# DYNAMICS OF CHARGED PARTICLES AND THEIR RADIATION FIELD

This book provides a self-contained and systematic introduction to classical electron theory and its quantization, nonrelativistic quantum electrodynamics. The first half of the book covers the classical theory. It discusses the well-defined Abraham model of extended charges in interaction with the electromagnetic field, and gives a study of the effective dynamics of charges under the condition that, on the scale given by the size of the charge distribution, they are far apart and the applied potentials vary slowly. The second half covers the quantum theory, leading to a coherent presentation of nonrelativistic quantum electrodynamics. Topics discussed include nonperturbative properties of the basic Hamiltonian, the structure of resonances, the relaxation to the ground state through emission of photons, the nonperturbative derivation of the $g$-factor of the electron, and the stability of matter.

Suitable as a supplementary text for graduate courses, this book will also be a valuable reference for researchers in mathematical physics, classical electrodynamics, quantum optics, and applied mathematics. This title, first published in 2004, has been reissued as an Open Access publication on Cambridge Core.

HERBERT SPOHN is Professor of Mathematical Physics at Zentrum Mathematik, Technische Universität München. He obtained his Ph.D. from Ludwig-Maximilians-Universität München in 1975. He has done research at universities and institutes throughout the world. His research interests are in statistical physics, particularly dynamics and nonequilibrium statistical mechanics, with one focus on the derivation of macroscopic evolution equations from the dynamics of atoms. He has had numerous publications in these areas. From 2000 to 2002 he has been the president of the International Association of Mathematical Physics.

# DYNAMICS OF CHARGED PARTICLES AND THEIR RADIATION FIELD

HERBERT SPOHN
*Technische Universität München*

CAMBRIDGE
UNIVERSITY PRESS

CAMBRIDGE
UNIVERSITY PRESS

Shaftesbury Road, Cambridge CB2 8EA, United Kingdom

One Liberty Plaza, 20th Floor, New York, NY 10006, USA

477 Williamstown Road, Port Melbourne, VIC 3207, Australia

314–321, 3rd Floor, Plot 3, Splendor Forum, Jasola District Centre, New Delhi – 110025, India

103 Penang Road, #05-06/07, Visioncrest Commercial, Singapore 238467

Cambridge University Press is part of Cambridge University Press & Assessment, a department of the University of Cambridge.

We share the University's mission to contribute to society through the pursuit of education, learning and research at the highest international levels of excellence.

www.cambridge.org
Information on this title: www.cambridge.org/9781009402279

DOI: 10.1017/9781009402286



First published 2004
Reissued as OA 2023

*A catalogue record for this publication is available from the British Library.*

ISBN 978-1-009-40227-9 Hardback
ISBN 978-1-009-40223-1 Paperback

To the memory of my parents Ortrud Knopp and Karl Spohn

# Contents

# Preface

Physical theories, while devised to model a particular range of phenomena, are evidently linked in a hierarchical fashion. It is this structure which keeps fascinating me. In statistical mechanics, my scientific home-town, the link between atomic and macroscopic properties is one central issue. There we are taught that the emergence of a more restricted theory from a more general one has a richer structure than merely letting some parameter tend to infinity. I understood at some point, by accident, that similar issues appear in the dynamics of classical charges coupled to the Maxwell field. Since I could not find a satisfactory discussion in the literature, I decided to write up my own account. The theory so covered is the classical electron theory, a subject which is commonly regarded as settled with some modest revival through astrophysical applications. On the other hand, the quantized version of this theory is more lively than ever through the amazing advances in atomic physics and quantum optics. It thus seemed to me a welcome opportunity to expand my enterprise and to cover also nonrelativistic quantum electrodynamics, stressing its classical counterpart more than is done usually.

The research which has led to this book goes back about seven years and in part much longer. I am grateful for the constant help from my collaborators Volker Betz, Brian Davies, Rolf Dümcke, Detlef Dürr, Christian Hainzl, Masao Hirokawa, Fumio Hiroshima, Frank Hövermann, Matthias Hübner, Valery Imaikin, Sasha Komech, Markus Kunze, Joel Lebowitz, József Lőrinczi, Robert Minlos, Gianluca Panati, and Stefan Teufel. In this list I also include Michael Kiessling for many illuminating observations. In addition I thank him for a careful reading of a draft of the book.

As the project expanded I received comments, criticisms, remarks, and questions which in their total sum shaped my understanding of the subject and the way things were written down eventually. All I can do here is to deeply thank Robert Alicki, Asao Arai, Volker Bach, Gernot Bauer, Jens Bolte, Thomas Chen, Stephan

De Bièvre, Jan Dereziński, Thomas Erber, László Erdös, Raffaele Esposito, Jürg Fröhlich, Luigi Galgani, Christian Gérard, Shelly Goldstein, Vittorio Gorini, Marcel Griesemer, Vojkan Jakšić, Caroline Lasser, Elliott Lieb, Michael Loss, Claude-Alain Pillet, Mario Pulvirenti, Markus Rauscher, Luc Rey-Bellet, Fritz Rohrlich, Wolfgang Schleicher, Michael Sigal, and Hong-Tzer Yau. In addition, I appreciate the help with the figures from Patrik Ferrari.

This book is dedicated to my parents in deep gratitude for a wonderful childhood. My father furnished stability and my mother cared for the three boys, encouraging our curiosity to learn about the marvellously complex world around us. This gift constitutes a lasting source of joy.

Herbert Spohn<br>München<br>May 2004

# Symbols

## Symbols for physical quantities

| | |
|---|---|
| $A$ | quantized vector potential |
| $\boldsymbol{A}$ | vector potential |
| $\mathcal{A}$ | action |
| $A_t(x)$ | fluctuating vector potential |
| $A_\parallel$, $A_\perp$, $E_\parallel$, $E_\perp$ | longitudinal, tranverse fields |
| $\boldsymbol{B}$ | magnetic field |
| $\boldsymbol{E}$ | electric field |
| $\mathcal{E}$ | total energy |
| $E(\boldsymbol{p})$ | energy–momentum relation |
| $\boldsymbol{E}^0$, $\boldsymbol{B}^0$, $\boldsymbol{q}^0$, $\boldsymbol{v}^0$ | initial conditions |
| $E_\mathrm{s}$ | soliton energy |
| $E_\mathrm{bin}$ | binding energy |
| $\boldsymbol{E}_\mathrm{ex}$, $\boldsymbol{B}_\mathrm{ex}$ | external fields |
| $\boldsymbol{E}_\mathrm{ini}$, $\boldsymbol{B}_\mathrm{ini}$ | initial fields |
| $\boldsymbol{E}_\mathrm{out}$, $\boldsymbol{B}_\mathrm{out}$ | outgoing fields |
| $\boldsymbol{E}_\mathrm{ret}$, $\boldsymbol{B}_\mathrm{ret}$ | retarded fields |
| $\boldsymbol{E}_\mathrm{sc}$, $\boldsymbol{B}_\mathrm{sc}$ | scattered fields |
| $E_\mathrm{self}$ | self-energy |
| $\boldsymbol{E}_{\boldsymbol{v}}$, $\boldsymbol{B}_{\boldsymbol{v}}$ | soliton fields |
| $\mathbf{F}$ | electromagnetic field tensor |
| $\boldsymbol{F}$ | force |
| $H$ | Hamiltonian |
| $H_\mathrm{f}$ | field Hamiltonian |
| $H_p$, $H(p)$ | Hamiltonian at fixed total momentum $p$ |
| $H_\mathrm{sp}$ | spin Hamiltonian |
| $I_\mathrm{b}$, $I_\mathrm{f}$ | moment of inertia |

| | |
|---|---|
| $\boldsymbol{J}$ | total angular momentum |
| $J_{\mathrm{f}}$ | field angular momentum |
| $\mathsf{L}$, $\mathsf{L}_{\mathrm{at}}$, $\mathsf{L}_{\mathrm{f}}$, $\mathsf{L}_{\mathrm{int}}$ | Liouvillean |
| $L$, $\mathcal{L}$ | Lagrangian |
| $\mathsf{L}_{\mathrm{D}}$ | Davies generator |
| $M_{\mathrm{e}}$ | electric moment |
| $M_{\mathrm{m}}$ | magnetic moment |
| $N$ | number of particles |
| $\boldsymbol{N}$ | torque |
| $\mathcal{P}$ | total momentum |
| $\boldsymbol{P}_{\mathrm{s}}$ | soliton momentum |
| $\boldsymbol{P}_{\mathrm{f}}$ | field momentum |
| $\mathcal{P}_{\mathrm{f}}$, $P_{\mathrm{f}}$ | field momentum |
| $\mathcal{S}$ | soliton manifold |
| $T$ | temperature |
| $V_{\mathrm{coul}}$ | Coulomb potential |
| $V_{\mathrm{dar}}$ | Darwin potential |
| $Z$ | partition function, nucleon charge |
| $a^*$, $a$ | creation, annihilation operators |
| $c$ | velocity of light |
| $e\varphi$ | charge distribution |
| $e$ | electric charge |
| $e_{\lambda}$ | polarization vectors |
| $\mathbf{f}$ | Minkowski force |
| $f_{\alpha}$ | distribution function |
| $g$ | $g$-factor |
| $g_{\mu\nu}$ | metric tensor |
| $\hbar$ | Planck's constant |
| $\boldsymbol{j}$ | current density |
| $\mathbf{j}$ | four-current |
| $\boldsymbol{k}$ | momentum |
| $k_{\mathrm{B}}$ | Boltzmann's constant |
| $m$ | mass |
| $m_{\mathrm{b}}$ | bare mass |
| $m_{\mathrm{f}}$ | field mass |
| $m_{\mathrm{g}}$ | gyrational mass |
| $m_{\mathrm{eff}}$ | effective mass |
| $\widehat{n}$ | unit vector |
| $\mathbf{p}$ | four-momentum |

| | |
|---|---|
| $p$, $\boldsymbol{p}$, $P$ | momentum |
| $q$, $\boldsymbol{q}$ | position |
| $\mathbf{q}(\tau)$ | world line |
| $\boldsymbol{r}$ | position |
| $r_{\mathrm{B}}$ | Bohr radius |
| $\mathbf{s}$ | spin angular momentum |
| $\mathbf{t}$ | Minkowski torque |
| $t$ | time |
| $\mathbf{u}$ | four-velocity |
| $u$, $\boldsymbol{u}$ | velocity |
| $v$, $\boldsymbol{v}$ | velocity |
| $\mathbf{x}$ | four-space vector |
| $x$, $\boldsymbol{x}$ | space |
| $\Delta$ | Laplacian |
| $\Lambda$ | ultraviolet cutoff |
| $\Omega$ | four-gyration |
| $\Omega^{\pm}$ | wave operator |
| $\alpha$ | fine structure constant |
| $\beta$ | inverse temperature |
| $\gamma$ | relativistic velocity factor |
| $\delta^{\perp}$ | transverse delta function |
| $\lambda_{\mathrm{c}}$ | Compton wavelength |
| $\mu$ | magnetic moment |
| $\rho$ | charge distribution |
| $\rho$ | density matrix |
| $\sigma$ | Pauli spin matrices |
| $\tau$ | eigentime |
| $\phi$ | electrostatic potential |
| $\phi$, $\pi$ | scalar field, scalar momentum field |
| $\phi_{\mathrm{ex}}$, $A_{\mathrm{ex}}$ | external potentials |
| $\widehat{\varphi}$ | form factor |
| $\psi$ | wave function |
| $\psi_{\mathrm{g}}$ | ground state wave function |
| $\omega$ | angular velocity |
| $\omega_{\mathrm{c}}$ | cyclotron frequency |
| $\omega_{\mathrm{s}}$ | spin precession frequency |
| $\omega$ | free-field dispersion relation |
| $\omega_{\beta}$ | KMS state |
| $\widehat{\omega}$ | unit vector |

## Mathematical symbols

| | |
|---|---|
| $A(q, p)$ | operator-valued function |
| $B(\mathcal{H})$ | bounded operators on $\mathcal{H}$ |
| $C, C(\mathbb{R}, \mathbb{R}^d)$ | continuous functions on $\mathbb{R}$ with values in $\mathbb{R}^d$ |
| $C^\infty$ | infinitely often differentiable functions |
| $C^k$ | $k$ times differentiable functions |
| $\mathbb{C}$ | complex numbers |
| $D(\cdot, \cdot)$ | Dirichlet form |
| $D(A)$ | domain of operator $A$ |
| $\mathbb{E}$ | expectation |
| $\mathcal{F}$ | Fock space |
| $\mathcal{H}_\mathrm{f}$ | field Hilbert space |
| $\mathcal{H}_\mathrm{p}$ | particle Hilbert space |
| $L^2, L^2(\mathbb{R}^3, \mathrm{d}^3x)$ | Hilbert space of square-integrable functions on $\mathbb{R}^3$ |
| $\mathcal{M}_N$ | algebra of $N \times N$ matrices |
| $\mathbb{N}$ | positive integer numbers |
| $\mathbb{P}$ | probability measure |
| $\mathbb{R}$ | real numbers |
| Ran $A$ | range of operator A |
| $\mathcal{T}_1(\mathcal{H})$ | trace class operators on $\mathcal{H}$ |
| $\mathcal{T}_2(\mathcal{H})$ | Hilbert–Schmidt operators on $\mathcal{H}$ |
| $\mathcal{W}_\varepsilon$ | Weyl quantization |
| $\mathbb{Z}$ | integer numbers |
| $d(\cdot, \cdot)$ | metric |
| $\widehat{f}$ | Fourier transform of $f$ |
| $\ell, r$ | left, right representation |
| tr | trace |
| $\Omega$ | Fock vacuum |
| $\sigma(H)$ | spectrum of operator $H$ |
| $\Vert \cdot \Vert$ | Hilbert space norm |
| $\Vert \cdot \Vert_1$ | $L^1$-norm |
| $\Vert \cdot \Vert_\infty$ | $L^\infty$-norm |
| $\Vert \cdot \Vert_R$ | local energy norm |
| $\langle \cdot, \cdot \rangle_\mathcal{H}$ | Hilbert space scalar product |
| $\langle \cdot \vert \cdot \rangle$ | scalar product for Hilbert–Schmidt operators |
| $: \ :$ | normal order, Wick order |
| $\{\cdot, \cdot\}$ | Poisson bracket |
| $[\cdot, \cdot]$ | commutator |
| $\int \mathrm{d}q_s$ | stochastic integration |
| $\sharp$ | Moyal product |
| $\nabla$ | nabla operator |

# 1

# Scope, motivation, and orientation

If one accepts gravitational forces on the Newtonian level of precision and ignores nuclear fission and fusion, then most physical phenomena on the scale of the Earth are accounted for by electrons, nuclei, and photons. Here photons play a double role: they mediate the interaction between charges, and appear freely propagating in the form of electromagnetic radiation. In their first role it often suffices to ignore all dynamical aspects and replace the photons by the effective electrostatic Coulomb interaction. Conversely, in the study of radiation phenomena, matter in the form of nuclei and electrons can mostly be replaced by prescribed macroscopic quantities like charge, current, and polarization densities. In our treatise we plan to dwell on the border area, where the interaction between photons and electrons, respectively nuclei, must be fully retained. Our goal is to discuss the dynamics of the coupled system, charges and their radiation field.

Although such a description might give the impression that we will deal with relativistic quantum electrodynamics (QED), in fact we will not even touch upon it. This theory has been devised for predicting a few very specific effects, like the anomalous $g$-factor of the electron, and it does so with astounding precision. Relativistic QED is, however, not well adapted to discuss, say, the fluorescence of the hydrogen atom. Thus the subject to be covered is what is commonly known as nonrelativistic quantum electrodynamics. In fact our enterprise also has a classical part. Just as in studying quantum mechanics a good grasp of classical mechanics is most useful, we believe that an understanding of classical electron theory, i.e. classical charges in interaction with the Maxwell field, serves as a solid basis for taking up the corresponding quantum theory. The classical models discussed will be semirelativistic with one exception, namely a fully relativistic theory of extended classical charges.

Classical electron theory was at the forefront of research in the early 1900s when the development of a dynamical theory of the then newly discovered electron was attempted. The basic prediction was an energy–momentum relation for the electron (compare with chapter 4), which, however, depended on the details of the particular electron model adopted. This enterprise came to a standstill because of the advent of the theory of special relativity, which, advancing with a totally different set of arguments, required a relativistically covariant link between energy and momentum for massive particles. Classical electron theory further deteriorated simply because it had become evident that for the investigation of radiation from atoms the newly born quantum mechanics had to be used. A brief revival occurred in the struggle to formulate a consistent relativistic quantum theory for the electron–positron field coupled to the photons. The hope was that a refined understanding of the classical theory should give a hint on how to quantize and how to handle correctly the ultraviolet infinities. But as the proper quantum field theory surfaced, classical considerations faded away. In fact the theory emerged in a worse state than before as summarized in the 1963 opinion of R. Feynman: "The classical theory of electromagnetism is an unsatisfactory theory all by itself. The electromagnetic theory predicts the existence of an electromagnetic mass, but it also falls on its face in doing so, because it does not produce a consistent theory."

Because of its peculiar history, classical electron theory never had any share in the good fortune of being rewritten, modernized, and rewritten again, as can be seen from a rapid sample of standard textbooks on electrodynamics. While the conventional chapters essentially follow the same intrinsic pattern, obviously with a lot of variations on details, once it comes to the chapter on radiation reaction, Pandora's box opens. As a student I was rather dissatisfied with such a state of affairs and promised myself to come back to it at some point. The first few chapters of this treatise are my own rewriting of the classical theory. It is based on two cornerstones:

- a well-defined dynamical theory of extended charges in interaction with the electromagnetic field;
- a study of the effective dynamics of charges under the condition that they are far apart and the external potentials vary slowly on the scale given by the size of the charge distribution. This is the *adiabatic limit*.

Our approach reflects the great progress which has taken place in the theory of dynamical systems. After all, charges coupled to their radiation field can be considered as one particular case, but with some rather special features. Perhaps the most unusual one is the appearance of a center manifold in the effective dynamics, in case friction through radiation is included.

For nonrelativistic QED the situation could hardly be more different. Through the efforts made in atomic physics and quantum optics a structured theory emerged which is well covered in textbooks and reviews. It would make little sense in trying to compete with them. However, almost exclusively this theory is based either on such drastic simplifications that an exact solution becomes possible or on second-order time-dependent perturbation theory. In recent years there has been substantial progress, mostly within the quarters of mathematical physicists, in gaining an understanding of *nonperturbative* properties of the full basic Hamiltonian, among others the structure of resonances, the relaxation to the ground state through emission of photons, the nonperturbative derivation of the $g$-factor of the electron, and the stability of matter when the quantized radiation field is included. These and other topics will be covered in the second half of the book. Readers less interested in the classical theory may jump ahead to chapter 12, where the conclusions of chapters 2–11 are summarized and the contents of the quantum part outlined.

A few words on the style are in order. First of all, I systematically develop the theory and discuss some of the most prominent applications. No review is intended. For a subject with a long history, such an attitude looks questionable. After all, what did the many physicists working in that area contribute? To compensate, I include one historical chapter, which as very often in physics is the history as viewed from our present understanding. Since there are excellent historical studies, I hope to be excused. Further, at the end of each chapter I add *Notes and References* intended as a guide to all the material which has been left out. The level of the book is perhaps best characterized as being an advanced textbook. I assume a basic knowledge of Maxwell's theory of electromagnetism and of nonrelativistic quantum mechanics. On the other hand, the central topics are explained in detail and, for the reader to follow the discussion, there is no need of further outside sources. This brings me to the issue of mathematical rigor. In the case of classical electron theory, many claims of uncertain status are in the literature, hardly any numerical work is available, and there are no quantitative experimental verifications, as yet, with the exception of the lifetime of an electron captured in a Penning trap. More than in other fields one has to rely on fixed points in the form of mathematical theorems, which seems to be the only way to disentangle hard facts from "truths" handed down by tradition. For the quantum theory we venture into the nonperturbative regime which by definition requires a certain mathematical sophistication. In a few cases I decided to provide the full proof of the mathematical theorem. Otherwise I usually indicate its basic idea to proceed then with the formal computation. To give always full details would overload the text on an unacceptable scale and, in addition, would be duplication, since mostly the complete argument can be found elsewhere in the literature. Of course, there are stretches, possibly even long stretches, where such a firm foundation is not available and one has to proceed on the basis of limited evidence.

Our introduction might give the impression that all basic problems are resolved, nonrelativistic quantum electrodynamics is in good shape, and one only has to turn to exciting applications. This would be a far too simplistic reading. What I hope is to bring the dynamics of charges and their radiation field properly into focus. Once this point is reached, there are many loose ends. On the theoretical side, to mention only a few of them: on the classical level, the comparison between the true microscopic and approximate particle dynamics could be more precise; a similar program for the relativistic theory of an extended charge is hardly tackled; in the quantum theory the removal of the ultraviolet cutoff at the expense of energy and mass renormalization is still not understood; and the dynamics of many charges remains largely unexplored. Also quantitative experimental confirmation of the effective dynamics of an electron, as given through the Lorentz–Dirac equation on its center manifold, remains on the agenda. The greatest reward would be if my notes encourage further research.

# Part I

## Classical theory

# 2

# A charge coupled to its electromagnetic field

We plan to study the dynamics of a well-localized charge, like an electron or a proton, when coupled to its own electromagnetic field. The case of several particles is reserved for chapter 11. In a first attempt, one models the particle as a point charge with a definite mass. If its world line is prescribed, then the fields are determined through the inhomogeneous Lorentz–Maxwell equations. On the other hand, if the electromagnetic fields are given, then the motion of the point charge is governed by Newton's equation of motion with the Lorentz force as force law. While it then seems obvious how to marry the two equations, such as to have a coupled dynamics for the charge and its electromagnetic field, ambiguities and inconsistencies arise due to the infinite electrostatic energy of the Coulomb field of the point charge. Thus one is forced to introduce a slightly smeared charge distribution, i.e. an extended charge model. Mathematically this means that the interaction between particle and field is cut off or regularized at short distances, which seems to leave a lot of arbitrariness. There are also strong constraints, however. In particular, local charge conservation must be satisfied, the theory should be of Lagrangian form, and it should reproduce the two limiting cases mentioned already. In addition, as expected from any decent physical model, the theory should be well defined and empirically accurate within its domain of validity. In fact, up to the present time only two models have been worked out in some detail: (i) the semirelativistic Abraham model of a rigid charge distribution; and (ii) the Lorentz model of a relativistically covariant extended charge distribution. The aim of this chapter is to introduce and explain both models at some length. On the way we recall a few properties of the inhomogeneous Lorentz–Maxwell equations for later use.

A short preamble on units and other conventions is in order. We use the Heaviside–Lorentz units. In particular, the Coulomb potential is simply the inverse of the Laplacian with no extra factor. The vacuum susceptibilities are $\varepsilon_0 = 1 = \mu_0$, which fixes the unit of charge. $c$ is the speed of light. Mostly we will set $c = 1$ for convenience, thereby linking the units of space and time. If needed, one can easily retrieve these natural constants in the conventional way. At some parts below we will do this without notice, so as to have the dimensions right and to keep better track of the order of magnitudes. In the nonrelativistic setting we use $\nabla\times$ for rotation, but switch to the more proper exterior derivative, $\nabla_{\mathrm{g}}\wedge$, with $\mathbf{g}$ the metric tensor, in the relativistic context. We will use standard notation as often as possible. Since a fairly broad spectrum of material is covered, double meaning cannot be avoided entirely. At the risk of some repetition we strive for minimal ambiguity within a given chapter. In the classical part of the book we use boldface italic letters, $\boldsymbol{x}$, for three-vectors and boldface roman letters, $\mathbf{x}$, for four-vectors. In the quantum section such a notation tends to be cumbersome and we use lightface letters, $x$, throughout.

## 2.1 The inhomogeneous Maxwell–Lorentz equations

We prescribe a charge density, $\rho(\boldsymbol{x},t)$, and an associated current, $\boldsymbol{j}(\boldsymbol{x},t)$, linked through the law of charge conservation

$$\partial_t \rho(\boldsymbol{x},t) + \nabla\cdot\boldsymbol{j}(\boldsymbol{x},t) = 0\,. \tag{2.1}$$

Of course, $\boldsymbol{x}\in\mathbb{R}^3$ and $t\in\mathbb{R}$, where we use $\mathbb{R}^3$ to describe physical space and $\mathbb{R}$ as the time axis. The Maxwell equations for the electric field $\boldsymbol{E}$ and the magnetic field $\boldsymbol{B}$ consist of the two evolution equations

$$\begin{aligned} c^{-1}\partial_t \boldsymbol{B}(\boldsymbol{x},t) &= -\nabla\times\boldsymbol{E}(\boldsymbol{x},t)\,,\\ c^{-1}\partial_t \boldsymbol{E}(\boldsymbol{x},t) &= \nabla\times\boldsymbol{B}(\boldsymbol{x},t) - c^{-1}\boldsymbol{j}(\boldsymbol{x},t) \end{aligned} \tag{2.2}$$

and the two constraints

$$\nabla\cdot\boldsymbol{E}(\boldsymbol{x},t) = \rho(\boldsymbol{x},t)\,,\quad \nabla\cdot\boldsymbol{B}(\boldsymbol{x},t) = 0\,. \tag{2.3}$$

◇ *How are the Maxwell equations written and named?* According to my survey, there seems to be no universally accepted standard. As indicated by the name "electromagnetic", the order $\boldsymbol{E}, \boldsymbol{B}$ is very common and also adopted here. In the Lagrangian version $\boldsymbol{B}$ is position-like and $-\boldsymbol{E}$ is velocity-like, which would suggest the opposite order, namely $(\boldsymbol{B}, -\boldsymbol{E})$. In the nineteenth century the time-derivative was written at the right side of the equation. By present standards, in evolution equations like the Boltzmann, Navier–Stokes, and Schrödinger equation, the time-derivative is always at the left, which is also our convention here.

The common practice is to call the first equation of (2.2) together with the second equation of (2.3) the "homogeneous Maxwell equations" and the remaining

Paragraphs indicated by ◇ give explanations of notation and names.

pair the "inhomogeneous Maxwell equations". We follow here the convention used in the context of wave equations and call (2.2) with $\boldsymbol{j} = 0$ the "homogeneous Maxwell–Lorentz equations" and (2.2) with $\boldsymbol{j} \neq 0$ the "inhomogeneous Maxwell–Lorentz equations". The constraints (2.3) are always understood. "Maxwell–Lorentz equations" and "Maxwell equations" are used synonymously. ◇

We solve the Maxwell equations as a Cauchy problem, i.e. by prescribing the fields at time $t = 0$. If the constraints (2.3) are satisfied at $t = 0$, then by the continuity equation (2.1) they are satisfied at all times. Thus the initial data are

$$\boldsymbol{E}(\boldsymbol{x}, 0)\,, \quad \boldsymbol{B}(\boldsymbol{x}, 0) \tag{2.4}$$

together with the constraints

$$\nabla \cdot \boldsymbol{E}(\boldsymbol{x}, 0) = \rho(\boldsymbol{x}, 0)\,, \quad \nabla \cdot \boldsymbol{B}(\boldsymbol{x}, 0) = 0\,. \tag{2.5}$$

The choice $t = 0$ is merely a convention. In some cases it is preferable to prescribe the fields either in the remote past or the distant future. We will only consider physical situations where the fields decay at spatial infinity and thus have the finite energy

$$\mathcal{E} = \frac{1}{2} \int \mathrm{d}^3 x \left( \boldsymbol{E}(\boldsymbol{x}, t)^2 + \boldsymbol{B}(\boldsymbol{x}, t)^2 \right) < \infty\,. \tag{2.6}$$

In a thermal state at nonzero temperature, typical fields fluctuate without decay and one would be forced to consider infinite-energy solutions.

The Maxwell equations (2.2), (2.3) are inhomogeneous wave equations and are thus easy to solve. This will be done in Fourier space first, where the Fourier transform is denoted by $\widehat{\ }$ and defined through

$$\widehat{f}(\boldsymbol{k}) = (2\pi)^{-n/2} \int \mathrm{d}^n x \, \mathrm{e}^{-\mathrm{i}\boldsymbol{k}\cdot\boldsymbol{x}} f(\boldsymbol{x})\,. \tag{2.7}$$

Then, setting $c = 1$, (2.2) becomes

$$\begin{aligned} \partial_t \widehat{\boldsymbol{B}}(\boldsymbol{k}, t) &= -\mathrm{i}\boldsymbol{k} \times \widehat{\boldsymbol{E}}(\boldsymbol{k}, t)\,, \\ \partial_t \widehat{\boldsymbol{E}}(\boldsymbol{k}, t) &= \mathrm{i}\boldsymbol{k} \times \widehat{\boldsymbol{B}}(\boldsymbol{k}, t) - \widehat{\boldsymbol{j}}(\boldsymbol{k}, t) \end{aligned} \tag{2.8}$$

with the constraints

$$\mathrm{i}\boldsymbol{k} \cdot \widehat{\boldsymbol{E}}(\boldsymbol{k}, t) = \widehat{\rho}(\boldsymbol{k}, t)\,, \quad \mathrm{i}\boldsymbol{k} \cdot \widehat{\boldsymbol{B}}(\boldsymbol{k}, t) = 0 \tag{2.9}$$

and the conservation law

$$\partial_t \widehat{\rho}(\boldsymbol{k}, t) + \mathrm{i}\boldsymbol{k} \cdot \widehat{\boldsymbol{j}}(\boldsymbol{k}, t) = 0\,. \tag{2.10}$$

To solve the inhomogeneous equations (2.8), we rely, as usual, on the solution of the homogeneous equations,

$$\widehat{E}_0(\boldsymbol{k},t) = \big(\cos|\boldsymbol{k}|t + (1-\cos|\boldsymbol{k}|t)\widehat{\boldsymbol{k}}\otimes\widehat{\boldsymbol{k}}\big)\widehat{\boldsymbol{E}}(\boldsymbol{k},0) + \Big(\frac{1}{|\boldsymbol{k}|}\sin|\boldsymbol{k}|t\Big)\mathrm{i}\boldsymbol{k}\times\widehat{\boldsymbol{B}}(\boldsymbol{k},0)\,,$$

$$\widehat{B}_0(\boldsymbol{k},t) = \big(\cos|\boldsymbol{k}|t + (1-\cos|\boldsymbol{k}|t)\widehat{\boldsymbol{k}}\otimes\widehat{\boldsymbol{k}}\big)\widehat{\boldsymbol{B}}(\boldsymbol{k},0) - \Big(\frac{1}{|\boldsymbol{k}|}\sin|\boldsymbol{k}|t\Big)\mathrm{i}\boldsymbol{k}\times\widehat{\boldsymbol{E}}(\boldsymbol{k},0)\,. \tag{2.11}$$

Here $\widehat{\boldsymbol{k}} = \boldsymbol{k}/|\boldsymbol{k}|$ is the unit vector along $\boldsymbol{k}$ and for any pair of vectors $\boldsymbol{a}, \boldsymbol{b}$, $\boldsymbol{a}\otimes\boldsymbol{b}$ is the tensor of rank 2 defined through $(\boldsymbol{a}\otimes\boldsymbol{b})\boldsymbol{c} = \boldsymbol{a}(\boldsymbol{b}\cdot\boldsymbol{c})$ as acting on the vector $\boldsymbol{c}$.

We insert (2.11) in the time-integrated version of (2.8). Taking account of the constraints, making a partial integration, and using charge conservation, we arrive at

$$\begin{aligned}\widehat{\boldsymbol{E}}(\boldsymbol{k},t) &= (\cos|\boldsymbol{k}|t)\widehat{\boldsymbol{E}}(\boldsymbol{k},0) + (|\boldsymbol{k}|^{-1}\sin|\boldsymbol{k}|t)\mathrm{i}\boldsymbol{k}\times\widehat{\boldsymbol{B}}(\boldsymbol{k},0)\\ &\quad + \int_0^t \mathrm{d}s\big(-(|\boldsymbol{k}|^{-1}\sin|\boldsymbol{k}|(t-s))\mathrm{i}\boldsymbol{k}\hat{\rho}(\boldsymbol{k},s) - (\cos|\boldsymbol{k}|(t-s))\widehat{\boldsymbol{j}}(\boldsymbol{k},s)\big)\\ &= \widehat{\boldsymbol{E}}_{\mathrm{ini}}(\boldsymbol{k},t) + \widehat{\boldsymbol{E}}_{\mathrm{ret}}(\boldsymbol{k},t)\,,\end{aligned} \tag{2.12}$$

$$\begin{aligned}\widehat{\boldsymbol{B}}(\boldsymbol{k},t) &= (\cos|\boldsymbol{k}|t)\widehat{\boldsymbol{B}}(\boldsymbol{k},0) - (|\boldsymbol{k}|^{-1}\sin|\boldsymbol{k}|t)\mathrm{i}\boldsymbol{k}\times\widehat{\boldsymbol{E}}(\boldsymbol{k},0)\\ &\quad + \int_0^t \mathrm{d}s(|\boldsymbol{k}|^{-1}\sin|\boldsymbol{k}|(t-s))\mathrm{i}\boldsymbol{k}\times\widehat{\boldsymbol{j}}(\boldsymbol{k},s)\\ &= \widehat{\boldsymbol{B}}_{\mathrm{ini}}(\boldsymbol{k},t) + \widehat{\boldsymbol{B}}_{\mathrm{ret}}(\boldsymbol{k},t)\,.\end{aligned} \tag{2.13}$$

The first terms are the initial fields propagated up to time $t$, while the second terms are the retarded fields. If one wanted to solve the Maxwell equations run into the past, then the retarded fields should be replaced by the advanced fields.

Next, let us introduce the fundamental propagator, $G_t(\boldsymbol{x})$, of the wave equation which is defined as the Fourier transform of $(2\pi)^{-3/2}\,|\boldsymbol{k}|^{-1}\sin|\boldsymbol{k}|t$ and satisfies

$$\partial_t^2 G - \Delta G = \delta(\boldsymbol{x})\delta(t)\,. \tag{2.14}$$

This means $G_t(\boldsymbol{x}) = (2\pi)^{-1}\,\delta(|\boldsymbol{x}|^2 - t^2)$ and in particular for $t > 0$

$$G_t(\boldsymbol{x}) = \frac{1}{4\pi t}\delta(|\boldsymbol{x}| - t)\,. \tag{2.15}$$

Then in physical space the solution (2.12), (2.13) of the inhomogeneous Maxwell–Lorentz equations reads as

$$\begin{aligned} \boldsymbol{E}(t) &= \partial_t G_t * \boldsymbol{E}(0) + \nabla \times G_t * \boldsymbol{B}(0) - \int_0^t \mathrm{d}s \big(\nabla G_{t-s} * \rho(s) + \partial_t G_{t-s} * \boldsymbol{j}(s)\big) \\ &= \boldsymbol{E}_{\mathrm{ini}}(t) + \boldsymbol{E}_{\mathrm{ret}}(t)\,, \end{aligned} \tag{2.16}$$

$$\begin{aligned} \boldsymbol{B}(t) &= \partial_t G_t * \boldsymbol{B}(0) - \nabla \times G_t * \boldsymbol{E}(0) + \int_0^t \mathrm{d}s \nabla \times G_{t-s} * \boldsymbol{j}(s) \\ &= \boldsymbol{B}_{\mathrm{ini}}(t) + \boldsymbol{B}_{\mathrm{ret}}(t)\,. \end{aligned} \tag{2.17}$$

Here $*$ denotes convolution, i.e. $f_1 * f_2(\boldsymbol{x}) = \int \mathrm{d}^n y f_1(\boldsymbol{x}-\boldsymbol{y}) f_2(\boldsymbol{y})$.

For later purposes it will be convenient to have a more concise notation. In matrix form, the solution of the homogeneous Maxwell–Lorentz equations can be written as

$$\frac{\mathrm{d}}{\mathrm{d}t}\begin{pmatrix} \boldsymbol{E}(t) \\ \boldsymbol{B}(t) \end{pmatrix} = \begin{pmatrix} 0 & \nabla\times \\ -\nabla\times & 0 \end{pmatrix}\begin{pmatrix} \boldsymbol{E}(t) \\ \boldsymbol{B}(t) \end{pmatrix}, \quad \frac{\mathrm{d}}{\mathrm{d}t}\mathsf{F}(t) = \mathsf{A}\mathsf{F}(t) \tag{2.18}$$

with the column vector $\mathsf{F} = (\boldsymbol{E}, \boldsymbol{B})$. They have the solution

$$\mathsf{F}(t) = \mathsf{U}(t)\mathsf{F}(0)\,, \quad \mathsf{U}(t) = \mathrm{e}^{\mathsf{A}t} \tag{2.19}$$

with $\mathsf{U}(t)$ given explicitly by the terms with subscripts 'ini' in (2.17), (2.16). If we set $\mathsf{g}(t) = (\boldsymbol{j}(t), 0)$ as a column vector, then

$$\frac{\mathrm{d}}{\mathrm{d}t}\mathsf{F}(t) = \mathsf{A}\mathsf{F}(t) - \mathsf{g}(t)\,, \quad \mathsf{F}(t) = \mathsf{U}(t)\mathsf{F}(0) - \int_0^t \mathrm{d}s\, \mathsf{U}(t-s)\mathsf{g}(s)\,. \tag{2.20}$$

The expressions (2.16), (2.17) remain meaningful even in case $\rho, \boldsymbol{j}$ are generated by the motion of a single point charge. Let us denote by $\boldsymbol{q}(t)$ the position and by $\boldsymbol{v}(t) = \dot{\boldsymbol{q}}(t)$ the velocity of the particle carrying charge $e$. Then

$$\rho(\boldsymbol{x}, t) = e\delta(\boldsymbol{x} - \boldsymbol{q}(t))\,, \quad \boldsymbol{j}(\boldsymbol{x}, t) = e\delta(\boldsymbol{x} - \boldsymbol{q}(t))\boldsymbol{v}(t)\,. \tag{2.21}$$

Upon inserting this in (2.16), (2.17) one arrives at the Liénard–Wiechert fields. Since their derivation is presented in most textbooks, we do not repeat the computation here and only discuss the result. We take the world line, $t \mapsto \boldsymbol{q}(t)$, of the particle to be given for all times. Since the particle is assumed to have a relativistic kinetic energy, $|\dot{\boldsymbol{q}}(t)| < 1$. Next we prescribe the initial data for the fields at time $t = t_0$ and take the limit $t_0 \to -\infty$ in (2.16), (2.17). Then, at a fixed space-time point $(\boldsymbol{x}, t)$, the contribution from the initial fields vanishes and the retarded fields become the Liénard–Wiechert fields. To describe them we introduce the retarded time $t_{\text{ret}}$, depending on $\boldsymbol{x}$, $t$, as the unique solution of

$$t_{\text{ret}} = t - |\boldsymbol{x} - \boldsymbol{q}(t_{\text{ret}})|\,. \tag{2.22}$$

$t_{\text{ret}}$ is then the uniquely defined time point at which the world line crosses the backward light cone with apex at $(\boldsymbol{x}, t)$. Furthermore, we introduce the unit vector

$$\widehat{\boldsymbol{n}} = \frac{\boldsymbol{x} - \boldsymbol{q}(t_{\text{ret}})}{|\boldsymbol{x} - \boldsymbol{q}(t_{\text{ret}})|}\,. \tag{2.23}$$

Then the electric field generated by the moving point charge is given by

$$\boldsymbol{E}(\boldsymbol{x}, t) = \frac{e}{4\pi}\left[\frac{(1 - \boldsymbol{v}^2)(\widehat{\boldsymbol{n}} - \boldsymbol{v})}{(1 - \boldsymbol{v}\cdot\widehat{\boldsymbol{n}})^3|\boldsymbol{x} - \boldsymbol{q}|^2} + \frac{\widehat{\boldsymbol{n}} \times [(\widehat{\boldsymbol{n}} - \boldsymbol{v}) \times \dot{\boldsymbol{v}}]}{(1 - \boldsymbol{v}\cdot\widehat{\boldsymbol{n}})^3|\boldsymbol{x} - \boldsymbol{q}|}\right]\bigg|_{t=t_{\text{ret}}} \tag{2.24}$$

and the corresponding magnetic field is

$$\boldsymbol{B}(\boldsymbol{x}, t) = \widehat{\boldsymbol{n}} \times \boldsymbol{E}(\boldsymbol{x}, t)\,. \tag{2.25}$$

Equations (2.24) and (2.25) are less explicit than they appear to be, since $t_{\text{ret}}$ depends through (2.22) on the reference point $(\boldsymbol{x}, t)$ and the particle trajectory. The first contribution in (2.24) is proportional to $|\boldsymbol{x} - \boldsymbol{q}|^{-2}$ and independent of the acceleration. This is the near field, which in a certain sense remains attached to the particle all through its motion. The second contribution is proportional to $|\boldsymbol{x} - \boldsymbol{q}|^{-1}$ as well as to the acceleration. This is the far field, which carries the information on the radiation field escaping to infinity. Whenever $\boldsymbol{q}(t)$ is smooth in $t$, the Liénard–Wiechert fields are also smooth functions except at $\boldsymbol{x} = \boldsymbol{q}(t)$, where they diverge as $|\boldsymbol{x} - \boldsymbol{q}(t)|^{-2}$. The corresponding potentials have a Coulomb singularity at the world line of the particle.

## 2.2 Newton's equations of motion

We take now the point of view that the electromagnetic fields $\boldsymbol{E}$, $\boldsymbol{B}$ are given. The motion of a charged particle, with charge $e$, position $\boldsymbol{q}(t)$, and velocity $\boldsymbol{v}(t)$, is then governed by Newton's equations of motion,

$$\frac{\mathrm{d}}{\mathrm{d}t}\big(m_0\gamma\boldsymbol{v}(t)\big) = e\big(\boldsymbol{E}(\boldsymbol{q}(t), t) + c^{-1}\boldsymbol{v}(t) \times \boldsymbol{B}(\boldsymbol{q}(t), t)\big)\,, \tag{2.26}$$

$\gamma(\boldsymbol{v}) = 1/\sqrt{1 - (\boldsymbol{v}/c)^2}$, which as an ordinary differential equation has to be supplemented with the initial conditions $\boldsymbol{q}(0)$, $\boldsymbol{v}(0)$. The force law is determined through the Lorentz force and thus (2.26) is also called the Newton–Lorentz equations. The particle is relativistic with rest mass $m_0$ as measured through the response to external forces. Once the particle is dynamically coupled to the Maxwell field, $m_0$ will attain a new meaning.

The $(\boldsymbol{E}, \boldsymbol{B})$ fields in (2.26) are not completely arbitrary. They are subject to the Maxwell equations with source $(\rho, \boldsymbol{j})$. In other words, we have divided all charges into a single charged particle whose motion is determined through (2.26) and the rest whose motion is taken to be known.

The Newton–Lorentz equations (2.26) are of Hamiltonian form. To see this we introduce vector potentials $\phi$, $\boldsymbol{A}$ such that

$$\boldsymbol{E}(\boldsymbol{x}, t) = -\nabla\phi(\boldsymbol{x}, t) - c^{-1}\partial_t \boldsymbol{A}(\boldsymbol{x}, t), \quad \boldsymbol{B}(\boldsymbol{x}, t) = \nabla \times \boldsymbol{A}(\boldsymbol{x}, t). \tag{2.27}$$

Then the Lagrangian associated with (2.26) is

$$L(\boldsymbol{q}, \dot{\boldsymbol{q}}, t) = -m_0 c^2 (1 - c^{-2}\dot{\boldsymbol{q}}^2)^{1/2} - e(\phi(\boldsymbol{q}, t) - c^{-1}\dot{\boldsymbol{q}} \cdot \boldsymbol{A}(\boldsymbol{q}, t)). \tag{2.28}$$

To switch to the Hamiltonian framework, one introduces the canonical momentum

$$\boldsymbol{p} = m_0 \gamma(\dot{\boldsymbol{q}})\dot{\boldsymbol{q}} + \frac{e}{c}\boldsymbol{A}(\boldsymbol{q}, t) \tag{2.29}$$

and obtains the Hamiltonian function

$$H(\boldsymbol{q}, \boldsymbol{p}, t) = \left((c\,\boldsymbol{p} - e\boldsymbol{A}(\boldsymbol{q}, t))^2 + m_0^2 c^4\right)^{1/2} + e\phi(\boldsymbol{q}, t). \tag{2.30}$$

In particular, whenever the fields are time independent, the energy

$$\mathcal{E}(\boldsymbol{q}, \boldsymbol{v}) = m_0 \gamma(\boldsymbol{v}) + e\phi(\boldsymbol{q}) \tag{2.31}$$

is conserved along the solution trajectories of (2.26).

It should be noted that in general the solutions to Newton's equations of motion (2.26) will have a complicated structure even for time-independent fields. This has been amply demonstrated for particular cases. Depending on how the external fields are chosen, the motion would range from regular to fully chaotic with a mixed phase space as a rule.

## 2.3 Coupled Maxwell's and Newton's equations

While for most practical purposes, barring a few exceptional cases, it suffices to use either Maxwell's equations with prescribed sources or Newton's equations with prescribed forces, from a more fundamental point of view such a procedure is unsatisfactory. Physically it would seem more natural to have a coupled system of equations for the time evolution of the charged particles together with their electromagnetic field and to regard the two cases discussed above as emerging limit situations. If for the moment we restrict ourselves to a single particle, it is obvious how to proceed. From (2.2), (2.3) we have

$$\partial_t \boldsymbol{B}(\boldsymbol{x},t) = -\nabla \times \boldsymbol{E}(\boldsymbol{x},t)\,,$$
$$\partial_t \boldsymbol{E}(\boldsymbol{x},t) = \nabla \times \boldsymbol{B}(\boldsymbol{x},t) - e\delta(\boldsymbol{x}-\boldsymbol{q}(t))\boldsymbol{v}(t) \tag{2.32}$$

with the constraints

$$\nabla \cdot \boldsymbol{E}(\boldsymbol{x},t) = e\delta(\boldsymbol{x}-\boldsymbol{q}(t))\,, \quad \nabla \cdot \boldsymbol{B}(\boldsymbol{x},t) = 0\,. \tag{2.33}$$

Moreover, from (2.26) we have

$$\frac{\mathrm{d}}{\mathrm{d}t}\big(m_0\gamma \boldsymbol{v}(t)\big) = e\big(\boldsymbol{E}_{\mathrm{ex}}(\boldsymbol{q}(t)) + \boldsymbol{E}(\boldsymbol{q}(t),t) + \boldsymbol{v}(t) \times (\boldsymbol{B}_{\mathrm{ex}}(\boldsymbol{q}(t)) + \boldsymbol{B}(\boldsymbol{q}(t),t))\big)\,. \tag{2.34}$$

We added the external electromagnetic fields $\boldsymbol{E}_{\mathrm{ex}}$, $\boldsymbol{B}_{\mathrm{ex}}$, which will play a prominent role later on. They are derived from potentials as

$$\boldsymbol{E}_{\mathrm{ex}} = -\nabla\phi_{\mathrm{ex}}\,, \quad \boldsymbol{B}_{\mathrm{ex}} = \nabla \times \boldsymbol{A}_{\mathrm{ex}}\,. \tag{2.35}$$

We assume the potentials to be time independent for simplicity, although a considerable part of the theory to be developed will work also for time-dependent fields. As before, (2.32)–(2.34) are to be solved as an initial value problem. Thus $\boldsymbol{E}(\boldsymbol{x},0)$, $\boldsymbol{B}(\boldsymbol{x},0)$, $\boldsymbol{q}(0)$, and $\boldsymbol{v}(0)$ are supposed to be given. Note that the continuity equation is satisfied by fiat.

Equations (2.32), (2.34) are the stationary points of a Lagrangian action, which strengthens our trust in these equations, since every microscopic classical evolution equation seems to be of that form. We continue to use the underlying electromagnetic potentials as in (2.27), (2.35). Then the action for (2.32), (2.34) reads

$$\begin{aligned}\mathcal{A}([\boldsymbol{q},\phi,\boldsymbol{A}]) = \int \mathrm{d}t\big[&-m_0(1-\dot{\boldsymbol{q}}(t)^2)^{1/2} - e\big(\phi_{\mathrm{ex}}(\boldsymbol{q}(t)) + \phi(\boldsymbol{q}(t),t)\\ &-\dot{\boldsymbol{q}}(t)\cdot(\boldsymbol{A}_{\mathrm{ex}}(\boldsymbol{q}(t)) + \boldsymbol{A}(\boldsymbol{q}(t),t))\big)\big]\\ &+\frac{1}{2}\int \mathrm{d}t \int \mathrm{d}^3x\big[(\nabla\phi(\boldsymbol{x},t) + \partial_t\boldsymbol{A}(\boldsymbol{x},t))^2 - (\nabla\times\boldsymbol{A}(\boldsymbol{x},t))^2\big]\,.\end{aligned} \tag{2.36}$$

The only difficulty is that (2.32) and (2.33) taken together with (2.34) make no proper mathematical sense. As explained, the solution of the Maxwell equations is singular at $\boldsymbol{x} = \boldsymbol{q}(t)$, and in the Lorentz force we are asked to evaluate the fields precisely at that point. One might be tempted to put the blame on the mathematics which refuses to handle equations as singular as (2.32)–(2.34). However before such a drastic conclusion is drawn, the physics should be properly understood. The point charge carries along with it a potential which at short distances diverges as the Coulomb potential, cf. (2.24), and which therefore has the electrostatic energy

$$\frac{1}{2} \int\limits_{\{|\boldsymbol{x}-\boldsymbol{q}(t)| \leq R\}} \mathrm{d}^3x\, \boldsymbol{E}(\boldsymbol{x}, t)^2 \simeq \int_0^R \mathrm{d}r r^2 (r^{-2})^2 = \int_0^R \mathrm{d}r r^{-2} = \infty\,. \tag{2.37}$$

Taken literally, such an object would have an infinite mass and hence would not respond to external forces. It would keep its velocity for ever, which is inconsistent with what is observed.

*Thus we are forced to regularize at short distances the coupled system consisting of the Maxwell equations and Newton's equation of motion with the Lorentz force.*

In carrying out such a program there are two, in part, complementary points of view. The first one, which we will *not* follow here, starts from the idea that regularization is a mathematical device with the sole purpose of making sense of a singular mathematical object through a suitable limiting procedure. To illustrate this approach we can think of the following prominent mathematical physics example. The free scalar field, $\phi(\boldsymbol{x})$, in Euclidean quantum field theory in $1 + 1$ dimensions fluctuates so wildly at short distances that an interaction such as $\int \mathrm{d}^2x\, V(\phi(\boldsymbol{x}))$ with $V(\phi) = \phi^2 + \lambda\phi^4$ cannot be properly defined. One way, not necessarily optimal, to regularize the theory is to introduce a spatial lattice with spacing $a$. Such a lattice field theory is well defined in any finite volume. On taking the limit $a \to 0$ along with a simultaneous readjustment of the interaction potential, $V(\phi) = V_a(\phi)$, a Euclidean-invariant, interacting quantum field theory is obtained. Ideally this limit theory should be independent of the regularization scheme. For instance one could start with the free scalar field in the continuum and regularize $\phi(\boldsymbol{x})$ as $\phi * g(\boldsymbol{x})$ with a suitable test function $g$ concentrated at 0. Then the regularized interaction is $\int \mathrm{d}^2x\, V(\phi * g(\boldsymbol{x}))$ and in the limit $g(y) \to \delta(y)$ a quantum field theory should be obtained identical to the one from the lattice regularization.

In the second approach one argues that there is a physical cutoff coming from a more refined theory, which is then modeled in a phenomenological way. While this is a standard procedure, it is worthwhile to illustrate it again with a concrete example. Consider a large number ($\cong 10^{23}$) of $He^4$ atoms in a container of adjustable size and suppose we are interested in computing their free energy according to the rules of statistical mechanics. The more refined theory is here nonrelativistic quantum mechanics which treats the electrons and nuclei as point particles carrying a spin $\frac{1}{2}$, respectively spin 0. As far as we can tell, this model approximately covers the temperature range $T = 0$ K to $T = 10^5$ K, i.e. way beyond dissociation, and the density range $\rho = 0$ to $\rho = \rho_{\mathrm{cp}}$, the density of close packing. Beyond these limits relativistic effects must be taken into account. However, there is a more limited range where we can get away with a model of classical point particles interacting through an effective potential of Lennard-Jones type. Once this pair potential is specified classical statistical mechanics makes well-defined predictions at *any* $T$, $\rho$. There is no limitation in principle. Only outside a certain range of parameters would the classical model lose the correspondence with the real world. Already from the way the physical cutoff is described, there is a considerable amount of vagueness. How much error should we allow in the free energy? What about more detailed properties like density correlations? An effective potential can be defined quantum mechanically, but it is temperature dependent and never strictly a pair potential. Despite all these imprecisions and shortcomings, the equilibrium theory of fluids relies heavily on the availability of a classical model.

In the same spirit we modify the coupled Maxwell and Newton equations by introducing an extended charge distribution as a phenomenological model for the omitted quantum electrodynamics. The charge distribution is stabilized by strong interactions which act outside the realm of electromagnetic forces. On the classical level, say, an electron appears as an extended charged object with a size roughly of the order of its Compton wavelength, i.e. $4 \times 10^{-11}$ cm. We impose the obvious condition that the extended charge distribution has to be adjusted such that, in the range where classical electrodynamics is applicable, the coupled Maxwell and Newton equations correctly reproduce the empirical observations.

Such general clauses seem to leave a lot of freedom in the construction of the theory. However, charge conservation and the Lagrangian form of the equations of motion severely limit the possibilities. In fact, essentially only two models of extended charge distribution have been investigated so far.

(i) *The semirelativistic Abraham model of a rigid charge distribution.* The charge $e$ is assumed to be smeared out over a ball of radius $R_\varphi$. This means that in (2.32)–(2.34) the $\delta$-function is replaced by a smooth charge distribution $e\varphi$. $\varphi(\boldsymbol{x})$ is taken to be radial, vanishing for $|\boldsymbol{x}| > R_\varphi$, and normalized as $\int d^3x\varphi(\boldsymbol{x}) = 1$. Equivalently, having (2.32)–(2.34) recast in Fourier space, the couplings between the field modes with $|\boldsymbol{k}| \gtrsim 1/R_\varphi$ and the particle become suppressed. This particular choice for the internal structure of the charge is called the Abraham model (for a single nonrotating charge). For zero coupling the model is relativistic. However, $\varphi$ is taken to be rigid, thus velocity independent in a prescribed coordinate frame, which breaks Lorentz invariance. The standard examples are that the charge is uniformly distributed either over the ball, $\varphi(\boldsymbol{x}) = (4\pi R_\varphi^2/3)^{-1}$ for $|\boldsymbol{x}| \leq R_\varphi$, $\varphi(\boldsymbol{x}) = 0$ otherwise, or over the sphere, $\varphi(\boldsymbol{x}) = (4\pi R_\varphi^2)^{-1}\delta(|\boldsymbol{x}| - R_\varphi)$. In the quantized version of the Abraham model, cf. chapter 13 below, often a sharp cutoff in Fourier space is adopted, i.e. $\widehat{\varphi}(\boldsymbol{k}) = (2\pi)^{-3/2}$ for $|\boldsymbol{k}| \leq \Lambda = R_\varphi^{-1}$, $\widehat{\varphi}(\boldsymbol{k}) = 0$ otherwise; this has the slight disadvantage of being oscillating and having slow decay in position space.

Once the charge distribution is extended, besides its center of charge, also rotational degrees of freedom must be taken into account. The Abraham model allowing for a spinning charge will be discussed in chapter 10. Since the dynamical behavior then becomes more complex, it is advisable to omit spin in the first round.

The Abraham model will be studied in considerable detail. While defined for all velocities $|\boldsymbol{v}(t)| < c$, it becomes empirically inaccurate at velocities close to $c$. Despite this drawback we hope that the Abraham model will serve as a blueprint towards a more realistic description of matter.

(ii) *The Lorentz model of a relativistically rigid charge distribution.* More in accord with special relativity is to require that $e\varphi$ is the charge distribution in the momentary rest frame of the particle. While such a principle was already stated by Lorentz and Poincaré, a satisfactory *dynamical* theory has been arrived at only very recently. As we will explain in section 2.5, in a relativistic theory translational and rotational degrees of freedom are intrinsically coupled. To gain an understanding of how relativistic invariance would modify the theory, we insert some features of the Lorentz model, although our understanding of its dynamical properties is far less developed than that of the Abraham model.

We emphasize that for extended charge models the diameter $R_\varphi$ of the charge distribution defines a length (and upon dividing by $c$ also a time) scale, relative to which the approximate validity of effective theories, like the Lorentz–Dirac equation, can be addressed quantitatively. In fact, apart from the external forces, $R_\varphi$ is the *only* natural length scale available.

## 2.4 The Abraham model

Following Abraham, we model the charged particle as a spherically symmetric, rigid body to which the charge elements are permanently attached. The charge distribution is prescribed and independent of the particle's velocity, which singles out the laboratory frame. In a relativistic theory the charge distribution would appear to be Lorentz contracted. To be specific the charge distribution $e\varphi$ is assumed to be smooth, radial, and supported in a ball of radius $R_\varphi$, and normalized to $e$, i.e.

*Condition* $(C)$:

$$\varphi \in C^\infty(\mathbb{R}^3)\,, \quad \varphi(\boldsymbol{x}) = \varphi_{\mathrm{r}}(|\boldsymbol{x}|)\,, \quad \varphi(\boldsymbol{x}) = 0 \text{ for } |\boldsymbol{x}| \geq R_\varphi\,, \quad \int \mathrm{d}^3x\varphi(\boldsymbol{x}) = 1\,. \tag{2.38}$$

◇ $e\varphi(\boldsymbol{x})$ is the charge distribution and $\widehat{\varphi}(\boldsymbol{k})$ is the form factor, since in Fourier space it multiplies the current as $(2\pi)^{3/2}\widehat{\varphi}(\boldsymbol{k})\widehat{\boldsymbol{j}}(\boldsymbol{k},t)$. ◇

Our goal is to set up the Abraham model as a well-defined dynamical system. Usually this point is taken for granted. Since the occurrence of ill-defined equations of motion was one of our main objections to the $\delta$-charge, it is worthwhile to understand why this objection is no longer valid for a smeared out $\delta$.

The equations of motion for the Abraham model are

$$\partial_t \boldsymbol{B}(\boldsymbol{x}, t) = -\nabla \times \boldsymbol{E}(\boldsymbol{x}, t),$$

$$\partial_t \boldsymbol{E}(\boldsymbol{x}, t) = \nabla \times \boldsymbol{B}(\boldsymbol{x}, t) - e\varphi(\boldsymbol{x} - \boldsymbol{q}(t))\boldsymbol{v}(t), \tag{2.39}$$

$$\nabla \cdot \boldsymbol{E}(\boldsymbol{x}, t) = e\varphi(\boldsymbol{x} - \boldsymbol{q}(t)), \quad \nabla \cdot \boldsymbol{B}(\boldsymbol{x}, t) = 0, \tag{2.40}$$

$$\frac{\mathrm{d}}{\mathrm{d}t}\big(m_\mathrm{b}\gamma \boldsymbol{v}(t)\big) = e\big(\boldsymbol{E}_\mathrm{ex}(\boldsymbol{q}(t)) + \boldsymbol{E}_\varphi(\boldsymbol{q}(t), t) + \boldsymbol{v}(t) \times (\boldsymbol{B}_\mathrm{ex}(\boldsymbol{q}(t)) + \boldsymbol{B}_\varphi(\boldsymbol{q}(t), t))\big), \tag{2.41}$$

where we have set $c = 1$. In (2.41) we use the shorthand $\boldsymbol{E}_\varphi(\boldsymbol{x}) = \boldsymbol{E} * \varphi(\boldsymbol{x})$ and $\boldsymbol{B}_\varphi(\boldsymbol{x}) = \boldsymbol{B} * \varphi(\boldsymbol{x})$ so as to resemble (2.34). Strictly speaking also $\boldsymbol{E}_\mathrm{ex}$, $\boldsymbol{B}_\mathrm{ex}$ should be smeared over $\varphi$; however, this would only amount to a redefinition of the external potentials. In contrast to Newton's equations of motion (2.26), for the Abraham model we denote the mechanical mass of the particle by $m_\mathrm{b}$ to emphasize that this bare mass will differ from the observed mass of the compound object "particle plus surrounding Coulomb field". The external potentials $\phi_\mathrm{ex}$, $\boldsymbol{A}_\mathrm{ex}$ can be fairly arbitrary. We only require them and their derivatives to be smooth and locally bounded, to avoid too strong local oscillations. No condition on the increase at infinity is needed, since $|\boldsymbol{v}(t)| \le 1$. However, it is convenient to have the energy, as defined in (2.44), uniformly bounded from below. To keep things simple we make the (unnecessarily strong) assumptions

*Condition* $(P)$:

$$\phi_\mathrm{ex} \in C^\infty(\mathbb{R}^3), \quad \boldsymbol{A}_\mathrm{ex} \in C^\infty(\mathbb{R}^3, \mathbb{R}^3), \quad \phi_\mathrm{ex} \ge \bar{\phi} > -\infty. \tag{2.42}$$

*Moreover, there exists a constant $C$ such that* $|\nabla \phi_\mathrm{ex}| \le C$, $|\nabla \boldsymbol{A}_\mathrm{ex}| \le C$.

The Abraham model is derived from the Lagrangian

$$L = -m_\mathrm{b}(1 - \dot{\boldsymbol{q}}^2)^{1/2} - e\big(\phi_\mathrm{ex}(\boldsymbol{q}) + \phi_\varphi(\boldsymbol{q}) - \dot{\boldsymbol{q}} \cdot \boldsymbol{A}_\mathrm{ex}(\boldsymbol{q}) - \dot{\boldsymbol{q}} \cdot \boldsymbol{A}_\varphi(\boldsymbol{q})\big) + \frac{1}{2}\int \mathrm{d}^3x\big((\nabla\phi + \partial_t \boldsymbol{A})^2 - (\nabla \times \boldsymbol{A})^2\big). \tag{2.43}$$

Correspondingly, the energy

$$\mathcal{E}(\boldsymbol{E}, \boldsymbol{B}, \boldsymbol{q}, \boldsymbol{v}) = m_\mathrm{b}\gamma(\boldsymbol{v}) + e\phi_\mathrm{ex}(\boldsymbol{q}) + \frac{1}{2}\int \mathrm{d}^3x\big(\boldsymbol{E}(\boldsymbol{x})^2 + \boldsymbol{B}(\boldsymbol{x})^2\big) \tag{2.44}$$

is conserved.

As for any dynamical system, the first step in dealing with (2.39)–(2.41) is to construct a suitable phase space. The dynamical variables are $(\boldsymbol{E}(\boldsymbol{x}), \boldsymbol{B}(\boldsymbol{x}), \boldsymbol{q}, \boldsymbol{v}) = Y$ which is called a state of the system. We have $\boldsymbol{q} \in \mathbb{R}^3$, $\boldsymbol{v} \in \mathbb{V} = \{\boldsymbol{v} \mid |\boldsymbol{v}| < 1\}$. In addition, the energy (2.44) should be bounded. Thus it is natural to introduce the (real) Hilbert space

$$L^2 = L^2(\mathbb{R}^3, \mathbb{R}^3) \tag{2.45}$$

with norm $\|\boldsymbol{E}\| = (\int \mathrm{d}^3x |\boldsymbol{E}(\boldsymbol{x})|^2)^{1/2}$ and to define $\mathcal{L}$ as the set of states satisfying

$$\|Y\|_{\mathcal{L}} = \|\boldsymbol{E}\| + \|\boldsymbol{B}\| + |\boldsymbol{q}| + |\gamma(\boldsymbol{v})\boldsymbol{v}| < \infty\,. \tag{2.46}$$

In particular for the field energy, $\frac{1}{2}(\|\boldsymbol{E}\|^2 + \|\boldsymbol{B}\|^2) < \infty$. The norm $\|\cdot\|_{\mathcal{L}}$ gives rise to the metric

$$d(Y_1, Y_2) = \|\boldsymbol{E}_1 - \boldsymbol{E}_2\| + \|\boldsymbol{B}_1 - \boldsymbol{B}_2\| + |\boldsymbol{q}_1 - \boldsymbol{q}_2| + |\gamma(\boldsymbol{v}_1)\boldsymbol{v}_1 - \gamma(\boldsymbol{v}_2)\boldsymbol{v}_2|\,. \tag{2.47}$$

In addition, the constraints (2.40) have to be satisfied. Thus the phase space, $\mathcal{M}$, for the Abraham model is the nonlinear submanifold of $\mathcal{L}$ defined through

$$\nabla \cdot \boldsymbol{E}(\boldsymbol{x}) = e\varphi(\boldsymbol{x} - \boldsymbol{q})\,, \quad \nabla \cdot \boldsymbol{B}(\boldsymbol{x}) = 0\,. \tag{2.48}$$

$\mathcal{M}$ inherits its metric from $\mathcal{L}$.

On various occasions below we will need the property that the system forgets its initial field data. For this purpose it is helpful to have a little bit of smoothness and some decay at infinity. Formally we introduce the "good" subset $\mathcal{M}^\sigma \subset \mathcal{M}$, $0 \le \sigma \le 1$, consisting of fields such that componentwise and outside a ball of radius $R_0$, $|\boldsymbol{x}| \ge R_0$, we have

$$|\boldsymbol{E}(\boldsymbol{x})| + |\boldsymbol{B}(\boldsymbol{x})| + |\boldsymbol{x}|(|\nabla\boldsymbol{E}(\boldsymbol{x})| + |\nabla\boldsymbol{B}(\boldsymbol{x})|) \le C\,|\boldsymbol{x}|^{-1-\sigma}\,. \tag{2.49}$$

The Liénard–Wiechert fields (2.24), (2.25) are included in $\mathcal{M}^0$; moreover, $\mathcal{M}^0$ is dense in $\mathcal{M}$. However $\mathcal{M}^\sigma = \emptyset$ for $\sigma > 1$, by Gauss's law (2.40) with $e \neq 0$.

The evolution equations (2.39)–(2.41) are of the general form

$$\frac{\mathrm{d}}{\mathrm{d}t} Y(t) = F(Y(t)) \tag{2.50}$$

with $Y(0) = Y^0 \in \mathcal{M}$. We turn to the question of the existence and uniqueness of solutions of the Abraham model (2.50).

**Theorem 2.1** (Existence of the dynamics for the Abraham model). *Let the conditions (C) and (P) hold and let* $Y^0 = (\boldsymbol{E}^0(\boldsymbol{x}), \boldsymbol{B}^0(\boldsymbol{x}), \boldsymbol{q}^0, \boldsymbol{v}^0) \in \mathcal{M}$. *Then the* *integral equation associated with (2.50),*

$$Y(t) = Y^0 + \int_0^t \mathrm{d}s\, F(Y(s)), \tag{2.51}$$

*has a unique solution* $Y(t) = (\boldsymbol{E}(\boldsymbol{x},t), \boldsymbol{B}(\boldsymbol{x},t), \boldsymbol{q}(t), \boldsymbol{v}(t)) \in \mathcal{M}$, *which is continuous in* $t$ *and satisfies* $Y(0) = Y^0$. *Along the solution trajectory*

$$\mathcal{E}(Y(t)) = \mathcal{E}(Y^0) \tag{2.52}$$

*for all* $t$, *i.e. the energy is conserved.*

For short times existence and uniqueness follow through the contraction mapping principle with constants depending only on the initial energy. For smooth initial data, energy conservation is verified directly and by continuity it extends to all finite-energy data. Thus we can construct iteratively the solution for all times.

We first summarize some properties of the Maxwell equations. They follow directly from the Fourier and convolution representations (2.12), (2.13), respectively (2.16), (2.17).

**Lemma 2.2** *In the Maxwell equations (2.2), (2.3), let* $e\varphi(\boldsymbol{x},t) = e\varphi(\boldsymbol{x} - \boldsymbol{q}(t))$, $\boldsymbol{j}(\boldsymbol{x},t) = e\varphi(\boldsymbol{x} - \boldsymbol{q}(t))\boldsymbol{v}(t)$, *with prescribed* $t \mapsto (\boldsymbol{q}(t), \boldsymbol{v}(t))$ *continuous. Then (2.2), (2.3) has a unique solution in* $C(\mathbb{R}, L^2 \oplus L^2)$. *The solution map* $(\boldsymbol{E}^0, \boldsymbol{B}^0) \mapsto (\boldsymbol{E}(t), \boldsymbol{B}(t))$ *depends continuously on* $(\boldsymbol{q}(t), \boldsymbol{v}(t))$.

*Proof of Theorem 2.1*: Let $b > 0$ be fixed and choose initial data such that $\mathcal{E}(Y^0) \le b$.
(i) There exists a unique solution $Y(t) \in C([0,\delta], \mathcal{M})$ for $\delta = \delta(b)$ sufficiently small.

We write (2.41) in the form

$$\frac{\mathrm{d}}{\mathrm{d}t}\big(m_{\mathrm{b}}\gamma\, \boldsymbol{v}(t)\big) = \boldsymbol{F}_{\mathrm{ex}}(t) + \boldsymbol{F}_{\mathrm{ini}}(t) + \boldsymbol{F}_{\mathrm{self}}(t) \tag{2.53}$$

by inserting $\boldsymbol{E}(\boldsymbol{x},t), \boldsymbol{B}(\boldsymbol{x},t)$ from the Maxwell equations according to (2.16), (2.17). Let

$$\begin{aligned} W_t(\boldsymbol{x}) &= e^2 \int \mathrm{d}^3k |\widehat{\varphi}(\boldsymbol{k})|^2\, \mathrm{e}^{\mathrm{i}\boldsymbol{k}\cdot\boldsymbol{x}} \frac{1}{|\boldsymbol{k}|} \sin|\boldsymbol{k}|t \\ &= (2\pi)^3 e^2 \int \mathrm{d}^3y \int \mathrm{d}^3y' \varphi(\boldsymbol{y})\varphi(\boldsymbol{y}') \frac{1}{4\pi t}\, \delta(|\boldsymbol{y} + \boldsymbol{x} - \boldsymbol{y}'| - t)\,. \end{aligned} \tag{2.54}$$

Then

$$F_{\mathrm{ex}}(t) = e^{-}\big(E_{\mathrm{ex}}(q(t)) + v(t) \times B_{\mathrm{ex}}(q(t))\big)\,, \tag{2.55}$$

$$F_{\mathrm{ini}}(t) = \int d^3x\, e\varphi(x - q(t))\big[\partial_t G_t * E^0(x) + \nabla \times G_t * B^0(x) + v(t) \times \partial_t G_t * B^0(x) - v(t) \times (\nabla \times G_t * E^0(x))\big]\,, \tag{2.56}$$

$$F_{\mathrm{self}}(t) = \int_0^t ds\big[ - \nabla W_{t-s}(q(t) - q(s)) - v(s)\partial_t W_{t-s}(q(t) - q(s)) + v(t) \times (\nabla \times v(s) W_{t-s}(q(t) - q(s)))\big]\,. \tag{2.57}$$

We now integrate both sides of (2.53) over the time interval $[0, t]$. The resulting expression is regarded as a map from the trajectory $t \mapsto (q(t), v(t))$, $0 \le t \le \delta$, to the trajectory $t \mapsto (\bar{q}(t), \bar{v}(t))$ and is defined by

$$\bar{q}(t) = q^0 + \int_0^t ds\, v(s)\,, \tag{2.58}$$

$$m_{\mathrm{b}}\gamma(\bar{v}(t))\bar{v}(t) = m_{\mathrm{b}}\gamma(v^0)v^0 + \int_0^t ds\big(F_{\mathrm{ex}}(s) + F_{\mathrm{ini}}(s) + F_{\mathrm{self}}(s)\big)\,,$$

where $F_{\mathrm{ex}}(s)$, $F_{\mathrm{ini}}(s)$, and $F_{\mathrm{self}}(s)$ are functionals of $q(\cdot), v(\cdot)$ according to (2.55)–(2.57). Since $\varphi$, $W$, $\phi_{\mathrm{ex}}$, and $A_{\mathrm{ex}}$ are smooth, this map is a contraction in $C([0, t], \mathbb{R}^3 \times \mathbb{V})$, i.e.

$$\sup_{0 \le s \le t} \big(|\bar{q}_2(s) - \bar{q}_1(s)| + |\bar{v}_2(s) - \bar{v}_1(s)|\big) \le c(t, b) \sup_{0 \le s \le t} \big(|q_1(s) - q_2(s)| + |v_1(s) - v_2(s)|\big)\,, \tag{2.59}$$

with a constant $c(t, b)$ depending on $b$ and $c(t, b) < 1$ for sufficiently small $t$. Such a map has a unique fixed point which is the desired solution $(q(t), v(t))$. By the Maxwell equations also $B(x, t)$, $E(x, t)$ are uniquely determined.

(ii) The solution map $Y^0 \mapsto Y(t)$ is continuous in $\mathcal{M}$.

This follows from Lemma 2.2 and the continuous dependence of $(q(t), v(t))$ on the initial data.

(iii) The energy is conserved.

We choose smooth initial fields such that $E, B \in C^\infty(\mathbb{R}^3)$ and

$$|\nabla^\alpha E(x)| + |\nabla^\alpha B(x)| \le C(1 + |x|)^{-(2+|\alpha|)}\,. \tag{2.60}$$

Here $\alpha = (\alpha_1, \alpha_2, \alpha_3)$ is a multi-index with $\alpha_i = 0, 1, 2, \ldots$. This subset is dense in $\mathcal{M}$. By the convolution representation (2.16), (2.17) of the solution to the Maxwell equations we have $\boldsymbol{E}(\boldsymbol{x}, t), \boldsymbol{B}(\boldsymbol{x}, t) \in C^1([0, \delta] \times \mathbb{R}^3)$ and $|\boldsymbol{E}(\boldsymbol{x}, t)| + |\boldsymbol{B}(\boldsymbol{x}, t)| \le C(1 + |\boldsymbol{x}|)^{-2}$. Also $\boldsymbol{v}(t) \in C^1([0, \delta])$. Thus we are allowed to differentiate,

$$\frac{\mathrm{d}}{\mathrm{d}t}\mathcal{E}(Y(t)) = \gamma^3 \boldsymbol{v} \cdot \dot{\boldsymbol{v}} + \boldsymbol{v} \cdot \nabla\phi_{\mathrm{ex}}(\boldsymbol{q}) + \int \mathrm{d}^3x (\boldsymbol{E} \cdot \partial_t \boldsymbol{E} + \boldsymbol{B} \cdot \partial_t \boldsymbol{B})$$
$$= \int \mathrm{d}^3x \big(\boldsymbol{E} \cdot (\nabla \times \boldsymbol{B}) - \boldsymbol{B} \cdot (\nabla \times \boldsymbol{E})\big) = 0\,, \qquad (2.61)$$

since the fields decay and hence the surface terms vanish. Thus $\mathcal{E}(Y(t)) = \mathcal{E}(Y^0)$ for $0 \le t \le \delta$. By continuity this equality extends to all of $\mathcal{M}$.

(iv) The global solution exists.

From (iii) we know that $\mathcal{E}(Y(\delta)) = \mathcal{E}(Y^0) \le b$. Thus we can repeat the previous argument for $\delta \le t \le 2\delta$, etc. Backwards in time we still have the solution (2.16), (2.17) of the Maxwell equations, only the retarded fields have to be replaced by the advanced fields. Thereby we obtain the solution for all times. □

Theorem 2.1 ensures the existence and uniqueness of solutions for the Abraham model. For initial data $Y^0 \in \mathcal{M}$ the solution trajectory $t \to Y(t)$ lies in the phase space $\mathcal{M}$, is continuous in $t$, and its energy is conserved. We have thus established the basis for further investigations on the dynamics of the Abraham model.

## 2.5 The relativistically covariant Lorentz model

To improve on the semirelativistic Abraham model, following Lorentz, it is natural to assume that when viewed in a momentary inertial rest frame the charge and mass distribution of the particle remain unchanged. This is what one would call a relativistically rigid extended charge. Our requirement fixes uniquely the four-current density. The equations of motion then follow from a relativistically covariant action.

For obvious reasons we will switch to relativistic notation, where we follow the conventions of Misner, Thorne, and Wheeler. Our arena is the Minkowski space-time $\mathbb{M}^4$. A Lorentz frame, $\mathcal{F}_{\mathrm{L}}$, in $\mathbb{M}^4$ is specified through the tetrad $\{\mathbf{e}_0, \mathbf{e}_1, \mathbf{e}_2, \mathbf{e}_3\}$ of fixed unit vectors. They have the inner product

$$\mathbf{e}_\mu \cdot \mathbf{e}_\nu = g_{\mu\nu}\,, \qquad (2.62)$$

where $g_{\mu\nu}$ is the *metric tensor* with $g_{00} = -1$, $g_{\mu\mu} = 1$, $\mu = 1, 2, 3$, and $g_{\mu\nu} = 0$ otherwise. Therefore $\mathbb{M}^4$ can be identified with $\mathbb{R}^{1,3}$. In the given basis, a vector $\mathbf{x} \in \mathbb{M}^4$ is expanded as

$$\mathbf{x} = x^\mu \mathbf{e}_\mu \qquad (2.63)$$

using the Einstein summation convention over repeated indices. We group $\mathbf{x} = (t, \boldsymbol{x})$ with $t \in \mathbb{R}$ the time and $\boldsymbol{x} \in \mathbb{R}^3$ the space coordinate. The scalar product is $\mathbf{x} \cdot \mathbf{y} = g_{\mu\nu} x^\mu y^\nu$ and $|\mathbf{x}|^2 = \mathbf{x} \cdot \mathbf{x}$.

The motion of a particle is specified through its world line $\tau \mapsto \mathbf{q}(\tau)$ parametrized in terms of the eigentime $\tau$, $\mathrm{d}\tau^2 = -\mathrm{dx} \cdot \mathrm{dx}$. Denoting by $\dot{\mathbf{q}}$ differentiation of $\mathbf{q}(\tau)$ with respect to $\tau$, the four-velocity is $\mathbf{u}(\tau) = \dot{\mathbf{q}}(\tau)$. $\mathbf{u}$ is time-like, $\mathbf{u} \cdot \mathbf{u} = -1$, and $u_0 > 0$ for a particle moving forward in time. In the given Lorentz frame we have

$$\mathbf{u} = (\gamma, \gamma \boldsymbol{v}), \quad \gamma = (1 - |\boldsymbol{v}|^2)^{-1/2} \tag{2.64}$$

with $\boldsymbol{v}$ the usual three-velocity.

If the charged particle is at rest, then, as before, its charge is smeared according to the charge distribution $e\varphi$. In addition we assume that now the bare mass, $m_\mathrm{b}$, is smeared also according to $\varphi$. In principle, one should distinguish between the charge and mass form factor. We suppress such a distinction, since it can be unambiguously recovered from the prefactors $e$ and $m_\mathrm{b}$. By the definition of a rigid charge, we require that in any momentary rest frame the mass, respectively charge, distribution are given by $m_\mathrm{b}\varphi$, respectively $e\varphi$.

Since our charged body is extended, in its kinematical description, besides $\mathbf{q}(\tau)$ and the velocity $\mathbf{u}(\tau) = \dot{\mathbf{q}}(\tau)$, we have to specify its state of rotation. Let us introduce the (noninertial) body frame $\mathcal{F}_\mathrm{body}$ through the tetrad $\{\mathbf{e}'_\mu\}_{\mu=0,\ldots,3}$ of unit vectors. $\mathcal{F}_\mathrm{body}$ is fixed in the charged body and thus comoving and corotating. We set $\mathbf{e}'_0 = \mathbf{u}(\tau)$. $\{\mathbf{e}'_1, \mathbf{e}'_2, \mathbf{e}'_3\}$ gives then the spatial orientation of $\mathcal{F}_\mathrm{body}$ in the momentary rest frame. In the course of time $\mathcal{F}_\mathrm{body}$ evolves according to

$$\frac{\mathrm{d}}{\mathrm{d}\tau}\mathbf{e}'_\mu = -\boldsymbol{\Omega} \cdot \mathbf{e}'_\mu, \quad \mu = 0, \ldots, 3, \tag{2.65}$$

where $\boldsymbol{\Omega}$ is the antisymmetric tensor of the instantaneous rate of four-gyration of $\mathcal{F}_\mathrm{body}$ as seen in the Lorentz frame $\mathcal{F}_\mathrm{L}$.

Even if there is no external torque acting on the rigid charged body, the frame $\mathcal{F}_\mathrm{body}$ rotates. This is the famous Thomas precession, determined by the Fermi–Walker transport equation

$$\frac{\mathrm{d}}{\mathrm{d}\tau}\bar{\mathbf{e}}_\mu = -\boldsymbol{\Omega}_\mathrm{FW} \cdot \bar{\mathbf{e}}_\mu, \quad \mu = 0, \ldots, 3, \tag{2.66}$$

where

$$\boldsymbol{\Omega}_\mathrm{FW} = \dot{\mathbf{u}} \wedge \mathbf{u}. \tag{2.67}$$

Here the exterior product of two vectors is defined by $\mathbf{a} \wedge \mathbf{b} = \mathbf{a} \otimes \mathbf{b} - \mathbf{b} \otimes \mathbf{a}$ or, as acting on a vector $\mathbf{c}$, $(\mathbf{a} \wedge \mathbf{b}) \cdot \mathbf{c} = \mathbf{a}(\mathbf{b} \cdot \mathbf{c}) - \mathbf{b}(\mathbf{a} \cdot \mathbf{c})$. Together with the initial conditions $\bar{\mathbf{e}}_0(0) = \mathbf{u}(0)$, $\bar{\mathbf{e}}_\mu(0) = \mathbf{e}_\mu$, $\mu = 1, 2, 3$, (2.66) defines the noninertial frame $\mathcal{F}_{\mathrm{FW}}$.

If there is an external torque acting, then $\mathcal{F}_{\mathrm{body}} \neq \mathcal{F}_{\mathrm{FW}}$ and it is natural to introduce the intrinsic (Eulerian) four-gyration by

$$\boldsymbol{\Omega}_{\mathrm{E}} = \boldsymbol{\Omega} - \boldsymbol{\Omega}_{\mathrm{FW}}\,. \tag{2.68}$$

As $\boldsymbol{\Omega}$, $\boldsymbol{\Omega}_{\mathrm{FW}}$, also $\boldsymbol{\Omega}_{\mathrm{E}}$ is antisymmetric and satisfies

$$\boldsymbol{\Omega}_{\mathrm{E}} \cdot \mathbf{u} = 0\,. \tag{2.69}$$

Therefore $\boldsymbol{\Omega}_{\mathrm{E}}$ has only three independent components and is dual to a space-like four-vector $\mathbf{w}_{\mathrm{E}}$ which satisfies

$$\boldsymbol{\Omega}_{\mathrm{E}} \cdot \mathbf{w}_{\mathrm{E}} = 0\,, \quad \mathbf{w}_{\mathrm{E}} \cdot \mathbf{u} = 0\,. \tag{2.70}$$

In $\mathcal{F}_{\mathrm{FW}}$, $\mathbf{w}_{\mathrm{E}}$ is of the form $(0, \boldsymbol{\omega}_{\mathrm{E}})$, where $\boldsymbol{\omega}_{\mathrm{E}}$ is the usual angular velocity vector which points along the instantaneous axis of body gyration in the space-like three-slice of $\mathcal{F}_{\mathrm{FW}}$. For zero torque $\boldsymbol{\omega}_{\mathrm{E}} = 0$.

We conclude that relative to $\mathcal{F}_{\mathrm{FW}}$ the rotational state is either given by $\boldsymbol{\Omega}_{\mathrm{E}}(\tau)$ or by $\mathbf{w}_{\mathrm{E}}(\tau)$. $\mathbf{w}_{\mathrm{E}}(\tau)$ is space-like, $|\mathbf{w}_{\mathrm{E}}(\tau)|^2 \geq 0$.

### *2.5.1 The four-current density*

Our task is to construct a relativistically covariant current density, which will serve both as the source term in Maxwell's equations and as the force, respectively torque, term in Newton's equations of motion.

For a given world line let $\mathcal{F}'_{\mathrm{L}}$ be the momentary rest frame at time $\tau$ centered at $\mathbf{q}(\tau)$ with spatial axes oriented as in $\mathcal{F}_{\mathrm{L}}$. In the coordinates of $\mathcal{F}'_{\mathrm{L}}$, by definition, the four-current density is given by

$$\mathbf{j}'(t', x') = e\varphi_{\mathrm{r}}(|x'|)\delta(t')(1, 0)\,. \tag{2.71}$$

Transformed to our laboratory frame $\mathcal{F}_{\mathrm{L}}$ the current density becomes

$$\mathbf{j}(\mathbf{x}) = e\varphi_{\mathrm{r}}(|\mathbf{x} - \mathbf{q}(\tau_0)|)\mathbf{u}(\tau_0)|_{\sigma(\tau_0)}\,. \tag{2.72}$$

Here $\sigma(\tau)$ is the hyperplane defined by $\sigma(\tau) = \{\mathbf{y}|\mathbf{u}(\tau) \cdot (\mathbf{y} - \mathbf{q}(\tau)) = 0\}$ and the subscript in (2.72) means that for given $\mathbf{x}$ we have to choose $\tau_0$ such that $\mathbf{x} \in \sigma(\tau_0)$, see figure 2.1. In general, there will be several such planes, see figure 2.2. Of course, they contribute to the current only if $\mathbf{x} - \mathbf{q}(\tau_0)$ is space-like and the distance $|\mathbf{x} - \mathbf{q}(\tau_0)|$ satisfies $|\mathbf{x} - \mathbf{q}(\tau_0)| \leq R_\varphi$. Let us assume for the moment

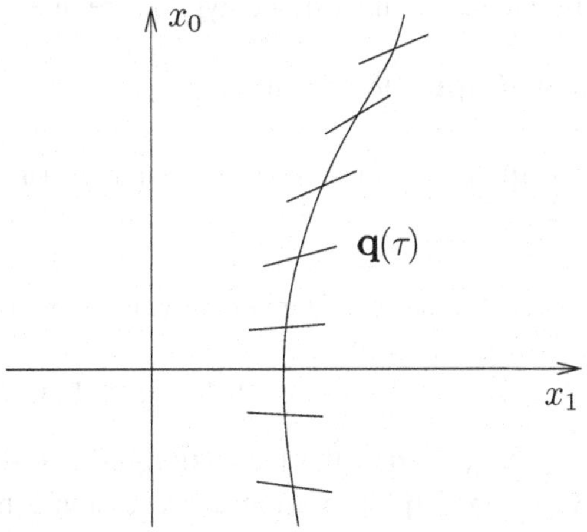


Figure 2.1: World line of an extended charge and the associated current density.

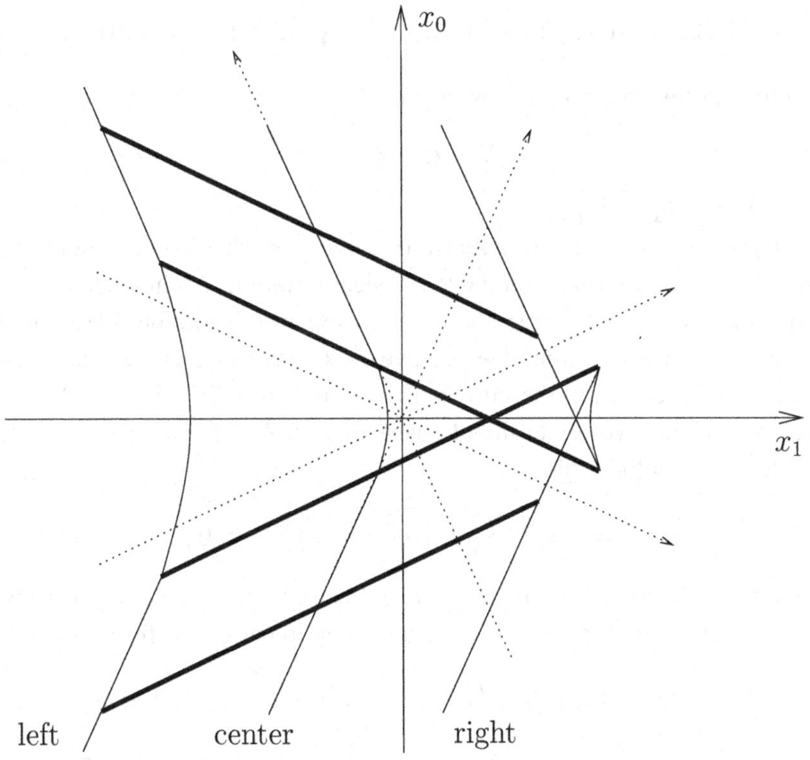


Figure 2.2: World line of an extended charge with large acceleration and backward currents.

that with this restriction there is only a single hyperplane intersecting $x$. Then

$$\begin{aligned}\mathbf{j}(\mathbf{x}) &= \int \mathrm{d}\tau e\varphi_{\mathrm{r}}(|\mathbf{x}-\mathbf{q}(\tau)|)\mathbf{u}(\tau)\delta(\tau-\tau_0)|_{\sigma(\tau_0)} \\ &= \int \mathrm{d}\tau e\varphi_{\mathrm{r}}(|\mathbf{x}-\mathbf{q}(\tau)|)\mathbf{u}(\tau)\big(1+\dot{\mathbf{u}}(\tau)\cdot(\mathbf{x}-\mathbf{q}(\tau))\big)\delta(\mathbf{u}(\tau)\cdot(\mathbf{x}-\mathbf{q}(\tau)))\,.\end{aligned} \tag{2.73}$$

The additional term comes from the change in the volume element, since

$$\frac{\mathrm{d}}{\mathrm{d}\tau}\mathbf{u}\cdot(\mathbf{x}-\mathbf{q}) = \dot{\mathbf{u}}\cdot(\mathbf{x}-\mathbf{q}) - \mathbf{u}\cdot\mathbf{u} = 1+\dot{\mathbf{u}}\cdot(\mathbf{x}-\mathbf{q})\,. \tag{2.74}$$

Note that, because of $\delta(\mathbf{u}\cdot(\mathbf{x}-\mathbf{q}))$, the factor $\mathbf{u}(1+\dot{\mathbf{u}}\cdot(\mathbf{x}-\mathbf{q}))$ in (2.73) may be replaced by $\mathbf{u}-\boldsymbol{\Omega}_{\mathrm{FW}}\cdot(\mathbf{x}-\mathbf{q})$. The Thomas precession generates a current in addition to that due to translations.

In general, the body-fixed frame will be rotated by $\boldsymbol{\Omega}$ and we arrive at the final form of the four-current density as

$$\mathbf{j}(\mathbf{x}) = \int \mathrm{d}\tau e\varphi_{\mathrm{r}}(|\mathbf{x}-\mathbf{q}|)\delta(\mathbf{u}\cdot(\mathbf{x}-\mathbf{q}))(\mathbf{u}-\boldsymbol{\Omega}\cdot(\mathbf{x}-\mathbf{q}))\,. \tag{2.75}$$

One readily verifies the charge conservation

$$\nabla_g\cdot\mathbf{j}(\mathbf{x}) = 0\,, \tag{2.76}$$

where $\nabla_g f = (-\partial_{x_0}f, \nabla f)$.

Before proceeding to the action for the dynamics, we should understand whether the current (2.75) conforms with naive physical intuition. An instructive example is a uniformly accelerated charge, the so-called hyperbolic motion. We assume that the particle is accelerated along the positive 1-axis starting from rest at the origin. In the orthogonal direction the current traces out a tube of diameter $2R_\varphi$ and it suffices to treat the two-dimensional space-time problem. The center, $C$, of the charge moves along the orbit

$$C = \Big(t, g^{-1}\Big(\sqrt{1+g^2t^2}-1\Big)\Big),\ t\geq 0\,, \tag{2.77}$$

where $g>0$ is the acceleration. The curves traced by the right and left ends, $C_+$ and $C_-$, are determined from (2.73) and are given in parameter form as

$$C_\pm = \Big((1\pm R_\varphi g)t, g^{-1}\Big((1+R_\varphi g)\sqrt{1+g^2t^2}-1\Big)\Big),\ t\geq 0\,. \tag{2.78}$$

The equal-time distance between the center and $C_+$ is $t^{-1}((R_\varphi g)^2+2R_\varphi g)/(2g^2(1+R_\varphi g))$ for large $t$ and is thus well bounded. However the left end motion depends crucially on the magnitude of $R_\varphi g$. If $R_\varphi g<1$, then the distance to the center is $t^{-1}((R_\varphi g)^2-2R_\varphi g)/(2g^2(1-R_\varphi g))$ for large $t$. On the other hand, for

$R_\varphi g > 1$, the left end moves into the past and the current density looks strangely distorted. To gain a feeling for the order of magnitudes involved we insert the classical electron radius. Then

$$g > \frac{c^2}{R_\varphi} = 10^{31}\,[\mathrm{m\,s^{-1}}]\,, \tag{2.79}$$

which is far beyond the domain of the validity of the theory. Of course, one would hope that for reasonable initial data such accelerations can never be reached. But the mere fact that charge elements may move backwards in time is an extra difficulty.

### 2.5.2 Relativistic action, equations of motion

For given current density, $\boldsymbol{j}$, the Maxwell equations read

$$\nabla_g \cdot^{*} \mathbf{F} = 0\,, \quad \nabla_g \cdot \mathbf{F} = \mathbf{j}\,, \tag{2.80}$$

where $\mathbf{F}$ is the antisymmetric electromagnetic field tensor of rank 2 and $^{*}\mathbf{F}$ its star dual. Equations (2.80) can be regarded as the Euler–Lagrange equations of an action functional $\mathcal{A}_\mathrm{f}$, which most conveniently is written in terms of a Lagrange density $\mathcal{L}_\mathrm{f}(\mathbf{x}) + \mathcal{L}_\mathrm{int}(\mathbf{x})$. The field part of the Lagrangian is given by

$$\mathcal{L}_\mathrm{f}(\mathbf{x}) = -\frac{1}{4}\mathrm{tr}[\mathbf{F}(\mathbf{x}) \cdot \mathbf{F}(\mathbf{x})]\,. \tag{2.81}$$

The interaction Lagrangian, $\mathcal{L}_\mathrm{int}(\mathbf{x})$, is defined through minimal coupling. We recall that (2.80) implies that $\mathbf{F}$ is the exterior derivative of a vector potential $\mathbf{A}$, $\mathbf{F} = \nabla_g \wedge \mathbf{A}$. If we adopt the Lorentz gauge $\nabla_g \cdot \mathbf{A} = 0$, then

$$\mathcal{L}_\mathrm{int}(\mathbf{x}) = \mathbf{A}(\mathbf{x}) \cdot \mathbf{j}(\mathbf{x})\,. \tag{2.82}$$

The variation of

$$\mathcal{A}_\mathrm{f} = \int (\mathcal{L}_\mathrm{f}(\mathbf{x}) + \mathcal{L}_\mathrm{int}(\mathbf{x}))\mathrm{d}^4\mathrm{x} \tag{2.83}$$

with respect to $\mathbf{A}$ yields indeed (2.80).

Thus we are left with writing down the particle Lagrangian. One might be tempted to simply take $-m_\mathrm{b}\int \mathrm{d}\tau$ from the relativistic mechanics of a single particle. This cannot be correct, unless all mass is concentrated at the center, i.e. $\varphi(\boldsymbol{x}) = \delta(\boldsymbol{x})$, since $-m_\mathrm{b}\int \mathrm{d}\tau$ ignores the energy stored in the inner rotation.

Including rotation the Lagrangian density for the particle becomes

$$\mathcal{L}_{\mathrm{p}}(\mathbf{x}) = -\int_{\tau_1}^{\tau_2} (1 - |\boldsymbol{\Omega}_{\mathrm{E}} \cdot (\mathbf{x} - \mathbf{q})|^2)^{1/2} m_{\mathrm{b}} \varphi_{\mathrm{r}}(|\mathbf{x} - \mathbf{q}|) \delta(\mathbf{u} \cdot (\mathbf{x} - \mathbf{q})) \mathrm{d}\tau, \quad (2.84)$$

where $\mathbf{q} = \mathbf{q}(\tau)$, $\mathbf{u} = \mathbf{u}(\tau)$, and $\boldsymbol{\Omega}_{\mathrm{E}} = \boldsymbol{\Omega}_{\mathrm{E}}(\tau)$ along the world line of the particle.

Let us check that (2.84) yields the physically correct equations of motion when $\mathbf{A}(\mathbf{x})$ is taken to be given. We have

$$\mathcal{A}_{\mathrm{p}} = \int (\mathcal{L}_{\mathrm{p}}(\mathbf{x}) + \mathcal{L}_{\mathrm{int}}(\mathbf{x})) \mathrm{d}^4 \mathrm{x} \quad (2.85)$$

and must work out the variation of the world line $\tau \mapsto \mathbf{q}(\tau)$ at fixed end points, $\delta\mathbf{q}(\tau_1) = 0 = \delta\mathbf{q}(\tau_2)$, which induces also a change in the Fermi–Walker frame. The second independent variation is the body-fixed frame $\mathcal{F}_{\mathrm{body}}$ relative to $\mathcal{F}_{\mathrm{FW}}$. Thereby we obtain two equations of motion, which we write as

$$\frac{\mathrm{d}}{\mathrm{d}\tau} \mathbf{p}(\tau) = \mathbf{f}(\tau), \quad (2.86)$$

$$\frac{\mathrm{d}}{\mathrm{d}\tau} \mathbf{s}(\tau) + \boldsymbol{\Omega}_{\mathrm{FW}} \cdot \mathbf{s}(\tau) = \mathbf{t}(\tau) . \quad (2.87)$$

Let us discuss each equation separately. $\mathbf{p}$ is the momentum of the particle, related to the velocity by

$$\mathbf{p} = m_{\mathrm{g}} \mathbf{u} . \quad (2.88)$$

$m_{\mathrm{g}}$ depends on $|\omega_{\mathrm{E}}|$ and is defined by

$$m_{\mathrm{g}} = \int_{\mathbb{R}^{1,3}} (1 - |\boldsymbol{\Omega}_{\mathrm{E}} \cdot \mathbf{x}|^2)^{-1/2} m_{\mathrm{b}} \varphi_{\mathrm{r}}(|\mathbf{x}|) \delta(\mathbf{u} \cdot \mathbf{x}) \mathrm{d}^4 \mathrm{x} . \quad (2.89)$$

$m_{\mathrm{g}}$ is the *bare gyrational mass*, a Lorentz scalar. For small gyration frequency it can be expanded as

$$m_{\mathrm{g}} = m_{\mathrm{b}} + \frac{1}{2} I_{\mathrm{nr}} |\omega_{\mathrm{E}}|^2 + \mathcal{O}(|\omega_{\mathrm{E}}|^4) \quad (2.90)$$

with

$$I_{\mathrm{nr}} = m_{\mathrm{b}} \frac{2}{3} \int \mathrm{d}^3 x \varphi(x) x^2 , \quad (2.91)$$

the moment of inertia in the nonrelativistic limit. $\mathbf{f}(\tau)$ in (2.86) is the Minkowski force

$$\mathbf{f}(\tau) = \int_{\mathbb{R}^{1,3}} \mathbf{F}(\mathbf{x}) \cdot (\mathbf{u} - \boldsymbol{\Omega} \cdot (\mathbf{x} - \mathbf{q})) e \varphi_{\mathrm{r}}(|\mathbf{x} - \mathbf{q}|) \delta(\mathbf{u} \cdot (\mathbf{x} - \mathbf{q})) \mathrm{d}^4 \mathrm{x} . \quad (2.92)$$

It reduces to the Lorentz force, $e\mathbf{F}\cdot\mathbf{u}$, in the case where $\mathbf{F}(\mathbf{x})$ is slowly varying on the scale of $R_\varphi$.

In the rotational equation (2.87), $\mathbf{s}$ is the four-vector of *spin angular momentum* and is related to the four-gyration by

$$\mathbf{s}_\mathrm{b} = I_\mathrm{b}\mathbf{w}_\mathrm{E} \tag{2.93}$$

with $I_\mathrm{b}$ the *relativistic moment of inertia* relative to $\mathbf{q}$,

$$I_\mathrm{b}(|\omega_\mathrm{E}|)\mathbf{g} = \int_{\mathbb{R}^{1,3}} (|\mathbf{x}|^2\mathbf{g} - \mathbf{x}\otimes\mathbf{x})(1 - |\boldsymbol{\Omega}_\mathrm{E}\cdot\mathbf{x}|^2)^{-1/2} m_\mathrm{b}\varphi_\mathrm{r}(|\mathbf{x}|)\delta(\mathbf{u}\cdot\mathbf{x})\mathrm{d}^4\mathrm{x}\,. \tag{2.94}$$

In (2.87) $\boldsymbol{s}$ is kinematically Fermi–Walker transported by $\boldsymbol{\Omega}_\mathrm{FW}$ and changed through the external Minkowski torque $\mathbf{t}(\tau)$. From the variation of (2.85) we obtain

$$\mathbf{t}(\tau) = \int_{\mathbb{R}^{1,3}} (\mathbf{x}-\mathbf{q})\wedge(\mathbf{F}(\mathbf{x})\cdot(\mathbf{u}-\boldsymbol{\Omega}\cdot(\mathbf{x}-\mathbf{q})))^{\perp} e\varphi_\mathrm{r}(|\mathbf{x}-\mathbf{q}|)\delta(\mathbf{u}\cdot(\mathbf{x}-\mathbf{q}))\mathrm{d}^4\mathrm{x}\,, \tag{2.95}$$

where by definition $\boldsymbol{a}^{\perp} = (\mathbf{g}+\mathbf{u}\otimes\mathbf{u})\cdot\mathbf{a}$. In the case of slow variation of $\mathbf{F}$, (2.95) becomes the BMT equation, cf. section 10.1.

We remark that through (2.86), (2.87) the translational and rotational motion are coupled in a rather complicated way with some simplification for a slowly varying external potential $\mathbf{A}_\mathrm{ex}$.

Having discussed the action (2.83) for the field at prescribed currents and the action (2.85) for the particle at prescribed fields, the action for the Lorentz model of an extended charge is inevitable. The Lagrangian density reads

$$\mathcal{L}(\mathbf{x}) = \mathcal{L}_\mathrm{p}(\mathbf{x}) + \mathcal{L}_\mathrm{int}(\mathbf{x}) + \mathcal{L}_\mathrm{f}(\mathbf{x}) \tag{2.96}$$

with the corresponding action

$$\mathcal{A} = \int_\Xi \mathcal{L}(\mathbf{x})\mathrm{d}^4\mathrm{x}\,. \tag{2.97}$$

To include an external potential, $\mathcal{L}_\mathrm{int}$ from (2.82) has to be merely modified to $\mathcal{L}_\mathrm{int}(\mathbf{x}) = \mathbf{A}(\mathbf{x})\cdot\mathbf{j}(\mathbf{x}) + \mathbf{A}_\mathrm{ex}(\mathbf{x})\cdot\mathbf{j}(\mathbf{x})$.

One has to be careful with the domain of integration, $\Xi$. It is a region of $\mathbb{M}^4$, which is bordered by two space-like surfaces, $\partial\Xi_i$, $i=1,2$. One first fixes an interval $[\tau_1,\tau_2]$ of eigentimes. Restricted to a ball of radius $R_\varphi$, $\partial\Xi_i = \{\mathbf{y}|\,\mathbf{u}(\tau_i)\cdot(\mathbf{y}-\mathbf{q}(\tau_i)) = 0\}$, $i=1,2$. $\partial\Xi_1$, $\partial\Xi_2$ are then smoothly extended to hypersurfaces such that they do not intersect each other, see figure 2.3. The variation is carried out at fixed end points, which means that $\mathbf{q}(\tau_1)$, $\mathbf{q}(\tau_2)$, $\boldsymbol{\Omega}_\mathrm{E}(\tau_1)$, $\boldsymbol{\Omega}_\mathrm{E}(\tau_2)$, and $\mathbf{A}$ on the hypersurfaces $\partial\Xi_i$, $i=1,2$, are prescribed. In addition we require a properly

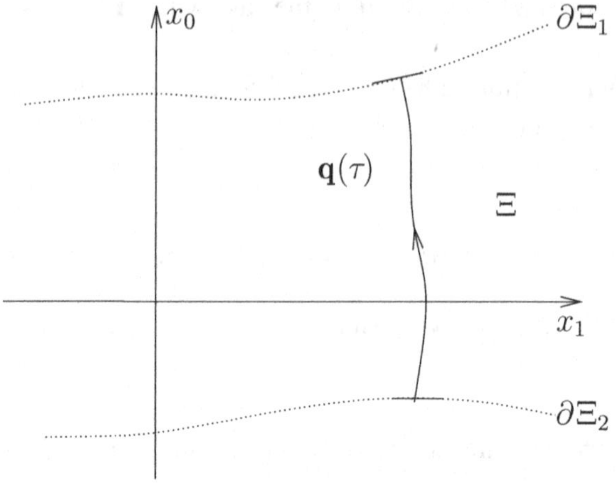


Figure 2.3: Space-like boundary surfaces in the variation of the action.

time-ordered history of momentary charge slices. Then the Euler–Lagrange equations for (2.97) are given by Maxwell's equations (2.80), by Newton's equations (2.86) for the translational degrees of freedom together with (2.88), (2.89), (2.92), and by Newton's equations (2.87) for the rotational degrees of freedom together with (2.93), (2.94), (2.95), as a coupled set of equations for the extended charge and the Maxwell field.

As for the Abraham model we should discuss the existence and uniqueness of solutions. This project is hampered by the fact that we have two constraints. The equator must have a subluminal speed of gyration, which is ensured by $|\boldsymbol{\omega}_{\mathrm{E}}|R_\varphi < 1$. In addition, the charge slices have to move forward in time, which is ensured by $|\ddot{\mathbf{q}}|R_\varphi < 1$. The difficulty is that, even if these conditions are met initially, there seems to be no mechanism which ensures their validity later on. At present, the general Cauchy problem is known to have a solution only for a finite interval of time, whose duration depends on the initial data.

## Notes and references

### *Sections 2.1 and 2.2*

The material discussed can be found in most textbooks. I find Landau and Lifshitz (1959), Panofsky and Phillips (1962), Jackson (1999), and Scharf (1994) particularly useful.

### *Section 2.3*

In our history chapter, chapter 3, we discuss the Wheeler–Feynman approach which cannot be subsumed under short distance regularization. In the literature the size of a classical electron, $r_{\mathrm{cl}}$, is usually determined through equating the rest mass with the Coulomb energy, $m_{\mathrm{e}}c^2 = e^2/r_{\mathrm{cl}}$, which gives $r_{\mathrm{cl}} = 3 \times 10^{-13}$cm. This is really a lower bound in the sense that an even smaller radius would be in contradiction to the experimentally observed mass of the electron (assuming a positive bare mass, cf. the discussion in section 6.3). Milonni (1994) argues that due to quantum fluctuations the electron appears to have a classical spread, which is given by its Compton wavelength $\lambda_{\mathrm{c}} = r_{\mathrm{cl}}/\alpha$, with $\alpha$ the fine structure constant. Renormalization in Euclidean quantum field theory is covered by Glimm and Jaffe (1987) and Huang (1998). Effective potentials for classical fluids are discussed, e.g., in Huang (1987).

### *Section 2.4*

The Abraham model was very popular in the early 1900s as studied by Abraham (1903, 1905), Lorentz (1892, 1915), Sommerfeld (1904a, 1904b, 1904c, 1905), and Schott (1912), among others. The extension to a rigid charge with rotation was already introduced in Abraham (1903) and further investigated by Herglotz (1903) and Schwarzschild (1903); compare with chapter 10. The dynamical systems point of view is stressed in Galgani *et al.* (1989). The proof of existence and uniqueness of the dynamics is taken from Komech and Spohn (2000), where a much wider class of external potentials is allowed. A somewhat different technique is used by Bauer and Dürr (2001). They also cover the case of a negative bare mass and discuss the smoothness of solutions in terms of the smoothness of initial data.

### *Section 2.5*

This section is based on Appel and Kiessling (2001). Amongst many other results they explain the somewhat tricky variation of the action (2.97). Global existence of solutions is available in the case where the charge moves with constant velocity (Appel and Kiessling 2002). Appel and Kiessling (2001) rely on the monumental work of Nodvik (1964), but differ in one crucial aspect. Nodvik assumes that the mass of the extended body is concentrated in its center, which implies $I_{\mathrm{b}} = 0$. Newton's equations for the torque degenerate then into a constraint, which makes the Cauchy problem singular. A discussion of the Nodvik model can be found in Rohrlich (1990), chapter 7-4. The relativistic Thomas precession is discussed in Thomas (1926, 1927), Møller (1952), and in Misner, Thorne and Wheeler (1973), which is an excellent source on relativistic electrodynamics. Another informative source is Thirring (1997).

Of course, relativistic theories were studied much earlier, e.g. Born (1909). I refer to Yaghjian (1992) for an exhaustive discussion. The early models use a continuum description of the extended charge where each charge element has a velocity. They are not dynamical models in our sense, simply because there are more unknowns than equations. Also inner rotation is neglected, which, as we discussed, is not admissible in a relativistic theory.

The current generated by a point charge can be written as

$$\mathbf{j}(\mathbf{x}) = e \int_{-\infty}^{\infty} d\tau \mathbf{u}(\tau) \delta(\mathbf{x} - \mathbf{q}(\tau)) . \tag{2.98}$$

McManus (1948) proposes to smear out the $\delta$-function as

$$\mathbf{j}(\mathbf{x}) = e \int_{-\infty}^{\infty} d\tau \mathbf{u}(\tau) \varphi_{\mathrm{MM}}((\mathbf{x} - \mathbf{q}(\tau))^2) , \tag{2.99}$$

which is to be inserted in the Lagrange density (2.82). He does not identify the conserved four-momentum, see also Peierls (1991) for illuminating explanations. Schwinger (1983) discusses the structure of the electromagnetic energy–momentum tensor in the case of rectilinear motion of the charge.

A more radical approach to a fully relativistic theory is to give up the notion of a material charged object and to regard electrons as point singularities of the Maxwell–Lorentz field. The guiding example are point vortices in a two-dimensional ideal Euler fluid, whose motion is governed by a closed set of differential equations which are of Hamiltonian form with the 1- and 2-component of the position as a canonically conjugate pair. In electrodynamics such a program was launched by Born (1933) and Born and Infeld (1933) and has not lost in attraction even now, mostly through activities in high-energy physics and string theory. Still, to have meaningful Newtonian equations of motion for the singularities is not so readily achieved. A recent proposal, based on the Hamilton–Jacobi equation, has been made by Kiessling (2003). He also provides a coherent overview of earlier attempts.

# 3

# Historical notes

## 3.1 Extended charge models (1897–1912)

When in 1897 J. J. Thomson identified the cathode rays as consisting of particles with charge $-e$, not only had he discovered the first elementary particle, but posed the theoretical challenge of computing the energy–momentum relation of this novel object. To put it concisely, we write the equations of motion in approximately uniform $\boldsymbol{E}$ and $\boldsymbol{B}$ fields as

$$m(\boldsymbol{v})\dot{\boldsymbol{v}} = e(\boldsymbol{E} + c^{-1}\boldsymbol{v} \times \boldsymbol{B}) \tag{3.1}$$

with $m(\boldsymbol{v})$ the velocity-dependent mass as a $3 \times 3$ matrix. The challenge was to predict the ratio $m(\boldsymbol{v})/e$. For small velocities it was well established that the mass is independent of $\boldsymbol{v}$. But for the electron with its tiny mass and unprecedented range of accessible velocities the case was wide open. In fact, Thomson (1881) himself had pointed out that, in analogy with a ball immersed in a fluid, the coupling to the self-generated electromagnetic field will induce a velocity dependence of the mass.

So which theory could be used to determine $m(\boldsymbol{v})$? In fact, there was little choice. Since the phenomenon under consideration is clearly electromagnetic, the Maxwell–Lorentz equations had to be used, and since the trajectory of a single charge was measured, one had to couple through Newton's equations of motion. Thus the electron was pictured as a tiny sphere charged with electricity. In the inhomogeneous Maxwell equations the current generated by that moving sphere had to be inserted. On the other hand the electromagnetic fields react back on the charge distribution through the Lorentz force. Thereby the so-called extended charge model was introduced. Abraham (1903, 1904) adopted a charge distribution which is rigid in the laboratory frame. The corresponding energy–momentum relation is discussed at length in the second volume of his book on electromagnetism (Abraham 1905), compare with section 4.1. For Abraham's model, Sommerfeld

(1904a, 1905) obtained an exact equation of motion for the electron. As a complicating and unfamiliar feature it contains memory terms through the integration over the retarded fields. Lorentz (1904a, b) proposed a charge distribution which is rigid in its momentary rest frame, and therefore, as seen from the laboratory frame, contracting parallel to its momentary velocity. It was left completely open by which forces this charge distribution would be kept in place. Poincaré (1905, 1906) developed nonelectromagnetic models where additional stresses counteracted the Coulomb repulsion. Bucherer (1904, 1905) and Langevin (1905) introduced a charge distribution Lorentz contracted under the constraint of constant volume.

Up to 1900 electromagnetism was dominated by mechanics, in the sense that physicists felt compelled to introduce mechanical models for electromagnetic fields. Light would propagate through a rather mysterious gas, called the ether, and not simply through vacuum. The great revolution of the young electrodynamicists of the day was to reverse this position and consider inertial mass to be of purely electromagnetic origin. This electromagnetic world picture was nourished by the fact that in all extended charge models the velocity-dependent mass has the additive structure $m(v) = m_{\mathrm{b}}\mathbb{1} + m_{\mathrm{f}}(v)$, as $3 \times 3$ matrices with $\mathbb{1}$ the unit matrix, where $m_{\mathrm{b}}$ is the bare mechanical mass of the particle, in accordance with Newtonian mechanics taken to be velocity independent, and $m_{\mathrm{f}}(v)$ is the mass due to the coupling to the field, which was to be computed from the model charge distribution. In the spirit of the electrodynamic world picture it was natural to set $m_{\mathrm{b}} = 0$. Then Lorentz predicted the standard relativistic velocity dependence, which only for $|v/c| > 0.3$ differed significantly from the results of Abraham and Bucherer.

While experiments were on the way to decide between the competing theories, the whole enterprise came to a sudden end, since Einstein (1905a, b) forcefully argued that just like electromagnetism in vacuum also the mechanical laws had to be Lorentz invariant. But if Einstein was right, then the energy–momentum relation of the electron had to be the relativistic one, as emphasized independently by Poincaré (1906). Thus the only free parameter was the rest mass of the electron which anyway could not be deduced from theory, since the actual charge distribution was not known. There was simply nothing left to compute. At the latest with the atomic model of Bohr, to say 1913, it became obvious that a theory based on classical electromagnetism could not account for the observed stability of atoms nor for the sharp spectral lines. Classical electron theory, as a tool for explaining properties of atoms, electrons, and nuclei, was abandoned.

The experimental status remained ambiguous for some time. Kaufmann (1901) favored Abraham's model up to 1906. Only through the experiments of Bucherer (1908, 1909) were the predictions of Einstein and Lorentz considered to be reasonably confirmed. Of course, by that time Einstein had already convinced the theoreticians, and any other outcome would have been in serious doubt. A repetition of these historical experiments dryly concludes that "it seems fair to say that the Bucherer–Neumann experiments proved very little, if anything more than the Kaufmann experiments, which indicated a large qualitative increase of mass with velocity", Zahn and Spees (1938).

The effective equation of motion for the electron as given by Eq. (3.1) could not possibly have been the full story. Through the work of Larmor it was already understood that a charge loses energy through radiation at a rate roughly proportional to $\dot{v}^2$. Lorentz observed that in the approximation of small velocities this loss could be accounted for by the friction or radiation reaction force

$$\boldsymbol{F}_{\mathrm{rr}} = \frac{e^2}{6\pi c^3}\,\ddot{\boldsymbol{v}}\,, \tag{3.2}$$

which had to be added to the Lorentz force in Eq. (3.1). In 1904 Abraham obtained this friction force for arbitrary velocities as

$$\boldsymbol{F}_{\mathrm{rr}} = \frac{e^2}{6\pi c^3}\left[\gamma^4 c^{-2}(\boldsymbol{v}\cdot\ddot{\boldsymbol{v}})\boldsymbol{v} + 3\gamma^6 c^{-4}(\boldsymbol{v}\cdot\dot{\boldsymbol{v}})^2\boldsymbol{v} + 3\gamma^4 c^{-2}(\boldsymbol{v}\cdot\dot{\boldsymbol{v}})\dot{\boldsymbol{v}} + \gamma^2\ddot{\boldsymbol{v}}\right]. \tag{3.3}$$

He argued that energy and momentum are transported to infinity through the far field. On that scale the charge distribution is like a point charge and the electromagnetic fields can be computed from the Liénard–Wiechert potentials. Using conservation of energy and momentum for the total system he showed that the loss at infinity could be balanced by the friction-like force (3.3). Von Laue (1909) realized that the radiation reaction (3.3) is relativistically covariant and can be written as

$$\mathbf{F}_{\mathrm{rr}} = \frac{e^2}{6\pi c^3}\left[\ddot{\mathbf{u}} - c^{-2}(\dot{\mathbf{u}}\cdot\dot{\mathbf{u}})\mathbf{u}\right], \tag{3.4}$$

with $\boldsymbol{u}$ the four-velocity. It is in this form that the radiation reaction appears in the famous 1921 review article of Pauli on relativity. But apparently there was no incentive to study properties of Newton's equations of motion (3.1) including the full radiation reaction correction (3.3). Using the data from the Kaufmann experiment Abraham estimated the radiation reaction to be down by a factor of $10^{-9}$ relative to the Hamiltonian motion. Schott (1912) after studying the motion in a uniform electric field concluded: "Hence the effect of the reaction due to radiation is quite inappreciable in this and probably in all practical cases."

The first chapter on the dynamics of classical electrons closes around 1912 with the relativistic version of elasticity theory for deformable bodies by Born (1909) and von Laue (1911a, b). In essence there were two results: (i) a relativistically covariant expression for the radiation reaction and (ii) an energy–momentum relation for the charged particle dependent on the particular model charge distribution.

Of these models only Lorentz's model of a charge distribution properly contracting along its instantaneous velocity is consistent with Einstein's theory of special relativity.

## 3.2 Nonrelativistic quantum electrodynamics

The time lapse was short: In late 1925 Heisenberg formulated his matrix mechanics and in early 1926 Schrödinger had come to wave mechanics. Through Dirac's transformation theory both approaches were shown to be equivalent. But more importantly in our context, Dirac clearly formulated the rules of canonical quantization, providing the tools for quantizing any Hamiltonian system including those with an infinite number of degrees of freedom. In 1928 Dirac discovered the relativistic generalization of the Schrödinger equation. From then on the theoretician's avant garde strived for creating a relativistic quantum electrodynamics understood as a specific quantum field theory – no small effort – which in a broad sense still continues with us today. The nonrelativistic theory, our concern here, was regarded as being settled. In fact, in its basic theoretical aspects, the research monograph of Heitler (1936) does not differ significantly from modern variants. But obviously, many fascinating phenomena and theoretical developments still lay ahead.

Let us briefly recall the major steps. Born, Heisenberg and Jordan (1926) quantized the wave equation by regarding it as corresponding to an infinite set of harmonic oscillators. They studied the energy fluctuations and derived Planck's law. On 2 February 1927 Dirac proudly reported to Bohr that, on the basis of the new quantum theory, he knew how to compute the lifetime and the line shape of an excited state of an atom in the approximation where only a single photon is emitted. A systematic quantum treatment of emission and absorption of radiation is Dirac (1927). Fermi (1930) recognized the importance of the Coulomb gauge and quantized a system with an arbitrary number of charges. His 1932 review article discusses the quantization of the (many-particle) Abraham model as we know it today; compare with chapter 13. With the theoretical foundations laid down, most physical processes of interest could be handled through second-order perturbation. Perturbation theory as applied to an isolated bound state had been well established. However, for radiation one has to deal with resonances, i.e. unperturbed energies embedded in the continuum energy of field modes. On a practical level Fermi's golden rule settled the issue. The reason why and in what sense this was the correct answer triggered a continuing theoretical effort. As the body of radiation phenomena explainable through quantum mechanics accumulated, the trust in the new theory increased. Divergences were of concern, but, according to Heitler, "it seems now that there is a certain limited field within which the present quantum electrodynamics is correct". High frequencies had to be cut off to taste. In this spirit Bethe arrived at his famous prediction for the Lamb shift of the 2S level of the hydrogen atom.

As well as ultraviolet divergence, nonrelativistic quantum electrodynamics is also infrared divergent, as discovered by Bloch and Nordsieck (1937) and more exhaustively studied by Pauli and Fierz (1938). Even today infrared divergence is a somewhat elusive physical phenomenon. It says that an accelerated charge radiates an infinite number of photons. Since their total energy is finite, by necessity these photons must have ever-increasing wavelengths.

## 3.3 The point charge

In the 1930s and early 1940s it was a fairly widespread belief that one way to overcome the difficulties of quantum electrodynamics is a better understanding of the classical theory of point charges coupled to their radiation field. Of course, this was to be understood only as an intermediate step to the final goal, namely a consistent quantized theory. Our third section deals with a single paper: "Classical theory of radiating electrons" submitted by P. A. M. Dirac on 15 March 1938. Dirac's paper was equally motivated by quantum electrodynamics; however, as such it is concerned only with classical electron theory.

We have to report the findings of Dirac in sufficient detail, since most later activities start from there. The formal argument in the original paper can be well followed and alternative versions can be found in Rohrlich (1990), Teitelbom *et al.* (1980), and Thirring (1997). Thus there is no need for repetition and we can focus on the conclusions. At first reading it is best to disregard all philosophical claims and concentrate on the equations. But before that, let us see how Dirac himself viewed the 1897–1912 period:

> The Lorentz model of the electron as a small sphere charged with electricity, possessing mass account of the energy of the electric field around it, has proved very valuable in accounting for the motion and radiation of electrons in a certain domain of problems, in which the electromagnetic field does not vary too rapidly and the accelerations of the electrons are not too great.

Dirac's goal was to construct quantum electrodynamics. There the electron is regarded as an elementary particle with, almost by definition, no internal structure. Thus Dirac had to dispense with model charges and develop a theory of *point-like* electrons.

What then did Dirac really accomplish? Of course, he assumes the validity of the inhomogeneous Maxwell equations. The current is generated by a point charge whose motion is yet to be determined. Mechanically this point charge is relativistic with bare mass $m_\mathrm{b}$. There is no explicit reaction of the field back onto the charge, since at no stage would Dirac invoke the Lorentz force. Instead conservation of energy and momentum should suffice to fix the true trajectory of the point charge. Note that this is very different from the extended charge models where the starting point is a closed system of equations for the particle and the Maxwell field. Dirac studies the flow of energy and momentum through a thin tube of radius $R$ around the world line of the particle. The computation simplifies by writing the retarded fields generated by the motion of the point charge as

$$\mathbf{F}_{\text{ret}} = \frac{1}{2}(\mathbf{F}_{\text{ret}} + \mathbf{F}_{\text{adv}}) + \frac{1}{2}(\mathbf{F}_{\text{ret}} - \mathbf{F}_{\text{adv}}) \tag{3.5}$$

in all of space-time. The difference term turns out to be finite on the world line of the charge and, through a balancing of energy and momentum, yields in the limit $R \to 0$, the relativistic radiation reaction (3.4).

The more delicate term in (3.5) is the sum, which is divergent on the world line of the particle. At the expense of ignoring other divergent terms, cf. Thirring (1997), Eq. (8.4.16), Dirac obtains the expected result, namely

$$-\frac{e^2}{4\pi Rc^2}\dot{\mathbf{u}} = -m_{\text{f}}\dot{\mathbf{u}}\,. \tag{3.6}$$

Adding the radiation reaction (3.4) and equating with the mechanical four-momentum, the final result is an equation of motion which determines the trajectory of the particle,

$$(m_{\text{b}} + m_{\text{f}})\dot{\mathbf{u}} = m_{\text{exp}}\dot{\mathbf{u}} = e\mathbf{F}_{\text{ex}} \cdot \mathbf{u} + \frac{e^2}{6\pi c^3}\left[\ddot{\mathbf{u}} - c^{-2}(\dot{\mathbf{u}} \cdot \dot{\mathbf{u}})\mathbf{u}\right] + \mathcal{O}(R) \tag{3.7}$$

with an error of the size of the tube, where we have added the prescribed electromagnetic field tensor $\mathbf{F}_{\text{ex}}$ of external fields.

To complete his argument, Dirac had to take the limit $R \to 0$. Since $m_{\text{f}} \to \infty$, this amounts to

$$m_{\text{b}} \to -\infty,\ m_{\text{f}} \to \infty,\ m_{\text{exp}} = m_{\text{b}} + m_{\text{f}} \quad \text{fixed}, \tag{3.8}$$

where $m_{\text{exp}}$ is adjusted such that it agrees with the experimentally determined mass of the charged particle. The combined limit (3.8) is the classical mass renormalization.

Dirac admits that "such a model is hardly a plausible one according to current physical ideas but this is not an objection to the theory provided we have a reasonable mathematical scheme."

Equation (3.7), dropping the terms $\mathcal{O}(R)$, is the Lorentz–Dirac equation. Within the framework of Dirac it makes no sense to ask whether the Lorentz–Dirac equation is "exact", since there is nothing to compare with. The Lorentz–Dirac equation comes as one package, so to speak. One could compare only with real experiments, which is difficult since the radiation reaction is very small, or one could compare with higher-level theories such as quantum electrodynamics. But this has never been seriously attempted, since, to begin with, it would require a well-defined relativistic quantum field theory which is a difficult task.

The Lorentz–Dirac equation is identical to the effective equations of motion obtained from extended charge models, if we ignore for a moment the possibility that the kinetic energy might come out differently depending on which model charge is used. In this sense Dirac has recovered the previous results through a novel approach. However, there is an important distinction. For extended charge models one has a true solution for the position of the charged particle, say $\tilde{\mathbf{q}}(t)$. One can then compare $\tilde{\mathbf{q}}(t)$ with a solution of the Lorentz–Dirac equation and hope for agreement in asymptotic regimes, like slowly varying potentials. In addition, for an extended charge model one can set the bare mass to some negative value and study the consequences.

Dirac continues with a remark which shattered the naive trust in classical electron theory. He observes that even for zero external fields Eq. (3.7) has solutions where $|v(t)/c| \to 1$ as $t \to \infty$ and $|\dot{v}(t)|$ increases beyond any bound. Such unphysical solutions he called runaway solutions. It is somewhat surprising that runaways apparently went completely unnoticed before, which only indicates that no attempt was made to apply the Lorentz–Dirac equation to a concrete physical problem. If one inserts numbers, then runaways grow very fast. For instance, for an electron $\dot{v}(t) = \dot{v}(0)e^{t/\tau}$ with $\tau = 10^{-23}$ s. Thus if the Lorentz–Dirac equation (3.7) is a valid approximation in an extended charge model, which after all was the general understanding of the 1897–1912 period, then this model must also have runaway solutions – a conclusion in obvious conflict with empirical evidence.

Dirac proposed to eliminate the runaway solutions by requiring the asymptotic condition

$$\lim_{t\to\infty} \dot{\mathbf{u}}(t) = 0 \,. \tag{3.9}$$

As a bonus the problem of the missing initial condition is resolved: since in (3.7) the third derivative appears, one has to know $\mathbf{q}(0)$, $\dot{\mathbf{q}}(0)$, as in any mechanical problem, and in addition $\ddot{\mathbf{q}}(0)$. If one accepts (3.9), the initial condition for $\dot{\mathbf{u}}$ is replaced by the asymptotic condition (3.9). Dirac checked that for zero external forces and for a spatially constant but time-dependent force the asymptotic condition singles out physically meaningful solutions. By the end of 1938 the classical electron theory was in an awkward shape, in fact in a much worse shape than by the end of 1912. Formal, but even by strict standards careful, derivations yielded an equation with unphysical solutions. How did they come into existence? While Dirac's asymptotic condition seemed to be physically sensible, it was very much ad hoc and imposed *post festum* to get rid of unwanted guests. Even physicists willing to accept the asymptotic condition as a new principle, like Haag (1955), could not be too happy. Solutions satisfying the asymptotic condition are acausal in the sense that the charge starts moving even before any force is acting. To be sure, the causality violation is on the time scale of $\tau = 10^{-23}$ s for an electron, and even shorter for a proton, and thus has no observable consequences. But acausality remains as a dark spot in relativistic theory. The clear recognition of runaway solutions generated a sort of consensus that the coupled Maxwell–Newton equations have internal difficulties.

In the preface of his book Rohrlich writes:

> Most applications treat electrons as point particles. At the same time, there was the widespread belief that the theory of point particles is beset with various difficulties such as infinite electrostatic self-energy, a rather doubtful equation of motion which admits physically meaningless solutions, violation of causality, and others. It is not surprising, therefore, that the very existence of a consistent classical theory of charged particles is often questioned.

In Chapter 28 of the Feynman Lectures we read:

> Classical mechanics is a mathematically consistent theory; it just doesn't agree with experience. It is interesting, though, that the classical theory of electromagnetism is an unsatisfactory theory all by itself. The electromagnetic theory predicts the existence of an electromagnetic mass, but it also falls on its face in doing so, because it does not produce a consistent theory.

And finally to quote from the textbook on mathematical physics by Thirring:

> Not all solutions to (3.7) are crazy. Attempts have been made to separate sense from nonsense by imposing special initial conditions. It is to be hoped that some day the real solution of the problem of the charge–field interaction will look differently, and the equations describing nature will not be so highly unstable that the balancing act can only succeed by having the system correctly prepared ahead of time by a convenient coincidence.

To be sure, these issues were of concern only to theoretical physicists in search of a secure foundation. Synchrotron radiation sources were built anyhow. The loss in energy of an electron during one revolution can be accounted for by Larmor's formula. This is then the amount of energy which has to be supplied in order to maintain a stationary electron current. The radiation emitted from the synchrotron source is computed from the inhomogeneous Lorentz–Maxwell equations with a point charge source, i.e. from the Liénard–Wiechert potentials. No problem.

### 3.4 Wheeler–Feynman electrodynamics

To avoid the infinities of self-interaction Wheeler and Feynman (1945, 1949) designed a radical solution, at least on the classical level, since the quantized version of their theory was never accomplished.

Their basic tenet is to have as dynamical degrees of freedom only the trajectories of the particles. As such there are no electromagnetic fields, even though one still uses them as a familiar and convenient notational device. As Wheeler (1998) puts it later on, the 1940s were his period of "all particles – no fields" and he wanted to understand how far this point of view could be pushed.

Wheeler–Feynman electrodynamics starts from an action which was first written down by Fokker (1929). Let us consider $N$ particles, where the $i$-th particle has mass $m_i$, charge $e_i$, and a motion given by the world line $\mathbf{q}_i(\tau_i)$, $i = 1, \dots, N$. The world line is parametrized by its eigentime $\tau_i$ and the dot '˙' denotes differentiation with respect to this eigentime. The action functional has the form

$$\mathsf{S} = -\sum_{i=1}^{N} m_i c^2 \int \mathrm{d}\tau_i + \frac{1}{2} \sum_{\substack{i,j=1\\ i\neq j}}^{N} e_i e_j \iint \delta((\mathbf{q}_i - \mathbf{q}_j)^2)(\dot{\mathbf{q}}_i \cdot \dot{\mathbf{q}}_j)\mathrm{d}\tau_i \mathrm{d}\tau_j \,. \tag{3.10}$$

A formal variation of $\mathsf{S}$ leads to the equations of motion

$$m_i \ddot{\mathbf{q}}_i = \frac{e_i}{c} \sum_{\substack{j=1\\ j\neq i}}^{N} \frac{1}{2}\big(\mathbf{F}_{\mathrm{ret}(j)}(\mathbf{q}_i) + \mathbf{F}_{\mathrm{adv}(j)}(\mathbf{q}_i)\big) \cdot \dot{\mathbf{q}}_i \,. \tag{3.11}$$

Here $\mathbf{F}_{\mathrm{ret}(j)}(\boldsymbol{q}_i)$ and $\mathbf{F}_{\mathrm{adv}(j)}(\boldsymbol{q}_i)$ are the retarded and advanced Liénard–Wiechert fields generated by the charge at $\boldsymbol{q}_j$ and evaluated at $\boldsymbol{q}_i$. They are derived from the retarded and advanced potentials

$$\mathbf{A}_{\mathrm{ret}(j)}(\mathbf{x}) = e_j \dot{\mathbf{q}}_j(\tau_{j\mathrm{ret}})[(\mathbf{x} - \mathbf{q}_j(\tau_{j\mathrm{ret}})) \cdot \dot{\mathbf{q}}_j(\tau_{j\mathrm{ret}})]^{-1} \,, \tag{3.12}$$

$$\mathbf{A}_{\mathrm{adv}(j)}(\mathbf{x}) = e_j \dot{\mathbf{q}}_j(\tau_{j\mathrm{adv}})[(\mathbf{x} - \mathbf{q}_j(\tau_{j\mathrm{adv}})) \cdot \dot{\mathbf{q}}_j(\tau_{j\mathrm{adv}})]^{-1} \tag{3.13}$$

with $\tau_{j\mathrm{ret}}$, respectively $\tau_{j\mathrm{adv}}$, the eigentime when the trajectory $\mathbf{q}_j$ crosses the backward, respectively the forward, light cone with apex at $\mathbf{x}$. Notationally (3.11) looks like a set of ordinary differential equations. In fact, the locations of the other particles have to be known both at the advanced and retarded times, a situation which is not covered by any of the standard techniques. Even if the existence of solutions is taken for granted, it is widely open which data would single out a specific one.

To transform (3.11) into a familiar form, we use the decomposition (3.5) and Dirac's observation that $(\mathbf{F}_{\mathrm{ret}} - \mathbf{F}_{\mathrm{adv}})/2$ at the trajectory of the particle yields the radiation reaction. Then

$$m_i \ddot{\mathbf{q}}_i = \frac{e_i}{c} \sum_{\substack{j=1 \\ j \neq i}}^{N} \mathbf{F}_{\mathrm{ret}(j)}(\mathbf{q}_i) \cdot \dot{\mathbf{q}}_i + \frac{e_i^2}{6\pi c^3} \left( \dddot{\mathbf{q}}_i - c^{-2} \left( \ddot{\mathbf{q}}_i \cdot \ddot{\mathbf{q}}_i \right) \dot{\mathbf{q}}_i \right)$$

$$+ \frac{e_i}{c} \sum_{j=1}^{N} \frac{1}{2} \big( \mathbf{F}_{\mathrm{adv}(j)}(\mathbf{q}_i) - \mathbf{F}_{\mathrm{ret}(j)}(\mathbf{q}_i) \big) \cdot \dot{\mathbf{q}}_i \, . \qquad (3.14)$$

Of course, being symmetric in time, we could have equally transformed to the advanced fields for the force and a radiation reaction with reversed sign.

As a specific example let us consider the scattering of two charges with all other charges far apart. In the framework of the Lorentz model one would start with two charges and their comoving Coulomb field, sufficiently far apart and with incoming velocities. If radiation reaction is neglected, the bare mass is renormalized, and the force on one particle is due to the other particle at the retarded time. In the Wheeler–Feynman theory for *two* particles, the mass is just the bare mass, the forces are the average of retarded and advanced, and there is no radiation reaction. The Wheeler–Feynman theory seems to be at variance with empirical observations.

The crucial new element of their theory is that even in the case of two-particle scattering, the motion of all other charges cannot be ignored. Thus in (3.14), we take only $i = 1, 2$, but sum over large $N$. Wheeler and Feynman spend a considerable amount of effort to argue that when averaged over the random-like motion of all other charges, the last term in (3.14) vanishes and they call this the condition of a perfect absorber. The exact cancellation is hard to check and one has to be satisfied with qualitative arguments. The perfect absorber granted, in the first sum of (3.14) only the terms $j = 1, 2$ contribute by assumption and one has achieved the reduction to a two-particle problem with retarded forces. In its 18-dimensional phase space there is a 12-dimensional submanifold of physical solutions; all others run away. Wheeler and Feynman discuss an energy-like quantity for the system of $N$ charges which seems to ensure that all solutions to (3.11) are well behaved. As a consequence, only the physical solutions to (3.14) with perfect absorber are a valid approximation to the motion of $N$ charges as governed by (3.11) and agreement with the conventional theory is accomplished.

## Notes and references

### *Section 3.1*

An authorative, highly recommended source on the history of the classical electron theory is Miller (1997), which should be augmented by Pais (1972, 1982), by Rohrlich (1973), and by the introductory chapters of Rohrlich (1990). For a discussion of the Kaufmann experiments I refer to Cushing (1981) and Miller (1997). The monograph by Schott (1912) is the most complete technical account. It contains lots of material which has become an integral part of our present-day textbooks on electrodynamics and discusses in detail properties of various electron models. Reviews of classical electron theory are Hönl (1952), Caldirola (1956), Erber (1961), Barut (1980), Teitelbom *et al.* (1980), Coleman (1982), and Pearle (1982). The interconnection with quantum electrodynamics before the 1947 Shelter Island conference is vividly described in Schweber (1994).

### *Section 3.2*

There are excellent studies of the historical development of quantum electrodynamics as culminating in the work of Dyson, Feynman, Schwinger, and Tomonaga, in which as one part also the nonrelativistic theory is discussed. The most complete coverage is Schweber (1994), where the mentioned letter by Dirac is reproduced. Miller (1994) covers the history up to 1938 and includes reprints of the most important papers. A somewhat different selection is Schwinger (1958) with a recommended introduction. A further source is the monumental work of Mehra and Rechenberg (2000) on *The Historical Development of Quantum Theory*. The relevant volume is no. 6, part 1. Modern textbooks and research monographs on nonrelativistic quantum electrodynamics are Heitler (1936, 1958), Power (1964), Louisell (1973), Healy (1982), Craig and Thirunamachandran (1984), Cohen-Tannoudji, Dupont-Roc and Grynberg (1989, 1992), Milonni (1994) among others. They all have a common core, but emphasize rather diverse aspects once it comes to applications.

### *Section 3.3*

Kramers' (1948) investigations on the mass renormalization in the classical theory were instrumental for a correct computation of the Lamb shift. We refer to Dresden (1987) and Schweber (1994).

### *Section 3.4*

The two-body problem in Wheeler–Feynman electrodynamics is discussed by Schild (1963). The existence and classification of solutions is studied by Bauer (1997). A few explicit solutions are listed in Stephas (1992).

The opposite extreme "no particles – all fields" is briefly mentioned in the Notes to section 2.5.

# 4

# The energy–momentum relation

If the external forces vanish, the equations of motion must have a solution, in which the particle travels at constant velocity $\boldsymbol{v}$ in the company of its electromagnetic fields. There seems to be no accepted terminology for this object. Since it will be used as a basic building block later on, we need a short descriptive name and we call this particular solution a *charge soliton*, or simply soliton, at velocity $\boldsymbol{v}$, in analogy to solitons of nonlinear wave equations. The soliton has an energy and a momentum which are linked through the energy–momentum relation.

For the Lorentz model, by Lorentz invariance, it suffices to determine the four-vector of total momentum in the rest frame, where it is of the form $(m_s, 0)$, $m_s$ being the rest mass of the soliton. $m_s$ depends on $|\omega_E|$. Through a Lorentz boost one obtains the charge soliton moving with velocity $\boldsymbol{v}$ and, of course, the relativistic energy–momentum relation. No such argument is available for the Abraham model and one simply has to compute its energy–momentum relation, which can be achieved along two equivalent routes. The first one is dynamic, as alluded to above, while the second one is static and directly determines the minimal energy at fixed total momentum. The minimizer is the charge soliton.

In the following two sections we compute the conserved energy and momentum, the charge solitons, and the energy–momentum relation for both the Abraham and the Lorentz model. $\phi_{ex} = 0$, $\boldsymbol{A}_{ex} = 0$ is assumed throughout.

## 4.1 The Abraham model

The mechanical momentum of the particle is given by

$$m_b \gamma \boldsymbol{v} \tag{4.1}$$

and the momentum of the field by

$$\mathcal{P}_f = \int d^3x \big(\boldsymbol{E}(\boldsymbol{x}) \times \boldsymbol{B}(\boldsymbol{x})\big) . \tag{4.2}$$

Thus we set the total momentum

$$\mathcal{P} = m_b \gamma v + \mathcal{P}_f \tag{4.3}$$

as a functional on $\mathcal{M}$. It is easily checked that $\mathcal{P}$ is conserved by the coupled Maxwell and Newton equations (2.39)–(2.41). To ensure that $\mathcal{P}$ corresponds physically to the total momentum we note that the Lagrangian (2.43) of the Abraham model is invariant under spatial translations. By Noether's theorem, this symmetry is linked with a conserved quantity which turns out to be $\mathcal{P}$.

We want to minimize the energy at fixed total momentum. One eliminates $v$ from (2.44) and (4.3) and thus has to minimize

$$\Big(m_b^2 + \Big(\mathcal{P} - \int d^3x (E \times B)\Big)^2\Big)^{1/2} + \frac{1}{2}\int d^3x (E^2 + B^2) \tag{4.4}$$

at fixed $\mathcal{P}$ and subject to the constraints $\nabla \cdot E = e\varphi$, $\nabla \cdot B = 0$. By translation invariance we may center $\varphi$ at an arbitrary $q \in \mathbb{R}^3$. For $q = 0$, say, the minimizer is unique and given by

$$\begin{aligned} E_v(x) &= -\nabla\phi_{v\varphi}(x) + v(v \cdot \nabla\phi_{v\varphi}(x)), \\ B_v(x) &= -v \times \nabla\phi_{v\varphi}(x) \end{aligned} \tag{4.5}$$

with $v \in \mathbb{V} = \{v \,|\, |v| < 1\}$. Here

$$\widehat{\phi}_v(k) = e[k^2 - (v \cdot k)^2]^{-1}, \tag{4.6}$$

or in physical space

$$\phi_v(x) = e(4\pi)^{-1}(\gamma^{-2}x^2 + (v \cdot x)^2)^{-1/2}, \tag{4.7}$$

and $\phi_{v\varphi}$ is shorthand for the convolution $\phi_v * \varphi$, i.e. $\widehat{\phi}_{v\varphi}(k) = (2\pi)^{3/2}\widehat{\varphi}(k)\widehat{\phi}_v(k)$. $v$ has to be adjusted such that $\mathcal{P} = P_s(v)$ with

$$\begin{aligned} P_s(v) &= m_b\gamma v + e^2 \int d^3k |\widehat{\varphi}(k)|^2 \big([k^2 - (k \cdot v)^2]^{-1} v \\ &\quad - \gamma^{-2}[k^2 - (k \cdot v)^2]^{-2}(k \cdot v)k\big) \\ &= v\big(m_b\gamma + m_f |v|^{-3}\big[-|v| + (1 + v^2)\operatorname{arctanh}|v|\big]\big), \end{aligned} \tag{4.8}$$

where $m_f$ is the electrostatic energy of the charge distribution $e\varphi$,

$$m_f = \frac{1}{2}e^2 \int d^3x\, d^3x' \varphi(x)\, \varphi(x')(4\pi|x - x'|)^{-1}. \tag{4.9}$$

The map $\mathbb{V} \ni v \mapsto P_s(v) \in \mathbb{R}^3$ is one-to-one and therefore $\mathcal{P} = P_s(v)$ has a unique solution. The minimizing energy is given by

$$E_{\mathrm{s}}(\boldsymbol{v}) = m_{\mathrm{b}}\gamma + \frac{1}{2}e^2 \int \mathrm{d}^3k|\widehat{\varphi}(\boldsymbol{k})|^2\,[\boldsymbol{k}^2 - (\boldsymbol{k}\cdot\boldsymbol{v})^2]^{-2}\big((1+\boldsymbol{v}^2)\boldsymbol{k}^2 - (3-\boldsymbol{v}^2)(\boldsymbol{v}\cdot\boldsymbol{k})^2\big)$$
$$= m_{\mathrm{b}}\gamma + m_{\mathrm{f}}\,|\boldsymbol{v}|^{-1}\big[-|\boldsymbol{v}| + 2\,\mathrm{arctanh}|\boldsymbol{v}|\big]. \tag{4.10}$$

Eliminating now $\boldsymbol{v}$ from $E_{\mathrm{s}}$ and $\boldsymbol{P}_{\mathrm{s}}$ yields the *energy–momentum relation*

$$E_{\mathrm{eff}}(\boldsymbol{p}) = E_{\mathrm{s}}(\boldsymbol{v}(\boldsymbol{p})) \tag{4.11}$$

with $\boldsymbol{v}(\boldsymbol{P}_{\mathrm{s}})$ the function inverse to $\boldsymbol{P}_{\mathrm{s}}(\boldsymbol{v})$. It is emphasized that $E_{\mathrm{eff}}$ depends on the charge distribution only through its electrostatic energy.

We note that

$$\boldsymbol{P}_{\mathrm{s}}(\boldsymbol{v}) = \nabla_{\boldsymbol{v}}T(\boldsymbol{v}), \tag{4.12}$$

where

$$T(\boldsymbol{v}) = -m_{\mathrm{b}}\gamma^{-1} + \frac{1}{2}e^2\,\gamma^{-2}\int \mathrm{d}^3k|\widehat{\varphi}(\boldsymbol{k})|^2\,[\boldsymbol{k}^2 - (\boldsymbol{k}\cdot\boldsymbol{v})^2]^{-1}$$
$$= -m_{\mathrm{b}}\gamma^{-1} - m_{\mathrm{f}}\,|\boldsymbol{v}|^{-1}(1-|\boldsymbol{v}|^2)\,\mathrm{arctanh}|\boldsymbol{v}|, \tag{4.13}$$

and that

$$E_{\mathrm{s}}(\boldsymbol{v}) = \boldsymbol{P}_{\mathrm{s}}(\boldsymbol{v})\cdot\boldsymbol{v} - T(\boldsymbol{v}). \tag{4.14}$$

This suggests that $T$ will play the role of the inertial term in an effective Lagrangian and $E_{\mathrm{s}}$ the role of an effective Hamiltonian as our notation in (4.11) indicates already. In particular,

$$\boldsymbol{v} = \nabla_{\boldsymbol{p}}\,E_{\mathrm{eff}}(\boldsymbol{p}) \tag{4.15}$$

and, equivalently,

$$\frac{\mathrm{d}\boldsymbol{P}_{\mathrm{s}}(\boldsymbol{v})}{\mathrm{d}\boldsymbol{v}}\,\boldsymbol{v} = \nabla_{\boldsymbol{v}}\,E_{\mathrm{s}}(\boldsymbol{v}) \tag{4.16}$$

which implies that $\boldsymbol{v}$ is to be interpreted as a velocity and $\mathrm{d}\boldsymbol{P}_{\mathrm{s}}/\mathrm{d}\boldsymbol{v}$, regarded as a $3\times 3$ matrix, as the velocity-dependent mass.

For a relativistic theory one expects that

$$E_{\mathrm{s}}(\boldsymbol{v}) = (m_{\mathrm{b}} + m_{\mathrm{f}})\gamma\,, \quad \boldsymbol{P}_{\mathrm{s}}(\boldsymbol{v}) = (m_{\mathrm{b}} + m_{\mathrm{f}})\gamma\boldsymbol{v}\,. \tag{4.17}$$

Since the Abraham model is semirelativistic, there is no reason for such a property to be satisfied. Still, as in the relativistic case, the energy–momentum relation depends on the charge distribution $e\varphi$ only through $m_{\mathrm{f}}$.

To gain a feeling for the field contributions to the mass we define

$$m_{\mathrm{f}}(v) = \frac{\mathrm{d}(P_{\mathrm{s}} - m_{\mathrm{b}}\gamma v)}{\mathrm{d}v} = m_{\mathrm{l}}(v)\widehat{v}\otimes\widehat{v} + m_{\mathrm{t}}(v)(\mathbb{1} - \widehat{v}\otimes\widehat{v}), \tag{4.18}$$

where $\widehat{v}$ is the unit vector along $v$; $m_{\mathrm{l}}(v)$ is the longitudinal and $m_{\mathrm{t}}(v)$ is the transverse field mass. Using (4.8) one obtains

$$m_{\mathrm{l}}(v) = m_{\mathrm{f}}|v|^{-3}\big(2|v|(1 - |v|^2)^{-1} - 2\,\mathrm{arctanh}|v|\big), \tag{4.19}$$

$$m_{\mathrm{t}}(v) = m_{\mathrm{f}}|v|^{-3}\big(-|v| + (1 + |v|^2)\mathrm{arctanh}|v|\big), \tag{4.20}$$

and by expanding in small $v$, i.e. small $|v|/c$,

$$m_{\mathrm{l}}(v) = \frac{4}{3}m_{\mathrm{f}}\Big(1 + \frac{6}{5}|v|^2 + \frac{9}{7}|v|^4 + \cdots\Big), \tag{4.21}$$

$$m_{\mathrm{t}}(v) = \frac{4}{3}m_{\mathrm{f}}\Big(1 + \frac{2}{5}|v|^2 + \frac{9}{35}|v|^4 + \cdots\Big). \tag{4.22}$$

In particular one has

$$E_{\mathrm{s}}(v) - E_{\mathrm{s}}(0) \cong \frac{1}{2}\Big(m_{\mathrm{b}} + \frac{4}{3}m_{\mathrm{f}}\Big)v^2, \quad P_{\mathrm{s}}(v) = \Big(m_{\mathrm{b}} + \frac{4}{3}m_{\mathrm{f}}\Big)v. \tag{4.23}$$

Thus the effective mass in the nonrelativistic approximation is

$$m_{\mathrm{eff}} = m_{\mathrm{b}} + \frac{4}{3}m_{\mathrm{f}}. \tag{4.24}$$

We compare (4.19)–(4.22) with a relativistic particle for small $v$ and of the same mass. Then

$$m^{\mathrm{rel}} = m_{\mathrm{l}}^{\mathrm{rel}}\widehat{v}\otimes\widehat{v} + m_{\mathrm{t}}^{\mathrm{rel}}(\mathbb{1} - \widehat{v}\otimes\widehat{v}) \tag{4.25}$$

with

$$m_{\mathrm{l}}^{\mathrm{rel}}(v) = \Big(m_{\mathrm{b}} + \frac{4}{3}m_{\mathrm{f}}\Big)\gamma^3 = \Big(m_{\mathrm{b}} + \frac{4}{3}m_{\mathrm{f}}\Big)\Big(1 + \frac{3}{2}|v|^2 + \frac{9}{8}|v|^4 + \cdots\Big), \tag{4.26}$$

$$m_{\mathrm{t}}^{\mathrm{rel}}(v) = \Big(m_{\mathrm{b}} + \frac{4}{3}m_{\mathrm{f}}\Big)\gamma = \Big(m_{\mathrm{b}} + \frac{4}{3}m_{\mathrm{f}}\Big)\Big(1 + \frac{1}{2}|v|^2 + \frac{3}{8}|v|^4 + \cdots\Big). \tag{4.27}$$

If one sets the bare mass to zero, $m_{\mathrm{b}} = 0$, even for $|v| = 0.5$ the error in the velocity-dependent mass is less than 5%. Only at speeds $|v| > 0.5$ will the Abraham model lose its empirical validity. The model could be partially saved by declaring the Compton wavelength as the characteristic size of the charge distribution. Then $m_{\mathrm{f}}/m_{\mathrm{b}} \cong 0.01$ and the relativistic dispersion would be violated only for speeds very close to one.

The energy minimizer has a simple dynamical interpretation. We look for a solution of (2.39)–(2.41) traveling at constant velocity. Let us first define

$$S_{q,v} = (\boldsymbol{E}_v(\boldsymbol{x}-\boldsymbol{q}),\, \boldsymbol{B}_v(\boldsymbol{x}-\boldsymbol{q}),\, \boldsymbol{q},\, \boldsymbol{v}) \tag{4.28}$$

with $\boldsymbol{v} \in \mathbb{V}$, $\boldsymbol{q} \in \mathbb{R}^3$, and $\boldsymbol{B}_v$, $\boldsymbol{E}_v$ from (4.5). Then the solution traveling at constant velocity is

$$Y(t) = S_{q+vt,v}\,. \tag{4.29}$$

The particular state (4.28) will play an important role and is called a *charge soliton*, labeled by its center $\boldsymbol{q}$ and its velocity $\boldsymbol{v}$. It has the energy $\mathcal{E}(S_{q,v}) = E_{\mathrm{s}}(\boldsymbol{v})$ and momentum $\mathcal{P}(S_{q,v}) = \boldsymbol{P}_{\mathrm{s}}(\boldsymbol{v})$. The set of all charge solitons is

$$\mathcal{S} = \{S_{q,v} |\, \boldsymbol{v} \in \mathbb{V}, \boldsymbol{q} \in \mathbb{R}^3\} \subset \mathcal{M}\,. \tag{4.30}$$

Sometimes we use the same words and symbols for the field configuration only.

There is an instructive alternate way to represent the charge soliton. We consider the inhomogeneous Maxwell–Lorentz equations (2.39) and prescribe the initial data at time $\tau$. We require that the particle travels along the straight line $\boldsymbol{q} = \boldsymbol{v}t$. If we let $\tau \to -\infty$ and consider the solution at time $t = 0$, then in (2.16), (2.17) the initial fields will have escaped to infinity and only the retarded fields survive. Using (2.16), (2.17) this leads to

$$\begin{aligned} \boldsymbol{E}_v(\boldsymbol{x}) = -\int_{-\infty}^{0} \mathrm{d}t \int \mathrm{d}^3y\, \big(\nabla G_{-t}(\boldsymbol{x}-\boldsymbol{y})\, e\varphi(\boldsymbol{y}-\boldsymbol{v}t) \\ + \partial_t G_{-t}(\boldsymbol{x}-\boldsymbol{y})\boldsymbol{v}e\varphi(\boldsymbol{y}-\boldsymbol{v}t)\big)\,, \end{aligned} \tag{4.31}$$

$$\boldsymbol{B}_v(\boldsymbol{x}) = \int_{-\infty}^{0} \mathrm{d}t \int \mathrm{d}^3y\, \nabla \times G_{-t}(\boldsymbol{x}-\boldsymbol{y})\, \boldsymbol{v}e\varphi(\boldsymbol{y}-\boldsymbol{v}t)\,, \tag{4.32}$$

which can be checked either in Fourier space or as being a solution of the Maxwell equations traveling at constant velocity $\boldsymbol{v}$.

## 4.2 The Lorentz model

We fix a Lorentz frame, $\mathcal{F}_{\mathrm{L}}$, and seek a solution with $\mathbf{q}(\tau) = 0$, $\mathbf{w}(\tau) = \mathbf{w}$ for all $\tau$. The corresponding four-current is

$$\mathbf{j}(\mathbf{x}) = e\varphi_{\mathrm{r}}(|\mathbf{x}|)\boldsymbol{\Omega}\cdot\mathbf{x} \tag{4.33}$$

and provides the source for the electromagnetic vector potential. The inhomogeneous Maxwell equations yield

$$\phi_{\mathbf{0},\omega}(\boldsymbol{x}) = \int d^3x' \frac{1}{4\pi|\boldsymbol{x}-\boldsymbol{x}'|} e\varphi(\boldsymbol{x}'), \tag{4.34}$$

$$\boldsymbol{A}_{\mathbf{0},\omega}(\boldsymbol{x}) = \int d^3x' \frac{1}{4\pi|\boldsymbol{x}-\boldsymbol{x}'|} \boldsymbol{\omega}_{\mathrm{E}} \times \boldsymbol{x}' e\varphi(\boldsymbol{x}'), \tag{4.35}$$

the index $\mathbf{0}$ standing for $\boldsymbol{v} = \mathbf{0}$.

Outside the support of the charge distribution, $\phi_{\mathbf{0},\omega}$ is the Coulomb potential,

$$\phi_{\mathbf{0},\omega}(\boldsymbol{x}) = \frac{e}{4\pi|\boldsymbol{x}|}, \quad |\boldsymbol{x}| \geq R_\varphi, \tag{4.36}$$

and $\boldsymbol{A}_{\mathbf{0},\omega}$ is the vector potential generated by the magnetic moment

$$\boldsymbol{\mu} = \frac{1}{2}\int d^3x \boldsymbol{x} \times (\boldsymbol{\omega}_{\mathrm{E}} \times \boldsymbol{x}) e\varphi(\boldsymbol{x}) = \mu\boldsymbol{\omega}_{\mathrm{E}} \text{ with } \mu = \frac{1}{3}e\int d^3x\varphi(\boldsymbol{x})\boldsymbol{x}^2, \tag{4.37}$$

which means

$$\boldsymbol{A}_{\mathbf{0},\omega}(\boldsymbol{x}) = \frac{\boldsymbol{\mu}\times\boldsymbol{x}}{4\pi|\boldsymbol{x}|^3}, \quad |\boldsymbol{x}| \geq R_\varphi. \tag{4.38}$$

To check the Lorentz force and torque we note that a well-defined momentum and angular momentum requires the equator to have subluminal speed, i.e.

$$\omega R_\varphi \leq 1, \quad \omega = |\boldsymbol{\omega}_{\mathrm{E}}|. \tag{4.39}$$

Inserting the fields (4.34), (4.35) in Eqs. (2.92), (2.95) we indeed find $\mathbf{f}(\tau) = 0$, $\mathbf{t}(\tau) = 0$ and thus (2.86), (2.87) are satisfied.

The family of charge solitons is obtained from (4.34), (4.35) through a Lorentz boost with velocity $\mathbf{u} = (\gamma, \gamma\boldsymbol{v})$. They are labeled by their center at $t = 0$, set equal to zero here, by the velocity $\boldsymbol{v}$, and by their angular velocity $\boldsymbol{\omega}$. Explicitly we have

$$\phi(\boldsymbol{x}, t) = \phi_{\boldsymbol{v},\omega}(\boldsymbol{x} - \boldsymbol{v}t), \quad \boldsymbol{A}(\boldsymbol{x}, t) = \boldsymbol{A}_{\boldsymbol{v},\omega}(\boldsymbol{x} - \boldsymbol{v}t). \tag{4.40}$$

Because of the convolution structure $\phi_{\boldsymbol{v},\omega}$, $\boldsymbol{A}_{\boldsymbol{v},\omega}$ are more easily written in Fourier space, where

$$\widehat{\phi}_{\boldsymbol{v},\omega}(\boldsymbol{k}) = \frac{e}{\boldsymbol{k}^2 - (\boldsymbol{k}\cdot\boldsymbol{v})^2}\left[\widehat{\varphi}(D^{-1}\boldsymbol{k}) + \boldsymbol{v}\cdot(\boldsymbol{\omega}\times i\nabla_{\boldsymbol{k}}\widehat{\varphi})(D^{-1}\boldsymbol{k})\right], \tag{4.41}$$

$$\widehat{\boldsymbol{A}}_{\boldsymbol{v},\omega}(\boldsymbol{k}) = \frac{e}{\boldsymbol{k}^2 - (\boldsymbol{k}\cdot\boldsymbol{v})^2}\Big[\boldsymbol{v}\widehat{\varphi}(D^{-1}\boldsymbol{k}) + \frac{1}{\gamma}(\boldsymbol{\omega}\times i\nabla_{\boldsymbol{k}}\widehat{\varphi})(D^{-1}\boldsymbol{k}) + \frac{\gamma\boldsymbol{v}}{1+\gamma}\boldsymbol{v}\cdot(\boldsymbol{\omega}\times i\nabla_{\boldsymbol{k}}\widehat{\varphi})(D^{-1}\boldsymbol{k})\Big] \tag{4.42}$$

with $D^{-1}\boldsymbol{k} = \boldsymbol{k} - (\gamma^{-1} - 1)(\widehat{\boldsymbol{v}} \cdot \boldsymbol{k})\widehat{\boldsymbol{v}}$. We note that (4.40), (4.42) coincide with (4.5), (4.6) for $\boldsymbol{\omega} = 0$ and $D = 1$. Put differently, (4.40) and (4.42) properly incorporate the Lorentz contraction of the charge distribution and the extra fields due to the nonvanishing magnetic moment. To obtain the energy–momentum relation we only have to compute the energy of the soliton in its rest frame. By rotation invariance, this energy depends on $\boldsymbol{\omega}$ through its absolute value $\omega = |\boldsymbol{\omega}|$. From (2.89) the bare gyrational mass of the particle is given by

$$\begin{aligned} m_{\mathrm{g}}(\omega) &= m_{\mathrm{b}} \int \mathrm{d}^3x \varphi(\boldsymbol{x})(1 - |\boldsymbol{\omega}_{\mathrm{E}} \times \boldsymbol{x}|^2)^{-1/2} \\ &= m_{\mathrm{b}} \int_0^\infty \mathrm{d}r 4\pi r^2 \varphi_{\mathrm{r}}(r) \frac{1}{\omega r} \operatorname{arctanh} \omega r \,. \end{aligned} \tag{4.43}$$

The field energy is defined through

$$m_{\mathrm{f}} = \frac{1}{2} \int \mathrm{d}^3x (\boldsymbol{E}^2 + \boldsymbol{B}^2) \,. \tag{4.44}$$

Inserting from (4.34), (4.35) results in

$$m_{\mathrm{f}}(\omega) = \frac{1}{2} e^2 \int \mathrm{d}^3k |\widehat{\varphi}|^2 \frac{1}{\boldsymbol{k}^2} + \frac{1}{3} \omega^2 e^2 \int \mathrm{d}^3k |\nabla_{\boldsymbol{k}} \widehat{\varphi}|^2 \frac{1}{\boldsymbol{k}^2} \,. \tag{4.45}$$

Thus the charge soliton carries the energy

$$m_{\mathrm{s}}(\omega) = m_{\mathrm{g}}(\omega) + m_{\mathrm{f}}(\omega) \tag{4.46}$$

and its energy–momentum relation is necessarily relativistic,

$$E = (\boldsymbol{p}^2 + m_{\mathrm{s}}^2)^{1/2} \,. \tag{4.47}$$

The rotational degrees of freedom are handled in the same spirit. The charge distribution carries the magnetic moment defined in (4.37). $\mu$ sets the rotational coupling to the electromagnetic field. Like the charge, it is not renormalized through the interaction with the field. According to (2.93), (2.94), the bare angular momentum of the particle is

$$\mathrm{s}_{\mathrm{b}} = I_{\mathrm{b}}(\omega) \mathbf{w}_{\mathrm{E}} \,, \tag{4.48}$$

where

$$I_{\mathrm{b}}(\omega) = m_{\mathrm{b}} \int_0^\infty \mathrm{d}r 4\pi r^2 \varphi_{\mathrm{r}}(r) \frac{1}{2\omega^2} \Big( -1 + \frac{1 + \omega^2 r^2}{\omega r} \operatorname{arctanh} \omega r \Big) \,. \tag{4.49}$$

In addition, the soliton carries a field angular momentum defined by

$$\boldsymbol{s}_{\mathrm{f}} = \int \mathrm{d}^3x \boldsymbol{x} \times (\boldsymbol{E} \times \boldsymbol{B}) \tag{4.50}$$

with $\boldsymbol{E}$, $\boldsymbol{B}$ in their rest frame inserted from (4.34), (4.35). One obtains

$$\mathbf{s}_{\mathrm{f}} = I_{\mathrm{f}}\mathbf{w}_{\mathrm{E}}\,, \quad I_{\mathrm{f}} = \frac{2}{3}e^2 \int \mathrm{d}^3k |\nabla_k \widehat{\varphi}|^2 \frac{1}{k^2}\,. \tag{4.51}$$

Thus the charge soliton carries the spin

$$\mathbf{s}_{\mathrm{s}} = \mathbf{s}_{\mathrm{b}} + \mathbf{s}_{\mathrm{f}} = (I_{\mathrm{b}}(\omega) + I_{\mathrm{f}})\mathbf{w}_{\mathrm{E}}\,. \tag{4.52}$$

## 4.3 The limit of zero bare mass

The bare mass seems to be an artifact of the theory, since there is no way to determine its value through experiments involving only electromagnetic forces (unless the charge distribution could be probed). Thus a natural and conceptually attractive proposal is to take $m_{\mathrm{b}} = 0$, thereby declaring all mass to be of electromagnetic origin. We discuss here the limit $m_{\mathrm{b}} \to 0_+$ on the level of the energy–momentum relation, whereas the correct procedure would be to study this limit on the level of a solution to the evolution equations. The problem remains unexplored, since for the equations of motion zero bare mass is rather singular.

(i) *Abraham model.* Since $m_{\mathrm{s}}$ is additive, the only choice is simply to set $m_{\mathrm{b}} = 0$. In particular, the kinetic energy equals $\frac{1}{2}(\frac{4}{3}m_{\mathrm{f}})\boldsymbol{v}^2$ for small velocities. If we equate $4m_{\mathrm{f}}/3$ with $m_{\mathrm{exp}}$, the experimental mass of the electron, we conclude that $R_\varphi \cong r_{\mathrm{cl}} = 3 \times 10^{-13}$ cm with a prefactor which depends on the choice of the form factor $\widehat{\varphi}$.

(ii) *Lorentz model.* Since $m_{\mathrm{g}}$ depends on $\omega$, the Lorentz model offers more variety. We recall Eq. (4.43). If the integral is bounded, which in particular is the case for $\varphi$ bounded and $\omega R_\varphi \leq 1$, then $m_{\mathrm{g}}$ vanishes in the limit $m_{\mathrm{b}} \to 0$. We conclude that $m_{\mathrm{s}} = m_{\mathrm{f}}(\omega)$ and $I_{\mathrm{s}} = I_{\mathrm{f}}$. A novel situation occurs if the integral in (4.43) can be made to diverge, for which we must choose $\varphi$ to be well concentrated at the sphere with radius $R_\varphi$. To be concrete let us set $R = R_\varphi$ and $\varphi(\boldsymbol{x}) = \delta(|\boldsymbol{x}| - R)(4\pi R^2)^{-1}$. We also reintroduce $c$. Then the integral in (4.43) becomes

$$m_{\mathrm{g}}(\omega) = m_{\mathrm{b}}\frac{c}{\omega R}\,\mathrm{arctanh}\,\frac{\omega R}{c}\,. \tag{4.53}$$

We let $\omega R/c \to 1$ and $m_{\mathrm{b}} \to 0$ such that

$$m_{\mathrm{b}}\,\mathrm{arctanh}\frac{\omega R}{c} \to \bar{m} \tag{4.54}$$

with $\bar{m} \geq 0$ still at our choice. Note that in this limit the equator rotates with the speed of light. For the mass, moment of inertia, and magnetic moment of the soliton, one obtains, respectively,

$$m_s = \bar{m} + \frac{11}{18}\frac{e^2}{c^2R}, \quad I_s = \frac{2}{3}\bar{m}cR + \frac{2}{9}\frac{e^2}{c}, \quad \mu = \frac{1}{3}eR, \tag{4.55}$$

which leaves us with $R$ and $\bar{m}$ as free parameters. They can be fitted through the experimentally determined mass and gyromagnetic ratio of the electron. While for the mass we simply set $m_s = m_{\mathrm{exp}}$, the $g$-factor requires a more elaborate discussion which will be taken up in section 10.1.

## Notes and references

### *Section 4.1*

Abraham (1905) computes the energy–momentum relation in essence along the same lines as outlined here (except for the variational characterization). Sommerfeld (1905) uses the expansion of the exact self-force, as will be explained in chapter 7. Lorentz (1904a) proposes a model charge which relativistically contracts parallel to its momentary velocity. Thus provisionally we replace the charge distribution $e\varphi(\boldsymbol{x})$ by its Lorentz contracted version

$$\varphi_{\mathrm{L}}(\boldsymbol{x}) = \gamma\varphi_{\mathrm{r}}([\boldsymbol{x}^2 + \gamma^2(\boldsymbol{x}\cdot\boldsymbol{v})^2]^{1/2}), \quad \widehat{\varphi}_{\mathrm{L}}(\boldsymbol{k}) = \widehat{\varphi}_{\mathrm{r}}([\boldsymbol{k}^2 - (\boldsymbol{v}\cdot\boldsymbol{k})^2]^{1/2}). \tag{4.56}$$

This expression is substituted in (4.5) and gives the electromagnetic fields comoving with the charge at velocity $\boldsymbol{v}$. Their energy and momentum are computed as before with the result

$$\boldsymbol{P}_{\mathrm{L}}(\boldsymbol{v}) = \boldsymbol{v}\Big(m_{\mathrm{b}}\gamma(\boldsymbol{v}) + \frac{4}{3}m_{\mathrm{f}}\gamma(\boldsymbol{v})\Big), \tag{4.57}$$

$$E_{\mathrm{L}}(\boldsymbol{v}) = m_{\mathrm{b}}\gamma(\boldsymbol{v}) + m_{\mathrm{f}}\gamma(\boldsymbol{v})\Big(1 + \frac{1}{3}\boldsymbol{v}^2\Big). \tag{4.58}$$

The momentum has the anticipated form, except for the factor 4/3 which should be 1. The energy has an unwanted $\boldsymbol{v}^2/3$. In particular the relation (4.16) does not hold, which implies that the power equation $\frac{d}{dt}E_{\mathrm{L}}(\boldsymbol{v})$ differs from the force equation $\boldsymbol{v}\cdot\frac{d}{dt}\boldsymbol{P}_{\mathrm{L}}(\boldsymbol{v})$. We refer to Yaghjian (1992) for a thorough discussion, which however somehow misses step zero, namely to specify a relativistically covariant model for an extended charge, as, e.g., in section 2.5. Schott (1912, 1915) employs a deformable elastic medium as a model charge. To compute the velocity-dependent mass he uses essentially the same method as Sommerfeld, an exact self-force and an expansion in the charge diameter. Schott considers also electron models different from those of Abraham and Lorentz. Reviews are Neumann (1914) and Richardson (1916).

There have been various attempts to improve on the oversimplistic version (4.56) of the Lorentz model. Fermi (1922) argues that in a relativistic theory energy and momentum have to be redefined. His argument has been rediscovered several times and is explained in Rohrlich (1990). Poincaré (1906) takes the elastic stresses into account. We refer to Rohrlich (1960) and Yaghjian (1992), and the instructive example by Schwinger (1983).

### *Section 4.2*

Since the Lorentz model is defined through a Lagrangian, the total energy and momentum are determined from Noether's theorem for space-time translations. The transformation as a four-vector is then automatically guaranteed, a property which we used in the computation of the soliton mass.

### *Section 4.3*

The limit of zero bare mass is discussed in Appel and Kiessling (2001).

# 5
# Long-time asymptotics

For any dynamical system one of the first qualitative issues is to understand whether there are general patterns governing the long-time behavior. In this spirit we plan to study the long-time asymptotics of the Abraham model with prescribed external potentials. The basic mechanism at work is the loss of energy radiated to infinity, which is proportional to $\dot{\boldsymbol{v}}(t)^2$ according to Larmor's formula. Since the energy is bounded from below, we expect

$$\lim_{t\to\infty} \dot{\boldsymbol{v}}(t) = 0 \tag{5.1}$$

under rather general conditions. In fact, one would also expect that the velocity tends to a definite limit,

$$\lim_{t\to\infty} \boldsymbol{v}(t) = \boldsymbol{v}_\infty \in \mathbb{V}\,, \tag{5.2}$$

which leaves us with two qualitatively rather different cases.

(i) $\boldsymbol{v}_\infty = 0$. The charged particle comes to rest confined by the external potentials.
(ii) $\boldsymbol{v}_\infty \neq 0$. The charge escapes into a region with zero external potentials and travels there with constant velocity.

If we take also the asymptotics for $t \to -\infty$ into account, then four familiar cases arise: excitation by incident radiation and subsequent relaxation, (i) → (i); ionization, (i) → (ii); capture through radiation losses, (ii) → (i); and scattering of light from a freely moving charged particle, (ii) → (ii).

There must be a corresponding long-time asymptotic for the radiation field. It consists of a part attached to the motion of the particle and a part scattered to infinity. Thus a more complete description of the long-time solution is

$$Y(t) \cong S_{\boldsymbol{q}(t),\boldsymbol{v}(t)} + (\boldsymbol{E}_{\mathrm{out}}(t), \boldsymbol{B}_{\mathrm{out}}(t), 0, 0) \tag{5.3}$$

for large $t$. Here $S_{q(t),v(t)}$ is the charge soliton at the current position and momentum and $\boldsymbol{E}_{\mathrm{out}}(t)$, $\boldsymbol{B}_{\mathrm{out}}(t)$ are the solution of the homogeneous Maxwell equations with appropriately adjusted initial conditions, the scattering data which depend on $Y(0)$.

At present two techniques are at hand for establishing a limit like (5.3). The first one exploits the fact that energy cannot be radiated to infinity forever. This route requires that all field modes are coupled to the particle as expressed by the

*Wiener condition (W):*

$$\widehat{\varphi}(\boldsymbol{k}) > 0 . \tag{5.4}$$

The second route is based on a contraction method. It needs no extra condition and gives explicit convergence rates. However, it requires $|e|$ to be sufficiently small, i.e. $|e| < \bar{e}$ with a suitable $\bar{e}$ depending only on the initial energy. Presumably $(W)$ and $\bar{e}$ are artifacts of our mathematical technique.

## 5.1 Radiation damping and the relaxation of the acceleration

We will establish the limit (5.1) under the Wiener condition, but otherwise in complete generality. The proof follows rather closely physical intuition and leads to an equation of convolution type which has a definite long-time limit.

Let us consider a ball of radius $R$ centered at the origin. At time $t$ the sum of the field energy in this ball and of the mechanical energy of the particle is given by

$$\mathcal{E}_R(t) = \mathcal{E}(t) - \frac{1}{2} \int_{\{|\boldsymbol{x}| \geq R\}} \mathrm{d}^3x \big(\boldsymbol{E}(\boldsymbol{x},t)^2 + \boldsymbol{B}(\boldsymbol{x},t)^2\big) \tag{5.5}$$

provided $R$ is sufficiently large. Using the conservation of total energy, $\mathcal{E}(t) = \mathcal{E}(0)$, $\mathcal{E}_R$ changes in time as

$$\frac{\mathrm{d}}{\mathrm{d}t} \mathcal{E}_R(t) = -R^2 \int \mathrm{d}^2\omega\, \widehat{\omega} \cdot [\boldsymbol{E}(R\widehat{\omega},t) \times \boldsymbol{B}(R\widehat{\omega},t)] , \tag{5.6}$$

where $\widehat{\omega}$ is a vector on the unit sphere, $\mathrm{d}^2\omega$ the surface measure normalized to $4\pi$, and $\boldsymbol{E} \times \boldsymbol{B}$ the Poynting vector for the flux in energy at the surface of the ball under consideration. Since the total energy is bounded from below, we conclude that

$$\mathcal{E}_R(R) - \mathcal{E}_R(R+t) = - \int_R^{R+t} \mathrm{d}s \frac{\mathrm{d}}{\mathrm{d}s} \mathcal{E}_R(s) \leq C \tag{5.7}$$

with the constant $C = \mathcal{E}(0) - \bar{\phi}$ independent of $R$ and $t$.

In (5.7) we first take the limit $R \to \infty$, which yields the energy radiated to infinity during the time interval $[0, t]$ through a large sphere centered at the origin. Subsequently we take the limit $t \to \infty$ to obtain the total radiated energy. To state the result let us define

$$E_\infty(\widehat{\omega}, t) = -\frac{e}{4\pi}\int \mathrm{d}^3y\, \varphi(y - q(t + \widehat{\omega}\cdot y)) \times \big[(1-\widehat{\omega}\cdot v)^{-1}\dot{v} + (1-\widehat{\omega}\cdot v)^{-2}(\widehat{\omega}\cdot\dot{v})(v-\widehat{\omega})\big]\big|_{t+\widehat{\omega}\cdot y} \tag{5.8}$$

which is a functional of the actual trajectory of the particle. Whatever its motion we must have

$$\int_0^\infty \mathrm{d}t \int \mathrm{d}^2\omega\, |E_\infty(\widehat{\omega}, t)|^2 \le C < \infty\,. \tag{5.9}$$

Note that the integrand in (5.9) is proportional to $\dot{v}(t)^2$, which therefore is expected to decay to zero for large $t$.

To establish (5.9) is somewhat tedious with pieces of the argument explained in the section below and in section 8.5. One imagines that the trajectory $t \mapsto q(t)$ is given and solves the inhomogeneous Maxwell–Lorentz equations according to (2.16), (2.17). If the time-zero fields are in $\mathcal{M}^\sigma, 0 < \sigma \le 1$, see the definition (2.49), then $E_{\mathrm{ini}}(t)$ and $B_{\mathrm{ini}}(t)$ decay as stated in (5.28). Therefore $|\frac{\mathrm{d}}{\mathrm{d}s}\mathcal{E}_R(s)| < CR^2(1+s)^{-2-2\sigma}$ and the contribution to (5.7) from the initial fields vanishes in the limit $R \to \infty$. Next one has to study the asymptotics of the retarded fields, which is carried out in section 8.5. There $\varepsilon$ is fixed, and for our purpose we may set $\varepsilon = 1$. In addition in (8.48) the sphere of radius $R$ is centered at $q^\varepsilon(t)$, rather than at the origin. This means, in the present context one can use the asymptotics (8.51), (8.52) as $R \to \infty$ with $q^\varepsilon(t)$ replaced by 0. Combining both arguments proves that (5.9) follows from (5.7) in the limit $R \to \infty$.

The real task is to extract from (5.9) that the acceleration vanishes for long times.

**Theorem 5.1** (Long-time limit of the acceleration). *For the Abraham model satisfying* $(C)$, $(P)$, *and the Wiener condition* $(W)$ *let the initial data be* $Y(0) = (E^0, B^0, q^0, v^0) \in \mathcal{M}^\sigma$ *with* $0 < \sigma \le 1$. *Then*

$$\lim_{t\to\infty} \dot{v}(t) = 0\,. \tag{5.10}$$

*Proof*: By energy conservation $|v(t)| \le \bar{v} < 1$. Inserting in (2.41) and using $(P)$ we conclude that $|\dot{v}(t)| \le C$. Differentiating (2.41) and using again $(P)$ also $|\ddot{v}(t)| \le C$ uniformly in $t$. Therefore $E_\infty(\widehat{\omega}, t)$ is Lipschitz continuous jointly in

$\widehat{\omega}, t$. Since the energy dissipation (5.9) is bounded, this implies

$$\lim_{t\to\infty} \boldsymbol{E}_\infty(\widehat{\omega}, t) = 0 \tag{5.11}$$

uniformly in $\widehat{\omega}$.

We analyze the structure of the integrand in (5.8). The retarded argument depends only on $y_\parallel = \widehat{\omega}\cdot\boldsymbol{y}$. Therefore the integration over $\boldsymbol{y}_\perp = \boldsymbol{y} - y_\parallel\widehat{\omega}$ can be carried out and we are left with a one-dimensional integral of convolution type. We set $\varphi_a(x_3) = \int dx_1 dx_2\, \varphi(\boldsymbol{x})$. Then

$$\begin{aligned}\boldsymbol{E}_\infty(\widehat{\omega}, t) &= \frac{e}{4\pi}\int dy_\parallel\, \varphi_a(y_\parallel - q_\parallel(t+y_\parallel))\\ &\quad\times\big[(1-\widehat{\omega}\cdot\boldsymbol{v})^{-2}\widehat{\omega}\times((\widehat{\omega}-\boldsymbol{v})\times\dot{\boldsymbol{v}})\big]\big|_{t+y_\parallel}\\ &= \frac{e}{4\pi}\int ds\varphi_a(t-(s-q_\parallel(s)))\\ &\quad\times\big[(1-\widehat{\omega}\cdot\boldsymbol{v})^{-2}\widehat{\omega}\times((\widehat{\omega}-\boldsymbol{v})\times\dot{\boldsymbol{v}})\big]\big|_{s}\,.\end{aligned} \tag{5.12}$$

Since $|\dot{q}_\parallel(s)| < 1$, we can substitute $\theta = s - q_\parallel(s)$ and obtain the convolution representation

$$\boldsymbol{E}_\infty(\widehat{\omega}, t) = \int d\theta\, \varphi_a(t-\theta)\boldsymbol{g}_{\widehat{\omega}}(\theta) = \varphi_a * \boldsymbol{g}_{\widehat{\omega}}(t)\,, \tag{5.13}$$

where

$$\boldsymbol{g}_{\widehat{\omega}}(\theta) = \frac{e}{4\pi}\big[(1-\widehat{\omega}\cdot\boldsymbol{v})^{-2}\widehat{\omega}\times((\widehat{\omega}-\boldsymbol{v})\times\dot{\boldsymbol{v}})\big]\big|_{s(\theta)}\,. \tag{5.14}$$

From (5.11) we know that $\lim_{t\to\infty} \varphi_a * \boldsymbol{g}_{\widehat{\omega}}(t) = 0$. If $\widehat{\varphi}(\boldsymbol{k}_0) = 0$ for some $\boldsymbol{k}_0$, hence $\widehat{\varphi}$ violating the Wiener condition, then we could choose $\boldsymbol{g}_{\widehat{\omega}}(\theta)$ periodic with frequency $|\boldsymbol{k}_0|$ and still have $\varphi_a * \boldsymbol{g}_{\widehat{\omega}}(t) = 0$. At this point no further progress seems to be possible. However under the Wiener condition $(W)$ and with the smoothness of $\boldsymbol{g}_{\widehat{\omega}}(\theta)$ already established, Pitt's extension to the Tauberian theorem of Wiener assures us that

$$\lim_{\theta\to\infty} \boldsymbol{g}_{\widehat{\omega}}(\theta) = 0\,, \tag{5.15}$$

which, since $\theta(t) \to \infty$ as $t \to \infty$, implies

$$\lim_{t\to\infty} \widehat{\omega}\times((\widehat{\omega}-\boldsymbol{v}(t))\times\dot{\boldsymbol{v}}(t)) = 0 \tag{5.16}$$

for every $\widehat{\omega}$ in the unit sphere. Replacing $\widehat{\omega}$ by $-\widehat{\omega}$ and summing both expressions yields $\widehat{\omega}\times(\widehat{\omega}\times\dot{\boldsymbol{v}}(t)) \to 0$ as $t \to \infty$. Since this is true for every $\widehat{\omega}$, the claim follows. □

Note that by fiat Theorem 5.1 avoids any claims as regards the convergence of $(\boldsymbol{q}(t), \boldsymbol{v}(t))$ as $t \to \infty$.

Since the acceleration vanishes for large times, the comoving electromagnetic fields will adjust locally to the appropriate charge soliton. We established already that $\boldsymbol{E}_{\text{ini}}(t)$ and $\boldsymbol{B}_{\text{ini}}(t)$ decay. Thus one only has to consider the retarded fields $\boldsymbol{E}_{\text{ret}}(\boldsymbol{x}+\boldsymbol{q}(t), t)$, $\boldsymbol{B}_{\text{ret}}(\boldsymbol{x}+\boldsymbol{q}(t), t)$ relative to the position of the particle and compare them with the soliton fields $\boldsymbol{E}_{\boldsymbol{v}(t)}(\boldsymbol{x})$, $\boldsymbol{B}_{\boldsymbol{v}(t)}(\boldsymbol{x})$ at the current velocity. For this purpose one uses the representations (4.31), (4.32) for the charge soliton and (2.16), (2.17) for the retarded fields. We insert the explicit form (2.15) of the propagator. This yields

$$
\begin{aligned}
\boldsymbol{E}_{\boldsymbol{v}}(\boldsymbol{x}) = e \int d^3y\, (4\pi|\boldsymbol{x}-\boldsymbol{y}|)^{-1}\big(&|\boldsymbol{x}-\boldsymbol{y}|^{-1}\varphi(\boldsymbol{y}-\boldsymbol{v}|\boldsymbol{x}-\boldsymbol{y}|)\widehat{\boldsymbol{n}} \\
&+\boldsymbol{v}\cdot\nabla\varphi(\boldsymbol{y}-\boldsymbol{v}|\boldsymbol{x}-\boldsymbol{y}|)(\boldsymbol{v}-\widehat{\boldsymbol{n}})\big), \qquad (5.17)
\end{aligned}
$$

$$
\begin{aligned}
\boldsymbol{B}_{\boldsymbol{v}}(\boldsymbol{x}) = e \int d^3y (4\pi|\boldsymbol{x}-\boldsymbol{y}|)^{-1}\widehat{\boldsymbol{n}} \times \big(&-|\boldsymbol{x}-\boldsymbol{y}|^{-1}\varphi(\boldsymbol{y}-|\boldsymbol{x}-\boldsymbol{y}|\boldsymbol{v})\boldsymbol{v} \\
&+\boldsymbol{v}\cdot\nabla\varphi(\boldsymbol{y}-|\boldsymbol{x}-\boldsymbol{y}|\boldsymbol{v})\boldsymbol{v}\big), \qquad (5.18)
\end{aligned}
$$

where $\widehat{\boldsymbol{n}} = (\boldsymbol{x}-\boldsymbol{y})/|\boldsymbol{x}-\boldsymbol{y}|$. Similarly for the retarded fields

$$
\begin{aligned}
\boldsymbol{E}_{\text{ret}}(\boldsymbol{x}+\boldsymbol{q}(t), t) = \int d^3y\, (4\pi|\boldsymbol{x}-\boldsymbol{y}|)^{-1}\big(&|\boldsymbol{x}-\boldsymbol{y}|^{-1}\varphi(\boldsymbol{y}+\boldsymbol{q}(t)-\boldsymbol{q}(\tau))\widehat{\boldsymbol{n}} \\
&+\boldsymbol{v}(\tau)\cdot\nabla\varphi(\boldsymbol{y}+\boldsymbol{q}(t)-\boldsymbol{q}(\tau))(\boldsymbol{v}(\tau)-\widehat{\boldsymbol{n}}) \\
&-\varphi(\boldsymbol{y}+\boldsymbol{q}(t)-\boldsymbol{q}(\tau))\dot{\boldsymbol{v}}(\tau)\big), \qquad (5.19)
\end{aligned}
$$

$$
\begin{aligned}
\boldsymbol{B}_{\text{ret}}(\boldsymbol{x}+\boldsymbol{q}(t), t) = \int d^3y (4\pi|\boldsymbol{x}-\boldsymbol{y}|)^{-1}\widehat{\boldsymbol{n}} \times \big(&-|\boldsymbol{x}-\boldsymbol{y}|^{-1}\varphi(\boldsymbol{y}+\boldsymbol{q}(t) \\
&-\boldsymbol{q}(\tau))\boldsymbol{v}(\tau)+\boldsymbol{v}(\tau)\cdot\nabla\varphi(\boldsymbol{y}+\boldsymbol{q}(t)-\boldsymbol{q}(\tau))\boldsymbol{v}(\tau) \\
&-\varphi(\boldsymbol{y}+\boldsymbol{q}(t)-\boldsymbol{q}(\tau))\dot{\boldsymbol{v}}(\tau)\big), \qquad (5.20)
\end{aligned}
$$

where $\tau = t - |\boldsymbol{x}-\boldsymbol{y}|$ and $t \ge t_\varphi = 2R_\varphi/(1-\bar{v})$.

We compare the fields locally and use the result that $\lim_{t\to\infty} \dot{\boldsymbol{v}}(t) = 0$. Then, for any fixed $R > 0$,

$$
\begin{aligned}
\lim_{t\to\infty} \int_{\{|\boldsymbol{x}|\le R\}} d^3x \Big(&\big(\boldsymbol{E}(\boldsymbol{x}+\boldsymbol{q}(t), t) - \boldsymbol{E}_{\boldsymbol{v}(t)}(\boldsymbol{x})\big)^2 \\
&+\big(\boldsymbol{B}(\boldsymbol{x}+\boldsymbol{q}(t), t) - \boldsymbol{B}_{\boldsymbol{v}(t)}(\boldsymbol{x})\big)^2\Big) = 0\,. \qquad (5.21)
\end{aligned}
$$

The scattered fields are not covered by (5.21) and will be studied in section 5.3.

## 5.2 Convergence to the soliton manifold

In the case of *zero* external potentials, in essence any solution $Y(t)$ rapidly converges to the soliton manifold $\mathcal{S}$ as $t \to \infty$, in particular $\boldsymbol{v}(t) \to \boldsymbol{v}_\infty$. Such behavior will be of importance in the discussion of the adiabatic limit, see chapter 6, where it will be explained that in the matching to a comparison dynamics one cannot use the naive $\boldsymbol{v}(0)$ but instead must take $\boldsymbol{v}_\infty$. For hydrodynamic boundary value problems such a property is known as the slip condition, since the extrapolation from the bulk does not coincide with the boundary conditions imposed externally.

To prove the envisaged behavior we need a little preparation. Firstly we must have some decay and smoothness of the initial fields at infinity. We already introduced such a set of "good" initial data, $\mathcal{M}^\sigma$, compare with (2.49), and therefore require here $Y(0) \in \mathcal{M}^\sigma, 0 < \sigma \leq 1$. Secondly, we need a notion for two field configurations being close to each other. At a given time and far away from the particle the fields are determined by their initial data. Only close to the particle are they Coulombic. Therefore it is natural to measure closeness in the *local energy norm* defined by

$$\|(\boldsymbol{E}, \boldsymbol{B})\|_R^2 = \frac{1}{2} \int_{\{|\boldsymbol{x}| \leq R\}} d^3x \big(\boldsymbol{E}(\boldsymbol{x})^2 + \boldsymbol{B}(\boldsymbol{x})^2\big) \tag{5.22}$$

for given radius $R$.

The true solution is $Y(t) = (\boldsymbol{E}(\boldsymbol{x}, t), \boldsymbol{B}(\boldsymbol{x}, t), \boldsymbol{q}(t), \boldsymbol{v}(t))$ which is to be compared with the charge soliton approximation $\big(\boldsymbol{E}_{\boldsymbol{v}(t)}(\boldsymbol{x} - \boldsymbol{q}(t)), \boldsymbol{B}_{\boldsymbol{v}(t)}(\boldsymbol{x} - \boldsymbol{q}(t)), \boldsymbol{q}(t), \boldsymbol{v}(t)\big)$. We set $\boldsymbol{Z}_1(\boldsymbol{x}, t) = \boldsymbol{E}(\boldsymbol{x}, t) - \boldsymbol{E}_{\boldsymbol{v}(t)}(\boldsymbol{x} - \boldsymbol{q}(t))$, $\boldsymbol{Z}_2(\boldsymbol{x}, t) = \boldsymbol{B}(\boldsymbol{x}, t) - \boldsymbol{B}_{\boldsymbol{v}(t)}(\boldsymbol{x} - \boldsymbol{q}(t))$, $Z = (\boldsymbol{Z}_1, \boldsymbol{Z}_2)$ and want to establish that $\|Z(\cdot + \boldsymbol{q}(t), t)\|_R \to 0$ for large times at fixed $R$.

**Proposition 5.2** (Long-time limit for the velocity). *For the Abraham model with zero external potentials and satisfying (C) let $|e| \leq \bar{e}$ with sufficiently small $\bar{e}$ and let the initial data be $Y(0) \in \mathcal{M}^\sigma$ for some $\sigma \in (0, 1]$. Then for every $R > 0$ we have*

$$\|Z(\cdot + \boldsymbol{q}(t), t)\|_R \leq C_R (1 + |t|)^{-1-\sigma} . \tag{5.23}$$

*In addition, the acceleration is bounded as*

$$|\dot{\boldsymbol{v}}(t)| \leq C(1 + |t|)^{-1-\sigma} \tag{5.24}$$

*and there exists a $\boldsymbol{v}_\infty \in \mathbb{V}$ such that*

$$\lim_{t \to \infty} \boldsymbol{v}(t) = \boldsymbol{v}_\infty . \tag{5.25}$$

*Proof*: Using the Maxwell equations together with the identities $(\boldsymbol{v}\cdot\nabla)\,\boldsymbol{E}_v = -\nabla\times\boldsymbol{B}_v + e\varphi\boldsymbol{v}$, $(\boldsymbol{v}\cdot\nabla)\boldsymbol{B}_v = \nabla\times\boldsymbol{E}_v$ one obtains

$$\frac{\mathrm{d}}{\mathrm{d}t}Z(t) = AZ(t) - g(t)\,, \tag{5.26}$$

where $A$ is defined in (2.18) and $g(t)$ has the components $(\dot{\boldsymbol{v}}(t)\cdot\nabla_v)\boldsymbol{E}_v(\boldsymbol{x}-\boldsymbol{q}(t))$, $(\dot{\boldsymbol{v}}(t)\cdot\nabla_v)\boldsymbol{B}_v(\boldsymbol{x}-\boldsymbol{q}(t))$, and therefore

$$Z(t) = U(t)Z(0) - \int_0^t \mathrm{d}s\, U(t-s)g(s) \tag{5.27}$$

with $U(t) = \mathrm{e}^{At}$.

For the first term we note that $\boldsymbol{Z}_1(\boldsymbol{x},0) = \boldsymbol{E}^0(\boldsymbol{x}) - \boldsymbol{E}_{v^0}(\boldsymbol{x}-\boldsymbol{q}^0)$, $\boldsymbol{Z}_2(\boldsymbol{x},0) = \boldsymbol{B}^0(\boldsymbol{x}) - \boldsymbol{B}_{v^0}(\boldsymbol{x}-\boldsymbol{q}^0) \in \mathcal{M}^\sigma$ by assumption. Using the solution of the inhomogeneous Maxwell–Lorentz equations in position space and the bound (2.49) one has

$$\begin{aligned}
|\boldsymbol{Z}_1(\boldsymbol{x},t)| + |\boldsymbol{Z}_2(\boldsymbol{x},t)| &\le C\,t^{-2}\int \mathrm{d}^3y\,\delta(|\boldsymbol{x}-\boldsymbol{y}|-t)\big(|\boldsymbol{Z}_1(\boldsymbol{y},0)| + |\boldsymbol{Z}_2(\boldsymbol{y},0)|\big)\\
&\quad + C\,t^{-1}\int \mathrm{d}^3y\,\delta(|\boldsymbol{x}-\boldsymbol{y}|-t)\big(|\nabla\boldsymbol{Z}_1(\boldsymbol{y},0)|\\
&\quad + |\nabla\boldsymbol{Z}_2(\boldsymbol{y},0)|\big)\\
&\le C\,t^{-2}\int \mathrm{d}^3y\,\delta(|\boldsymbol{x}-\boldsymbol{y}|-t)(1+|\boldsymbol{y}|)^{-1-\sigma}\\
&\quad + C\,t^{-1}\int \mathrm{d}^3y\,\delta(|\boldsymbol{x}-\boldsymbol{y}|-t)(1+|\boldsymbol{y}|)^{-2-\sigma}\\
&\le C\,(1+t)^{-1-\sigma}\,.
\end{aligned} \tag{5.28}$$

The integrand in the second term of (5.27) will be estimated in section 7.3 with the bound

$$\|U(t-s)g(s)\|_{R_\varphi} \le C(\bar{v})e^2(1+(t-s)^2)^{-1}\|Z(\cdot+\boldsymbol{q}(s),s)\|_{R_\varphi}\,; \tag{5.29}$$

compare with (7.36).

We choose $R \ge R_\varphi$. From (5.29) and (5.28)

$$\begin{aligned}
\|Z(\cdot+\boldsymbol{q}(t),t)\|_R &\le C(1+t)^{-1-\sigma}\\
&\quad + C(\bar{v})e^2\int_0^t \mathrm{d}s\,(1+(t-s)^2)^{-1}\|Z(\cdot+\boldsymbol{q}(s),s)\|_R\,.
\end{aligned} \tag{5.30}$$

Let $\kappa = \sup_{t\geq 0} (1+t)^{1+\sigma} \|Z(\cdot + q(t), t)\|_R$. Then

$$\kappa \leq C + C(\bar{v})e^2 \Big( \int_0^t \mathrm{d}s\, (1+(t-s)^2)^{-1}(1+s)^{-1-\sigma} \Big)\kappa\,, \tag{5.31}$$

which implies $\kappa < \infty$ provided $C(\bar{v})\, e^2$ is sufficiently small.

To estimate the decay rate for the acceleration we start from Newton's equations of motion in the form

$$\frac{\mathrm{d}}{\mathrm{d}t}\big(m_{\mathrm{b}}\gamma \boldsymbol{v}(t)\big) = e\big(\boldsymbol{E}_\varphi(\boldsymbol{q}(t)) - \boldsymbol{E}_{\boldsymbol{v}(t)\varphi}(0) + \boldsymbol{v} \times (\boldsymbol{B}_\varphi(\boldsymbol{q}(t)) - \boldsymbol{B}_{\boldsymbol{v}(t)\varphi}(0))\big)\,, \tag{5.32}$$

which uses the fact that the force from the soliton field vanishes. By energy conservation $|\boldsymbol{v}(t)| \leq \bar{v} < 1$. Therefore (5.32) implies

$$|\dot{\boldsymbol{v}}(t)| \leq C\, e\, \|Z(\cdot + \boldsymbol{q}(t), t)\|_{R_\varphi} \tag{5.33}$$

and (5.24) follows from (5.23). Since $\boldsymbol{v}(t) = \boldsymbol{v}(0) + \int_0^t \mathrm{d}s\, \dot{\boldsymbol{v}}(s)$, one has $|\boldsymbol{v}(t) - \boldsymbol{v}_\infty| \leq C\,(1+|t|)^{-\sigma}$. □

## 5.3 Scattering theory

We still have to provide an analysis of the scattered wave. Our results are somewhat fragmentary and we start with an easy and sufficient integrability condition.

**Theorem 5.3** (Existence of scattering solutions). *For the Abraham model satisfying (C) and (P) let $Y(t) \in \mathcal{M}$ be a solution. If*

$$\int_0^\infty \mathrm{d}t\, |\dot{\boldsymbol{v}}(t)| < \infty\,, \tag{5.34}$$

*then there exist scattering data $(\boldsymbol{E}_{\mathrm{sc}}, \boldsymbol{B}_{\mathrm{sc}})$ such that*

$$\lim_{t\to\infty} \big(\|\boldsymbol{E}(t) - \boldsymbol{E}_{\boldsymbol{v}(t)}(\cdot - \boldsymbol{q}(t)) - \boldsymbol{E}_{\mathrm{sc}}(t)\| + \|\boldsymbol{B}(t) - \boldsymbol{B}_{\boldsymbol{v}(t)}(\cdot - \boldsymbol{q}(t)) - \boldsymbol{B}_{\mathrm{sc}}(t)\|\big) = 0\,, \tag{5.35}$$

*where $(\boldsymbol{E}_{\mathrm{sc}}(t), \boldsymbol{B}_{\mathrm{sc}}(t)) = U(t)(\boldsymbol{E}_{\mathrm{sc}}, \boldsymbol{B}_{\mathrm{sc}})$ propagate according to the homogeneous Maxwell–Lorentz equations.*

Note that in (5.35) the difference is in the global energy norm and therefore carries the information on the scattered wave.

*Proof*: The difference in (5.35) is $Z(t)$ by definition. (5.26) remains valid in the presence of external forces, which means that

$$Z(t) = U(t)\big(Z(0) - \int_0^t \mathrm{d}s\, U(-s)g(s)\big)\,. \tag{5.36}$$

We set

$$\begin{aligned} \boldsymbol{E}_{\mathrm{sc}}(\boldsymbol{x}) &= \boldsymbol{E}^0(\boldsymbol{x}) - \boldsymbol{E}_{v^0}(\boldsymbol{x}-\boldsymbol{q}^0) - \int_0^\infty \mathrm{d}t\,(\dot{\boldsymbol{v}}(t)\cdot\nabla_v)\boldsymbol{E}_v(\boldsymbol{x}-\boldsymbol{q}(t))\,,\\ \boldsymbol{B}_{\mathrm{sc}}(\boldsymbol{x}) &= \boldsymbol{B}^0(\boldsymbol{x}) - \boldsymbol{B}_{v^0}(\boldsymbol{x}-\boldsymbol{q}^0) - \int_0^\infty \mathrm{d}t\,(\dot{\boldsymbol{v}}(t)\cdot\nabla_v)\boldsymbol{B}_v(\boldsymbol{x}-\boldsymbol{q}(t))\,. \end{aligned} \tag{5.37}$$

Since $|\boldsymbol{v}(t)| \le \bar{v} < 1$, the integrands have uniformly bounded energy norm. Thus by assumption (5.34) the integrals converge in $\mathcal{M}$ and define $(\boldsymbol{E}_{\mathrm{sc}}, \boldsymbol{B}_{\mathrm{sc}}) \in \mathcal{M}$. Hence (5.35) follows. □

There are two cases of interest for which the integrability condition (5.34) can be checked.

(i) *Compton scattering (zero external potential).* If $|e| \le \bar{e}$, then by (5.24) $|\dot{\boldsymbol{v}}(t)| \le C(1+|t|)^{-1-\sigma}$ which implies (5.34). For a freely moving charge the asymptotic motion is rectilinear and the scattered waves propagate according to the free Maxwell equations. Such a result also applies to a charge reaching an essentially potential-free region. The standard example is a charge scattered by an infinitely heavy nucleus. For sufficiently long times the incident fields have decayed already and we assume that the charge has reached, with its velocity pointing outwards, a large sphere centered at the nucleus. Then the external force decays as $1/t^2$ which combined with Theorem 5.3 yields the desired asymptotics.

(ii) *Rayleigh scattering (bounded motion).* Under the Wiener condition $(W)$ we already know that $\lim_{t\to\infty} \dot{\boldsymbol{v}}(t) = 0$. If in addition the motion is bounded,

$$|\boldsymbol{q}(t)| \le \bar{q} \tag{5.38}$$

for all $t$, then necessarily

$$\lim_{t\to\infty} \boldsymbol{v}(t) = 0\,, \tag{5.39}$$

i.e. the particle comes to rest. Inserting in Newton's equations of motion (2.34) and using the fact that the fields become locally soliton-like, we infer that

$$\lim_{t\to\infty} \nabla\phi_{\mathrm{ex}}(\boldsymbol{q}(t)) = 0\,. \tag{5.40}$$

Let us define the set $\mathcal{A}$ of critical points for the potential $\phi_{\mathrm{ex}}$, $\mathcal{A} = \{\boldsymbol{q} \mid \nabla\phi_{\mathrm{ex}}(\boldsymbol{q}) = 0\}$. By (5.40), $\boldsymbol{q}(t)$ approaches $\mathcal{A}$ as a set. If $\mathcal{A}$ happens to be a discrete set, then, by the continuity of solutions in $t$, $\boldsymbol{q}(t)$ has to converge to some definite $\boldsymbol{q}^* \in \mathcal{A}$.

Such reasoning yields no rate of convergence. The situation improves in the case where $\boldsymbol{q}^*$ is a stable local minimum of $\phi_{\text{ex}}$. We linearize the Maxwell equations at $Y^* = S_{q^*,\mathbf{0}}$. The solution to the linearized equations converges exponentially fast to zero. Therefore, once $\boldsymbol{q}(t)$ is in the vicinity of $\boldsymbol{q}^*$, the velocity decays exponentially ensuring (5.34). In particular, if $\phi_{\text{ex}}$ is strictly convex and if $(W)$ holds, then the asymptotics (5.35) of Theorem 5.3 hold for every $Y(0) \in \mathcal{M}$.

A standard situation not covered by (i) and (ii) is the motion in a uniform magnetic field. Even if one assumes that the motion is bounded, one can only conclude that $\boldsymbol{v}(t) \to 0$. The attractor $\mathcal{A}$ equals $\mathbb{R}^3$. Physically one would expect the charge to spiral inwards and to come to rest at its center of gyration. Another instructive example is the motion in a confining $\phi_{\text{ex}}$ with a flat bottom, say $\{\boldsymbol{x} \mid |\boldsymbol{x}| \le 1\}$ and $A_{\text{ex}} = 0$. Each time the particle is reflected by the confining potential, it loses energy. Thus $\boldsymbol{v}(t) \to 0$ as $t \to \infty$, but $\boldsymbol{q}(t)$ has no limit.

## Notes and references

### *Section 5.1*

The long-time asymptotics are studied in Komech and Spohn (2000), where the details of the proof can be found. See also Komech, Spohn and Kunze (1997). Pitt's version of the Wiener theorem is proved in Rudin (1977), Theorem 9.7(b). We remark that Theorem 5.1 provides no rate of convergence. Thus to investigate the asymptotics of the velocity and position requires extra considerations.

Theorem 5.1 can also be read that under the Wiener condition the Abraham model admits no periodic solution. In the literature, Bohm and Weinstein (1948), Eliezer (1950), and in particular the review by Pearle (1982), periodic solutions of the Abraham model have been reported repeatedly for the case of a charged sphere, i.e. $\varphi(\boldsymbol{x}) = (4\pi a^2)^{-1}\delta(|\boldsymbol{x}| - a)$, which is not covered by Theorem 5.1 since $(W)$ is violated. These computations invoke certain approximations and it is not clear whether the full model, as defined by (2.39)–(2.41), has periodic solutions. Pearle (1977) argues that in the Nodvik model there are no periodic solutions. Kunze (1998) proves that if there is a periodic solution, its frequency is determined by the zeros of the radial part of the form factor $\widehat{\varphi}$, which under $(C)$ form a discrete set. If $\widehat{\varphi}$ has a zero, then the linearized system admits a periodic solution. However, the full nonlinear equations have no periodic solution, at least in a small neighborhood of the linearized periodic solution.

As will be explained in chapter 11, the Abraham model extends in the obvious way to the dynamics of many charges. The argument of Theorem 5.1 applied to this case yields that the acceleration of the center of mass relaxes to zero. One would expect particles to form neutral lumps, each of which is traveling at constant velocity for large $t$. No argument towards a proof is in sight.

### *Section 5.2*

The contraction method was first developed in Komech, Kunze and Spohn (1999). Komech and Spohn (1998) prove the convergence to the soliton manifold in the case of a scalar wave field requiring only $(W)$ and not the restriction $|e| < \bar{e}$. No convergence rates are obtained. Their result is extended to the Abraham model by Imaikin, Komech and Mauser (2003). Orbital stability was established before by Bambusi and Galgani (1993). Bambusi (1994) investigates the long-time stability in the case of an attractive central potential.

### *Section 5.3*

Our results are based on Imaikin, Komech and Spohn (2002). The linearization argument is fully carried out in Komech, Spohn and Kunze (1997).

# 6
# Adiabatic limit

If we assume that the mass of an electron is purely electromagnetic, then by equating its rest energy and electrostatic Coulomb energy the charge distribution must be concentrated in a ball of radius

$$r_{\mathrm{cl}} = \frac{e^2}{m_{\mathrm{e}}c^2} = 3 \times 10^{-13} \text{ cm} \tag{6.1}$$

which is the so-called classical electron radius. Quantum mechanically one argues that on the basis of light scattering the electron appears to have an effective size of the order of the Compton wavelength $\lambda_{\mathrm{c}} = \hbar m_{\mathrm{e}}/c = (e^2/\hbar c)^{-1} r_{\mathrm{cl}} = 137\, r_{\mathrm{cl}}$. Thus empirically $R_\varphi$ is limited to $r_{\mathrm{cl}} \leq R_\varphi \leq 137 r_{\mathrm{cl}}$. Electromagnetic fields which can be manipulated in the laboratory vary little over that length scale. $r_{\mathrm{cl}}$ defines a time scale through the time span for light to travel across the diameter of the charge distribution,

$$t_{\mathrm{cl}} = r_{\mathrm{cl}}/c = 10^{-23} \text{ s}, \quad \text{equivalently a frequency } \omega_{\mathrm{cl}} = 10^{23} \text{ Hz}. \tag{6.2}$$

Again, manufactured frequencies are much smaller than $\omega_{\mathrm{cl}}$. Space-time variations as fast as (6.1) and (6.2) lead us deeply into the quantum regime. Thus it is natural and physically compelling to study the dynamics of a charged particle under external potentials which vary *slowly* on the scale of the charge radius $R_\varphi$, which is the only length scale available. This means we have to introduce a scale of potentials and enquire about an approximately autonomous particle dynamics with an error depending on the scale under consideration. We will introduce such a scheme in the following section. The resulting problem has many similarities with the derivation of hydrodynamics from Newtonian particle dynamics – with the most welcome bonus that it is simpler mathematically by many orders of magnitude. Still, the comparison is instructive.

### 6.1 Scaling limit for external potentials of slow variation

For the Abraham model, see Eq. (2.41), the Lorentz force has in addition to the dynamical fields $\boldsymbol{E}(\boldsymbol{x}, t)$, $\boldsymbol{B}(\boldsymbol{x}, t)$ also prescribed external fields, which are the gradients of the external potentials $\phi_{\text{ex}}(\boldsymbol{x})$, $\boldsymbol{A}_{\text{ex}}(\boldsymbol{x})$.

We want to impose the condition that $\phi_{\text{ex}}$ and $\boldsymbol{A}_{\text{ex}}$ are slowly varying on the scale of $R_\varphi$. Formally we introduce a small *dimensionless* parameter $\varepsilon$ and consider the potentials

$$\phi_{\text{ex}}(\varepsilon \boldsymbol{x}), \quad \boldsymbol{A}_{\text{ex}}(\varepsilon \boldsymbol{x}), \tag{6.3}$$

which are slowly varying in the limit $\varepsilon \to 0$. Most of our results extend to potentials which vary also slowly in time. For simplicity we restrict ourselves to time-independent potentials here. Clearly, $\varepsilon$ appears as a parameter of the potential, just like $\omega_0$ is a parameter of the harmonic oscillator potential $\frac{1}{2} m\omega_0^2 x^2$. But $\varepsilon$ should really be thought of as a bookkeeping device which orders the magnitude of the various terms and the space-time scales according to the powers of $\varepsilon$. Such a scheme is familiar from very diverse contexts and appears whenever one has to deal with a problem involving scale separation.

So how small is $\varepsilon$? From the discussion above one might infer that if $\phi_{\text{ex}}$, $\boldsymbol{A}_{\text{ex}}$ vary over a scale of 1 mm, then $\varepsilon = 10^{-12}$. This is a totally meaningless statement, because $e\phi_{\text{ex}}$, $e\boldsymbol{A}_{\text{ex}}$ have the dimension of energy and thus the variation depends on the adopted energy scale. In (6.3) we merely stretch the spatial axes by a factor $\varepsilon^{-1}$ and fix the energy scale. Since from experience this point is likely to be confusing, let us consider the specific example of a charge revolving in the uniform magnetic field $\boldsymbol{B}_{\text{ex}} = (0, 0, B_0)$. The corresponding vector potential is linear in $\boldsymbol{x}$, and to introduce $\varepsilon$ as in (6.3) just means that the magnetic field strength equals $\varepsilon B_0$. The limit $\varepsilon \to 0$ is a limit of small magnetic field strength *relative* to some reference field $B_0$. Thus to obtain $\varepsilon$ we first have to determine the reference field and compare it with the magnetic field of interest. This shows that in order to fix $\varepsilon$ we have to specify the physical situation in detail, in particular the external potentials, the mass of the particle, the charge of the particle, $\gamma(\boldsymbol{v})$, and the time span of interest.

The scaling scheme (6.3) has the great advantage that the analysis can be carried out in generality. In a second step one has to figure out $\varepsilon$ for a *concrete* situation, which leads to a quantitative estimate of the error terms. For instance, if in the case above we consider an electron with velocities such that $\gamma \leq 10$, then, by comparing the Hamiltonian term and the friction term, the reference field turns out to be $B_0 = 10^{17}$ gauss. Laboratory magnetic fields are less than $10^5$ gauss and thus $\varepsilon < 10^{-12}$. In this and many other concrete examples, $\varepsilon$ is very small, less than $10^{-10}$, which implies that, firstly, all corrections *beyond* radiation reaction are negligible. Secondly, we do not have to go each time through the scheme indicated above and may as well set $\varepsilon = 1$ thereby returning to conventional units. Still on a theoretical level the use of the scale parameter $\varepsilon$ is very convenient. In an appendix to this section we will work out the example of a constant magnetic field more explicitly. If the reader feels uneasy about the scaling limit, (s)he should consult this example first.

Adopting (6.3), Newton's equations of motion now read

$$\frac{\mathrm{d}}{\mathrm{d}t}\big(m_{\mathrm{b}}\gamma \boldsymbol{v}(t)\big) = e\big(\boldsymbol{E}_{\varphi}(\boldsymbol{q}(t),t) + \varepsilon \boldsymbol{E}_{\mathrm{ex}}(\varepsilon \boldsymbol{q}(t)) + \boldsymbol{v}(t) \times \big(\boldsymbol{B}_{\varphi}(\boldsymbol{q}(t),t) + \varepsilon \boldsymbol{B}_{\mathrm{ex}}(\varepsilon \boldsymbol{q}(t))\big)\big)\,, \tag{6.4}$$

where

$$\boldsymbol{E}_{\mathrm{ex}} = -\nabla\phi_{\mathrm{ex}}\,, \quad \boldsymbol{B}_{\mathrm{ex}} = \nabla \times \boldsymbol{A}_{\mathrm{ex}}\,. \tag{6.5}$$

Note that if $\boldsymbol{E}_{\mathrm{ex}}, \boldsymbol{B}_{\mathrm{ex}}$ are smeared by $\varphi$, as would be proper, the resulting error in (6.4) is of order $\varepsilon^3$, which can be ignored for our purposes.

Equation (6.4) has to be supplemented with Maxwell's equations (2.39), (2.40). Our goal is to understand the structure of the solution for small $\varepsilon$, and as a first qualitative step one should discuss the rough order of magnitudes in powers of $\varepsilon$. But before that we have to specify the initial data. We give ourselves $\boldsymbol{q}^0$, $\boldsymbol{v}^0$ as the initial position and velocity of the charge. The initial fields are assumed to be Coulombic, i.e. of the form of a charge soliton centered at $\boldsymbol{q}^0$ with velocity $\boldsymbol{v}^0$, compare with (4.28), which we formalize as

*Condition* ($I$):

$$Y(0) = S_{\boldsymbol{q}^0,\boldsymbol{v}^0}\,. \tag{6.6}$$

Equivalently, according to (4.31), (4.32), we may say that the particle has traveled freely with velocity $\boldsymbol{v}^0$ for the infinite time span $(-\infty, 0]$. At time $t = 0$ the external potentials are turned on. Geometrically, our initial data are exactly on the soliton manifold $\mathcal{S}$ considered as a submanifold of the phase space $\mathcal{M}$. If there are no external forces, the solution stays on $\mathcal{S}$ and moves along a straight line. For slowly varying external potentials as in (6.3) we will show that the solution stays $\varepsilon$-close to $\mathcal{S}$ in the local energy distance.

On general grounds one may wonder whether such specific initial data are really required. In analogy to hydrodynamics, we call this the initial slip problem. In times of order $t_{\varphi}$ $(= R_{\varphi}/c)$, the fields close to the charge acquire their Coulombic form while the external forces are still negligible; compare with figure 6.1. However, during that period the particle might gain or lose in momentum and energy through the interaction with its own field and the data at time $t_{\varphi}$ close to the particle

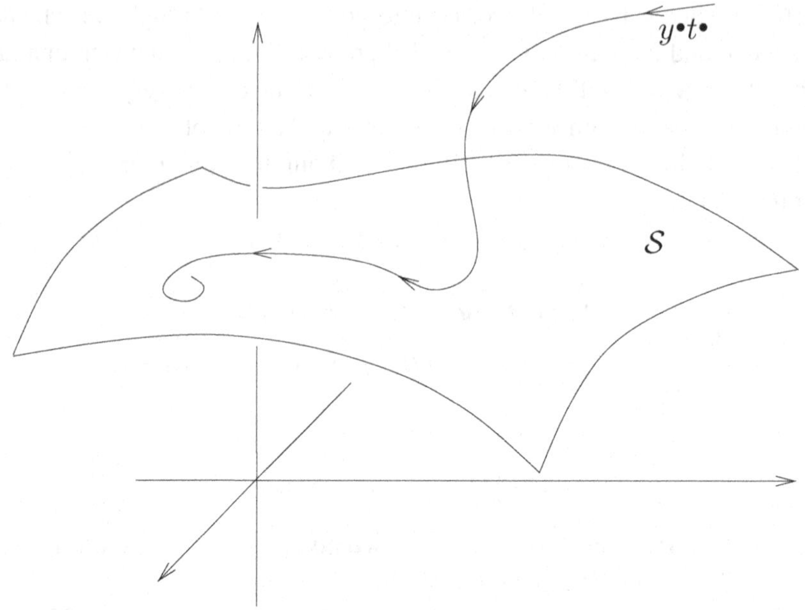


Figure 6.1: Schematic phase space with attractive soliton manifold $\mathcal{S}$. Away from $\mathcal{S}$ the motion is fast, on $\mathcal{S}$ it is slow.

are approximately of the form $S_{\tilde{\boldsymbol{q}},\tilde{\boldsymbol{v}}}$, where $\tilde{\boldsymbol{q}}$ and $\tilde{\boldsymbol{v}}$ are to be computed from the full solution. Of course, at a distance $ct$ away from the charge, the field still remembers its $t = 0$ data. Thus we see that the initial slip problem translates into the long-time asymptotics of a charge at *zero* external potentials but with general initial field data. We refer to section 5.2, where this point has been studied in detail. At the moment we just circumvent the initial slip by fiat.

Let us discuss the three relevant time scales, where we recall that $t_\varphi = R_\varphi/c$.

(i) *Microscopic scale*, $t = \mathcal{O}(t_\varphi)$, $\boldsymbol{q} = \mathcal{O}(R_\varphi)$. On this scale the particle moves along an essentially straight line. The electromagnetic fields adjust themselves to their comoving Coulombic form. As we will see, they do this with a precision $\mathcal{O}(\varepsilon)$ in the energy norm.

(ii) *Macroscopic potential scale*, $t = \mathcal{O}(\varepsilon^{-1}t_\varphi)$, $\boldsymbol{q} = \mathcal{O}(\varepsilon^{-1}R_\varphi)$. This scale is defined by the variation of the potentials, i.e. on this scale the potentials are $\phi_{\text{ex}}(\boldsymbol{x})$, $\boldsymbol{A}_{\text{ex}}(\boldsymbol{x})$. The particle follows the external forces. Since it is in company with almost Coulombic fields, the particle responds to the forces according to the effective energy–momentum relation, which we determined in chapter 4. On the macroscopic scale the motion is Hamiltonian up to errors of order $\varepsilon$. There is no dissipation of energy and momentum.

(iii) *Macroscopic friction scale*. Accelerated charges lose energy through radiation, which means that there must be friction corrections to the effective

Hamiltonian motion. According to Larmor's formula the radiation losses are proportional to $\dot{v}(t)^2$. Since the external forces are of the order $\varepsilon$, these losses are proportional to $\varepsilon^2$ when measured in microscopic units. Integrated over a time span $\varepsilon^{-1}t_\varphi$ the friction results in an effect of order $\varepsilon$. Thus we expect order $\varepsilon$ dissipative corrections to the conservative motion on the macroscopic scale. Followed over the even longer time scale $\varepsilon^{-2}t_\varphi$, the radiation reaction results in $\mathcal{O}(1)$ deviations from the Hamiltonian trajectory.

On the friction time scale the motion either comes to a standstill or stays uniform. In addition, as will be shown, the dissipative effective equation has the same long-time behavior as the true solution. Thus we expect no further qualitatively distinct time scale beyond the friction scale.

From our description, in a certain sense, the most natural scale is the macroscopic scale and we transform Maxwell's and Newton's equations to this new scale by setting

$$t' = \varepsilon t\,, \quad \boldsymbol{x}' = \varepsilon \boldsymbol{x}\,. \tag{6.7}$$

We have the freedom of how to scale the amplitudes of the dynamic part of the electromagnetic fields. We require that their energy is independent of $\varepsilon$. Then

$$\boldsymbol{E}'(\boldsymbol{x}', t') = \varepsilon^{-3/2}\,\boldsymbol{E}(\boldsymbol{x}, t)\,, \quad \boldsymbol{B}'(\boldsymbol{x}', t') = \varepsilon^{-3/2}\,\boldsymbol{B}(\boldsymbol{x}, t)\,. \tag{6.8}$$

Finally the new position and velocity are

$$\boldsymbol{q}'(t') = \varepsilon \boldsymbol{q}(t)\,, \quad \boldsymbol{v}'(t') = \boldsymbol{v}(t)\,, \tag{6.9}$$

so that $\frac{\mathrm{d}}{\mathrm{d}t'}\boldsymbol{q}' = \boldsymbol{v}'$. There is little risk of confusion in omitting the prime. We then denote

$$\boldsymbol{q}^\varepsilon(t) = \varepsilon \boldsymbol{q}(\varepsilon^{-1}t)\,, \quad \boldsymbol{v}^\varepsilon(t) = \boldsymbol{v}(\varepsilon^{-1}t)\,, \quad \varphi_\varepsilon(\boldsymbol{x}) = \varepsilon^{-3}\,\varphi(\varepsilon^{-1}\boldsymbol{x})\,, \tag{6.10}$$

which means that $\int \mathrm{d}^3x\, \varphi_\varepsilon(\boldsymbol{x}) = 1$ independent of $\varepsilon$ and that $\varphi_\varepsilon$ is supported in a ball of radius $\varepsilon R_\varphi$. In the macroscopic coordinates the coupled Maxwell's and Newton's equations read

$$\begin{aligned}
&\partial_t \boldsymbol{B}(\boldsymbol{x}, t) = -\nabla \times \boldsymbol{E}(\boldsymbol{x}, t)\,,\\
&\partial_t \boldsymbol{E}(\boldsymbol{x}, t) = \nabla \times \boldsymbol{B}(\boldsymbol{x}, t) - \sqrt{\varepsilon}\, e\varphi_\varepsilon(\boldsymbol{x} - \boldsymbol{q}^\varepsilon(t))\boldsymbol{v}^\varepsilon(t)\,,\\
&\frac{\mathrm{d}}{\mathrm{d}t}\big(m_\mathrm{b}\gamma \boldsymbol{v}^\varepsilon(t)\big) = e\big(\boldsymbol{E}_\mathrm{ex}(\boldsymbol{q}^\varepsilon(t)) + \boldsymbol{v}^\varepsilon(t) \times \boldsymbol{B}_\mathrm{ex}(\boldsymbol{q}^\varepsilon(t))\big)\\
&\qquad\qquad + \sqrt{\varepsilon}\, e\big(\boldsymbol{E}_{\varphi_\varepsilon}(\boldsymbol{q}^\varepsilon(t), t) + \boldsymbol{v}^\varepsilon(t) \times \boldsymbol{B}_{\varphi_\varepsilon}(\boldsymbol{q}^\varepsilon(t), t)\big)
\end{aligned} \tag{6.11}$$

together with the constraints

$$\nabla \cdot \boldsymbol{E} = \sqrt{\varepsilon}\, e\varphi_\varepsilon(\cdot - \boldsymbol{q}^\varepsilon(t))\,, \quad \nabla \cdot \boldsymbol{B} = 0\,. \tag{6.12}$$

On the macroscopic scale the conserved energy is

$$\mathcal{E}_{\text{mac}} = m_{\text{b}}\gamma(\boldsymbol{v}) + e\phi_{\text{ex}}(\boldsymbol{q}) + \frac{1}{2}\int \mathrm{d}^3x\,\big(\boldsymbol{E}(\boldsymbol{x})^2 + \boldsymbol{B}(\boldsymbol{x})^2\big)\,. \tag{6.13}$$

Also the initial data have to be transformed and become

*Condition* $(I_\varepsilon)$:

$$Y^\varepsilon(0) = S^\varepsilon_{\boldsymbol{q}^0,\boldsymbol{v}^0} = (\boldsymbol{E}^\varepsilon_{\boldsymbol{v}^0}(\boldsymbol{x}-\boldsymbol{q}^0), \boldsymbol{B}^\varepsilon_{\boldsymbol{v}^0}(\boldsymbol{x}-\boldsymbol{q}^0), \boldsymbol{q}^0, \boldsymbol{v}^0) \tag{6.14}$$

with

$$\boldsymbol{E}^\varepsilon_{\boldsymbol{v}} = -\nabla\phi^\varepsilon_{\boldsymbol{v}} + \boldsymbol{v}(\boldsymbol{v}\cdot\nabla\phi^\varepsilon_{\boldsymbol{v}})\,, \quad \boldsymbol{B}^\varepsilon_{\boldsymbol{v}} = -\boldsymbol{v}\times\nabla\phi^\varepsilon_{\boldsymbol{v}}\,, \tag{6.15}$$

where now

$$\widehat{\phi}^\varepsilon_{\boldsymbol{v}}(\boldsymbol{k}) = \frac{\sqrt{\varepsilon}\,e\widehat{\varphi}(\varepsilon\boldsymbol{k})}{\boldsymbol{k}^2 - (\boldsymbol{v}\cdot\boldsymbol{k})^2}\,. \tag{6.16}$$

On the macroscopic scale, the scaling parameter $\varepsilon$ can be absorbed into the "effective" charge distribution $\sqrt{\varepsilon}e\widehat{\varphi}_\varepsilon$. Its electrostatic energy,

$$m_{\text{e}} = \frac{1}{2}e^2\int \mathrm{d}^3k\,\varepsilon|\widehat{\varphi}_\varepsilon(\boldsymbol{k})|^2\frac{1}{\boldsymbol{k}^2} = \frac{1}{2}\int \mathrm{d}^3k|\widehat{\varphi}(\boldsymbol{k})|^2\,\frac{1}{\boldsymbol{k}^2}\,, \tag{6.17}$$

is independent of $\varepsilon$, while its charge

$$e\int \mathrm{d}^3x\,\sqrt{\varepsilon}\,\varphi_\varepsilon(\boldsymbol{x}) = \sqrt{\varepsilon}\,e \tag{6.18}$$

vanishes as $\sqrt{\varepsilon}$. Recall that $\varepsilon$ is a "bookkeeping device".

We argued that on the macroscopic scale the response to external potentials in the motion of the charges is of order one. We thus expect that $\boldsymbol{q}^\varepsilon(t)$ tends to a nondegenerate limit as $\varepsilon \to 0$, i.e.

$$\lim_{\varepsilon\to 0}\boldsymbol{q}^\varepsilon(t) = \boldsymbol{r}(t)\,, \quad \lim_{\varepsilon\to 0}\boldsymbol{v}^\varepsilon(t) = \boldsymbol{u}(t)\,. \tag{6.19}$$

The position $\boldsymbol{r}(t)$ and velocity $\boldsymbol{u}(t)$ should be governed by an effective Lagrangian. In section 4.1 we determined the effective inertial term. If the potentials add in as usual, one has

$$L_{\text{eff}}(\boldsymbol{q},\dot{\boldsymbol{q}}) = T(\dot{\boldsymbol{q}}) - e\big(\phi_{\text{ex}}(\boldsymbol{q}) - \dot{\boldsymbol{q}}\cdot\boldsymbol{A}_{\text{ex}}(\boldsymbol{q})\big)\,, \tag{6.20}$$

which results in the equations of motion

$$\dot{\boldsymbol{r}} = \boldsymbol{u}\,, \quad m(\boldsymbol{u})\dot{\boldsymbol{u}} = e(\boldsymbol{E}_{\text{ex}}(\boldsymbol{r}) + \boldsymbol{u}\times\boldsymbol{B}_{\text{ex}}(\boldsymbol{r}))\,. \tag{6.21}$$

The velocity-dependent mass $m(\boldsymbol{u})$ has a bare and a field contribution. From (4.12) we conclude that

$$m(\boldsymbol{u}) = \frac{\mathrm{d}P_{\mathrm{s}}(\boldsymbol{u})}{\mathrm{d}\boldsymbol{u}} \tag{6.22}$$

as a $3 \times 3$ matrix. If instead of the velocity we introduce the canonical momentum, $\boldsymbol{p}$, then the effective Hamiltonian reads

$$H_{\mathrm{eff}}(\boldsymbol{r}, \boldsymbol{p}) = E_{\mathrm{eff}}(\boldsymbol{p} - e\boldsymbol{A}_{\mathrm{ex}}(\boldsymbol{r})) + e\phi_{\mathrm{ex}}(\boldsymbol{r}) \tag{6.23}$$

with Hamilton's equations of motion

$$\dot{\boldsymbol{r}} = \nabla_{\boldsymbol{p}} H_{\mathrm{eff}}\,, \quad \dot{\boldsymbol{p}} = -\nabla_{\boldsymbol{r}} H_{\mathrm{eff}}\,. \tag{6.24}$$

Our plan is to establish the limit (6.19) and to investigate the corrections due to radiation losses.

### 6.1.1 Appendix 1: How small is $\varepsilon$?

We consider an electron moving in an external magnetic field oriented along the $z$-axis, $\boldsymbol{B}_{\mathrm{ex}} = (0, 0, B_0)$. The corresponding vector potential is $\boldsymbol{A}_{\mathrm{ex}}(\boldsymbol{x}) = \frac{1}{2} B_0(-x_2, x_1, 0)$. According to our convention the slowly varying vector potential is given by $\boldsymbol{A}_{\mathrm{ex}}(\varepsilon \boldsymbol{x}) = \frac{1}{2}\varepsilon B_0(-x_2, x_1, 0)$. Thus $B_0$ is a reference field strength, which is to be determined, and $B = \varepsilon B_0$ is the physical field strength in the laboratory. The motion of the electron is assumed to be in the 1–2 plane and we set $\boldsymbol{v} = (\boldsymbol{u}, 0)$. According to section 9.2, example (iii), within a good approximation the motion of the electron is governed by

$$\gamma \dot{\boldsymbol{u}} = \omega_{\mathrm{c}}(\boldsymbol{u}^{\perp} - \beta \omega_{\mathrm{c}} \boldsymbol{u})\,. \tag{6.25}$$

Here $\boldsymbol{u}^{\perp} = (-u_2, u_1)$, $\omega_{\mathrm{c}} = eB/m_0 c$ is the cyclotron frequency, and $\beta = e^2/6\pi c^3 m_0$. The first term is the Lorentz force and the second term accounts for the radiation reaction.

We now choose the reference field $B_0$ such that the two terms balance, i.e.

$$B_0 = (\beta e/m_0 c)^{-1}\,. \tag{6.26}$$

For electrons

$$B_0 = 1.1 \times 10^{17} \text{ gauss} \tag{6.27}$$

and even larger by a factor $(1836)^2$ for protons. For a laboratory field of $10^5$ gauss this yields

$$\varepsilon = 10^{-12}\,. \tag{6.28}$$

Written in units of $B_0$, (6.25) becomes

$$\gamma \dot{\boldsymbol{u}} = \varepsilon \omega_{\mathrm{c}}^0 (\boldsymbol{u}^{\perp} - \varepsilon \boldsymbol{u}) \tag{6.29}$$

with $\beta \omega_{\mathrm{c}}^0 = 1$, i.e. $\omega_{\mathrm{c}}^0 = e B_0 / m_0 c = 1.6 \times 10^{28}\ \mathrm{s}^{-1}$. Thus friction is of relative order $\varepsilon$ and higher-order corrections would be of relative order $\varepsilon^2$. As will be demonstrated, the dimensionless scaling parameter $\varepsilon$ serves as a bookkeeping device to track the relative order of the various terms contributing to the dynamics.

### 6.1.2 Appendix 2: Adiabatic protection

The adiabatic limit, as discussed above, relies on the fact that photons have zero mass. If they had finite mass, radiation damping would be hindered. This point can be most easily argued in the context of a scalar wave field. Moreover, rather than having a particle interacting with the field, it suffices to have a source fixed at the origin.

The scalar wave field is denoted by $\phi$ with canonically conjugate momentum field $\pi$. They are governed by

$$\partial_t \phi(\boldsymbol{x}, t) = \pi(\boldsymbol{x}, t)\,, \quad \partial_t \pi(\boldsymbol{x}, t) = \Delta\phi(\boldsymbol{x}, t) - \kappa^2 \phi(\boldsymbol{x}, t) + \alpha(t)\delta(\boldsymbol{x})\,. \tag{6.30}$$

$\alpha(t)$ is a smooth function vanishing outside the interval $[0, T]$. Assuming that $\phi = 0$, $\pi = 0$ initially we want to determine how much energy is radiated in the long-time limit.

The local field energy is given by

$$e(\boldsymbol{x}, t) = \frac{1}{2}\big(\pi(\boldsymbol{x}, t)^2 + (\nabla\phi(\boldsymbol{x}, t))^2 + \kappa^2 \phi(\boldsymbol{x}, t)^2\big) \tag{6.31}$$

from which, using (6.30), the energy current

$$\boldsymbol{j}_{\mathrm{e}} = -\pi \nabla \phi \tag{6.32}$$

follows. The energy flow through a sphere of radius $R$ is given by

$$\begin{aligned} -\int_0^\infty \mathrm{d}t\, R^2 \int \mathrm{d}^2\omega\, \pi(\boldsymbol{\omega} R, t)\boldsymbol{\omega} \cdot \nabla\phi(\boldsymbol{\omega} R, t) &\\ = -4\pi R^2 \int_0^\infty \mathrm{d}t\, \pi(R, t)\phi'(R, t) &\\ = -4\pi \int_0^\infty \mathrm{d}t\, R\pi(R, R + t) R\phi'(R, R + t)\,. & \end{aligned} \tag{6.33}$$

The first step uses radial symmetry of the solution to (6.30), while retaining the notation for the radial fields and setting $\phi'(R, t) = \partial_R \phi(R, t)$, and the second step uses the condition that the solution is supported inside the light cone. To separate between near and far field one still has to take the limit $R \to \infty$ in (6.33). Thus

$$E_{\text{diss}} = \lim_{R\to\infty} -4\pi \int_0^\infty \mathrm{d}t\, R\pi(R, R+t) R\phi'(R, R+t)\,. \tag{6.34}$$

The fundamental solution of (6.30) is

$$\begin{pmatrix}\phi(t)\\ \pi(t)\end{pmatrix} = \begin{pmatrix}\partial_t G & G\\ \partial_t^2 G & \partial_t G\end{pmatrix}\begin{pmatrix}\phi\\ \pi\end{pmatrix}. \tag{6.35}$$

$G$ is the propagator for $t \geq 0$,

$$G(\boldsymbol{x}, t) = \frac{1}{4\pi|\boldsymbol{x}|}\delta(t - |\boldsymbol{x}|) - \frac{1}{4\pi}\kappa^2 F\big(\kappa\sqrt{t^2 - \boldsymbol{x}^2}\big)\chi(\boldsymbol{x}^2 \leq t^2) \tag{6.36}$$

with

$$F(z) = \frac{1}{z} J_1(z) \tag{6.37}$$

and $J_1$ the integer Bessel function of order 1. For the initial conditions $\phi = 0, \pi = 0$ the solution to (6.30) is then

$$\phi(\boldsymbol{x}, t) = \int_0^t \mathrm{d}s\, G(\boldsymbol{x}, t-s)\alpha(s)\,, \quad \pi(\boldsymbol{x}, t) = \int_0^t \mathrm{d}s\, \partial_t G(\boldsymbol{x}, t-s)\alpha(s)\,. \tag{6.38}$$

Before inserting them in (6.34) both terms have to be somewhat simplified through partial integrations using the condition that $\alpha(0) = 0$. For the momentum field one obtains

$$4\pi R\pi(R, R+t) = \dot{\alpha}(t) - \kappa^2 R \int_0^t \mathrm{d}s\, F\Big(\kappa\sqrt{(t-s)(2R+t-s)}\Big)\dot{\alpha}(s)\,. \tag{6.39}$$

For the scalar field there are two subleading contributions, which vanish as $R \to \infty$, and the leading term

$$4\pi R\phi'(R, R+t) = -\dot{\alpha}(t) + \kappa^2 R \int_0^t \mathrm{d}s\, F\Big(\kappa\sqrt{(t-s)(2R+t-s)}\Big) \times \frac{R}{R+t-s}\dot{\alpha}(s) + \mathcal{O}\Big(\frac{1}{R}\Big)\,. \tag{6.40}$$

We insert (6.39) and (6.40) into (6.33), which results in four terms. The first one is clearly $(4\pi)^{-1}\int_0^\infty \mathrm{d}t\dot{\alpha}(t)^2$. For the cross-term the integral involving $F$ converges to $\dot{\alpha}(t)$ as $R \to \infty$. Thus the cross-terms add up to $-(2\pi)^{-1}\int_0^\infty \mathrm{d}t\dot{\alpha}(t)^2$. The fourth term requires more work. The $t$-integration of (6.33) is split into $[0, T]$ and $[T, \infty]$. The first integral yields $(4\pi)^{-1}\int_0^\infty \mathrm{d}t\dot{\alpha}(t)^2$, thereby cancelling terms

1 to 3. The remainder is

$$E_{\text{diss}} = \lim_{R\to\infty} \frac{1}{4\pi} \int_0^T \mathrm{d}s\dot{\alpha}(s) \int_0^T \mathrm{d}s'\dot{\alpha}(s') \int_T^\infty \mathrm{d}t\kappa^4 RF\Big(\kappa\sqrt{(t-s)(2R+t-s)}\Big)$$
$$\times RF\Big(\kappa\sqrt{(t-s')(2R+t-s')}\Big)\frac{R}{R+t-s'}. \tag{6.41}$$

At this point one can use the asymptotics of $J_1$ for large arguments leading through oscillating integrands to

$$E_{\text{diss}} = \frac{1}{2\pi}\int_\kappa^\infty \mathrm{d}\omega \frac{1}{\omega}\sqrt{\omega^2-\kappa^2}|\omega\widehat{\alpha}(\omega)|^2 . \tag{6.42}$$

In the limit $\kappa \to 0$ one obtains the familiar analog of the Larmor formula as

$$E_{\text{diss}} = \frac{1}{4\pi}\int \mathrm{d}t\dot{\alpha}(t)^2 . \tag{6.43}$$

If $\alpha$ has slow time variation, incorporated as $\alpha(\varepsilon t)$, $\varepsilon \ll 1$, then

$$E_{\text{diss}} = \varepsilon\frac{1}{4\pi}\int \mathrm{d}t\dot{\alpha}(t)^2 , \tag{6.44}$$

which in our working example would determine the time scale for radiation damping. On the other hand, for $\kappa > 0$

$$E_{\text{diss}} = \varepsilon\frac{1}{2\pi}\int_{\kappa/\varepsilon}^\infty \mathrm{d}\omega \frac{1}{\omega}\sqrt{\omega^2-(\kappa/\varepsilon)^2}|\omega\widehat{\alpha}(\omega)|^2 . \tag{6.45}$$

If $\widehat{\alpha}$ has exponential decay, $\widehat{\alpha}(\omega) \cong \mathrm{e}^{-\gamma|\omega|}$ for large $|\omega|$, then $E_{\text{diss}} = \varepsilon \mathrm{e}^{-\gamma\kappa/\varepsilon}$. The low frequencies of the source do not couple to the medium.

If photons were massive, the adiabatic motion of charges would be protected in the sense that radiation damping is of order $\mathrm{e}^{-1/\varepsilon}$ rather than of order $\varepsilon^2$ as is the case for photons with dispersion $\omega(\boldsymbol{k}) = c|\boldsymbol{k}|$.

## 6.2 Comparison with the hydrodynamic limit

In hydrodynamics one assumes that a small droplet of fluid with center $\boldsymbol{r}$ has its intrinsic velocity, $\boldsymbol{u}(\boldsymbol{r})$, and that relative to the moving frame the particles are distributed according to thermal equilibrium with density $\rho(\boldsymbol{r})$ and temperature $T(\boldsymbol{r})$. For such notions to be reasonably well defined, the hydrodynamic fields $\rho, \boldsymbol{u}, T$ must be slowly varying on the scale of the typical interparticle distance. This is how the analogy with the Maxwell–Newton equations arises. As for them we have three characteristic space-time scales.

(i) *Microscopic scale*. The microscopic scale is measured in units of a collision time, respectively interatomic distance. On that scale the hydrodynamic fields are frozen. Possible deviations from local equilibrium relax through collisions. To prove such behavior one has to establish a sufficiently fast relaxation to equilibrium. For Newtonian particles no general method is available. For the Maxwell field the situation is much simpler. Local deviations from the Coulomb field are transported off to infinity and are no longer seen.

(ii) *Macroscopic Euler scale.* The macroscopic space-time scale is defined by the variation of the hydrodynamic fields. If, as before, we introduce the dimensionless scaling parameter $\varepsilon$, then space-time is $\mathcal{O}(\varepsilon^{-1})$ in microscopic units. On the macroscopic scale the time between collisions is $\mathcal{O}(\varepsilon)$, the interparticle distance $\mathcal{O}(\varepsilon)$, and the pair potential for the particle at position $\boldsymbol{q}_i$ and the one at $\boldsymbol{q}_j$ is $V(\varepsilon^{-1}(\boldsymbol{q}_i - \boldsymbol{q}_j))$. On the macroscopic scale the hydrodynamic fields evolve according to the Euler equations. These are first-order equations, which must be so, since space and time are scaled in the same way. The Euler equations are of Hamiltonian form. There is no dissipation, and no entropy is produced. In fact, there is a slight complication here. Even for smooth initial data the Euler equations develop shock discontinuities. There the assumption of slow variation fails and shocks are a source of entropy.

(iii) *Macroscopic friction scale.* In a real fluid there are frictional forces which are responsible for the relaxation to global equilibrium. One adds to the Euler equations diffusive-like terms, which are second order in spatial derivatives, and obtains the compressible Navier–Stokes equations incorporating the shear and volume viscosity resulting from friction in momentum transport and thermal conductivity resulting from friction in energy transport. On the macroscopic scale these corrections are of order $\varepsilon$. In the same spirit, based on the full Maxwell–Newton equations, there will be dissipative terms of order $\varepsilon$ which have to be added to (6.21). Of course, in this context one has to deal only with ordinary differential equations as effective dynamics.

## 6.3 Point-charge limit, negative bare mass

The conventional point-charge limit is to let the diameter of the charge distribution tend to zero under the condition that the total charge remains fixed. Accordingly, let us consider now $R_\varphi$ as a reference scale and let $R/R_\varphi \to 0$. Then for the point charge one sets

$$\varphi_R(\boldsymbol{x}) = R^{-3}\varphi(\boldsymbol{x}/R) \tag{6.46}$$

and takes the limit $R \to 0$. This means that the charge diameter is small in units of the variation of the external potential, since this is the only other length scale available. At first sight, one just seems to say that the potentials vary slowly on the scale set by the charge diameter and that hence the point-charge limit and the adiabatic limit coincide. To see the difference let us consider the electrostatic energy

$$\frac{1}{2}e^2 \int d^3k|\widehat{\varphi}_R(\boldsymbol{k})|^2 \frac{1}{\boldsymbol{k}^2} = \frac{1}{R} m_f \,. \tag{6.47}$$

In particular, the ratio of field mass to bare mass grows as $R^{-1}$ in the point-charge limit and remains constant in the adiabatic limit.

To display the order of magnitude of the various dynamical contributions we resort again to our standard example of an electron in a uniform magnetic field $\boldsymbol{B}_{ex} = B\widehat{\boldsymbol{n}}$, $\widehat{\boldsymbol{n}} = (0, 0, 1)$ with $B$ of the order of 1 tesla $= 10^4$ gauss, say. It suffices to consider small velocities. In the adiabatic limit we set $B = \varepsilon B_0$ where the reference field is $B_0 = 1.1 \times 10^{17}$ gauss; compare with appendix 1 to section 6.1. Up to higher-order corrections, the motion of the electron is then governed by

$$\left(m_b + \frac{4}{3} m_f\right)\dot{\boldsymbol{v}} = \frac{e}{c}\,\varepsilon B_0(\boldsymbol{v} \times \widehat{\boldsymbol{n}}) + \frac{e^2}{6\pi c^3}\,\ddot{\boldsymbol{v}} + \mathcal{O}(\varepsilon^3) \tag{6.48}$$

on the microscopic scale. Going over to the macroscopic time scale, $t' = \varepsilon^{-1} t$, (6.48) becomes

$$\left(m_b + \frac{4}{3} m_f\right)\dot{\boldsymbol{v}} = \frac{e}{c}\, B_0(\boldsymbol{v} \times \widehat{\boldsymbol{n}}) + \frac{e^2}{6\pi c^3}\,\varepsilon\ddot{\boldsymbol{v}} + \mathcal{O}(\varepsilon^2)\,. \tag{6.49}$$

Setting $m_0 = m_b + \frac{4}{3} m_f$, $\omega_c^0 = e\,B_0/m_0 c$, $\beta = e^2/6\pi c^3 m_0$, and restricting to the motion on the critical manifold, as will be explained in chapter 9, Eq. (6.49) becomes

$$\dot{\boldsymbol{v}} = \omega_c^0\big(\boldsymbol{v} \times \widehat{\boldsymbol{n}} + \varepsilon\beta\omega_c^0\,(\boldsymbol{v} \times \widehat{\boldsymbol{n}}) \times \widehat{\boldsymbol{n}}\big) + \mathcal{O}(\varepsilon^2)\,, \tag{6.50}$$

equivalently, on the microscopic time scale

$$\dot{\boldsymbol{v}} = \omega_c\big(\boldsymbol{v} \times \widehat{\boldsymbol{n}} + \beta\omega_c(\boldsymbol{v} \times \widehat{\boldsymbol{n}}) \times \widehat{\boldsymbol{n}}\big) + \mathcal{O}(\varepsilon^3) \tag{6.51}$$

with the cyclotron frequency $\omega_c = e\,\varepsilon B_0/m_0 c = eB/m_0 c$.

For the point-charge limit we rely on the Taylor expansion of section 7.2. Then, for small velocities,

$$\left(m_b + R^{-1}\frac{4}{3} m_f\right)\dot{\boldsymbol{v}} = \frac{e}{c} B(\boldsymbol{v} \times \widehat{\boldsymbol{n}}) + \frac{e^2}{6\pi c^3}\,\ddot{\boldsymbol{v}} + \mathcal{O}(R)\,. \tag{6.52}$$

Since based on the same expansion, as long as no limit is taken, of course, we can switch back and forth between (6.52) and (6.48), respectively (6.49), provided the appropriate units are used. This can be seen more easily if we accept momentarily the differential–difference equation

$$m_{\mathrm{b}}\dot{\boldsymbol{v}}(t) = e\big(\boldsymbol{E}_{\mathrm{ex}}(\boldsymbol{q}(t)) + c^{-1}\boldsymbol{v}(t) \times \boldsymbol{B}_{\mathrm{ex}}(\boldsymbol{q}(t))\big) + \frac{e^2}{12\pi c R^2}\big(\boldsymbol{v}(t - 2c^{-1}R) - \boldsymbol{v}(t)\big), \tag{6.53}$$

which is exact for a uniformly charged sphere at small velocities, see section 7.1. If we expand in the charge diameter $R$, then

$$\Big(m_{\mathrm{b}} + \frac{e^2}{6\pi R c^2}\Big)\dot{\boldsymbol{v}} = e(\boldsymbol{E}_{\mathrm{ex}} + c^{-1}\boldsymbol{v} \times \boldsymbol{B}_{\mathrm{ex}}) + \frac{e^2}{6\pi c^3}\ddot{\boldsymbol{v}} + \mathcal{O}(R), \tag{6.54}$$

which is the analog of (6.52). On the other hand, if we assume that the external fields are slowly varying, as explained in section 6.1, then on the macroscopic scale

$$\varepsilon m_{\mathrm{b}}\dot{\boldsymbol{v}}(t) = \varepsilon\, e\big(\boldsymbol{E}_{\mathrm{ex}}(\boldsymbol{q}(t)) + c^{-1}\boldsymbol{v}(t) \times \boldsymbol{B}_{\mathrm{ex}}(\boldsymbol{q}(t))\big) + \frac{e^2}{12\pi c R_\varphi^2}\big(\boldsymbol{v}(t - 2\varepsilon c^{-1}R_\varphi) - \boldsymbol{v}(t)\big), \tag{6.55}$$

where $R_\varphi$ is now regarded as fixed. Taylor expansion in $\varepsilon$ yields

$$\Big(m_{\mathrm{b}} + \frac{e^2}{6\pi R_\varphi c^2}\Big)\dot{\boldsymbol{v}} = e(\boldsymbol{E}_{\mathrm{ex}} + c^{-1}\boldsymbol{v} \times \boldsymbol{B}_{\mathrm{ex}}) + \varepsilon\frac{e^2}{6\pi c^3}\ddot{\boldsymbol{v}} + \mathcal{O}(\varepsilon^2) \tag{6.56}$$

which is the analog of (6.49).

As can be seen from (6.52), in the point-charge limit the total mass becomes so large that the particle hardly responds to the magnetic field. The only way out seems to formally compensate the diverging $R^{-1}\,(4/3)m_{\mathrm{f}}$ by setting

$$m_{\mathrm{b}} = -R^{-1}\,(4/3)\,m_{\mathrm{f}} + m_{\mathrm{exp}} \tag{6.57}$$

with $m_{\mathrm{exp}}$ the experimental mass of the charged particle. But this is asking for trouble, since the energy (2.44) is no longer bounded from below and potential energy can be transferred to kinetic mechanical energy without limit. To see this mechanism in detail we consider the Abraham model with $\boldsymbol{B}_{\mathrm{ex}} = 0$ and $\phi_{\mathrm{ex}}$ varying only along the 1-axis. The bare mass of the particle is now $-m_{\mathrm{b}}$, with $m_{\mathrm{b}} > 0$ as before. We set $\boldsymbol{q}(t) = (q_t, 0, 0)$, $\boldsymbol{v}(t) = (v_t, 0, 0)$, $\boldsymbol{E}_{\mathrm{ex}} = (-\phi'(q), 0, 0)$. $\phi$ is assumed to be strictly convex with a minimum at $q = 0$. Initially the particle is at rest at the minimum of the potential. Thus $\boldsymbol{E}(\boldsymbol{x}, 0) = \boldsymbol{E}_0(\boldsymbol{x})$ from (4.5) and $\boldsymbol{B}(\boldsymbol{x}, 0) = 0$. We now give the particle a slight kick to the right, which means $q_0 = 0$, $v_0 > 0$. By conservation of energy

$$\begin{aligned}
-m_{\mathrm{b}}c^2\,\gamma(v_t) + e\,\phi(q_t) + \frac{1}{2}\int \mathrm{d}^3x\,\big(\boldsymbol{E}(\boldsymbol{x},t)^2 + \boldsymbol{B}(\boldsymbol{x},t)^2\big) \\
= -m_{\mathrm{b}}c^2\gamma(v_0) + e\phi(q_0) + \frac{1}{2}\int \mathrm{d}^3x\,\boldsymbol{E}(\boldsymbol{x},0)^2\,. \qquad (6.58)
\end{aligned}$$

We split $\boldsymbol{E}$ into longitudinal and transverse components, $\boldsymbol{E} = \boldsymbol{E}_{\parallel} + \boldsymbol{E}_{\perp}$, $\widehat{\boldsymbol{E}}_{\parallel} = \widehat{\boldsymbol{k}}(\widehat{\boldsymbol{k}}\cdot\widehat{\boldsymbol{E}})$. Clearly $\int \mathrm{d}^3x\,\boldsymbol{E}_{\parallel}\cdot\boldsymbol{E}_{\perp} = 0$ and therefore

$$\begin{aligned}
\int \mathrm{d}^3x\boldsymbol{E}(\boldsymbol{x},t)^2 \ge \int \mathrm{d}^3x\boldsymbol{E}_{\parallel}(\boldsymbol{x},t)^2 = \int \mathrm{d}^3k(\widehat{\boldsymbol{k}}\cdot\widehat{\boldsymbol{E}}(\boldsymbol{k},t))^2 \\
= e^2\int \mathrm{d}^3k\,|\boldsymbol{k}|^{-2}\,|\widehat{\varphi}(\boldsymbol{k})|^2 = \int \mathrm{d}^3x\,\boldsymbol{E}(\boldsymbol{x},0)^2\,, \qquad (6.59)
\end{aligned}$$

since the initial field has zero transverse component. Inserting in (6.58) yields

$$\dot{q}_t^2 \ge 1 - \big[\gamma(v_0) + (e/m_{\mathrm{b}}c^2)(\phi(q_t) - \phi(q_0))\big]^{-2}\,. \qquad (6.60)$$

Since $\gamma(v_0) > 1$, $\dot{q}_t > 0$ for short times. As the particle moves to the right, $(\phi(q_t) - \phi(q_0))$ is increasing and therefore $\dot{q}_t \to 1$ and $q_t \to \infty$ as $t \to \infty$. Note that $v_0$ and $m_{\mathrm{b}}$ can be arbitrarily small. Not surprisingly, the Abraham model with a negative bare mass behaves rather unphysically. A tiny initial kick suffices to generate a runaway solution.

The point-charge limit is honored by a long tradition, which however seems to have constantly overlooked that physically it is more appropriate to have the external potentials slowly varying on the scale of a fixed-size charge distribution. Then there is no need to introduce a negative bare mass and there are no runaway solutions.

## Notes and references

### *Section 6*

The importance of slowly varying external potentials has been emphasized repeatedly. In the early literature slow variation appears as the quasi-stationary hypothesis and quasi-stationary motion (Miller 1997). Such principles remain vague and, interestingly enough, in more mathematical considerations the size $R_\varphi$ of the charge distribution is taken as an expansion parameter rather than the appropriate parameter in the potential. To me it is rather surprising that, apparently, there is no systematic study of the equations of motion with external potentials of slow variation. We use the notion "adiabatic limit" to correspond to the adiabatic theorem in classical and quantum mechanics which refers to a Hamiltonian with slow time-dependence. More appropriately we should speak of "space-adiabatic limit", since the slow variation is in space, the slow variation in time being a consequence.

### Section 6.1

In the context of charges coupled to the Maxwell field the adiabatic limit was first introduced in Komech, Kunze and Spohn (1999) and in Kunze and Spohn (2000a). The fundamental solution (6.36) of the Klein–Gordon equation is discussed in Morse and Feshbach (1953). De Bièvre (private communication) points out that the dissipated energy (6.42) can be guessed also from elementary considerations. In Fourier space the wave equation becomes

$$\partial_t^2 \widehat{\phi}_t(\boldsymbol{k}) = -\omega(\boldsymbol{k})^2 \widehat{\phi}_t(\boldsymbol{k}) + (2\pi)^{-3/2} \alpha(t) \tag{6.61}$$

with $\omega(\boldsymbol{k})^2 = \boldsymbol{k}^2 + \kappa^2$. For a forced harmonic oscillator the equation of motion reads $\ddot{x} = -\omega^2 x + f(t)$ and the energy transferred by the forcing is $\pi |\widehat{f}(\omega)|^2$. Inserting in (6.61) and integrating over all $\boldsymbol{k}$ yields (6.42). Schwinger (1949) uses a similar argument for the radiated energy.

### Section 6.2

A more detailed discussion of the hydrodynamic limit can be found in Spohn (1991).

### Section 6.3

In the early days of classical electron theory, one simply expanded in $R_\varphi$. $R_\varphi$ was considered to be small, but finite, roughly of the order of the classical electron radius. Schott (1912) pushes the expansion to include the radiation reaction. Apparently, the notion of a point charge is first stated explicitly by Frenkel (1925). The difficulties resulting from the point charge were clearly understood by P. Ehrenfest as stressed by Pauli in his 1933 obituary. The point-charge limit is at the core of the famous Dirac (1938) paper, cf. section 3.3. Since then the limit $m_\mathrm{b} \to -\infty$ has become a standard piece of the theory, reproduced in textbooks and survey articles. The negative bare mass was soon recognized as a source of instability. We refer to the review by Erber (1961). On a linearized level stability is studied by Wildermuth (1955) and by Moniz and Sharp (1977) and Levine, Moniz and Sharp (1977). Bambusi (1996), Bambusi and Noja (1996), and Noja and Posilicano (1998, 1999) discuss the point-charge limit in the dipole approximation and show that then the true solution is well approximated by the linear Lorentz–Dirac equation with the full, both physical and unphysical, solution manifold explored. An extension to the nonlinear theory is attempted by Marino (2002). The bound (6.60) is taken from Bauer and Dürr (2001), which seems to be the only quantitative handling of the instability for the full nonlinear problem.

# 7

# Self-force

The inhomogeneous Maxwell equations have been solved in (2.16), (2.17). Thus it is natural to insert them into the Lorentz force in order to obtain a closed, albeit memory equation for the position of the particle.

According to (2.16), (2.17) the Maxwell fields are a sum of initial and retarded terms. We discuss first the contribution from the initial fields. By our specific choice of initial conditions they have the representation, for $t \geq 0$,

$$\begin{aligned} \boldsymbol{E}_{\text{ini}}(\boldsymbol{x},t) = -\int_{-\infty}^{0} \mathrm{d}s \int \mathrm{d}^3y \,\big(\nabla G_{t-s}(\boldsymbol{x}-\boldsymbol{y})\, e\varphi(\boldsymbol{y}-\boldsymbol{q}^0-\boldsymbol{v}^0 s) \\ + \partial_t G_{t-s}(\boldsymbol{x}-\boldsymbol{y})\boldsymbol{v}^0\, e\varphi(\boldsymbol{y}-\boldsymbol{q}^0-\boldsymbol{v}^0 s)\big)\,, \end{aligned} \tag{7.1}$$

$$\boldsymbol{B}_{\text{ini}}(\boldsymbol{x},t) = \int_{-\infty}^{0} \mathrm{d}s \int \mathrm{d}^3y\, \nabla \times G_{t-s}(\boldsymbol{x}-\boldsymbol{y})\,\boldsymbol{v}^0 e\varphi(\boldsymbol{y}-\boldsymbol{q}^0-\boldsymbol{v}^0 s)\,; \tag{7.2}$$

compare with (4.31), (4.32). Since $G_t$ is concentrated on the light cone, one concludes from (7.1), (7.2) that $\boldsymbol{E}_{\text{ini}}(\boldsymbol{x},t) = 0$, $\boldsymbol{B}_{\text{ini}}(\boldsymbol{x},t) = 0$ for $|\boldsymbol{q}^0 - \boldsymbol{x}| \leq t - R_\varphi$. If we had allowed for more general initial data, such a property would hold only asymptotically for large $t$.

Next we note that constrained by energy conservation the particle cannot travel too far. Using the bound on the potential, one can find a $\bar{v} < 1$ such that

$$\sup_{t\in\mathbb{R}} |\boldsymbol{v}(t)| < \bar{v} < 1\,, \tag{7.3}$$

cf. Eq. (7.26). The charge distribution vanishes for $|\boldsymbol{x} - \boldsymbol{q}(t)| \geq R_\varphi$. Therefore, once

$$t \geq \bar{t}_\varphi = 2R_\varphi/(1-\bar{v})\,, \tag{7.4}$$

the initial fields and the charge distribution have no overlap. We conclude that for $t > \bar{t}_\varphi$ the initial fields make no contribution to the self-force and it remains to discuss the effect of the retarded fields.

We insert (2.12), (2.13) into the Lorentz force for which purpose it is convenient to use the scaled version (6.11). The external potentials are set equal to zero for a while. Then on the macroscopic scale, for $t \geq \varepsilon \bar{t}_\varphi$,

$$\frac{\mathrm{d}}{\mathrm{d}t}\big(m_\mathrm{b}\gamma\, \boldsymbol{v}^\varepsilon(t)\big) = \boldsymbol{F}^\varepsilon_{\mathrm{self}}(t) \tag{7.5}$$

with the self-force

$$\begin{aligned}\boldsymbol{F}^\varepsilon_{\mathrm{self}}(t) = e^2 \int_0^t \mathrm{d}s\, \varepsilon \int \mathrm{d}^3k\, |\widehat{\varphi}(\varepsilon \boldsymbol{k})|^2\, \mathrm{e}^{-\mathrm{i}\boldsymbol{k}\cdot(\boldsymbol{q}^\varepsilon(t)-\boldsymbol{q}^\varepsilon(s))}\big((|\boldsymbol{k}|^{-1} \sin|\boldsymbol{k}|(t-s))\mathrm{i}\boldsymbol{k} \\ - (\cos|\boldsymbol{k}|(t-s))\boldsymbol{v}^\varepsilon(s) - (|\boldsymbol{k}|^{-1}\sin|\boldsymbol{k}|(t-s))\,\boldsymbol{v}^\varepsilon(t)\times(\mathrm{i}\boldsymbol{k}\times\boldsymbol{v}^\varepsilon(s))\big),\end{aligned} \tag{7.6}$$

which in position space for $\varepsilon = 1$ was already written down in Eq. (2.57).

Equation (7.5) is exact under the stated conditions on the initial fields. No information has been discarded. The interaction with the field has been merely transcribed into a memory term. To make further progress we have to use a suitable approximation which exploits the assumption that the external forces are slowly varying. Since this corresponds to small $\varepsilon$, we just have to Taylor-expand $\boldsymbol{F}^\varepsilon_{\mathrm{self}}(t)$, which is carried out in section 7.2 with the proper justification left for section 7.3. But before that, and to make contact with previous work, we take a closer look at the memory term.

## 7.1 Memory equation

Equation (7.6) can be simplified, for which it is convenient to set $\varepsilon = 1$. By partial integration

$$\begin{aligned}&\int_0^t \mathrm{d}s \int \mathrm{d}^3k\, |\widehat{\varphi}(\boldsymbol{k})|^2\, \mathrm{e}^{-\mathrm{i}\boldsymbol{k}\cdot(\boldsymbol{q}(t)-\boldsymbol{q}(s))}\boldsymbol{v}(s)\,\frac{\mathrm{d}}{\mathrm{d}s}\,|\boldsymbol{k}|^{-1}\sin|\boldsymbol{k}|(t-s) \\ &\quad = -\int \mathrm{d}^3k\, |\widehat{\varphi}(\boldsymbol{k})|^2\, \mathrm{e}^{-\mathrm{i}\boldsymbol{k}\cdot(\boldsymbol{q}(t)-\boldsymbol{q}(0))}\boldsymbol{v}(0)|\boldsymbol{k}|^{-1}\sin|\boldsymbol{k}|t \\ &\qquad - \int_0^t \mathrm{d}s \int \mathrm{d}^3k |\widehat{\varphi}(\boldsymbol{k})|^2\, \mathrm{e}^{-\mathrm{i}\boldsymbol{k}\cdot(\boldsymbol{q}(t)-\boldsymbol{q}(s))}(|\boldsymbol{k}|^{-1}\sin|\boldsymbol{k}|(t-s))\big(\dot{\boldsymbol{v}}(s) \\ &\qquad + \mathrm{i}(\boldsymbol{k}\cdot\boldsymbol{v}(s))\boldsymbol{v}(s)\big).\end{aligned} \tag{7.7}$$

Since $t \geq \bar{t}_\varphi$, the boundary term vanishes. Inserting (7.7) into (7.6), returning to physical space, and setting $t - s = \tau$, one has for $t \geq \bar{t}_\varphi$

$$\boldsymbol{F}_{\text{self}}(t) = -e^2 \int_0^\infty \mathrm{d}\tau \, \big[\dot{\boldsymbol{v}}(t-\tau) + (1 - \boldsymbol{v}(t) \cdot \boldsymbol{v}(t-\tau))\nabla_{\boldsymbol{x}} + \boldsymbol{v}(t-\tau)(\boldsymbol{v}(t) - \boldsymbol{v}(t-\tau)) \cdot \nabla_{\boldsymbol{x}}\big] W_t(\boldsymbol{x})|_{\boldsymbol{x}=\boldsymbol{q}(t)-\boldsymbol{q}(t-\tau)}, \tag{7.8}$$

where

$$W_t(\boldsymbol{x}) = \int \mathrm{d}^3k \, |\widehat{\varphi}(\boldsymbol{k})|^2 \, \mathrm{e}^{-\mathrm{i}\boldsymbol{k}\cdot\boldsymbol{x}} |\boldsymbol{k}|^{-1} \sin |\boldsymbol{k}| t \,. \tag{7.9}$$

In (7.8) we have extended the integration to $\infty$, since the integrand vanishes anyway for $\tau \geq \bar{t}_\varphi$. Carrying out the integrations on the angles in (7.9) one obtains

$$W_t(\boldsymbol{x}) = |\boldsymbol{x}|^{-1}\big(h(|\boldsymbol{x}| + t) - h(|\boldsymbol{x}| - t)\big), \tag{7.10}$$

$$h(w) = 2\pi \int_0^\infty \mathrm{d}k \, g(k) \cos kw \tag{7.11}$$

with $g(|\boldsymbol{k}|) = |\widehat{\varphi}(\boldsymbol{k})|^2$. Since $\varphi$ vanishes for $|\boldsymbol{x}| \geq R_\varphi$, $h(w) = 0$ for $|w| \geq 2R_\varphi$. Note that $|\boldsymbol{q}(t) - \boldsymbol{q}(t-\tau)| \leq \bar{v}\,\tau$. Thus for $t \geq \bar{t}_\varphi$ we indeed have $W_t(\boldsymbol{q}(t) - \boldsymbol{q}(t-\tau)) = 0$, as claimed before. $\boldsymbol{F}_{\text{self}}(t)$ has a finite memory extending backwards in time up to $t - \bar{t}_\varphi$.

To go beyond (7.10) one has to use a specific form factor $\widehat{\varphi}$. Two choices, popular at the time, are $\varphi_{\mathrm{s}}(\boldsymbol{x}) = (4\pi R_\varphi^2)^{-1}\,\delta(|\boldsymbol{x}| - R_\varphi)$ and $\varphi_{\mathrm{b}}(\boldsymbol{x}) = e\,(4\pi R_\varphi^3/3)^{-1}$ for $|\boldsymbol{x}| \leq R_\varphi$, $\varphi_{\mathrm{b}}(\boldsymbol{x}) = 0$ for $|\boldsymbol{x}| \geq R_\varphi$. For the uniformly charged sphere one finds

$$h(R_\varphi w) = \begin{cases} (8\pi R_\varphi)^{-1}(1 - |w|/2) & \text{for } |w| \leq 2\,, \\ 0 & \text{for } |w| \geq 2\,, \end{cases} \tag{7.12}$$

and for the uniformly charged ball

$$h(R_\varphi w) = \begin{cases} (8\pi R_\varphi)^{-1}\,\frac{9}{8}\,\widetilde{h} * \widetilde{h}(w) & \text{for } |w| \leq 2\,, \\ 0 & \text{for } |w| \geq 2\,, \end{cases} \tag{7.13}$$

with $\widetilde{h}(w) = (1 - w^2)$ for $|w| \leq 1$ and $\widetilde{h}(w) = 0$ otherwise.

For the charged sphere $W_t(\boldsymbol{x})$ is piecewise linear and, by first taking the gradient of $W$, the time integrations simplify. In the approximation of small velocities the motion of the charged particle is then governed by the differential–difference equation

$$m_{\mathrm{b}}\dot{v}(t) = e\big(E_{\mathrm{ex}}(q(t)) + v(t) \times B_{\mathrm{ex}}(q(t))\big) + \frac{e^2}{12\pi R_\varphi^2}\big(v(t-2R_\varphi) - v(t)\big), \tag{7.14}$$

where we have reintroduced the external fields.

The memory equation (7.14) is of suggestive simplicity. To have a well-defined dynamics one has to prescribe $q(0)$ and $v(t)$ for $-2R_\varphi \le t \le 0$ as initial data. Of course, the coupled system determines these data completely. However, the supporters of differential–difference equations regard (7.14) as the starting point with no instruction for the choice of initial data. Their claim is that solutions to (7.14) are not very sensitive to this choice. While there is some evidence on the linearized level, the dependence on the initial data for the full nonlinear problem remains to be studied.

## 7.2 Taylor expansion

We return to Eq. (7.5). As will be explained in section 7.3, one knows that there exists a constant $C$, independent of $\varepsilon$ for $\varepsilon < \varepsilon_0$, such that

$$\begin{aligned} &|\ddot{q}^\varepsilon(t)| \le C\,, \quad |\dddot{q}^\varepsilon(t)| \le C\big(1 + \varepsilon(\varepsilon + |t|)^{-2}\big)\,, \\ &|\ddddot{q}^\varepsilon(t)| \le C\big(1 + \varepsilon(\varepsilon + |t|)^{-2} + \varepsilon(\varepsilon + |t|)^{-3}\big) \end{aligned} \tag{7.15}$$

for all $t$, provided the total charge $e$ is sufficiently small. This smallness condition merely reflects the fact that at present we do not know how to do better mathematically. Physically we expect (7.15) to hold no matter how large $e$.

Note that in higher time derivatives the mismatch of the initial conditions becomes visible. Only if the charge is allowed to move for a time span of order $\varepsilon^{1/3}$, which is short on the macroscopic scale but long as $\mathcal{O}(\varepsilon^{-2/3})$ on the microscopic scale, do the derivatives become uniformly bounded.

Because of (7.15) we are allowed to Taylor-expand in (7.6). To simplify notation we set $v^\varepsilon(t) = v$ and $t - s = \tau$. Then

$$v^\varepsilon(s) = v^\varepsilon(t-\tau) = v - \dot{v}\tau + \frac{1}{2}\ddot{v}\tau^2 + \mathcal{O}(\tau^3)\,, \tag{7.16}$$

$$\begin{aligned} \mathrm{e}^{-\mathrm{i}k\cdot(q^\varepsilon(t)-q^\varepsilon(s))} &= \mathrm{e}^{-\mathrm{i}k\cdot(q^\varepsilon(t)-q^\varepsilon(t-\tau))} = \mathrm{e}^{-\mathrm{i}(k\cdot v)\tau}\Big(1 + \frac{1}{2}\tau^2\mathrm{i}(k\cdot\dot{v}) - \frac{1}{6}\tau^3\,\mathrm{i}(k\cdot\ddot{v}) \\ &\quad - \frac{1}{2}\Big(\frac{1}{2}\tau^2(k\cdot\dot{v}) - \frac{1}{6}\tau^3(k\cdot\ddot{v})\Big)^2 + \mathcal{O}((|k|\tau^2)^3)\Big)\,. \end{aligned} \tag{7.17}$$

Inserting in (7.6) and substituting $s' = \varepsilon^{-1}s,\ \boldsymbol{k}' = \varepsilon\boldsymbol{k}$ yields

$$\begin{aligned}
\boldsymbol{F}^{\varepsilon}_{\text{self}}(t) = e^2 \int_0^{\varepsilon^{-1}t} \mathrm{d}\tau \varepsilon^{-1} \int \mathrm{d}^3k |\widehat{\varphi}(\boldsymbol{k})|^2 \mathrm{e}^{-\mathrm{i}(\boldsymbol{k}\cdot\boldsymbol{v})\tau} \Big\{ (|\boldsymbol{k}|^{-1} \sin |\boldsymbol{k}|\tau)\mathrm{i}\boldsymbol{k} \\
- (\cos |\boldsymbol{k}|\tau)\Big(\boldsymbol{v} - \varepsilon\tau\dot{\boldsymbol{v}} + \frac{1}{2}\varepsilon^2\tau^2\ddot{\boldsymbol{v}}\Big) - (|\boldsymbol{k}|^{-1} \sin |\boldsymbol{k}|\tau)(\boldsymbol{v} \times (\mathrm{i}\boldsymbol{k} \times \boldsymbol{v}) \\
- \boldsymbol{v} \times (\mathrm{i}\boldsymbol{k} \times \varepsilon\tau\dot{\boldsymbol{v}}) + \frac{1}{2}\boldsymbol{v} \times (\mathrm{i}\boldsymbol{k} \times \varepsilon^2\tau^2\ddot{\boldsymbol{v}})) + \frac{1}{2}\varepsilon\tau^2\mathrm{i}(\boldsymbol{k}\cdot\dot{\boldsymbol{v}}) \\
\times \big((|\boldsymbol{k}|^{-1} \sin |\boldsymbol{k}|\tau)\mathrm{i}\boldsymbol{k} - (\cos |\boldsymbol{k}|\tau)(\boldsymbol{v} - \varepsilon\tau\dot{\boldsymbol{v}}) - (|\boldsymbol{k}|^{-1} \sin |\boldsymbol{k}|\tau)(\boldsymbol{v} \times (\mathrm{i}\boldsymbol{k} \times \boldsymbol{v}) \\
- \boldsymbol{v} \times (\mathrm{i}\boldsymbol{k} \times \varepsilon\tau\dot{\boldsymbol{v}}))\big) + \Big(-\frac{1}{6}\varepsilon^2\tau^3\mathrm{i}(\boldsymbol{k}\cdot\ddot{\boldsymbol{v}}) - \frac{1}{8}\varepsilon^2\tau^4(\boldsymbol{k}\cdot\dot{\boldsymbol{v}})^2\Big) \\
\times \big((|\boldsymbol{k}|^{-1} \sin |\boldsymbol{k}|\tau)\mathrm{i}\boldsymbol{k} - (\cos |\boldsymbol{k}|\tau)\boldsymbol{v} - (|\boldsymbol{k}|^{-1} \sin |\boldsymbol{k}|\tau)(\boldsymbol{v} \times (\mathrm{i}\boldsymbol{k} \times \boldsymbol{v}))\big)\Big\} \\
+ \mathcal{O}(\varepsilon^2)\,. \qquad (7.18)
\end{aligned}$$

The terms proportional to $\varepsilon^{-1}$ cancel by symmetry. We sort all other terms,

$$\begin{aligned}
\boldsymbol{F}^{\varepsilon}_{\text{self}}(t) = e^2 \int \mathrm{d}^3k\, |\widehat{\varphi}(\boldsymbol{k})|^2 \Big\{ \big( - (\boldsymbol{v}\cdot\dot{\boldsymbol{v}})\nabla_v + \dot{\boldsymbol{v}}(\boldsymbol{v}\cdot\nabla_v)\big) \int_0^{\varepsilon^{-1}t} \mathrm{d}\tau \mathrm{e}^{-\mathrm{i}(\boldsymbol{k}\cdot\boldsymbol{v})\tau} (|\boldsymbol{k}|^{-1} \sin |\boldsymbol{k}|\tau) \\
+ \big(\dot{\boldsymbol{v}} + \frac{1}{2}\boldsymbol{v}(\dot{\boldsymbol{v}}\cdot\nabla_v)\big) \int_0^{\varepsilon^{-1}t} \mathrm{d}\tau\, \tau \mathrm{e}^{-\mathrm{i}(\boldsymbol{k}\cdot\boldsymbol{v})\tau} (\cos |\boldsymbol{k}|\tau) + \varepsilon\Big(\frac{1}{2}\big[ - (\boldsymbol{v}^2 - 1) \\
\times (\dot{\boldsymbol{v}}\cdot\nabla_v)\nabla_v + \boldsymbol{v}(\boldsymbol{v}\cdot\nabla_v)(\dot{\boldsymbol{v}}\cdot\nabla_v) + (\boldsymbol{v}\cdot\ddot{\boldsymbol{v}})\nabla_v - \ddot{\boldsymbol{v}}(\boldsymbol{v}\cdot\nabla_v)\big] \\
+ \frac{1}{6}\big[ - (1 - \boldsymbol{v}^2)(\ddot{\boldsymbol{v}}\cdot\nabla_v)\nabla_v - \boldsymbol{v}(\boldsymbol{v}\cdot\nabla_v)(\ddot{\boldsymbol{v}}\cdot\nabla_v) + 3(\boldsymbol{v}\cdot\dot{\boldsymbol{v}})(\dot{\boldsymbol{v}}\cdot\nabla_v)\nabla_v \\
- 3\dot{\boldsymbol{v}}(\boldsymbol{v}\cdot\nabla_v)(\dot{\boldsymbol{v}}\cdot\nabla_v)\big] + \frac{1}{8}\big[(\boldsymbol{v}^2 - 1)(\dot{\boldsymbol{v}}\cdot\nabla_v)^2\nabla_v \\
- \boldsymbol{v}(\boldsymbol{v}\cdot\nabla_v)(\dot{\boldsymbol{v}}\cdot\nabla_v)^2\big]\Big) \int_0^{\varepsilon^{-1}t} \mathrm{d}\tau\, \tau \mathrm{e}^{-\mathrm{i}(\boldsymbol{k}\cdot\boldsymbol{v})\tau} (|\boldsymbol{k}|^{-1} \sin |\boldsymbol{k}|\tau) \\
+ \varepsilon\Big( - \ddot{\boldsymbol{v}} - \frac{1}{6}\big[\boldsymbol{v}(\ddot{\boldsymbol{v}}\cdot\nabla_v) + 3\dot{\boldsymbol{v}}(\dot{\boldsymbol{v}}\cdot\nabla_v)\big]\Big) \int_0^{\varepsilon^{-1}t} \mathrm{d}\tau\, \tau^2 \mathrm{e}^{-i(\boldsymbol{k}\cdot\boldsymbol{v})\tau} \cos |\boldsymbol{k}|\tau \Big\} \\
+ \mathcal{O}(\varepsilon^2)\,. \qquad (7.19)
\end{aligned}$$

To take the limit $\varepsilon \to 0$ we go back to position space and use the fundamental solution of the wave equation. Then

$$\lim_{\varepsilon\to 0} \int_0^{\varepsilon^{-1}t} \mathrm{d}\tau \int \mathrm{d}^3k |\widehat{\varphi}(\boldsymbol{k})|^2 \, \mathrm{e}^{-\mathrm{i}(\boldsymbol{k}\cdot\boldsymbol{v})\tau} (|\boldsymbol{k}|^{-1} \sin|\boldsymbol{k}|\tau)\, \tau^p$$

$$= \int_0^{\infty} \mathrm{d}t \int \mathrm{d}^3x \int \mathrm{d}^3y \, \varphi(\boldsymbol{x})\varphi(\boldsymbol{y}) \frac{1}{4\pi t} \delta(|\boldsymbol{x} + \boldsymbol{v}t - \boldsymbol{y}| - t)\, t^p$$

$$= \begin{cases} \int \mathrm{d}^3k\, |\widehat{\varphi}(\boldsymbol{k})|^2 [\boldsymbol{k}^2 - (\boldsymbol{k}\cdot\boldsymbol{v})^2]^{-1} & \text{for } p = 0, \\ \int \mathrm{d}^3x\, \varphi(\boldsymbol{x}) \int \mathrm{d}^3y \varphi(\boldsymbol{y})\, (\gamma^2/4\pi) & \text{for } p = 1. \end{cases} \tag{7.20}$$

By the same method

$$\lim_{\varepsilon\to 0} \int_0^{\varepsilon^{-1}t} \mathrm{d}\tau \int \mathrm{d}^3k\, |\widehat{\varphi}(\boldsymbol{k})|^2 \, \mathrm{e}^{-\mathrm{i}(\boldsymbol{k}\cdot\boldsymbol{v})\tau}\, \tau^{1+p} \frac{\mathrm{d}}{\mathrm{d}\tau} (|\boldsymbol{k}|^{-1} \sin|\boldsymbol{k}|\tau)$$

$$= -\big(1 + p + (\boldsymbol{v}\cdot\nabla_{\boldsymbol{v}})\big) \int_0^{\infty} \mathrm{d}t \int \mathrm{d}^3k\, |\widehat{\varphi}(\boldsymbol{k})|^2 \, \mathrm{e}^{-\mathrm{i}(\boldsymbol{k}\cdot\boldsymbol{v})t} (|\boldsymbol{k}|^{-1} \sin|\boldsymbol{k}|t) t^p$$

$$= \begin{cases} -\int \mathrm{d}^3k\, |\widehat{\varphi}(\boldsymbol{k})|^2 (\boldsymbol{k}^2 + (\boldsymbol{k}\cdot\boldsymbol{v})^2)[\boldsymbol{k}^2 - (\boldsymbol{k}\cdot\boldsymbol{v})^2]^{-2} & \text{for } p = 0, \\ -\int \mathrm{d}^3x\, \varphi(\boldsymbol{x}) \int \mathrm{d}^3y \varphi(\boldsymbol{y})\, (2\gamma^4/4\pi) & \text{for } p = 1. \end{cases} \tag{7.21}$$

Collecting all terms the final result reads

$$\begin{aligned} \boldsymbol{F}^{\varepsilon}_{\mathrm{self}}(t) = &- m_{\mathrm{f}}(\boldsymbol{v})\dot{\boldsymbol{v}} + \varepsilon(e^2/6\pi)\big[\gamma^4(\boldsymbol{v}\cdot\ddot{\boldsymbol{v}})\boldsymbol{v} + 3\gamma^6(\boldsymbol{v}\cdot\dot{\boldsymbol{v}})^2\boldsymbol{v} \\ &+ 3\gamma^4(\boldsymbol{v}\cdot\dot{\boldsymbol{v}})\dot{\boldsymbol{v}} + \gamma^2\ddot{\boldsymbol{v}}\big] + \mathcal{O}(\varepsilon^2) \end{aligned} \tag{7.22}$$

for $t > 0$ with

$$\begin{aligned} m_{\mathrm{f}}(\boldsymbol{v}) = m_{\mathrm{e}}\Big[&\big(|\boldsymbol{v}|^{-2}\gamma^2(3 - \boldsymbol{v}^2) - |\boldsymbol{v}|^{-3}(3 + \boldsymbol{v}^2)\operatorname{arctanh}|\boldsymbol{v}|\big)\widehat{\boldsymbol{v}} \otimes \widehat{\boldsymbol{v}} \\ &+ \big(-|\boldsymbol{v}|^{-2} + |\boldsymbol{v}|^{-3}(1 + \boldsymbol{v}^2)\operatorname{arctanh}|\boldsymbol{v}|\big)\mathbb{1}\Big]. \end{aligned} \tag{7.23}$$

Note that $m_{\mathrm{f}}(\boldsymbol{v}) = \mathrm{d}(\boldsymbol{P}_{\mathrm{s}} - m_{\mathrm{b}}\gamma\boldsymbol{v})/\mathrm{d}\boldsymbol{v}$ as a $3 \times 3$ matrix.

Up to order $\varepsilon$, $\boldsymbol{F}^{\varepsilon}_{\mathrm{self}}(t)$ consists of two parts of a rather different character. The term $-m_{\mathrm{f}}(\boldsymbol{v})\dot{\boldsymbol{v}}$ is the contribution from the electromagnetic field to the change in total momentum. We computed this term already in section 4.1 via a completely different route. As emphasized there, since the Abraham model is semirelativistic, the velocity dependence of $m_{\mathrm{f}}$ has no reason to be of relativistic form and indeed it is not. The term proportional to $\varepsilon$ in (7.22) is the *radiation reaction*. Again there is no a priori reason to expect it to be relativistic, but in fact it is. Using the four-vector notation of section 2.5, the radiation reaction can be rewritten as

$$\varepsilon(e^2/6\pi)(\ddot{\mathbf{u}} - (\dot{\mathbf{u}} \cdot \dot{\mathbf{u}})\mathbf{u}) = \varepsilon(e^2/6\pi)(\mathbf{g} + \mathbf{u} \otimes \mathbf{u}) \cdot \ddot{\mathbf{u}} \,. \tag{7.24}$$

The space part is the term proportional to $e^2$ of (7.22), i.e. the radiation reaction force, and the time part is the work done by this force per unit time.

## 7.3 How can the acceleration be bounded?

We return to the microscopic time scale. From the conservation of energy together with condition $(P)$, we have

$$\begin{aligned} E_{\mathrm{s}}(\boldsymbol{v}^0) + e\phi(\varepsilon \boldsymbol{q}^0) &= \mathcal{E}(\boldsymbol{E}^0, \boldsymbol{B}^0, \boldsymbol{q}^0, \boldsymbol{v}^0) = \mathcal{E}(\boldsymbol{E}(t), \boldsymbol{B}(t), \boldsymbol{q}(t), \boldsymbol{v}(t)) \\ &\geq m_{\mathrm{b}}\gamma(\boldsymbol{v}(t)) + e\,\bar{\phi} \end{aligned} \tag{7.25}$$

and therefore

$$\sup_{t\in\mathbb{R}} |\boldsymbol{v}(t)| \leq \bar{v} < 1 \,. \tag{7.26}$$

In (6.4) the external forces are of order $\varepsilon$. Superficially the self-force is of order one. However for a Coulombic charge soliton field the self-force vanishes. Thus if we could show that the deviations from the appropriate local soliton field are of order $\varepsilon$, then the acceleration would satisfy

$$\sup_{t\in\mathbb{R}} |\dot{\boldsymbol{v}}(t)| \leq C\,\varepsilon \tag{7.27}$$

with $C$ a suitable constant. This is what we want to prove. We will not keep track of the constants, and the value of $C$ changes from equation to equation. We make sure, however, that the $e$-dependence is explicit and that $C$ depends only on $\overline{v}$, and thus is determined by the initial conditions. Of course, to justify the Taylor expansion of section 7.2, we also need analogous estimates of higher derivatives, which can be obtained with considerably more effort through the same scheme. Here we want to explain how to get (7.27) and why we need $e$ to be sufficiently small, at least for the moment.

From the equations of motion one has

$$\dot{\boldsymbol{v}} = m_0(\boldsymbol{v})^{-1}\big[\varepsilon e\big(\boldsymbol{E}_{\mathrm{ex}}(\varepsilon\boldsymbol{q}) + \boldsymbol{v}\times\boldsymbol{B}_{\mathrm{ex}}(\varepsilon\boldsymbol{q})\big) + e\big(\boldsymbol{E}_{\varphi}(\boldsymbol{q}) + \boldsymbol{v}\times\boldsymbol{B}_{\varphi}(\boldsymbol{q})\big)\big], \tag{7.28}$$

where $m_0^{-1}(\boldsymbol{v}) = (m_{\mathrm{b}}\gamma)^{-1}(\mathbb{1} - \widehat{\boldsymbol{v}}\otimes\widehat{\boldsymbol{v}})$ is the matrix inverse of $m_0(\boldsymbol{v})$. Clearly by (7.26) we have $\|m_0(\boldsymbol{v})^{-1}\| \leq C$ and, by condition $(P)$, the first term is bounded as

$$\varepsilon\,\big|e\big(\boldsymbol{E}_{\mathrm{ex}}(\varepsilon\boldsymbol{q}) + \boldsymbol{v}\times\boldsymbol{B}_{\mathrm{ex}}(\varepsilon\boldsymbol{q})\big)\big| \leq C|e|\varepsilon \,. \tag{7.29}$$

On the other hand, the self-force looks to be of order one. To reduce it in order we have to exploit the fact that $\boldsymbol{E}$, $\boldsymbol{B}$ deviate only slightly from $\boldsymbol{E}_v$, $\boldsymbol{B}_v$ close to the charge distribution, i.e. we subtract zero and rewrite the self-force as

$$e\big(\boldsymbol{E}_\varphi(\boldsymbol{q}) - \boldsymbol{E}_{v\varphi}(\boldsymbol{q}) + \boldsymbol{v} \times (\boldsymbol{B}_\varphi(\boldsymbol{q}) - \boldsymbol{B}_{v\varphi}(\boldsymbol{q}))\big)\,. \tag{7.30}$$

Our goal is to show that this difference is of order $\varepsilon$.

Let us define then

$$Z(\boldsymbol{x}, t) = \begin{pmatrix} \boldsymbol{E}(\boldsymbol{x}, t) - \boldsymbol{E}_{v(t)}(\boldsymbol{x} - \boldsymbol{q}(t)) \\ \boldsymbol{B}(\boldsymbol{x}, t) - \boldsymbol{B}_{v(t)}(\boldsymbol{x} - \boldsymbol{q}(t)) \end{pmatrix}. \tag{7.31}$$

Using Maxwell's equations and the relations $(\boldsymbol{v} \cdot \nabla)\, \boldsymbol{E}_v = -\nabla \times \boldsymbol{B}_v + e\varphi \boldsymbol{v}$, $(\boldsymbol{v} \cdot \nabla)\boldsymbol{B}_v = \nabla \times \boldsymbol{E}_v$ one obtains

$$\frac{\mathrm{d}}{\mathrm{d}t} Z(t) = A Z(t) - g(t)\,, \tag{7.32}$$

where $A$ is defined in (2.18) and

$$g(\boldsymbol{x}, t) = \begin{pmatrix} (\dot{\boldsymbol{v}}(t) \cdot \nabla_v) \boldsymbol{E}_v(\boldsymbol{x} - \boldsymbol{q}(t)) \\ (\dot{\boldsymbol{v}}(t) \cdot \nabla_v) \boldsymbol{B}_v(\boldsymbol{x} - \boldsymbol{q}(t)) \end{pmatrix}. \tag{7.33}$$

Therefore (7.32) has again the structure of the inhomogeneous Maxwell equations. Since $Z(0) = 0$ by our assumption on the initial data, one has

$$Z(t) = -\int_0^t \mathrm{d}s\, U(t-s) g(s)\,. \tag{7.34}$$

In terms of $Z(t)$, using (7.28), (7.30), the acceleration is bounded through

$$|\dot{\boldsymbol{v}}(t)| \le C(\varepsilon + |e|) \int \mathrm{d}^3 x \varphi(\boldsymbol{x}) |\boldsymbol{Z}_1(\boldsymbol{x} + \boldsymbol{q}(t), t) + \boldsymbol{v}(t) \times \boldsymbol{Z}_2(\boldsymbol{x} + \boldsymbol{q}(t), t)|\,. \tag{7.35}$$

Let us set $W(t, s) = U(t-s) g(s)$. Below we will prove that

$$|\boldsymbol{W}_1(t, s, \boldsymbol{q}(t) + \boldsymbol{x})| + |\boldsymbol{W}_2(t, s, \boldsymbol{q}(t) + \boldsymbol{x})| \le |e| C |\dot{\boldsymbol{v}}(s)| (1 + (t-s)^2)^{-1} \tag{7.36}$$

for $|\boldsymbol{x}| \le R_\varphi$. Therefore inserting (7.36) in (7.35) one obtains

$$|\dot{\boldsymbol{v}}(t)| \le |e| C \left( \varepsilon + |e| \int_0^t \mathrm{d}s\, (1 + (t-s)^2)^{-1}\, |\dot{\boldsymbol{v}}(s)| \right). \tag{7.37}$$

Let $\kappa = \sup_{t\geq 0} |\dot{v}(t)|$. Then (7.37) reads

$$\kappa \leq |e|\,C\Big(\varepsilon + |e|\kappa \int_0^\infty \mathrm{d}s\,(1+s^2)^{-1}\Big), \quad \kappa \leq \frac{|e|\,C}{1-e^2\,C}\,\varepsilon\,. \tag{7.38}$$

From the computation below we will see that $C$ depends on $\bar{v}$ (and on model parameters like the form factor $\widehat{\varphi}$), but not on $e$. Thus taking $|e|$ sufficiently small one can ensure $e^2\,C < 1$ and therefore $\kappa \leq C\varepsilon$ as claimed.

We still have to establish (7.36). $U(t)$ is given in Eqs. (2.12), (2.13). Since $\nabla \cdot \boldsymbol{g}_1(s) = 0 = \nabla \cdot \boldsymbol{g}_2(s)$, the term proportional to $\boldsymbol{k} \otimes \boldsymbol{k}$ drops out. In real space $|\boldsymbol{k}|^{-1} \sin |\boldsymbol{k}|t$ becomes $G_t$ from (2.15) and $\cos |\boldsymbol{k}|t$ becomes $\partial_t G_t$. Therefore

$$\begin{aligned}
\boldsymbol{W}_1(t,s,\boldsymbol{x}) &= \frac{1}{4\pi(t-s)^2}\int \mathrm{d}^3y\,\delta(|\boldsymbol{x}-\boldsymbol{y}|-(t-s))\\
&\quad\times\big[(t-s)\nabla\times \boldsymbol{g}_2(\boldsymbol{y},s) + \boldsymbol{g}_1(\boldsymbol{y},s) - (\boldsymbol{x}-\boldsymbol{y})\cdot\nabla \boldsymbol{g}_1(\boldsymbol{y},s)\big],\\
\boldsymbol{W}_2(t,s,\boldsymbol{x}) &= \frac{1}{4\pi(t-s)^2}\int \mathrm{d}^3y\,\delta(|\boldsymbol{x}-\boldsymbol{y}|-(t-s))\\
&\quad\times\big[-(t-s)\nabla\times \boldsymbol{g}_1(\boldsymbol{y},s) + \boldsymbol{g}_2(\boldsymbol{y},s) - (\boldsymbol{x}-\boldsymbol{y})\cdot\nabla \boldsymbol{g}_2(\boldsymbol{y},s)\big].
\end{aligned} \tag{7.39}$$

We insert $g$ from (7.33). $\boldsymbol{E}_v$ and $\boldsymbol{B}_v$ are first-order derivatives of the function $\phi_{v\varphi}$ which according to (4.7) is given by

$$\phi_{v\varphi}(\boldsymbol{x}) = e\int \mathrm{d}^3y\varphi(\boldsymbol{x}-\boldsymbol{y})(4\pi)^{-1}\big[((1-v^2)\boldsymbol{y}^2 + (v\cdot\boldsymbol{y})^2)\big]^{-1/2}\,. \tag{7.40}$$

Using (4.5) one has componentwise

$$\begin{aligned}
&|\nabla_v \boldsymbol{E}_v(\boldsymbol{x})| + |\nabla_v \boldsymbol{B}_v(\boldsymbol{x})| \leq C\,(\,|\nabla\phi_v(\boldsymbol{x})| + |\nabla\nabla_v\phi_v(\boldsymbol{x})|)\,,\\
&|\nabla\nabla_v \boldsymbol{E}_v(\boldsymbol{x})| + |\nabla\nabla_v \boldsymbol{B}_v(\boldsymbol{x})| \leq C\,(\,|\nabla\nabla_v\phi_v(\boldsymbol{x})| + |\nabla\nabla\nabla_v\phi_v(\boldsymbol{x})|)
\end{aligned} \tag{7.41}$$

and taking successive derivatives in (7.40) one obtains the bounds

$$\begin{aligned}
&|\nabla\phi_v(\boldsymbol{x})| + |\nabla\nabla_v\phi_v(\boldsymbol{x})| \leq e\,C\,(1+|\boldsymbol{x}|)^{-2}\,,\\
&|\nabla\nabla\phi_v(\boldsymbol{x})| + |\nabla\nabla\nabla_v\phi_v(\boldsymbol{x})| \leq e\,C\,(1+|\boldsymbol{x}|)^{-3}\,,
\end{aligned} \tag{7.42}$$

which imply

$$\begin{aligned}
&|\boldsymbol{g}_1(\boldsymbol{x},s)| + |\boldsymbol{g}_2(\boldsymbol{x},s)| \leq e\,C|\dot{v}(s)|(1+|\boldsymbol{x}-\boldsymbol{q}(s)|^2)^{-1}\,,\\
&|\nabla \boldsymbol{g}_1(\boldsymbol{x},s)| + |\nabla \boldsymbol{g}_2(\boldsymbol{x},s)| \leq e\,C|\dot{v}(s)|(1+|\boldsymbol{x}-\boldsymbol{q}(s)|^3)^{-1}\,.
\end{aligned} \tag{7.43}$$

We insert the bound (7.43) in (7.39) which results in an upper bound on $W(t, s, \boldsymbol{q}(t) + \boldsymbol{x})$. Using the condition that $|\boldsymbol{x}| \le R_\varphi$ and $|\boldsymbol{q}(t) - \boldsymbol{q}(s)| \le \bar{v}|t - s|$ finally yields (7.36).

We summarize our findings as

**Theorem 7.1** (Bounds on the velocity and its derivatives). *For the Abraham model satisfying conditions $(C)$, $(P)$, and $(I)$ there exist constants $C$, depending through $\bar{v}$ only on the initial conditions, and $\bar{e}$ such that*

$$|\boldsymbol{v}(t)| \le \bar{v} < 1\,, \quad |\dot{\boldsymbol{v}}(t)| \le C\varepsilon\,, \quad |\ddot{\boldsymbol{v}}(t)| \le C\big(\varepsilon^2 + \varepsilon(1 + |t|)^{-2}\big)\,,$$
$$|\dddot{\boldsymbol{v}}(t)| \le C\big(\varepsilon^3 + \varepsilon^2(1 + |t|)^{-2} + \varepsilon(1 + |t|)^{-3}\big) \tag{7.44}$$

*for all $t$ on the microscopic time scale, provided the charge is sufficiently small, i.e. $|e| < \bar{e}$.*

By keeping track of the constant $C$, one could get a bound on the charge admissible in Theorem 7.1. Since we believe this restriction to be an artifact of the method anyhow, there is no point in the effort.

## Notes and references

### *Section 7.1*

Sommerfeld (1904a, 1905) systematically uses memory equations. In fact he considers the Abraham model with the kinetic energy $m_b v^2/2$ for the particle and wants to understand what happens when $v(0) > c$. He argues that the particle rapidly loses its energy to become slower than $c$ by emitting what we now call Čerenkov radiation. The differential–difference equation (7.14) is derived by Page (1918) and its relativistic generalization by Caldirola (1956). For reviews we refer to Erber (1961) and Pearle (1982). Moniz and Sharp (1974, 1977) supply a linear stability analysis and show that the solutions to (7.14) are stable provided $R_\varphi$ is not too small. For that reason Rohrlich (1997) regards (7.14) and its relativistic sister as the fundamental starting point for the classical dynamics of extended charges. We take the Abraham model as the basic dynamical theory. Memory equations are a useful tool in analyzing its properties.

### *Section 7.2*

The Taylor expansion is taken from Kunze and Spohn (2000a). Such an expansion was already used in Sommerfeld (1904a, 1905), to be repeated in various disguises. The traditional expansion parameter is the size of the charge distribution, which in our context is replaced by the scaling parameter $\varepsilon$ controlling the variation of the potentials.

## *Section 7.3*

The contraction argument appears in Komech, Kunze and Spohn (1999). The bound on $\dot{v}(t)$ is taken from Kunze and Spohn (2000a), where also higher derivatives are discussed. It is claimed that $|\ddot{v}(t)| \leq C\varepsilon^2$ and $|\dddot{v}(t)| \leq C\varepsilon^3$. In the argument some initial terms are overlooked and the correct bounds are as given in (7.44).

# 8
# Comparison dynamics

The expansion of the self-force suggests that if we are willing to accept an error of order $\varepsilon^2$, the trajectory of the charged particle is governed by an autonomous equation – a substantial simplification of the hitherto coupled problem. An error of order $\varepsilon^2$ in the equation does *not* imply an error of the same order in the solution. This point must be discussed, but let us proceed for a while in good faith and simply ignore the error in Eq. (7.22). Then we obtain the following approximate equation for the motion of the charge,

$$\begin{aligned} \dot{\boldsymbol{q}} &= \boldsymbol{v}\,, \\ m(\boldsymbol{v})\dot{\boldsymbol{v}} &= e\big(\boldsymbol{E}_{\mathrm{ex}}(\boldsymbol{q}) + \boldsymbol{v}\times\boldsymbol{B}_{\mathrm{ex}}(\boldsymbol{q})\big) + \varepsilon(e^2/6\pi)\big[\gamma^4(\boldsymbol{v}\cdot\ddot{\boldsymbol{v}})\boldsymbol{v} \\ &\quad + 3\gamma^6(\boldsymbol{v}\cdot\dot{\boldsymbol{v}})^2\boldsymbol{v} + 3\gamma^4(\boldsymbol{v}\cdot\dot{\boldsymbol{v}})\dot{\boldsymbol{v}} + \gamma^2\ddot{\boldsymbol{v}}\big]\,. \end{aligned} \tag{8.1}$$

Here $m(\boldsymbol{v})$ is the effective velocity-dependent mass. It is the sum of the bare mass and the mass (7.23) induced by the field,

$$m(\boldsymbol{v}) = m_{\mathrm{b}}(\gamma\mathbb{1} + \gamma^3\boldsymbol{v}\otimes\boldsymbol{v}) + m_{\mathrm{f}}(\boldsymbol{v})\,. \tag{8.2}$$

As anticipated in section 4.1, via a distinct route, the leading contribution to (8.1) is derived from the effective Lagrangian

$$L_{\mathrm{eff}}(\boldsymbol{q},\dot{\boldsymbol{q}}) = T(\dot{\boldsymbol{q}}) - e\big(\phi_{\mathrm{ex}}(\boldsymbol{q}) - \dot{\boldsymbol{q}}\cdot\boldsymbol{A}_{\mathrm{ex}}(\boldsymbol{q})\big)\,, \tag{8.3}$$

or equivalently from the Hamiltonian

$$H_{\mathrm{eff}}(\boldsymbol{q},\boldsymbol{p}) = E_{\mathrm{eff}}\big(\boldsymbol{p} - e\boldsymbol{A}_{\mathrm{ex}}(\boldsymbol{q})\big) + e\phi_{\mathrm{ex}}(\boldsymbol{q})\,. \tag{8.4}$$

For later purposes it is more convenient to work with the energy function

$$H(\boldsymbol{q},\boldsymbol{v}) = E_{\mathrm{s}}(\boldsymbol{v}) + e\phi_{\mathrm{ex}}(\boldsymbol{q})\,, \tag{8.5}$$

which is conserved by the solutions to (8.1) with $\varepsilon = 0$; compare with (4.14).

The term of order $\varepsilon$ in (8.1) describes the radiation reaction. If included, the energy of the particle fails to be conserved and the energy balance becomes

$$\frac{\mathrm{d}}{\mathrm{d}t} H(\boldsymbol{q}, \boldsymbol{v}) - \frac{\mathrm{d}}{\mathrm{d}t} \varepsilon\, (e^2/6\pi)\, \gamma^4 (\boldsymbol{v} \cdot \dot{\boldsymbol{v}}) = -\varepsilon\, (e^2/6\pi) \left[\gamma^4 \dot{\boldsymbol{v}}^2 + \gamma^6 (\boldsymbol{v} \cdot \dot{\boldsymbol{v}})^2\right]. \quad (8.6)$$

The term $-\varepsilon(e^2/6\pi)\gamma^4(\boldsymbol{v} \cdot \dot{\boldsymbol{v}}) = E_{\mathrm{schott}}(\boldsymbol{v}, \dot{\boldsymbol{v}})$ is the *Schott energy*. It has no definite sign. The Schott energy is stored in the near field and can be reversibly exchanged with the mechanical energy of the charge. The right-hand side of (8.6) is the irreversible loss of energy through radiation; compare with section 8.4. Equation (8.6) is analogous to the balance equations in hydrodynamics and a familiar way to rewrite it is

$$e\boldsymbol{v} \cdot E_{\mathrm{ex}}(\boldsymbol{q}) = \frac{\mathrm{d}}{\mathrm{d}t} \big(E_{\mathrm{s}}(\boldsymbol{v}) + E_{\mathrm{schott}}(\boldsymbol{v}, \dot{\boldsymbol{v}})\big) + \varepsilon\, (e^2/6\pi) \left[\gamma^4 \dot{\boldsymbol{v}}^2 + \gamma^6 (\boldsymbol{v} \cdot \dot{\boldsymbol{v}})^2\right]. \quad (8.7)$$

In other words, the work done by the external electric field acting on the charge is divided up into the change in its kinetic energy, the change of the Schott energy, and radiation.

If we set $G_\varepsilon = E_{\mathrm{s}} + E_{\mathrm{schott}}$, then $G_\varepsilon$ is decreasing in time, and integrating both sides of (8.6) yields

$$\begin{aligned} &-G_\varepsilon\, (\boldsymbol{q}(t), \boldsymbol{v}(t), \dot{\boldsymbol{v}}(t)) + G_\varepsilon\, (\boldsymbol{q}(0), \boldsymbol{v}(0), \dot{\boldsymbol{v}}(0)) \\ &= \varepsilon\, (e^2/6\pi) \int_0^t \mathrm{d}s \left[\gamma^4\, \dot{\boldsymbol{v}}(s)^2 + \gamma^6\, (\boldsymbol{v}(s) \cdot \dot{\boldsymbol{v}}(s))^2\right]. \end{aligned} \quad (8.8)$$

The mechanical energy is bounded from below, but the Schott energy does not have a definite sign. *If* (!) the Schott energy remains bounded in the course of time, then

$$\int_0^\infty \mathrm{d}t \left[\gamma^4\, \dot{\boldsymbol{v}}(t)^2 + \gamma^6\, (\boldsymbol{v}(t) \cdot \dot{\boldsymbol{v}}(t))^2\right] < \infty\,, \quad (8.9)$$

which implies

$$\lim_{t \to \infty} \dot{\boldsymbol{v}}(t) = 0\,. \quad (8.10)$$

The same conclusion was already reached for the Abraham model in Theorem 5.1, with no adiabatic limit there. Instead of (8.9) we used the bounded energy dissipation (5.9). Since both the approximate and the true solutions have the same long-time asymptotics, we expect no further time scale, i.e. higher corrections to (8.1) should not change the qualitative behavior of solutions and merely increase in precision. One important difference must be stressed, however: Theorem 5.1 holds for every solution, whereas (8.10) holds only for those with bounded Schott energy.

Unfortunately, the energy balance (8.7) by itself does not tell the full story. As noticed apparently first by Dirac (1938), Eq. (8.1) has solutions which run away exponentially fast. This does not contradict (8.8). $G_\varepsilon(t)$ diverges to $-\infty$ and the time integral diverges to $+\infty$ as $t \to \infty$. The occurrence of runaway solutions can be seen most easily in the approximation of small velocities, setting $\boldsymbol{B}_{\rm ex} = 0$, and linearizing $\phi_{\rm ex}$ around a stable minimum, say at $\boldsymbol{q} = 0$. Then (8.1) becomes

$$\dot{\boldsymbol{q}} = \boldsymbol{v}\,, \quad m\dot{\boldsymbol{v}} = -m\,\omega_0^2 \boldsymbol{q} + \varepsilon\, km\, \ddot{\boldsymbol{v}} \tag{8.11}$$

with $km = e^2/6\pi$. The three components of the linear equation (8.11) decouple and for each component there are three modes of the form $\mathrm{e}^{zt}$. The characteristic equation is $z^2 = -\omega_0^2 + \varepsilon k z^3$ and to leading order the eigenvalues are $z_\pm = \pm\,\mathrm{i}\omega_0 - \varepsilon(k\omega_0^2/2)$, $z_3 = (1/\varepsilon k) + \mathcal{O}(1)$. Thus in the nine-dimensional phase space for (8.11) there is a stable six-dimensional hyperplane, $\mathcal{C}_\varepsilon$. On $\mathcal{C}_\varepsilon$ the motion is weakly damped, with friction coefficient $\varepsilon\,(k\omega_0^2/2)$, and relaxes as $t \to \infty$ to rest at $\boldsymbol{q} = 0$. Transversal to $\mathcal{C}_\varepsilon$ the solution runs away as $\mathrm{e}^{(t/\varepsilon k)}$.

Clearly such runaway solutions violate the stability estimate (7.15). Thus the full Maxwell–Newton equations do not have runaways. They somehow appear as an artifact of the Taylor expansion of $\boldsymbol{F}^\varepsilon_{\rm self}(t)$ from (7.6). Dirac simply postulated that physical solutions must satisfy the *asymptotic condition*

$$\lim_{t\to\infty} \dot{\boldsymbol{v}}(t) = 0\,. \tag{8.12}$$

In the linearized version (8.11) this means that the initial conditions have to lie in $\mathcal{C}_\varepsilon$. In Theorem 5.1 we proved the asymptotic condition to hold for the Abraham model. Thus only those solutions to (8.1) satisfying the asymptotic condition can serve as a comparison dynamics to the true solution. We then have to understand how the asymptotic condition arises, even more so the global structure of the solution flow to (8.1).

We note that in (8.1) the highest derivative is multiplied by a small prefactor. Such equations have been studied in great detail under the header of (geometric) singular perturbation theory. The main conclusion is that the structure found for the linear equation (8.11) persists for the nonlinear equation (8.1). Of course the hyperplane $\mathcal{C}_\varepsilon$ is now deformed into some manifold, called the critical (or center) manifold. We explain a standard example in the following section and then apply the theory to (8.1).

## 8.1 An example for singular perturbation theory

As a purely mathematical example we consider the coupled system

$$\dot{x} = f(x, y)\,, \quad \varepsilon\,\dot{y} = y - h(x)\,. \tag{8.13}$$

$h$ and $f$ are bounded, smooth functions. The phase space is $\mathbb{R}^2$. The question we address is to understand how the solutions to (8.13) behave for small $\varepsilon$. If we set $\varepsilon = 0$, then $y = h(x)$ and we obtain the autonomous equation

$$\dot{x} = f(x, h(x))\,. \tag{8.14}$$

Geometrically this means that the two-dimensional phase space has been squeezed to the line $y = h(x)$ and the base point, $x(t)$, is governed by (8.14). $\{(x, h(x))\,|\, x \in \mathbb{R}\}$ is the critical manifold to zeroth order in $\varepsilon$.

To see some motion appear in the phase space ambient to $\mathcal{C}_0$ we change from $t$ to the fast time scale $\tau = \varepsilon^{-1}t$. Denoting differentiation with respect to $\tau$ by $'$, (8.13) goes over to

$$x' = \varepsilon\, f(x, y)\,, \quad y' = y - h(x)\,. \tag{8.15}$$

In the limit $\varepsilon \to 0$ we now have $x' = 0$, i.e. $x(\tau) = x_0$ and $y' = y - h(x_0)$ with solution $y(\tau) = (y_0 - h(x_0))\mathrm{e}^{\tau} + h(x_0)$. Thus on this time scale, $\mathcal{C}_0$ consists exclusively of repelling fixed points. This is why $\mathcal{C}_0$ is called critical. The linearization at $\mathcal{C}_0$ has the eigenvalue 1 transverse and the eigenvalue 0 tangential to $\mathcal{C}_0$. In the theory of dynamical systems zero eigenvalues in the linearization turn out to be linked to center manifolds, and thus $\mathcal{C}_0$ is also called the center manifold (at $\varepsilon = 0$). The basic result of singular perturbation theory is that for small $\varepsilon$ the critical manifold deforms smoothly into $\mathcal{C}_\varepsilon$; compare with figure 8.1. Thus $\mathcal{C}_\varepsilon$ is invariant under the solution flow to (8.13). Its linearization at $(x, y) \in \mathcal{C}_\varepsilon$ has an eigenvalue of $\mathcal{O}(1)$ with eigenvector tangential to $\mathcal{C}_\varepsilon$ and an eigenvalue $1/\varepsilon$ with eigenvector transverse to $\mathcal{C}_\varepsilon$. Thus for an initial condition slightly away from $\mathcal{C}_\varepsilon$ the solution very rapidly diverges to infinity. Since $\mathcal{C}_0$ is deformed by order $\varepsilon$, also $\mathcal{C}_\varepsilon$ is of the form $\{(x, h_\varepsilon(x))\,|\, x \in \mathbb{R}\}$. According to (8.13) the base point evolves as

$$\dot{x} = f(x, h_\varepsilon(x))\,. \tag{8.16}$$

Since $h_\varepsilon$ is smooth in $\varepsilon$, it can be Taylor-expanded as

$$h_\varepsilon(x) = \sum_{j=0}^{m} \varepsilon^j\, h_j(x) + \mathcal{O}(\varepsilon^{m+1})\,. \tag{8.17}$$

By (8.13) and (8.16) we have the identity

$$\varepsilon\, \partial_x h_\varepsilon(x)\, f(x, h_\varepsilon(x)) = h_\varepsilon(x) - h(x)\,. \tag{8.18}$$

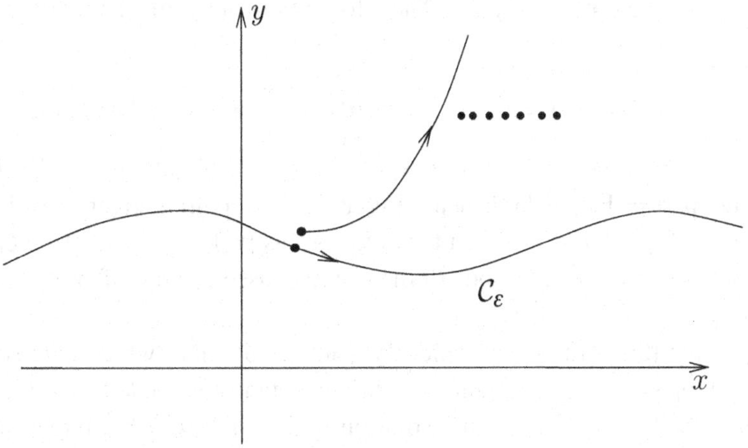


Figure 8.1: Repulsive center manifold $\mathcal{C}_\varepsilon$. The motion on $\mathcal{C}_\varepsilon$ is slow and the motion away from $\mathcal{C}_\varepsilon$ is fast.

Substituting into (8.17) and comparing powers of $\varepsilon$ one can thus determine $h_j(x)$ recursively. To lowest order we obtain

$$h_0(x) = h(x)\,, \quad h_1(x) = h'(x)\, f(x, h(x)) \tag{8.19}$$

and to order $\varepsilon$ the base point is governed by

$$\dot{x} = f(x, h(x)) + \varepsilon\, \partial_y f(x, h(x))\, h'(x)\, f(x, h(x))\,. \tag{8.20}$$

Given the geometric picture of the center manifold, the stable (i.e. not runaway) solutions to (8.13) can be determined to any required precision.

## 8.2 The critical manifold

Our task is to cast (8.1) into the canonical form used in singular perturbation theory. We set $(\boldsymbol{x}_1, \boldsymbol{x}_2) = \boldsymbol{x} = (\boldsymbol{q}, \boldsymbol{v}) \in \mathbb{R}^3 \times \mathbb{V}$, $\boldsymbol{y} = \dot{\boldsymbol{v}} \in \mathbb{R}^3$,

$$\boldsymbol{f}(\boldsymbol{x}, \boldsymbol{y}) = (\boldsymbol{x}_2, \boldsymbol{y}) \in \mathbb{V} \times \mathbb{R}^3 \tag{8.21}$$

and

$$\begin{aligned}\boldsymbol{g}(\boldsymbol{x}, \boldsymbol{y}, \varepsilon) = \gamma^{-2}\kappa(\boldsymbol{x}_2)^{-1}\big((6\pi/e^2)\left[m(\boldsymbol{x}_2)\boldsymbol{y} - \boldsymbol{F}_{\mathrm{ex}}(\boldsymbol{x})\right] \\ -\varepsilon\left[3\gamma^6(\boldsymbol{x}_2 \cdot \boldsymbol{y})^2 \boldsymbol{x}_2 + 3\gamma^4 (\boldsymbol{x}_2 \cdot \boldsymbol{y})\boldsymbol{y}\right]\big)\,,\end{aligned} \tag{8.22}$$

where $\gamma = (1 - \boldsymbol{x}_2^2)^{-1/2}$ as before, $\boldsymbol{F}_{\mathrm{ex}}(\boldsymbol{x}) = e(\boldsymbol{E}_{\mathrm{ex}}(\boldsymbol{x}_1) + \boldsymbol{x}_2 \times \boldsymbol{B}_{\mathrm{ex}}(\boldsymbol{x}_1))$, and $\kappa(\boldsymbol{v})$ is the $3 \times 3$ matrix $\kappa(\boldsymbol{v}) = \mathbb{1} + \gamma^2\, \boldsymbol{v} \otimes \boldsymbol{v}$ with inverse matrix $\kappa(\boldsymbol{v})^{-1} = \mathbb{1} - \boldsymbol{v} \otimes \boldsymbol{v}$. With this notation Eq. (8.1) reads

$$\dot{\boldsymbol{x}} = \boldsymbol{f}(\boldsymbol{x}, \boldsymbol{y})\,, \quad \varepsilon\, \dot{\boldsymbol{y}} = \boldsymbol{g}(\boldsymbol{x}, \boldsymbol{y}, \varepsilon)\,. \tag{8.23}$$

We set $\boldsymbol{h}(\boldsymbol{x}) = m(\boldsymbol{x}_2)^{-1} \boldsymbol{F}_{\rm ex}(\boldsymbol{x})$. Then for $\varepsilon = 0$ the critical manifold, $\mathcal{C}_0$, is given by

$$\mathcal{C}_0 = \{(\boldsymbol{x}, \boldsymbol{h}(\boldsymbol{x})) \,|\, \boldsymbol{x} \in \mathbb{R}^3 \times \mathbb{V}\} = \{(\boldsymbol{q}, \boldsymbol{v}, \dot{\boldsymbol{v}}) \,|\, m(\boldsymbol{v})\dot{\boldsymbol{v}} = \boldsymbol{F}_{\rm ex}(\boldsymbol{q}, \boldsymbol{v})\}, \quad (8.24)$$

which means that, for $\varepsilon = 0$, it is spanned by the solutions of the leading Hamiltonian part of Eq. (8.1). Linearizing at $\mathcal{C}_0$ the repelling eigenvalue is dominated by $\gamma^{-2}\kappa(\boldsymbol{x}_2)^{-1} m(\boldsymbol{x}_2)$ which tends to zero as $|\boldsymbol{x}_2| \to 1$. Therefore $\mathcal{C}_0$ is not uniformly hyperbolic, which is one of the standard assumptions of singular perturbation theory.

To overcome this difficulty we modify $\boldsymbol{g}$ to $\boldsymbol{g}_\delta$, $\delta$ small, which agrees with $\boldsymbol{g}$ on $\mathbb{R}^3 \times \{\boldsymbol{v} \,|\, |\boldsymbol{v}| \le 1 - \delta\} \times \mathbb{R}^3$ and which is constantly extended to values $|\boldsymbol{v}| \ge 1 - \delta$. Thus for $|\boldsymbol{x}_2(t)| \le 1 - \delta$ the solution to $\dot{\boldsymbol{x}} = \boldsymbol{f}$, $\varepsilon\dot{\boldsymbol{y}} = \boldsymbol{g}_\delta$ agrees with the solution to $\dot{\boldsymbol{x}} = \boldsymbol{f}$, $\varepsilon\dot{\boldsymbol{y}} = \boldsymbol{g}$. For sufficiently small $\varepsilon$ the modified equation then has a critical manifold $\mathcal{C}_\varepsilon$ with the properties discussed in the example of section 8.1. We only have to make sure that the modification is never seen by the solution. Thus, for the initial condition $|\boldsymbol{v}(0)| \le \bar{v}$, we have to find a $\delta = \delta(\bar{v})$ such that $|\boldsymbol{v}(t)| \le 1 - \delta$ for all times. To do so, one needs the energy balance (8.7).

We consider the modified evolution with vector field $(\boldsymbol{f}, \boldsymbol{g}_\delta)$ and choose the initial velocity such that $|\boldsymbol{v}(0)| \le \bar{v} < 1$. For $\varepsilon$ small enough this dynamics has a critical manifold of the form $\dot{\boldsymbol{v}} = \boldsymbol{h}_\varepsilon(\boldsymbol{q}, \boldsymbol{v})$ and $|\boldsymbol{h}_\varepsilon(\boldsymbol{q}, \boldsymbol{v})| \le c_1 = c_1(\delta)$. We start the dynamics on $\mathcal{C}_\varepsilon$. According to (8.8), for all $t \ge 0$,

$$\begin{aligned} G_\varepsilon\,(\boldsymbol{q}(t), \boldsymbol{v}(t), \boldsymbol{h}_\varepsilon(t)) \le G_\varepsilon(0) &= H(\boldsymbol{q}(0), \boldsymbol{v}(0)) - \varepsilon(e^2/6\pi)(\boldsymbol{v}(0) \cdot \boldsymbol{h}_\varepsilon(0)) \\ &\le E_{\rm s}(\bar{v}) + e\phi_{\rm ex}\,(\boldsymbol{q}(0)) + \varepsilon c_1\,. \end{aligned} \quad (8.25)$$

We now choose $\delta$ such that $\bar{v} \le 1 - 2\delta$. Since the initial conditions are on $\mathcal{C}_\varepsilon$, the solution will stay for a while on $\mathcal{C}_\varepsilon$ until the first time, $\tau$, when $|\boldsymbol{v}(\tau)| = 1 - \delta$ occurs. After that time the modification becomes visible. At time $\tau$ we have, using the lower bound on the energy and (8.25),

$$\begin{aligned} E_{\rm s}(\boldsymbol{v}(\tau)) + e\bar{\phi} \le H(\boldsymbol{q}(\tau), \boldsymbol{v}(\tau)) &= G_\varepsilon(\tau) + \varepsilon(e^2/6\pi)\,\gamma^4\,(\boldsymbol{v}(\tau) \cdot \boldsymbol{h}_\varepsilon(\tau)) \\ &\le E_{\rm s}(\bar{v}) + e\phi_{\rm ex}\,(\boldsymbol{q}(0)) + 2\varepsilon c_1 \end{aligned} \quad (8.26)$$

and therefore

$$E_{\rm s}(1 - \delta) \le E_{\rm s}(1 - 2\delta) + e\,(\phi_{\rm ex}\,(\boldsymbol{q}(0)) - \bar{\phi}) + 2\varepsilon c_1\,. \quad (8.27)$$

$E_{\rm s}(1 - \delta) \cong 1/\sqrt{\delta}$ for small $\delta$, which implies

$$\frac{1}{\sqrt{\delta}} \le c_2 + 4\,\varepsilon c_1 \quad (8.28)$$

with $c_2 = 2e\,(\phi_{\text{ex}}(\boldsymbol{q}(0)) - \bar{\phi})$. We now choose $\delta$ so small that $1/\sqrt{\delta} > c_2 + 1$ and then $\varepsilon$ so small that $4\varepsilon c_1 < 1$. Then (8.28) is a contradiction to the assumption that $|v(\tau)| = 1 - \delta$. We thus conclude that $\tau = \infty$ and the solution trajectory stays on $\mathcal{C}_\varepsilon$ for all times.

Equipped with this information we have for small $\varepsilon$ the critical manifold

$$\dot{v} = \boldsymbol{h}_\varepsilon\,(\boldsymbol{q}, v)\,. \tag{8.29}$$

On the critical manifold the Schott energy is bounded and from the argument leading to (8.10) we conclude that Dirac's asymptotic condition holds on $\mathcal{C}_\varepsilon$. On the other hand, slightly off $\mathcal{C}_\varepsilon$ the solution diverges with a rate of order $1/\varepsilon$. Therefore the asymptotic condition singles out, for given $\boldsymbol{q}(0)$, $v(0)$, the *unique* $\dot{v}(0)$ on $\mathcal{C}_\varepsilon$.

The motion on the critical manifold is governed by an effective equation which can be determined approximately following the scheme of section 8.1. We define

$$\boldsymbol{h}(\boldsymbol{q}, v) = m(v)^{-1}\, e\big(\boldsymbol{E}_{\text{ex}}(\boldsymbol{q}) + v \times \boldsymbol{B}_{\text{ex}}(\boldsymbol{q})\big)\,. \tag{8.30}$$

Then, up to errors of order $\varepsilon^2$,

$$\begin{aligned} m(v)\dot{v} = {} & e\,\big(\boldsymbol{E}_{\text{ex}}(\boldsymbol{q}) + v \times \boldsymbol{B}_{\text{ex}}(\boldsymbol{q})\big) \\ & + \varepsilon\,(e^2/6\pi)\,\big[\gamma^2 \kappa(v)\big((v \cdot \nabla_q)\boldsymbol{h} + (\boldsymbol{h} \cdot \nabla_v)\boldsymbol{h} + 3\gamma^2 (v \cdot \boldsymbol{h})\boldsymbol{h}\big)\big]\,. \end{aligned} \tag{8.31}$$

The physical solutions of (8.1), in the sense of satisfying the asymptotic condition, are governed by Eq. (8.31). Thus it, and *not* Eq. (8.1), must be regarded as the correct comparison dynamics to the true microscopic evolution equations (6.11). Note that the error accumulated in going from (8.1) to (8.31) is of the same order as the error made in the derivation of Eq. (8.1).

Because of the special structure of (8.1), on a formal level the final result (8.31) can be deduced without the help of geometric perturbation theory. We regard $m(v)\dot{v} = e\,(\boldsymbol{E}_{\text{ex}}(\boldsymbol{q}) + v \times \boldsymbol{B}_{\text{ex}}(\boldsymbol{q}))$ as the "unperturbed" equation and substitute for the terms inside the square bracket, which means replacing $\dot{v}$ by $\boldsymbol{h}$ and $\ddot{v}$ by $\dot{\boldsymbol{h}} = (v \cdot \nabla_q)\boldsymbol{h} + (\boldsymbol{h} \cdot \nabla_v)\boldsymbol{h}$. While yielding the correct answer, one misses the geometrical picture of the critical manifold and the associated motion in phase space.

## 8.3 Tracking of the true solution

From (6.11) we have the true solution $\boldsymbol{q}^\varepsilon(t)$, $v^\varepsilon(t)$ with initial conditions $\boldsymbol{q}^0$, $v^0$ and correspondingly adapted field data. We face the problem of how well this solution is tracked by the comparison dynamics (8.1) on its critical manifold. Let us first disregard the radiation reaction. From our a priori estimates we know that

$$\dot{\boldsymbol{q}}^\varepsilon = v^\varepsilon, \quad m(v^\varepsilon)\dot{v}^\varepsilon = e\,\big(\boldsymbol{E}_{\text{ex}}(\boldsymbol{q}^\varepsilon) + v^\varepsilon \times \boldsymbol{B}_{\text{ex}}(\boldsymbol{q}^\varepsilon)\big) + \mathcal{O}(\varepsilon) \tag{8.32}$$

which should be compared to

$$\dot{\boldsymbol{r}} = \boldsymbol{u}\,, \quad m(\boldsymbol{u})\dot{\boldsymbol{u}} = e\left(\boldsymbol{E}_{\mathrm{ex}}(\boldsymbol{r}) + \boldsymbol{u} \times \boldsymbol{B}_{\mathrm{ex}}(\boldsymbol{u})\right). \tag{8.33}$$

We switched to the variables $\boldsymbol{r}, \boldsymbol{u}$ instead of $\boldsymbol{q}, \boldsymbol{v}$ so as to distinguish more clearly between the true and comparison dynamics.

**Theorem 8.1** (Adiabatic limit, conservative tracking dynamics). *For the Abraham model satisfying the conditions (C), (P), and (I) let $|e| \leq \bar{e}$ and $\varepsilon \leq \varepsilon_0$ be sufficiently small. Let $\boldsymbol{r}(t), \boldsymbol{u}(t)$ be the solution to the comparison dynamics (8.33) with initial conditions $\boldsymbol{r}(0) = \boldsymbol{q}^0$, $\boldsymbol{u}(0) = \boldsymbol{v}^0$. Then for every $\tau > 0$ there exist constants $c(\tau)$ such that*

$$|\boldsymbol{q}^{\varepsilon}(t) - \boldsymbol{r}(t)| \leq c(\tau)\varepsilon\,, \quad |\boldsymbol{v}^{\varepsilon}(t) - \boldsymbol{u}(t)| \leq c(\tau)\varepsilon \tag{8.34}$$

*for $0 \leq t \leq \tau$.*

*Proof*: Let $\delta(t) = |\boldsymbol{q}^{\varepsilon}(t) - \boldsymbol{r}(t)| + |\boldsymbol{v}^{\varepsilon}(t) - \boldsymbol{u}(t)|$. Converting the differential equations (8.32), (8.33) into their integral form, one obtains

$$\begin{aligned}\delta(t) &\leq \delta(0) + C\int_0^t \mathrm{d}s\,\delta(s) + \varepsilon\int_0^t \mathrm{d}s\,C\left(1 + \varepsilon(\varepsilon + s)^{-2}\right)\\ &\leq \delta(0) + \varepsilon C(t+1) + C\int_0^t \mathrm{d}s\,\delta(s)\,.\end{aligned} \tag{8.35}$$

Since $\delta(0) = 0$ by assumption, Gronwall's lemma yields the bound $\delta(t) \leq \varepsilon C e^{Ct}$. □

Theorem 8.1 states that, up to an error of order $\varepsilon$, the true solution is well approximated by the Hamiltonian dynamics (8.33).

In the next order the comparison dynamics reads

$$\begin{aligned}\dot{\boldsymbol{r}} &= \boldsymbol{u}\,,\\ m(\boldsymbol{u})\dot{\boldsymbol{u}} &= e\left(\boldsymbol{E}_{\mathrm{ex}}(\boldsymbol{r}) + \boldsymbol{u} \times \boldsymbol{B}_{\mathrm{ex}}(\boldsymbol{r})\right)\\ &\qquad + \varepsilon(e^2/6\pi)\left[\gamma^4\,(\boldsymbol{u}\cdot\ddot{\boldsymbol{u}})\boldsymbol{u} + 3\gamma^6(\boldsymbol{u}\cdot\dot{\boldsymbol{u}})^2\,\boldsymbol{u} + 3\gamma^4(\boldsymbol{u}\cdot\dot{\boldsymbol{u}})\dot{\boldsymbol{u}} + \gamma^2\ddot{\boldsymbol{u}}\right]\end{aligned} \tag{8.36}$$

restricted to its critical manifold $\mathcal{C}_{\varepsilon}$. Since the radiation reaction is proportional to $\varepsilon$, the solution $\boldsymbol{r}(t), \boldsymbol{u}(t)$ depends now on $\varepsilon$, a dependence which is suppressed in our notation. Naively one would expect that improving the equation by a term of order $\varepsilon$ increases the precision to order $\varepsilon^2$, i.e.

$$|\boldsymbol{q}^{\varepsilon}(t) - \boldsymbol{r}(t)| + |\boldsymbol{v}^{\varepsilon}(t) - \boldsymbol{u}(t)| = \mathcal{O}(\varepsilon^2)\,. \tag{8.37}$$

An alternative option to keeping track of the $\varepsilon$-correction is to consider longer times, of the order $\varepsilon^{-1}\tau$ on the macroscopic time scale. Then the radiative effects add up to deviations of order one from the Hamiltonian trajectory. Thus

$$|\boldsymbol{q}^{\varepsilon}(t)-\boldsymbol{r}(t)|=\mathcal{O}(\varepsilon) \quad \text{for} \quad 0\le t\le\varepsilon^{-1}\tau\,. \tag{8.38}$$

One should be somewhat careful here. In a scattering situation the charged particle reaches the force-free region after a finite macroscopic time. According to (8.37) the error in the velocity is then $\mathcal{O}(\varepsilon^2)$, which builds up an error in the position of order $\varepsilon$ over a time span $\varepsilon^{-1}\tau$. Thus we cannot hope to do better than (8.38). On the other hand, when the motion remains bounded, as e.g. in a uniform external magnetic field, the charge comes to rest at some point $\boldsymbol{q}^*$ in the long-time limit and the rest point $\boldsymbol{q}^*$ is the same for the true and the comparison dynamics. At least, for an external electrostatic potential with a discrete set of critical points we have already established such behavior and presumably it holds in general. Thus for small $\varepsilon$ we have $\boldsymbol{q}^{\varepsilon}(\varepsilon^{-1}\tau)\cong\boldsymbol{q}^*$ and also $\boldsymbol{r}^{\varepsilon}(\varepsilon^{-1}\tau)\cong\boldsymbol{q}^*$. Therefore, in the case of bounded motion, we conjecture that (8.38) holds for *all* times.

**Conjecture 8.2** (Adiabatic limit including friction). *For the Abraham model satisfying (C), (P), and (I) let $\boldsymbol{q}(t)$ be bounded, i.e. $|\boldsymbol{q}(t)|\le C$ for all $t\ge 0$, and $\varepsilon\le\varepsilon_0$. Then there exists $(\boldsymbol{r}(0)\,,\boldsymbol{u}(0),\dot{\boldsymbol{u}}(0))\in\mathcal{C}_\varepsilon$ such that*

$$\sup_{t\ge 0}|\boldsymbol{q}^{\varepsilon}(t)-\boldsymbol{r}(t)|=\mathcal{O}(\varepsilon)\,, \tag{8.39}$$

*where $\boldsymbol{r}(t)$ is the solution to (8.36) with the initial conditions given before.*

In a more descriptive mode, the true solution $\boldsymbol{q}^{\varepsilon}(t)$ is $\varepsilon$-shadowed for all times by one solution (and thus by many solutions) of the comparison dynamics.

At present we are far from such strong results. The problem is that an error of order $\varepsilon^2$ in (8.36) is generically amplified as $\varepsilon^2 e^{t/\varepsilon}$. Although such an increase violates the a priori bounds, it renders a proof of (8.39) difficult. We seem to be back to (8.34) which carries no information on the radiation reaction. Luckily the radiation correction in (8.36) can be seen in the energy balance.

**Theorem 8.3** (Adiabatic limit including friction). *Under the assumptions of Theorem 8.1 one has*

$$\Big|[E_{\mathrm{s}}(\boldsymbol{v}^{\varepsilon}(t))+e\,\phi_{\mathrm{ex}}(\boldsymbol{q}^{\varepsilon}(t))]-[E_{\mathrm{s}}(\boldsymbol{u}(t))+e\phi_{\mathrm{ex}}(\boldsymbol{r}(t))]\Big|\le Cc(\tau)\varepsilon^2 \tag{8.40}$$

*for $t_\varepsilon\le t\le\tau$. Here $(\boldsymbol{r}(t),\boldsymbol{u}(t))$ is the solution to (8.36) with initial data $\boldsymbol{r}(t_\varepsilon)=\boldsymbol{q}^{\varepsilon}(t_\varepsilon)$, $\boldsymbol{u}(t_\varepsilon)=\boldsymbol{v}^{\varepsilon}(t_\varepsilon)$, $\dot{\boldsymbol{u}}^{\varepsilon}(t_\varepsilon)=h_\varepsilon(\boldsymbol{q}^{\varepsilon}(t_\varepsilon),\boldsymbol{v}^{\varepsilon}(t_\varepsilon))$ and $t_\varepsilon=\varepsilon^{1/3}$.*

To achieve a precision of order $\varepsilon^2$, the initial slip in (7.15) does not allow one to match the true and comparison dynamics at $t=0$. One needs $|\dddot{\boldsymbol{q}}^{\varepsilon}(t)|$ uniformly bounded, which is ensured only for $t \geq C\varepsilon^{1/3}$, i.e. $t \geq t_\varepsilon$ with the arbitrary choice $C = 1$.

*Proof*: Let us use the estimate (7.22) on the self-force and denote the error term by $\boldsymbol{f}^\varepsilon(t)$. Then $|\boldsymbol{f}^\varepsilon(t)| \leq C\varepsilon^2$ for $t_\varepsilon \leq t$. As in (8.7),

$$\frac{\mathrm{d}}{\mathrm{d}t} G_\varepsilon(\boldsymbol{q}^\varepsilon, \boldsymbol{v}^\varepsilon, \dot{\boldsymbol{v}}^\varepsilon) = \boldsymbol{f}^\varepsilon(t) \cdot \boldsymbol{v}^\varepsilon - \varepsilon\,(e^2/6\pi)\big[\gamma^4(\dot{\boldsymbol{v}}^\varepsilon)^2 + \gamma^6(\boldsymbol{v}^\varepsilon \cdot \dot{\boldsymbol{v}}^\varepsilon)^2\big] \tag{8.41}$$

and therefore

$$\begin{aligned} |H(\boldsymbol{q}^\varepsilon, \boldsymbol{v}^\varepsilon) - H(\boldsymbol{r}, \boldsymbol{u})| \leq{}& \varepsilon\,(e^2/6\pi)|\gamma(\boldsymbol{v}^\varepsilon)^4(\boldsymbol{v}^\varepsilon \cdot \dot{\boldsymbol{v}}^\varepsilon) - \gamma(\boldsymbol{u})^4(\boldsymbol{u} \cdot \dot{\boldsymbol{u}})| \\ &+ \int_{t_\varepsilon}^{t} \mathrm{d}s\big(|\boldsymbol{f}^\varepsilon \cdot \boldsymbol{v}^\varepsilon| + \varepsilon(e^2/6\pi)|\gamma(\boldsymbol{v}^\varepsilon)^4(\dot{\boldsymbol{v}}^\varepsilon)^2 \\ &+ \gamma(\boldsymbol{v}^\varepsilon)^6(\boldsymbol{v}^\varepsilon \cdot \dot{\boldsymbol{v}}^\varepsilon)^2 - \gamma(\boldsymbol{u})^4(\dot{\boldsymbol{u}})^2 - \gamma(\boldsymbol{u})^6(\boldsymbol{u} \cdot \dot{\boldsymbol{u}})^2|\big)\,. \end{aligned} \tag{8.42}$$

Since $|\boldsymbol{v}^\varepsilon|$, $|\boldsymbol{u}|$ remain bounded away from 1, the $\gamma$-factors are uniformly bounded, and it suffices to estimate the difference on the Hamiltonian level of precision. From Theorem 8.1 one has the bound $|\boldsymbol{v}^\varepsilon(t) - \boldsymbol{u}(t)| \leq c(\tau)\varepsilon$. Inserting (8.34) into (8.32) and (8.33), we obtain the same bound for the first derivative, $|\dot{\boldsymbol{v}}^\varepsilon(t) - \dot{\boldsymbol{u}}(t)| \leq c(\tau)\varepsilon$. Moreover $\int_{t_\varepsilon}^{t} \mathrm{d}s\,|\boldsymbol{f}^\varepsilon(s)| \leq Ct\varepsilon^2$. Working out the differences in (8.42), one concludes

$$|H(\boldsymbol{q}^\varepsilon(t), \boldsymbol{v}^\varepsilon(t)) - H(\boldsymbol{r}(t), \boldsymbol{u}(t))| \leq C(t + c(t))\varepsilon^2\,, \tag{8.43}$$

as claimed. □

## 8.4 Electromagnetic fields in the adiabatic limit

So far we have concentrated on the Lorentz force with retarded fields and have obtained approximate evolution equations for the charged particle. Such an approximate solution can be reinserted into the inhomogeneous Maxwell–Lorentz equations in order to obtain the electromagnetic fields in the adiabatic limit.

As before, let $(\boldsymbol{q}^\varepsilon(t), \boldsymbol{v}^\varepsilon(t))$, $t \geq 0$, be the true solution. We extend it to $\boldsymbol{q}^\varepsilon(t) = \boldsymbol{q}^0 + \boldsymbol{v}^0 t$, $\boldsymbol{v}^\varepsilon(t) = \boldsymbol{v}^0$ for $t \leq 0$. According to (4.31), (4.32) and using the scaled fields as in (6.8), one has

$$\frac{1}{\sqrt{\varepsilon}}\boldsymbol{E}(t) = -\int_{-\infty}^{t} \mathrm{d}s\big(\nabla G_{t-s} * \rho_\varepsilon(s) + \partial_t G_{t-s} * \boldsymbol{j}_\varepsilon(s)\big) \tag{8.44}$$

with $\rho_\varepsilon(\boldsymbol{x}, t) = e\varphi_\varepsilon(\boldsymbol{x} - \boldsymbol{q}^\varepsilon(t))$, $\boldsymbol{j}_\varepsilon(\boldsymbol{x}, t) = e\varphi_\varepsilon(\boldsymbol{x} - \boldsymbol{q}^\varepsilon(t))\boldsymbol{v}^\varepsilon(t)$. Inserting from (2.15) and by partial integration,

$$
\begin{aligned}
\frac{1}{\sqrt{\varepsilon}}\boldsymbol{E}(\boldsymbol{x}, t) = &- \int_{-\infty}^{t} \mathrm{d}s \int \mathrm{d}^3 y \frac{1}{4\pi(t-s)} \delta(|\boldsymbol{x} - \boldsymbol{y}| - (t-s)) \nabla \rho_\varepsilon(\boldsymbol{y}, s) \\
&- \int_{-\infty}^{t} \mathrm{d}s \int \mathrm{d}^3 y \frac{1}{4\pi(t-s)^2} \delta(|\boldsymbol{x} - \boldsymbol{y}| - (t-s)) \\
&\times [(\boldsymbol{y} - \boldsymbol{x}) \cdot \nabla \boldsymbol{j}_\varepsilon(\boldsymbol{y}, s) + \boldsymbol{j}_\varepsilon(\boldsymbol{y}, s)] \\
= &- e \int \mathrm{d}^3 y \Big( \frac{1}{4\pi|\boldsymbol{x} - \boldsymbol{y}|} \nabla \varphi_\varepsilon(\boldsymbol{y} - \boldsymbol{q}^\varepsilon(t - |\boldsymbol{x} - \boldsymbol{y}|)) \boldsymbol{v}^\varepsilon(t - |\boldsymbol{x} - \boldsymbol{y}|) \\
&+ \frac{1}{4\pi|\boldsymbol{x} - \boldsymbol{y}|^2} \boldsymbol{v}^\varepsilon(t - |\boldsymbol{x} - \boldsymbol{y}|)(1 + (\boldsymbol{y} - \boldsymbol{x}) \cdot \nabla) \\
&\varphi_\varepsilon(\boldsymbol{y} - \boldsymbol{q}^\varepsilon(t - |\boldsymbol{x} - \boldsymbol{y}|)) \Big) . \qquad (8.45)
\end{aligned}
$$

In the same fashion

$$
\frac{1}{\sqrt{\varepsilon}}\boldsymbol{B}(\boldsymbol{x}, t) = -e \int \mathrm{d}^3 y \frac{1}{4\pi|\boldsymbol{x} - \boldsymbol{y}|} \boldsymbol{v}^\varepsilon(t - |\boldsymbol{x} - \boldsymbol{y}|) \times \nabla \varphi_\varepsilon(\boldsymbol{y} - \boldsymbol{q}^\varepsilon(t - |\boldsymbol{x} - \boldsymbol{y}|)) . \qquad (8.46)
$$

In the limit $\varepsilon \to 0$ one has $\varphi_\varepsilon(\boldsymbol{x}) \to \delta(\boldsymbol{x})$ and, by Theorem 8.1, $\boldsymbol{q}^\varepsilon(t) \to \boldsymbol{r}(t)$, $\boldsymbol{v}^\varepsilon(t) \to \boldsymbol{u}(t)$, where $\boldsymbol{r}(t) = \boldsymbol{q}^0 + \boldsymbol{v}^0 t$, $\boldsymbol{u}(t) = \boldsymbol{v}^0$ for $t \le 0$. We substitute $\boldsymbol{y}' = \boldsymbol{y} - \boldsymbol{q}^\varepsilon(t - |\boldsymbol{x} - \boldsymbol{y}|)$ with volume element $\det(\mathrm{d}y/\mathrm{d}y') = [1 - \boldsymbol{v}^\varepsilon(t - |\boldsymbol{x} - \boldsymbol{y}|) \cdot (\boldsymbol{x} - \boldsymbol{y})/|\boldsymbol{x} - \boldsymbol{y}|]^{-1}$. Then $\delta(\boldsymbol{y}')$ leads to the constraint $0 = \boldsymbol{y} - \boldsymbol{r}(t - |\boldsymbol{x} - \boldsymbol{y}|)$ which has the unique solution $\boldsymbol{y} = \boldsymbol{r}(t_{\mathrm{ret}})$; compare with (2.22). In particular the volume element $\det(\mathrm{d}y/\mathrm{d}y')$ becomes $[1 - \widehat{\boldsymbol{n}} \cdot \boldsymbol{u}(t_{\mathrm{ret}})]^{-1}$ in the limit, with $\widehat{\boldsymbol{n}} = \widehat{\boldsymbol{n}}(\boldsymbol{x}, t) = (\boldsymbol{x} - \boldsymbol{r}(t_{\mathrm{ret}}))/|\boldsymbol{x} - \boldsymbol{r}(t_{\mathrm{ret}})|$.

We conclude that

$$
\lim_{\varepsilon \to 0} \frac{1}{\sqrt{\varepsilon}} \boldsymbol{E}(\boldsymbol{x}, t) = \bar{\boldsymbol{E}}(\boldsymbol{x}, t) , \quad \lim_{\varepsilon \to 0} \frac{1}{\sqrt{\varepsilon}} \boldsymbol{B}(\boldsymbol{x}, t) = \bar{\boldsymbol{B}}(\boldsymbol{x}, t) , \qquad (8.47)
$$

where $\bar{\boldsymbol{E}}, \bar{\boldsymbol{B}}$ are the Liénard–Wiechert fields (2.24), (2.25) generated by a point charge moving along the trajectory $t \mapsto \boldsymbol{r}(t)$. The convergence in (8.47) is pointwise, except for the Coulomb singularity at $\boldsymbol{x} = \boldsymbol{r}(t)$.

## 8.5 Larmor's formula

We want to determine the energy per unit time radiated to infinity and consider, for this purpose, a ball of radius $R$ centered at $\boldsymbol{q}^\varepsilon(t)$. At time $t + R$ the energy in this ball is

$$\mathcal{E}_{R,q^\varepsilon(t)}(t+R) = \mathcal{E}(0) - \frac{1}{2}\int_{\{|x-q^\varepsilon(t)|\geq R\}} \mathrm{d}^3x \left(E(x,t+R)^2 + B(x,t+R)^2\right) \tag{8.48}$$

using conservation of total energy. The radiation emitted from the charge at time $t$ reaches the surface of the ball at time $t+R$, and the energy loss per unit time is given by

$$\begin{aligned}
I_{R,\varepsilon}(t) &= \frac{\mathrm{d}}{\mathrm{d}t}\mathcal{E}_{R,q^\varepsilon(t)} \\
&= \int \mathrm{d}^3x\, \delta(|x - q^\varepsilon(t)| - R)\left(\frac{1}{2}(n(x)\cdot v^\varepsilon(t))\left(E(x,t+R)^2\right.\right. \\
&\quad \left.\left. + B(x,t+R)^2\right) + E(x,t+R)\cdot\left(n(x)\times B(x,t+R)\right)\right) \\
&= \frac{1}{2}R^2\int \mathrm{d}^2\omega\left((\widehat{\omega}\cdot v^\varepsilon(t))\left(E(q^\varepsilon(t)+R\widehat{\omega},t+R)^2\right.\right. \\
&\quad \left. + B(q^\varepsilon(t)+R\widehat{\omega},t+R)^2\right) + 2E(q^\varepsilon(t)+R\widehat{\omega},t+R) \\
&\quad \left.\cdot\left(\widehat{\omega}\times B(q^\varepsilon(t)+R\widehat{\omega},t+R)\right)\right),
\end{aligned} \tag{8.49}$$

where $n(x)$ is the outer normal of the ball and $|\widehat{\omega}| = 1$, with $\mathrm{d}^2\omega$ the integration over the unit sphere. Equation (8.49) holds for sufficiently large $R$, since we used $\{x|\ |x - q^\varepsilon(t)| \geq R\} \cap \{x|\ |x - q^\varepsilon(t+R)| \leq \varepsilon R_\varphi\} = \emptyset$, which is the case for $(1-\bar{v})R \geq \varepsilon R_\varphi$.

Equation (8.49) still contains the reversible energy transport between the considered ball and its complement. To isolate that part of the energy which is irreversibly lost one has to take the limit $R \to \infty$. For this purpose we first partially integrate in (8.45), (8.46) by using the identity

$$\nabla\varphi = \nabla_y\varphi - \frac{y-x}{|y-x|}\left(1 + \frac{(y-x)\cdot v^\varepsilon}{|y-x|}\right)^{-1}(v^\varepsilon\cdot\nabla_y)\varphi \tag{8.50}$$

at the argument $y - q^\varepsilon(t - |y-x|)$. For large $R$ the fields in (8.49) then become

$$\begin{aligned}
RE(q^\varepsilon(t)+R\widehat{\omega},t+R) &\cong \sqrt{\varepsilon}\int \mathrm{d}^3y \frac{e}{4\pi}\varphi_\varepsilon(y-q^\varepsilon)\left[-(1-\widehat{\omega}\cdot v^\varepsilon)^{-1}\dot{v}^\varepsilon\right. \\
&\quad \left.-(1-\widehat{\omega}\cdot v^\varepsilon)^{-2}(\widehat{\omega}\cdot\dot{v}^\varepsilon)(v^\varepsilon-\widehat{\omega})\right]\Big|_{t+\widehat{\omega}\cdot(y-q^\varepsilon(t))},
\end{aligned} \tag{8.51}$$

$$
\begin{aligned}
RB(q^\varepsilon(t)+R\widehat{\omega},t+R) &\cong \sqrt{\varepsilon}\int \mathrm{d}^3y\frac{e}{4\pi}\varphi_\varepsilon(y-q^\varepsilon)\big[-(1-\widehat{\omega}\cdot v^\varepsilon)^{-1}(\widehat{\omega}\times\dot{v}^\varepsilon) \\
&\quad -(1-\widehat{\omega}\cdot v^\varepsilon)^{-2}(\widehat{\omega}\cdot\dot{v}^\varepsilon)(\widehat{\omega}\times v^\varepsilon)\big]\big|_{t+\widehat{\omega}\cdot(y-q^\varepsilon(t))} \\
&= \widehat{\omega}\times RE(q^\varepsilon(t)+R\widehat{\omega},t+R)\,,
\end{aligned}
\tag{8.52}
$$

where we used the property that $t+R-|q^\varepsilon(t)+R\widehat{\omega}-y| = t+\widehat{\omega}\cdot(y-q^\varepsilon(t))+\mathcal{O}(1/R)$ for large $R$. Inserting in (8.49) yields

$$
\begin{aligned}
\lim_{R\to\infty} I_{R,\varepsilon}(t) &= I_\varepsilon(t) \\
&= -\lim_{R\to\infty}\int \mathrm{d}^2\omega(1-\widehat{\omega}\cdot v^\varepsilon(t))\big(RE(q^\varepsilon(t)+R\widehat{\omega},t+R)\big)^2 \qquad (8.53)\\
&= -\varepsilon\int \mathrm{d}^2\omega(1-\widehat{\omega}\cdot v^\varepsilon(t)) \\
&\quad\times\bigg(\Big[\frac{e}{4\pi}\int \mathrm{d}^3y\varphi_\varepsilon(y-q^\varepsilon)(1-\widehat{\omega}\cdot v^\varepsilon)^{-2}(\widehat{\omega}\cdot\dot{v}^\varepsilon)\Big]^2 \\
&\quad -\Big[\frac{e}{4\pi}\int \mathrm{d}^3y\varphi_\varepsilon(y-q^\varepsilon)(1-\widehat{\omega}\cdot v^\varepsilon)^{-1}\dot{v}^\varepsilon \\
&\quad +(1-\widehat{\omega}\cdot v^\varepsilon)^{-2}(\widehat{\omega}\cdot\dot{v}^\varepsilon)v^\varepsilon\Big]^2\bigg)\bigg|_{t+\widehat{\omega}\cdot(y-q^\varepsilon(t))}. \qquad (8.54)
\end{aligned}
$$

$I_\varepsilon(t)$ is the energy radiated per unit time at $\varepsilon$ fixed. As argued before, it is indeed of order $\varepsilon$. From the expression (8.53) it can be seen that $I_\varepsilon(t)\le 0$.

Equation (8.54) is not yet Larmor's formula. To obtain it we have to go to the adiabatic limit $\varepsilon\to 0$. Then $q^\varepsilon(t)\to r(t)$. Since $\varphi_\varepsilon(x)\to\delta(x)$, we have $y\cong q^\varepsilon(t)\cong r(t)$ in (8.54). From the $\mathrm{d}^3y$ volume element we get an additional factor of $(1-\widehat{\omega}\cdot v^\varepsilon)^{-1}$. Thus

$$
\begin{aligned}
\lim_{\varepsilon\to 0}\varepsilon^{-1}I_\varepsilon(t) = I(t) &= -e^2\int \mathrm{d}^2\omega(1-\widehat{\omega}\cdot u(t))\big(4\pi(1-\widehat{\omega}\cdot u(t))^{-3}\big)^2 \\
&\quad\times\big((\widehat{\omega}\cdot\dot{u}(t))^2-[(1-\widehat{\omega}\cdot u(t))\dot{u}(t)+(\widehat{\omega}\cdot\dot{u}(t))u(t)]^2\big) \\
&= -(e^2/6\pi)\big[\gamma^4\dot{u}(t)^2+\gamma^6(u(t)\cdot\dot{u}(t))^2\big] \\
&= -(e^2/6\pi)\gamma^6\big[\dot{u}(t)^2-(u(t)\times\dot{u}(t))^2\big],
\end{aligned}
\tag{8.55}
$$

which is the standard textbook formula of Larmor. Note that the same energy loss per unit time was obtained already in (8.6) using only the energy balance for the comparison dynamics.

Starting from (8.49) one could alternatively first take the limit $\varepsilon^{-1} I_{R,\varepsilon}(t) \to I_{R,0}(t)$, which is the change of energy in a ball of radius $R$ centered at the particle's position $\boldsymbol{r}(t)$ in the adiabatic limit. As before the irreversible energy loss is isolated through

$$\lim_{R\to\infty} I_{R,0}(t) = I(t)\,. \tag{8.56}$$

The energy loss does not depend on the order of limits, as it should be.

We recall that in Larmor's treatment the trajectory of the charge, taken as a point charge, is prescribed. In our case the charged particle is guided by external fields and interacts with its own Maxwell field, which is physically somewhat more realistic. Since the charge distribution is extended, by necessity, Larmor's formula holds only in the adiabatic approximation.

## Notes and references

### *Section 8*

The radiation damped harmonic oscillator is discussed in Jackson (1999) with a variety of physical applications. The asymptotic condition was first stated in Dirac (1938). It has been reemphasized by Haag (1955) in analogy to a similar condition in quantum field theory.

### *Section 8.1*

Singular, or geometric, perturbation theory is a standard tool in the theory of dynamical systems. Sakamoto (1990) presents the theory at the level of generality needed here. We refer to Jones (1995) for a review with many applications. In the context of synergetics (Haken 1983) one talks of slow and fast variables and the slaving principle, which means that fast variables are enslaved by the slow ones. Within our context this would correspond to an attractive critical manifold. The renormalization group flows in critical phenomena have a structure similar to that discovered here: the critical surface corresponds to critical couplings which then flow to some fixed point governing the universal critical behavior. The critical surface is repelling, and slightly away from that surface the trajectory moves towards either the high- or low-temperature fixed points.

### *Section 8.2*

Particular cases have been studied before, most extensively the one-dimensional potential step of finite width and with linear interpolation (Haag 1955; Baylis and

Huschilt 1976; Carati and Galgani 1993; Carati *et al.* 1995; Blanco 1995; Ruf and Srikanth 2000), head-on collision in the two-body problem (Huschilt and Baylis 1976), the motion in a uniform magnetic field (Endres 1993), and motion in an attractive Coulomb potential (Marino 2003). These authors emphasize that there can be several solutions to the asymptotic condition. From the point of view of singular perturbation theory such behavior is generic. If $\varepsilon$ is increased, then the critical manifold is strongly deformed and is no longer given as a graph of a function. For specified $\boldsymbol{q}(0)$, $\boldsymbol{v}(0)$ there are then several $\dot{\boldsymbol{v}}(0)$ on $\mathcal{C}_\varepsilon$, which means that the solution to the asymptotic condition is not unique. However, these authors fail to emphasize that the nonuniqueness in the examples occurs only at such high field strengths that a classical theory has long lost its empirical validity. At moderate field strengths the worked-out examples confirm our findings. The general applicability of singular perturbation theory was first recognized in Spohn (1998).

### *Sections 8.3, 8.4, and 8.5*

The discussion is adapted from Kunze and Spohn (2000a, 2000b).

# 9

# The Lorentz–Dirac equation

We return to the Lorentz model and add slowly varying external potentials. On a formal level one can carry out the expansion in $\varepsilon$ just as for the Abraham model. The net result is that the rotational degrees of freedom decouple from the translational degrees of freedom, and the latter are governed by

$$m_0 \dot{\mathbf{u}} = (e/c)\mathbf{F} \cdot \mathbf{u} + (e^2/6\pi c^3)(\ddot{\mathbf{u}} - c^{-2}(\dot{\mathbf{u}} \cdot \dot{\mathbf{u}})\mathbf{u}), \tag{9.1}$$

which includes radiation reaction. Equation (9.1) is the Lorentz–Dirac equation, written in microscopic units. $m_0$ is the experimental rest mass of the particle. We reintroduced the speed of light, $c$. $\mathbf{F}$ is the electromagnetic field tensor of the *external* fields, where for better readability we omit the subscript "ex" in this section. The scaling parameter $\varepsilon$ has been reabsorbed into the definition of $\mathbf{F}$, which amounts to setting $\varepsilon = 1$. It should be kept in mind that the radiation reaction is a small correction to the Hamiltonian part.

If one fixes an inertial frame of reference and goes over to three-vectors, then the time component of the Lorentz–Dirac equation reads

$$\frac{\mathrm{d}}{\mathrm{d}t}\big(m_0 c^2 \gamma(\boldsymbol{v}) + e\phi(\boldsymbol{q}) - (e^2/6\pi c^3)\,\gamma^4(\boldsymbol{v}\cdot\dot{\boldsymbol{v}})\big) = -(e^2/6\pi c^3)\,\gamma^4(\dot{\boldsymbol{v}}\cdot\kappa(\boldsymbol{v})\dot{\boldsymbol{v}}), \tag{9.2}$$

and the space part becomes

$$\begin{aligned} m_0\gamma\,\kappa(\boldsymbol{v})\dot{\boldsymbol{v}} &= e(\boldsymbol{E}(\boldsymbol{q}) + c^{-1}\boldsymbol{v}\times\boldsymbol{B}(\boldsymbol{q})) \\ &\quad + (e^2/6\pi c^3)\,\gamma^2\kappa(\boldsymbol{v})\,[\ddot{\boldsymbol{v}} + 3\gamma^2 c^{-2}\,(\boldsymbol{v}\cdot\dot{\boldsymbol{v}})\,\dot{\boldsymbol{v}}], \end{aligned} \tag{9.3}$$

where as before $\kappa(\boldsymbol{v}) = \mathbb{1} + c^{-2}\,\gamma^2\,\boldsymbol{v}\otimes\boldsymbol{v}$. Equation (9.3) differs from its semirelativistic sister (8.1) only through the proper relativistic kinetic energy. Equation (9.2) is identical to the energy balance (8.6), again with proper adjustment of the kinetic energy. Thus we can follow the blueprint of section 8.2 to establish the existence of the critical manifold and to derive an effective second-order equation for the motion on the critical manifold.

The Lorentz–Dirac equation makes definite predictions about the orbit of a charged particle, including the effects of radiation losses, and one would expect that these predictions can be verified experimentally. Of course, if radiation damping is neglected, there is a multitude of laboratory set-ups. The real challenge is to observe quantitatively the minute changes in the Hamiltonian orbit due to radiation losses. We will discuss two proposals in section 9.3. The first one is the motion of an electron in a Penning trap. In the quadratic approximation for the quadrupole field, this problem can still be handled analytically, which is done in section 9.2 along with a few other examples of independent interest. The second proposal is the motion of an electron when hit by an ultrastrong laser pulse. In this case the external potentials are time dependent and one has to rely on a numerical integration of the effective second-order equation.

## 9.1 Critical manifold, the Landau–Lifshitz equation

We write (9.3) in the standard form of singular perturbation theory; compare with section 8.2. Then

$$\dot{\boldsymbol{x}} = \boldsymbol{f}(\boldsymbol{x}, \boldsymbol{y}), \quad \varepsilon \dot{\boldsymbol{y}} = \boldsymbol{g}(\boldsymbol{x}, \boldsymbol{y}, \varepsilon) \tag{9.4}$$

with

$$\boldsymbol{f}(\boldsymbol{x}, \boldsymbol{y}) = (\boldsymbol{x}_2, \boldsymbol{y}), \tag{9.5}$$

$$\begin{aligned} \boldsymbol{g}(\boldsymbol{x}, \boldsymbol{y}, \varepsilon) = {}& (6\pi c^3/e^2)\big(m_0\gamma^{-1}\boldsymbol{y} - e\gamma^{-2}\kappa(\boldsymbol{x}_2)^{-1}(\boldsymbol{E}(\boldsymbol{x}_1) + c^{-1}\boldsymbol{x}_2 \times \boldsymbol{B}(\boldsymbol{x}_1))\big) \\ & - 3\varepsilon\gamma^2 c^{-2}(\boldsymbol{x}_2 \cdot \boldsymbol{y})\boldsymbol{y}. \end{aligned} \tag{9.6}$$

To conform with (8.1) we reintroduced the small parameter $\varepsilon$. At zeroth order the critical manifold is $\{(\boldsymbol{x}, \boldsymbol{h}(\boldsymbol{x}))|\boldsymbol{x} \in \mathbb{R}^3 \times \mathbb{V}\}$ with $\boldsymbol{h}(\boldsymbol{q}, \boldsymbol{v}) = (e/m_0)\gamma^{-1}\kappa(\boldsymbol{v})^{-1}\big(\boldsymbol{E}(\boldsymbol{q}) + c^{-1}\boldsymbol{v} \times \boldsymbol{B}(\boldsymbol{q})\big)$. Linearizing (9.5), (9.6) at $\boldsymbol{y} = \boldsymbol{h}(\boldsymbol{x})$ the repelling eigenvalue is $(6\pi c^3/e^2)\, m_0\gamma^{-1} + \mathcal{O}(\varepsilon)$, which vanishes as $|\boldsymbol{v}|/c \to 1$. Thus we have to rely on the construction of section 8.2, which ensures that for given maximal velocity $\bar{v}$ one can choose $\varepsilon$ small enough such that the orbit remains on the critical manifold for all times.

To order $\varepsilon$ the effective second-order equation is given by (8.31), except that now $m(\boldsymbol{v}) = m_0\gamma\kappa(\boldsymbol{v})$. We work out the various terms and switch back to microscopic units. Then the motion on the critical manifold of the Lorentz–Dirac equation is governed by

$$\begin{aligned}
\dot{\boldsymbol{q}} &= \boldsymbol{v},\\
m_0\gamma\,\kappa(\boldsymbol{v})\dot{\boldsymbol{v}} &= e(\boldsymbol{E}+c^{-1}\boldsymbol{v}\times\boldsymbol{B}) + \frac{e^2}{6\pi c^3}\Big[\frac{e}{m_0}\gamma\,(\boldsymbol{v}\cdot\nabla)(\boldsymbol{E}+c^{-1}\boldsymbol{v}\times\boldsymbol{B})\\
&\quad +\Big(\frac{e}{m_0}\Big)^2 c^{-1}\Big((\boldsymbol{E}\times\boldsymbol{B})+c^{-1}(\boldsymbol{v}\cdot\boldsymbol{E})\boldsymbol{E}+c^{-1}(\boldsymbol{v}\cdot\boldsymbol{B})\boldsymbol{B}\\
&\quad +\big(-\boldsymbol{E}^2-\boldsymbol{B}^2+c^{-2}(\boldsymbol{v}\cdot\boldsymbol{E})^2+c^{-2}(\boldsymbol{v}\cdot\boldsymbol{B})^2\\
&\quad +2c^{-1}\boldsymbol{v}\cdot(\boldsymbol{E}\times\boldsymbol{B})\big)\gamma^2c^{-1}\boldsymbol{v}\Big)\Big]. \qquad (9.7)
\end{aligned}$$

While singular perturbation theory provides a systematic method, Eq. (9.7) can also be derived formally. In (9.3) we regard $m_0\gamma\,\kappa(\boldsymbol{v})\dot{\boldsymbol{v}} = e\,(\boldsymbol{E}+c^{-1}\boldsymbol{v}\times\boldsymbol{B})$ as the unperturbed equation, differentiate it once, and substitute $\ddot{\boldsymbol{v}}$ inside the square brackets of (9.3). Resubstituting $\dot{\boldsymbol{v}}$ from the unperturbed equation results in Eq. (9.7). This argument is carried out more easily and in greater generality, because it allows for time-dependent potentials, in the covariant form of the Lorentz–Dirac equation. The unperturbed equation is

$$m_0\dot{\mathbf{u}} = (e/c)\mathbf{F}(\mathbf{q})\cdot\mathbf{u}. \qquad (9.8)$$

One differentiates with respect to the eigentime,

$$(m_0c/e)\ddot{\mathbf{u}} = (\mathbf{u}\cdot\nabla_g)\mathbf{F}(\mathbf{q})\cdot\mathbf{u} + \mathbf{F}(\mathbf{q})\cdot\dot{\mathbf{u}}. \qquad (9.9)$$

Substituting (9.9) in (9.1) and resubstituting (9.8) yields

$$\begin{aligned}
m_0\dot{\mathbf{u}} &= (e/c)\mathbf{F}\cdot\mathbf{u} + \frac{e^2}{6\pi c^3}\big[(e/m_0c)(\mathbf{u}\cdot\nabla_g)\mathbf{F}\cdot\mathbf{u} + (e/m_0c)^2\mathbf{F}\cdot\mathbf{F}\cdot\mathbf{u}\\
&\quad -(e/m_0c^2)^2(\mathbf{F}\cdot\mathbf{u})\cdot(\mathbf{F}\cdot\mathbf{u})\mathbf{u}\big]. \qquad (9.10)
\end{aligned}$$

In three-vector notation the space part of Eq. (9.10) coincides with (9.7), except for the additional term $(e/m_0)\gamma(\partial_t\boldsymbol{E}+c^{-1}\boldsymbol{v}\times\partial_t\boldsymbol{B})$ because of the time dependence of the fields. As usual, the time component of (9.10) provides the energy balance.

Equation (9.10) and its formal derivation appeared for the first time in the second volume of the *Course in Theoretical Physics* by Landau and Lifshitz. Hence it seems to be appropriate to call Eq. (9.10) the Landau–Lifshitz equation. The error in going from (9.1) to (9.10) is of the same order as that in the derivation of the Lorentz–Dirac equation itself. *Thus we regard the Landau–Lifshitz equation as the effective equation governing the motion of a charged particle in the adiabatic limit.*

## 9.2 Some applications

(i) *Zero magnetic field, one-dimensional motion.* We assume $\boldsymbol{B} = 0$ and $\phi_{\text{ex}}$ to vary only along the 1-axis. Setting $\boldsymbol{v} = (v, 0, 0)$, $\boldsymbol{q} = (x, 0, 0)$, and $\boldsymbol{E} = (-\phi', 0, 0)$, the Landau–Lifshitz equation becomes

$$m_0 \gamma^3 \dot{v} = -e\phi'(x) - \frac{e^2}{6\pi c^3} \frac{e}{m_0} \gamma \phi''(x) v. \tag{9.11}$$

The radiation reaction is proportional to $-\phi''(x)v$, which can be regarded as a spatially varying friction coefficient. For a convex potential, $\phi'' > 0$, such as an oscillator potential, this friction coefficient is strictly positive and the resulting motion is damped until the minimum of $\phi$ is reached. In general, however, $\phi''$ will not have a definite sign, like in the case of the double well potential $\phi(x) \simeq (x^2 - 1)^2$ or the washboard potential $\phi(x) \simeq -\cos x$. At locations where $\phi''(x) < 0$ one has antifriction and the mechanical energy increases. This gain is always dominated by losses as can be seen from the energy balance

$$\begin{aligned} &\frac{\mathrm{d}}{\mathrm{d}t}\left[m_0 \gamma + e\phi + \frac{e^2}{6\pi c^3} \frac{e}{m_0} \gamma \phi' v\right] \\ &\quad = -\frac{e^2}{6\pi c^3}\left(\frac{e}{m_0}\right)^2 \phi'^2 - \frac{1}{m_0}\left(\frac{e^2}{6\pi c^3} \frac{e}{m_0}\right)^2 \gamma \phi' \phi'' v\,. \end{aligned} \tag{9.12}$$

The last term in (9.12) does not have a definite sign. But its prefactor is down by the factor $e^2/m_0c^3$ and therefore it is outweighed by $-\phi'^2$.

Equation (9.11) has one peculiar feature. If $\phi(x) = -Ex$, $E > 0$, over some interval $[a_-, a_+]$, then $\phi'' = 0$ over that interval and the friction term vanishes. The particle entering at $a_-$ is uniformly accelerated to the right until it reaches $a_+$. From Larmor's formula we know that the energy radiated per unit time equals $(e^2/6\pi c^3)(e/m_0)^2 E^2$. Since the mechanical energy is conserved, the radiated energy must come entirely from the Schott energy stored in the near field. The same behavior is found for the Lorentz–Dirac equation. If, locally, $\boldsymbol{E} = \text{const.}$ and $\boldsymbol{B} = 0$, then the Hamiltonian part is solved by hyperbolic motion, i.e. a constantly accelerated relativistic particle. For this solution the radiation reaction vanishes which means that locally the critical manifold happens to be independent of $\varepsilon$. The radiated energy originates exclusively from the Schott energy.

(ii) *Zero magnetic field, central potential.* For zero magnetic field the Landau–Lifshitz equation simplifies to

$$m_0 \gamma \kappa(\boldsymbol{v})\dot{\boldsymbol{v}} = e\boldsymbol{E} + \frac{e^2}{6\pi c^3}\Big[\frac{e}{m_0}\gamma(\boldsymbol{v}\cdot\nabla)\boldsymbol{E} + \Big(\frac{e}{m_0 c}\Big)^2\big((\boldsymbol{v}\cdot\boldsymbol{E})\boldsymbol{E} - \gamma^2\boldsymbol{E}^2\boldsymbol{v} + \gamma^2 c^{-2}(\boldsymbol{v}\cdot\boldsymbol{E})^2\boldsymbol{v}\big)\Big]. \quad (9.13)$$

We take $\boldsymbol{E} = -\nabla\phi_{\mathrm{ex}}$ and assume that $\phi_{\mathrm{ex}}$ is central. Let us set $\boldsymbol{q} = \boldsymbol{r}$, $|\boldsymbol{r}| = r$, $\hat{\boldsymbol{r}} = \boldsymbol{r}/|\boldsymbol{r}|$, $\phi_{\mathrm{ex}}(\boldsymbol{q}) = \phi(r)$ which implies $\boldsymbol{E} = -\phi'\hat{\boldsymbol{r}}$. Then (9.13) becomes

$$m_0 \gamma \kappa(\boldsymbol{v})\dot{\boldsymbol{v}} = -e\phi'\hat{\boldsymbol{r}} + \frac{e^2}{6\pi c^3}\Big[\frac{e}{m_0}\gamma\Big(-(\boldsymbol{v}\cdot\hat{\boldsymbol{r}})\phi''\hat{\boldsymbol{r}} - \frac{1}{r}(\boldsymbol{v} - (\boldsymbol{v}\cdot\hat{\boldsymbol{r}})\hat{\boldsymbol{r}})\phi'\Big) + \Big(\frac{e}{m_0 c}\Big)^2\phi'^2\Big((\boldsymbol{v}\cdot\hat{\boldsymbol{r}})\hat{\boldsymbol{r}} - \gamma^2\boldsymbol{v} + \gamma^2 c^{-2}(\boldsymbol{v}\cdot\hat{\boldsymbol{r}})^2\boldsymbol{v}\Big)\Big]. \quad (9.14)$$

The angular momentum $\boldsymbol{L} = \boldsymbol{r}\times m_0\gamma\boldsymbol{v}$ satisfies

$$\dot{\boldsymbol{L}} = \frac{e^2}{6\pi c^3}\Big[-\frac{e}{m_0}\frac{1}{r}\phi' - \Big(\frac{e}{m_0 c}\Big)^2\gamma^2\big(1 - c^{-2}(\boldsymbol{v}\cdot\hat{\boldsymbol{r}})^2\big)\phi'^2\Big]\boldsymbol{L}. \quad (9.15)$$

Thus the orientation of $\boldsymbol{L}$ is conserved and the motion lies in the plane perpendicular to $\boldsymbol{L}$. No further reduction seems to be possible and one would have to rely on numerical integration. Only for the harmonic oscillator, $\phi(r) = \frac{1}{2}m_0\omega_0^2 r^2$, can a closed form solution be achieved.

(iii) *Zero electrostatic field and constant magnetic field.* We set $\boldsymbol{B} = (0, 0, B)$ with constant $B$. Then (9.7) simplifies to

$$m_0 \gamma \kappa(\boldsymbol{v})\dot{\boldsymbol{v}} = \frac{e}{c}(\boldsymbol{v}\times\boldsymbol{B}) + \frac{e^2}{6\pi c^3}\Big(\frac{e}{m_0 c}\Big)^2\big[(\boldsymbol{v}\cdot\boldsymbol{B})\boldsymbol{B} - \gamma^2\boldsymbol{B}^2\boldsymbol{v} + \gamma^2 c^{-2}(\boldsymbol{v}\cdot\boldsymbol{B})^2\boldsymbol{v}\big]. \quad (9.16)$$

We multiply by $\kappa(\boldsymbol{v})^{-1}$ to obtain

$$m_0\gamma\dot{\boldsymbol{v}} = \frac{e}{c}(\boldsymbol{v}\times\boldsymbol{B}) + \frac{e^2}{6\pi c^3}\Big(\frac{e}{m_0 c}\Big)^2\big[(\boldsymbol{v}\cdot\boldsymbol{B})\boldsymbol{B} - \boldsymbol{B}^2\boldsymbol{v}\big]. \quad (9.17)$$

The motion parallel to $\boldsymbol{B}$ decouples with $\dot{v}_3 = 0$. We set $v_3 = 0$ and $\boldsymbol{v} = (\boldsymbol{u}, 0)$, $\boldsymbol{u}^{\perp} = (-u_2, u_1)$. Then the motion in the plane orthogonal to $\boldsymbol{B}$ is governed by

$$\gamma\dot{\boldsymbol{u}} = \omega_c(\boldsymbol{u}^{\perp} - \beta\omega_c\boldsymbol{u}), \quad (9.18)$$

with the cyclotron frequency $\omega_c = eB/m_0 c$ and $\beta = e^2/6\pi c^3 m_0$. Equation (9.18) holds over the entire velocity range. For an electron $\beta\omega_c = 8.8\times 10^{-18}B$ [gauss]. Thus even for very strong fields the friction is small compared to the inertial terms.

Equation (9.18) can be integrated as

$$\frac{d}{dt}\gamma = -\beta\omega_c^2(\gamma^2 - 1) \tag{9.19}$$

with solution

$$\gamma(t) = \big[\gamma_0 + 1 + (\gamma_0 - 1)e^{-2\beta\omega_c^2 t}\big]\big[\gamma_0 + 1 - (\gamma_0 - 1)e^{-2\beta\omega_c^2 t}\big]^{-1}, \tag{9.20}$$

which tells us how $\boldsymbol{u}(t)^2$ shrinks to zero. To determine the angular dependence we introduce polar coordinates as $\boldsymbol{u} = u(\cos\varphi, \sin\varphi)$. Then

$$\frac{du}{d\varphi} = -\beta\omega_c u, \quad \frac{d\varphi}{dt} = \gamma^{-1}\omega. \tag{9.21}$$

Thus $u(\varphi)$ shrinks exponentially,

$$u(\varphi) = u(0)\,e^{-\beta\omega_c\varphi}. \tag{9.22}$$

Since $\beta\omega_c = 8.8 \times 10^{-18} B$ [gauss] for an electron, even for strong fields the change of $u$ over one revolution is tiny.

To obtain the evolution of the position $\boldsymbol{q} = (\boldsymbol{r}, 0)$, $|\boldsymbol{r}| = r$, we use the fact that for zero radiation reaction, $\beta = 0$,

$$r = \frac{u}{\omega_c}\gamma\,. \tag{9.23}$$

By (9.22) this relation remains approximately valid for non-zero $\beta$. Inserting $u(t)$ from (9.20) one obtains

$$r(t) = r_0\,e^{-\beta\omega_c^2 t}\big[1 + ((\gamma_0 - 1)/2)(1 - e^{-2\beta\omega_c^2 t})\big]^{-1} \tag{9.24}$$

with $r_0$ the initial radius and $u(0)/c = (\gamma_0 - 1)^{1/2}/\gamma_0$ the initial speed which are related through (9.23). In the ultrarelativistic regime, $\gamma_0 \gg 1$, and for times such that $\beta\omega_c^2 t \ll 1$, (9.24) simplifies to

$$r(t) = r_0 \frac{1}{1 + \gamma_0\beta\omega_c^2 t} \tag{9.25}$$

and the initial decay is according to the power law $t^{-1}$ rather than exponential.

For an electron $\beta\omega_c^2 = 1.6 \times 10^{-6}(B\text{ [gauss]})^2\ \mathrm{s}^{-1}$. Therefore if one chooses a field strength $B = 10^3$ gauss and an initial radius of $r_0 = 10$ cm, which corresponds to the ultrarelativistic case of $\gamma = 6 \times 10^4$, then the radius shrinks within 0.9 s to $r(t) = 1\ \mu$m by which time the electron has made $2 \times 10^{14}$ revolutions.

(iv) *The Penning trap.* An electron can be trapped for a very long time in the combination of a homogeneous magnetic field and an electrostatic quadrupole potential, which has come to be known as a Penning trap. Its design has been optimized towards high-precision measurements of the gyromagnetic $g$-factor of the electron. Our interest here is that the motion in the plane orthogonal to the magnetic field consists of two coupled modes, which means that the damping cannot be guessed by pure energy considerations using Larmor's formula. One really needs the full power of the Landau–Lifshitz equation.

An ideal Penning trap has the electrostatic quadrupole potential

$$e\phi(\boldsymbol{x}) = \frac{1}{2} m\omega_z^2\Big( -\frac{1}{2}x_1^2 - \frac{1}{2}x_2^2 + x_3^2\Big), \tag{9.26}$$

which satisfies $\Delta\phi = 0$, superimposed with the uniform magnetic field

$$\boldsymbol{B} = (0, 0, B). \tag{9.27}$$

The quadrupole field provides an axial restoring force whereas the magnetic field is responsible for the radial restoring force, which however could be outweighed by the inverted part of the harmonic electrostatic potential.

We insert $\boldsymbol{E} = -\nabla\phi$ and $\boldsymbol{B}$ in the Landau–Lifshitz equation. The terms proportional to $(\boldsymbol{v}\cdot\nabla)\boldsymbol{E},\ \boldsymbol{E}\times\boldsymbol{B},\ (\boldsymbol{v}\cdot\boldsymbol{B})\boldsymbol{B}$, and $\boldsymbol{B}^2\boldsymbol{v}$ are linear in $\boldsymbol{v}$, respectively $\boldsymbol{q}$. The remaining terms are either cubic or quintic and will be neglected. This is justified provided

$$\frac{|\boldsymbol{v}|}{c} \ll 1 \tag{9.28}$$

and

$$(m_0\omega_z^2/e)\, r_{\max} \ll B, \ \text{i.e.} \quad r_{\max} \ll c(\omega_c/\omega_z^2), \tag{9.29}$$

if $r_{\max}$ denotes the maximal distance from the trap center. With these assumptions the Landau–Lifshitz equation decouples into an in-plane motion and an axial motion governed by

$$\dot{\boldsymbol{u}} = \frac{1}{2}\omega_z^2\boldsymbol{r} + \omega_c\boldsymbol{u}^\perp - \beta\Big[\Big(\omega_c^2 - \frac{1}{2}\omega_z^2\Big)\boldsymbol{u} + \frac{1}{2}\omega_c\omega_z^2\boldsymbol{r}^\perp\Big], \tag{9.30}$$

$$\ddot{z} = -\omega_z^2 z - \beta\omega_z^2\dot{z}. \tag{9.31}$$

Here $\boldsymbol{q} = (\boldsymbol{r}, z),\ \boldsymbol{v} = (\boldsymbol{u}, \dot{z}),\ (x_1, x_2)^\perp = (-x_2, x_1)$.

The axial motion is just a damped harmonic oscillation with frequency $\omega_z$ and friction coefficient

$$\gamma_z = \beta\omega_z^2. \tag{9.32}$$

The in-plane motion can be written in matrix form as

$$\frac{\mathrm{d}}{\mathrm{d}t}\psi = (A + \beta V)\psi \tag{9.33}$$

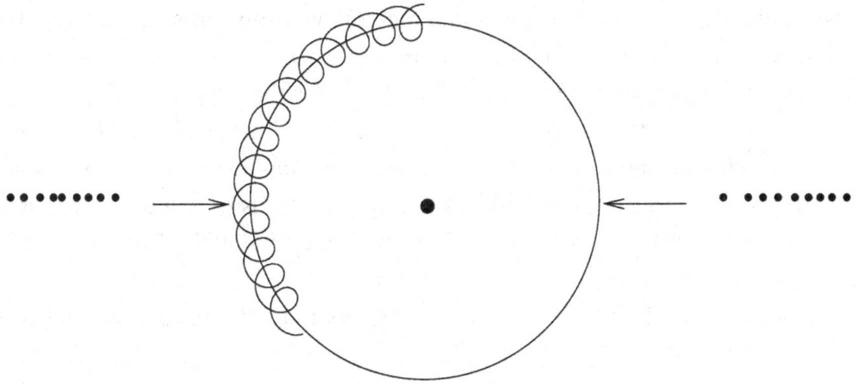

Figure 9.1: Orbit of an electron in a Penning trap seen from above.

with $\psi = (\boldsymbol{r}, \boldsymbol{u})$ and $A_{11} = 0$, $A_{12} = \mathbb{1}$, $A_{21} = \omega_z^2 \mathbb{1}$, $A_{22} = \mathrm{i}\omega_z \sigma_y$, $V_{11} = 0$, $V_{12} = 0$, $V_{21} = \mathrm{i}\omega_c \omega_z^2 \sigma_y$, $V_{22} = (\omega_z^2 - \omega_c^2)\mathbb{1}$, where $\sigma_y$ is the Pauli spin matrix with eigenvectors $\chi_\pm$, $\sigma_y \chi_\pm = \pm\chi_\pm$. The unperturbed motion is governed by the $4 \times 4$ matrix $A$. It has the eigenvectors $\psi_{+,\pm} = (\pm\mathrm{i}(1/\omega_+)\chi_\mp, \chi_\mp)$ with eigenvalues $\pm\mathrm{i}\omega_+$ and $\psi_{-,\pm} = (\pm\mathrm{i}(1/\omega_-)\chi_\mp, \chi_\mp)$ with eigenvalues $\pm\mathrm{i}\omega_-$, where

$$\omega_\pm = \frac{1}{2}\left(\omega_c \pm \sqrt{\omega_c^2 - 2\omega_z^2}\,\right). \tag{9.34}$$

The mode with frequency $\omega_+$ is called the cyclotron mode and that with $\omega_-$ is called the magnetron mode. Experimentally $\omega_c \gg \omega_z$ and therefore $\omega_+ \ll \omega_-$. The orbit is then an epicycle with rapid cyclotron and slow magnetron motion, as shown in figure 9.1. The adjoint matrix $A^*$ has eigenvectors orthogonal to the $\psi$'s. They are given by $\varphi_{+,\pm} = (\mp\mathrm{i}(\omega_z^2/\omega_+)\chi_\mp, \chi_\mp)$ with eigenvalues $\pm\mathrm{i}\omega_+$ and $\varphi_{-,\pm} = (-(\omega_z^2/\omega_-)\chi_\mp, \chi_\mp)$ with eigenvalues $\mp\mathrm{i}\omega_-$.

Since $\beta$ is small, the eigenfrequencies of $A + \beta V$ can be computed in first-order perturbation. The cyclotron mode attains a negative real part corresponding to the friction coefficient

$$\gamma_+ = \frac{e^2}{6\pi c^3 m_0} \frac{\omega_+^3}{\omega_+ - \omega_-} \tag{9.35}$$

and the magnetron mode attains a positive real part corresponding to the antifriction coefficient

$$\gamma_- = \frac{e^2}{6\pi c^3 m_0} \frac{\omega_-^3}{\omega_- - \omega_+}. \tag{9.36}$$

As the electron radiates, it lowers its potential energy by increasing the magnetron radius.

Experimentally $B = 6 \times 10^4$ gauss and the voltage drop across the trap is 10 V. This corresponds to $\omega_z = 4 \times 10^8$ Hz, $\omega_+ = 1.1 \times 10^{12}$ Hz, $\omega_- = 7.4 \times 10^4$ Hz. The conditions (9.28), (9.29) are easily satisfied. For the lifetimes $(1/\gamma_z) = 5 \times 10^8$ s, $(1/\gamma_+) = 8 \times 10^{-2}$ s, and $(1/\gamma_-) = -2 \times 10^{23}$ s are obtained. Thus the magnetron motion is stable, as observed through keeping a single electron trapped over weeks. The cyclotron motion decays within fractions of a second. The axial motion is in fact damped by coupling to the external circuit and decays also within a second.

The variation with the magnetic field can be more clearly discussed in terms of the dimensionless ratio $(\omega_c/\omega_z) = \lambda$. Then

$$\omega_\pm = \omega_z \frac{1}{2}\left(\lambda \pm \sqrt{\lambda^2 - 2}\right), \quad \gamma_\pm = \pm\beta\omega_z^2\left(\lambda \pm \sqrt{\lambda^2 - 2}\right)^3 / 8\sqrt{\lambda^2 - 2}. \tag{9.37}$$

For large $\lambda$, $\omega_+ \cong \lambda$, $\omega_- \cong \lambda^{-1}$, while $\gamma_+ \cong \lambda^2$, $\gamma_- \cong \lambda^{-4}$. As $\lambda \to \sqrt{2}$, we have $\omega_+ = \omega_- = \omega_z/\sqrt{2}$. However, the friction coefficients diverge as $(\lambda - \sqrt{2})^{-1/2}$. Let us call $B_c$ the critical field at which the mechanical motion becomes unstable. For $B > B_c$, one still has periodic motion with frequency $\omega_z/\sqrt{2}$, but the onsetting instability is revealed through the vanishing lifetime. In the mentioned experiment $\lambda = 2.7 \times 10^3$ and for fixed $\omega_z$ the critical field strength is $B_c = 30$ gauss.

## 9.3 Experimental status of the Lorentz–Dirac equation

Energy loss through radiation is a well-established phenomenon. Indeed, in synchrotron sources electrons slow down because of radiation losses, and energy has to be supplied to maintain a steady electron current. The supplied power is computed on the basis of Larmor's formula, and synchrotron sources are one prominent example to confirm its validity. On the other hand, the Lorentz–Dirac equation goes way beyond mere energy balancing and claims to predict the orbit of an electron. Here synchrotron sources provide no test, since the modification of the orbit due to radiation damping is lost in the noise of experimental uncertainties. As a fair summary, thus we can say that while qualitative aspects of radiation damping are well tested, there is no single experiment which probes quantitatively the predictions of the electron motion by the Lorentz–Dirac equation. We propose and discuss here two experiments which are within the reach of present-day techniques.

To cope with the smallness of the radiation reaction, in essence, only two approaches seem feasible. The first one is to wait long enough until the effects accumulate to something which may be detected, a route followed in the Penning-trap experiment. The other option is to use ultrastrong fields. In either case, there is no way to monitor directly the electron orbit and one has to rely on indirect evidence, like lifetimes or emission spectra.

(i) *Penning trap.* In the previous section we discussed the electron orbits for the Penning trap with the quadrupole potential in the quadratic approximation. The Lorentz–Dirac equation predicts, in particular, the lifetime of the cyclotron mode. For the field strengths used in the high-precision measurement of the $g$-factor, this lifetime is measured to 0.8 s in good agreement with the theoretical result. To have a more stringent test what would be needed is a systematic determination of how the lifetime depends on the magnetic field strength. Another option of interest is to turn the $\boldsymbol{B}$-field out off the symmetry axis. For this case we have not computed the cyclotron lifetime, but could have done so by the scheme explained, with the welcome complication that all three modes couple. The dependence of the cyclotron lifetime on the orientation of the $\boldsymbol{B}$-field would be a valuable test of the validity of the Lorentz–Dirac equation.

(ii) *Ultrastrong laser pulse.* A strong laser pulse hits a bound electron. Since the atom ionizes instantaneously, the electron is subject only to the time-dependent laser field. Thus we set $\boldsymbol{q}^0 = 0$, $\boldsymbol{v}^0 = 0$, and for the external fields

$$\begin{aligned} \boldsymbol{E}(\boldsymbol{x}, t) &= h(\omega t - \boldsymbol{k}\cdot\boldsymbol{x})\boldsymbol{E}_0 \cos(\omega t - \boldsymbol{k}\cdot\boldsymbol{x}), \\ \boldsymbol{B}(\boldsymbol{x}, t) &= h(\omega t - \boldsymbol{k}\cdot\boldsymbol{x})\boldsymbol{B}_0 \cos(\omega t - \boldsymbol{k}\cdot\boldsymbol{x}), \\ |\boldsymbol{E}_0| &= |\boldsymbol{B}_0|, \quad \boldsymbol{E}_0\cdot\boldsymbol{k} = 0 = \boldsymbol{B}_0\cdot\boldsymbol{k}, \quad \boldsymbol{E}_0\cdot\boldsymbol{B}_0 = 0. \end{aligned} \tag{9.38}$$

$h$ is a shape function. The motion of the electron is governed by the Landau–Lifshitz equation (9.7) augmented by the term

$$\frac{e^2}{6\pi c^3}\frac{e}{m_0}\gamma\frac{\partial}{\partial t}(\boldsymbol{E} + c^{-1}\boldsymbol{v}\times\boldsymbol{B}) \tag{9.39}$$

because of the time dependence of the external fields. Our dynamical problem is in fact two dimensional with the motion of the electron lying in the plane spanned by $\boldsymbol{E}_0$ and $\boldsymbol{k}$. Nevertheless one has to rely on numerical integration, and we discuss the example from Keitel *et al.* (1998).

The ultra-intense laser field has an intensity of $10^{22}$ W cm$^{-2}$. The frequency is chosen to be $\omega = 3.54 \times 10^{15}$ s$^{-1}$, which is in the near-infrared regime. We follow the motion of the electron up to 3000 laser cycles, i.e. up to the final time $t = 3000(2\pi/3.54 \times 10^{15})$ s $= 0.53 \times 10^{-11}$ s. Over that time span the shape function is assumed to interpolate linearly between zero and the full field strength.

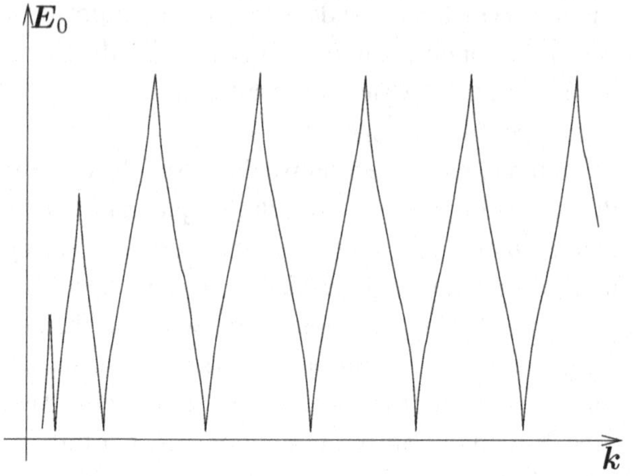


Figure 9.2: Orbit of an electron when hit by an ultrastrong laser pulse.

The electron motion is highly relativistic, as can be seen from the strong redshift corresponding to only the seven electron cycles displayed in figure 9.2. The electron is displaced by 0.1586 cm in the propagation direction and has a maximal amplitude of $0.795 \times 10^{-3}$ cm in the electric field direction.

The effects of radiation damping are minute. In the propagation direction the distance is *in*creased by the fraction $7 \times 10^{-7}$ and in the electric field direction it is *de*creased by the fraction $10^{-2}$. Thus a direct verification of the radiation reaction is out of reach. However, in the emission spectrum the radiation damping results in a roughly 1% change as compared to the frictionless solution with the Lorentz force from the external fields of (9.38). In an experiment the radiation spectrum has to be measured with such precision that, after the theoretical spectrum, computed without radiation reaction, has been subtracted, there is still a significant background which allows for a quantitative comparison with the emission spectrum predicted by the Lorentz–Dirac equation.

## Notes and references

### *Section 9*

The name Lorentz–Dirac is standard but historically inaccurate. Some authors, e.g. Rohrlich (1997), therefore propose Abraham–Lorentz–Dirac instead. The radiation reaction term was originally derived in Abraham (1905); compare with chapters 7 and 8. Von Laue (1909) realized its covariant form. In the Pauli (1921) encyclopaedia article on relativity the equation is stated as in (9.1). Dirac's contribution is explained in section 3.3. Plass (1961) is a summary of exact solutions of the Lorentz–Dirac equation.

### Section 9.1

Detailed case studies of the Lorentz–Dirac equation, including its center manifold, are listed in the Notes to section 8.2. Baylis and Huschilt (2002) critically explore the relation to the Landau–Lifshitz equation. The substitution trick seems to have been common knowledge. For example, without further comment it is used by Pauli (1929) and Heitler (1936) in the particular case of a harmonic oscillator. In its full generality the Landau–Lifshitz equation (9.10) appears already in the first edition of *Volume II: The Classical Theory of Fields* of the Landau–Lifshitz *Course in Theoretical Physics*. At no point is the reader given a hint on the geometrical picture of the solution flow and on the errors involved in the approximation. To me it is rather surprising that the contribution of Landau and Lifshitz is ignored in essentially all discussions of radiation reaction, one notable exception being Teitelbom *et al.* (1980). Spohn (1998, 2000a) uses singular perturbation theory to rederive the Landau–Lifshitz equation. The appearance of singular perturbation theory is difficult to track. For a particular application it is clearly stated by Burke (1970). There have been attempts to replace the Lorentz–Dirac equation by a second-order equation (Mo and Papas 1971; Shen 1972b; Bonnor 1974; Parrot 1987; Valentini 1988; Ford and O'Connell 1991, 1993). Based on Ford and O'Connell (1991), Jackson (1999) uses the substitution trick for a radiation damped harmonic oscillator and discusses several applications. In the case of arbitrary time-dependent potentials, only Landau and Lifshitz provide the correct center manifold equation. The structure discussed here reappears whenever a low-dimensional system is coupled to a wave equation; for an application in acoustics see Templin (1999).

### Section 9.2

Uniform acceleration is discussed in Fulton and Rohrlich (1960) and Rohrlich (1990). A constant magnetic field is of importance for synchrotron sources. Since the electron is maintained on a circular orbit, Larmor's formula is precise enough. Landau and Lifshitz (1959) give a brief discussion. The power law for the ultra-relativistic case is noted in Spohn (1998). Shen (1972a, 1978) discusses at which field strengths quantum corrections will become important. His results are only partially reliable, since his starting point is not the Landau–Lifshitz equation. The Penning trap is reviewed by Brown and Gabrielse (1986), which includes a discussion of the classical orbits and their lifetimes. They state the results (9.35), (9.36) as based on a quantum resonance computation. Since the final answer does not contain $\hbar$, it must follow from the Landau–Lifshitz equation (Spohn 2000a). In the classical framework, general trap potentials can be handled through numerical integration routines for ordinary differential equations. The self-force in the case of synchroton radiation is studied by Burko (2000).

### *Section 9.3*

The Penning-trap experiment is proposed in Spohn (2000a). The numerical results on ultrastrong laser pulses are taken from Keitel *et al.* (1998). Another proposal, which apparently never received the proper funding, is to measure the mega-gauss magnetic bremsstrahlung for ultrarelativistic electrons (Erber 1971; Shen 1970).

# 10
# Spinning charges

The Lorentz model includes by necessity the inner rotation of charges and, beyond the translational degrees of freedom, one has to determine its effective dynamics. This will lead to a derivation of the Bargmann–Michel–Telegdi (BMT) equation from a microscopic basis including an expression for the gyromagnetic ratio. We will also discuss the Abraham model with spin, a little-explored territory, since it is more easily controlled mathematically and it teaches us how the BMT equation is modified when Lorentz invariance is no longer available.

## 10.1 Effective spin dynamics of the Lorentz model

Let us recall the equations of motion for an extended charge, where for the moment the interaction with the self-field is ignored,

$$\dot{\mathbf{p}} = \mathbf{f}, \quad \dot{\mathbf{s}} + \boldsymbol{\Omega}_{\mathrm{FW}} \cdot \mathbf{s} = \mathbf{t}\,. \tag{10.1}$$

Here the external force $f$, respectively the external torque $t$, are defined through (2.92), respectively (2.95). Equation (10.1) must be supplemented by

$$\mathbf{p} = m_{\mathrm{g}}\mathbf{u}, \quad \mathbf{s} = I_{\mathrm{b}}\mathbf{w}, \tag{10.2}$$

which define the bare gyrational mass $m_{\mathrm{g}}$ and the bare moment of inertia $I_{\mathrm{b}}$. Both depend on $|\mathbf{w}|$.

We assume now that the external field tensor is slowly varying, by replacing $\mathbf{F}(\mathbf{q})$ by the scaled field tensor $\varepsilon\mathbf{F}(\varepsilon\mathbf{q})$ in (2.92), (2.95). Note that this prescription automatically includes slow variation in time. **f** and **t** simplify in the limit of small $\varepsilon$ and, on the macroscopic scale, (10.1) becomes

$$\dot{\mathbf{p}} = e\mathbf{F} \cdot \mathbf{u}, \quad \dot{\mathbf{s}} + \boldsymbol{\Omega}_{\mathrm{FW}} \cdot \mathbf{s} = \mu(\mathbf{F} \cdot \mathbf{w})^{\perp} \tag{10.3}$$

with the magnetic moment

$$\mu = \frac{1}{3} e \int d^3x \varphi(\boldsymbol{x}) \boldsymbol{x}^2 \tag{10.4}$$

and $\mathbf{a}^\perp = (\mathbf{g} + \mathbf{u} \otimes \mathbf{u}) \cdot \mathbf{a}$. Since $|\mathbf{w}|$ is conserved, the translational motion is autonomous, whereas the spin follows the local fields as they are encountered.

As a next step we have to include the coupling to the self-field. In principle the scheme of chapter 7 has to be repeated, but we prefer to take the static short-cut. The energy–momentum relation for the Lorentz model was computed in chapter 4. Thus we stipulate that the bare gyrational mass $m_g$ is renormalized to $m_g + m_f$ and the bare moment of inertia to $I_b + I_f$; see (4.43), (4.45), respectively (4.49), (4.51). This means that instead of (10.2) we have

$$\mathbf{p} = (m_g + m_f)\mathbf{u}, \quad \mathbf{s} = (I_b + I_f)\mathbf{w}\,. \tag{10.5}$$

Equation (10.3) together with (10.5) is the effective dynamics in the adiabatic limit on the Hamiltonian level neglecting radiation damping.

We want to compare our spin dynamics with the BMT equation which reads

$$\dot{\mathbf{w}} + \boldsymbol{\Omega}_{\mathrm{FW}} \cdot \mathbf{w} = \frac{g}{2} \frac{e}{m} (\mathbf{F} \cdot \mathbf{w})^\perp\,, \tag{10.6}$$

where $m$ is the experimental mass and $g$ the gyromagnetic ratio, which like the charge is an intrinsic property of the particle. Using the fact that $\boldsymbol{\Omega}_{\mathrm{FW}}$ is determined by Newton's translational equations of motion one arrives at the perhaps more familiar three-vector form for the angular velocity,

$$\begin{aligned} \dot{\omega} = \frac{e}{mc} \omega \times \Big[ \Big( \frac{g}{2} - 1 + \frac{1}{\gamma} \Big) \boldsymbol{B}_{\mathrm{ex}} - \Big( \frac{g}{2} - 1 \Big) \frac{\gamma}{1+\gamma} c^{-2} (\boldsymbol{v} \cdot \boldsymbol{B}_{\mathrm{ex}}) \boldsymbol{v} \\ - \Big( \frac{g}{2} - \frac{\gamma}{1+\gamma} \Big) c^{-1} \boldsymbol{v} \times \boldsymbol{E}_{\mathrm{ex}} \Big]\,. \end{aligned} \tag{10.7}$$

Here $\boldsymbol{v}$, $\boldsymbol{E}_{\mathrm{ex}}$, $\boldsymbol{B}_{\mathrm{ex}}$ are to be evaluated along the given orbit. To compare (10.6) with (10.3) one uses (10.6) and notes that, since $|\boldsymbol{w}|$ is a constant of motion,

$$\dot{\mathbf{w}} + \boldsymbol{\Omega}_{\mathrm{FW}} \cdot \mathbf{w} = \frac{\mu}{I_b + I_f} (\mathbf{F} \cdot \mathbf{w})^\perp\,. \tag{10.8}$$

Therefore the gyromagnetic ratio of the Lorentz model is given by

$$g = \frac{2\mu}{e} \frac{m_g + m_f}{I_b + I_f}\,. \tag{10.9}$$

The magnetic moment $\mu$ depends on the charge distribution, all other terms in (10.9) on the mass distribution. Through their variation any value of $g$ can be realized, unless the charge and mass form factors are equal to each other, as assumed already. In the case of a uniformly charged sphere [ball] of radius $R$ the integrals in (10.9) can be evaluated with the result (the first term refers to a sphere and [ ... ] to a ball)

$$\mu = \frac{1}{3}eR^2\,, \quad \Big[= \frac{1}{5}eR^2\Big]\,, \tag{10.10}$$

$$m_{\mathrm{g}} = m_{\mathrm{b}}\frac{1}{\omega R}\mathrm{arctanh}\omega R\,, \quad \Big[= m_{\mathrm{b}}\frac{3}{2(\omega R)^3}\big(\omega R - (1-(\omega R)^2)\mathrm{arctanh}\omega R\big)\Big]\,, \tag{10.11}$$

$$m_{\mathrm{f}} = \frac{1}{2}\frac{e^2}{4\pi R}\Big(1 + \frac{2}{9}(\omega R)^2\Big)\,, \quad \Big[= \frac{1}{2}\frac{e^2}{4\pi R}\Big(\frac{6}{5} + \frac{4}{35}(\omega R)^2\Big)\Big]\,, \tag{10.12}$$

$$I_{\mathrm{b}} = m_{\mathrm{b}}\frac{1}{2\omega^2}\Big(-1 + \frac{1+(\omega R)^2}{\omega R}\mathrm{arctanh}\omega R\Big)\,,$$
$$\Big[= m_{\mathrm{b}}\frac{1}{2\omega^2}\frac{3}{4(\omega R)^3}\big(3\omega R - (\omega R)^3 + (-3 + 2(\omega R)^2 + (\omega R)^4)\mathrm{arctanh}\omega R\big)\Big]\,, \tag{10.13}$$

$$I_{\mathrm{f}} = \frac{2}{9}\frac{e^2}{4\pi R}\,, \quad \Big[= \frac{4}{35}\frac{e^2}{4\pi R}\Big]\,. \tag{10.14}$$

In the limit $e \to 0$, $g_{\mathrm{sphere}}$ decreases from 1 to 2/3 and $g_{\mathrm{ball}}$ from 1 to 2/5 as $\omega R$ increases from 0 to 1. In the opposite limit $m_{\mathrm{b}} \to 0$, one obtains

$$g_{\mathrm{sphere}} = \frac{3}{2} + \frac{1}{3}(\omega R)^2\,, \quad g_{\mathrm{ball}} = \frac{21}{10} + \frac{1}{5}(\omega R)^2\,. \tag{10.15}$$

## 10.2 The Abraham model with spin

Abraham models the charge as a nonrelativistic rigid body with mass distribution $m_{\mathrm{b}}\varphi$ and charge distribution $e\varphi$, which for notational simplicity we take to be proportional to each other. A complete mechanical description must specify both the center of mass, $\boldsymbol{q}(t)$, and the angular velocity, $\boldsymbol{\omega}(t) \in \mathbb{R}^3$, relative to the center. The spinning charge generates the current

$$\boldsymbol{j}(\boldsymbol{x},t) = \big(\boldsymbol{v}(t) + \boldsymbol{\omega}(t) \times (\boldsymbol{x} - \boldsymbol{q}(t))\big)e\varphi(\boldsymbol{x} - \boldsymbol{q}(t))\,, \tag{10.16}$$

which satisfies charge conservation, since $\varphi$ is radial. Therefore the Maxwell equations have a modified source term and read

$$\partial_t \boldsymbol{B}(\boldsymbol{x},t) = -\nabla \times \boldsymbol{E}(\boldsymbol{x},t)\,,$$
$$\partial_t \boldsymbol{E}(\boldsymbol{x},t) = \nabla \times \boldsymbol{B}(\boldsymbol{x},t) - \big(\boldsymbol{v}(t) + \boldsymbol{\omega}(t) \times (\boldsymbol{x} - \boldsymbol{q}(t))\big)e\varphi(\boldsymbol{x} - \boldsymbol{q}(t))\,,$$
$$\nabla \cdot \boldsymbol{E}(\boldsymbol{x},t) = e\varphi(\boldsymbol{x} - \boldsymbol{q}(t)), \quad \nabla \cdot \boldsymbol{B}(\boldsymbol{x},t) = 0. \tag{10.17}$$

The momentum of the center of mass is $m_b v(t)$ and the angular momentum relative to $q(t)$ is

$$s = I_b \omega \quad \text{with} \quad I_b = \frac{2}{3} m_b \int d^3x \varphi(x) x^2 . \tag{10.18}$$

Therefore Newton's equations of motion for the translational degrees of freedom become

$$\frac{d}{dt} m_b v(t) = \int d^3x e\varphi(x - q(t)) \big[ E(x, t) + \big( v(t) + \omega(t) \times (x - q(t)) \big) \times B(x, t) \big] \tag{10.19}$$

and for the rotational degrees of freedom

$$\begin{aligned} \frac{d}{dt} I_b \, \omega(t) = \int & d^3x e\varphi(x - q(t))(x - q(t)) \\ & \times \big[ E(x, t) + \big( v(t) + \omega(t) \times (x - q(t)) \big) \times B(x, t) \big] . \end{aligned} \tag{10.20}$$

If in addition there are external forces acting on the charge, then $E$ and $B$ in (10.19), (10.20) would have to be replaced by $E + E_{ex}$ and $B + B_{ex}$, respectively.

The Abraham model of section 2.4 is obtained by formally setting $\omega(t) = 0$. Note that this is not consistent with Newton's torque equation (10.20), since $\dot{\omega}(t) \neq 0$, in general, even for $\omega(t) = 0$.

The Abraham model with spin conserves the energy

$$\mathcal{E} = \frac{1}{2} m_b v^2 + \frac{1}{2} I_b \omega^2 + \frac{1}{2} \int d^3x (E^2 + B^2) , \tag{10.21}$$

the linear momentum

$$\mathcal{P} = m_b v + \int d^3x E \times B , \tag{10.22}$$

and in addition the total angular momentum

$$\mathcal{J} = q \times m_b v + I_b \omega + \int d^3x x \times (E \times B) . \tag{10.23}$$

Of course, also the spinless Abraham model is invariant under rotations and there must exist a correspondingly conserved quantity, only it does not have the standard form of a total angular momentum, which from a somewhat different perspective indicates that inner rotations must be included.

In the by now established tradition, we assume that the external forces are slowly varying and want to derive in this adiabatic limit an effective equation of motion for the particle including its spin. As a first step of this program we have to determine the charge solitons. We set

$$\boldsymbol{q}(t) = \boldsymbol{v}t, \quad \boldsymbol{\omega}(t) = \boldsymbol{\omega}, \quad \boldsymbol{E}(\boldsymbol{x}, t) = \boldsymbol{E}(\boldsymbol{x} - \boldsymbol{v}t), \quad \boldsymbol{B}(\boldsymbol{x}, t) = \boldsymbol{B}(\boldsymbol{x} - \boldsymbol{v}t) \tag{10.24}$$

and have to determine the solutions of

$$\begin{aligned} &-\boldsymbol{v} \cdot \nabla \boldsymbol{B} = -\nabla \times \boldsymbol{E}, \quad -\boldsymbol{v} \cdot \nabla \boldsymbol{E} = \nabla \times \boldsymbol{B} - (\boldsymbol{v} + \boldsymbol{\omega} \times \boldsymbol{x}) e\varphi, \\ &\nabla \cdot \boldsymbol{E} = e\varphi, \quad \nabla \cdot \boldsymbol{B} = 0, \end{aligned} \tag{10.25}$$

$$0 = \int d^3x e\varphi(\boldsymbol{x}) \big[\boldsymbol{E}(\boldsymbol{x}) + (\boldsymbol{v} + \boldsymbol{\omega} \times \boldsymbol{x}) \times \boldsymbol{B}(\boldsymbol{x})\big], \tag{10.26}$$

$$0 = \int d^3x e\varphi(\boldsymbol{x}) \boldsymbol{x} \times \big[\boldsymbol{E}(\boldsymbol{x}) + (\boldsymbol{v} + \boldsymbol{\omega} \times \boldsymbol{x}) \times \boldsymbol{B}(\boldsymbol{x})\big], \tag{10.27}$$

for which we turn to Fourier space. The inhomogeneous Maxwell equations (10.25) are then solved by

$$\widehat{\boldsymbol{E}} = \widehat{\boldsymbol{E}}_1 + \widehat{\boldsymbol{E}}_2, \quad \widehat{\boldsymbol{B}} = \widehat{\boldsymbol{B}}_1 + \widehat{\boldsymbol{B}}_2 \tag{10.28}$$

with

$$\widehat{\boldsymbol{E}}_1(\boldsymbol{k}) = -\mathrm{i}\big[\boldsymbol{k}^2 - (\boldsymbol{k} \cdot \boldsymbol{v})^2\big]^{-1} (\boldsymbol{k} - (\boldsymbol{k} \cdot \boldsymbol{v})\boldsymbol{v}) e\widehat{\varphi}(\boldsymbol{k}), \tag{10.29}$$

$$\widehat{\boldsymbol{E}}_2(\boldsymbol{k}) = -\big[\boldsymbol{k}^2 - (\boldsymbol{k} \cdot \boldsymbol{v})^2\big]^{-1} (\boldsymbol{\omega} \times \boldsymbol{k})(\boldsymbol{v} \cdot \nabla_k) e\widehat{\varphi}(\boldsymbol{k}), \tag{10.30}$$

and

$$\widehat{\boldsymbol{B}}_1(\boldsymbol{k}) = \mathrm{i}\big[\boldsymbol{k}^2 - (\boldsymbol{k} \cdot \boldsymbol{v})^2\big]^{-1} (\boldsymbol{k} \times \boldsymbol{v}) e\widehat{\varphi}(\boldsymbol{k}), \tag{10.31}$$

$$\widehat{\boldsymbol{B}}_2(\boldsymbol{k}) = -\big[\boldsymbol{k}^2 - (\boldsymbol{k} \cdot \boldsymbol{v})^2\big]^{-1} (\boldsymbol{k} \times (\boldsymbol{\omega} \times \nabla_k)) e\widehat{\varphi}(\boldsymbol{k}). \tag{10.32}$$

Note that $\widehat{\boldsymbol{E}}_1$, $\widehat{\boldsymbol{B}}_1$ are odd, and $\widehat{\boldsymbol{E}}_2$, $\widehat{\boldsymbol{B}}_2$ are even in $\boldsymbol{k}$.

Since the integral over an odd term vanishes, a zero Lorentz force results in the condition

$$\begin{aligned} &-\int d^3k \widehat{\varphi}^* \big[\boldsymbol{k}^2 - (\boldsymbol{v} \cdot \boldsymbol{k})^2\big]^{-1} (\boldsymbol{\omega} \times \boldsymbol{k})(\boldsymbol{v} \cdot \nabla_k) \widehat{\varphi} \\ &-\int d^3k \widehat{\varphi}^* \big[\boldsymbol{k}^2 - (\boldsymbol{v} \cdot \boldsymbol{k})^2\big]^{-1} \boldsymbol{v} \times (\boldsymbol{k} \times (\boldsymbol{\omega} \times \nabla_k)) \widehat{\varphi} \\ &+\int d^3k \big[\boldsymbol{k}^2 - (\boldsymbol{v} \cdot \boldsymbol{k})^2\big]^{-1} ((\boldsymbol{\omega} \times \nabla_k) \widehat{\varphi}^*) \times (\boldsymbol{k} \times \boldsymbol{v}) \widehat{\varphi} \\ &\quad = -\int d^3k \widehat{\varphi}^* \big[\boldsymbol{k}^2 - (\boldsymbol{v} \cdot \boldsymbol{k})^2\big]^{-1} |\boldsymbol{k}|^{-1} \widehat{\varphi}_r' \\ &\qquad \times \big((\boldsymbol{\omega} \times \boldsymbol{k})(\boldsymbol{v} \cdot \boldsymbol{k}) + \boldsymbol{v} \times (\boldsymbol{k} \times (\boldsymbol{\omega} \times \boldsymbol{k})) - ((\boldsymbol{\omega} \times \boldsymbol{k}) \cdot \boldsymbol{v})\boldsymbol{k}\big) = 0 \end{aligned} \tag{10.33}$$

for every $\boldsymbol{v}$ and $\boldsymbol{\omega}$, using the fact that $\widehat{\varphi}$ is radial.

The Lorentz torque requires more work. Using again the fact that the integral over an odd term vanishes, we have

$$\begin{aligned} \mathrm{i}\int \mathrm{d}^3 k e\widehat{\varphi}^* \big(\nabla_k \times \widehat{\boldsymbol{E}}_1 + \nabla_k \times (\boldsymbol{v} \times \widehat{\boldsymbol{B}}_1) + \nabla_k \times ((\boldsymbol{\omega} \times i\nabla_k) \times \widehat{\boldsymbol{B}}_2)\big) \quad (10.34) \\ = -e^2 \int \mathrm{d}^3 k |\boldsymbol{k}|^{-1} \widehat{\varphi}_r^{*\prime} \big[\boldsymbol{k}^2 - (\boldsymbol{k}\cdot\boldsymbol{v})^2\big]^{-1} \widehat{\varphi}\big(\boldsymbol{k} \times (\boldsymbol{k} - (\boldsymbol{k}\cdot\boldsymbol{v})\boldsymbol{v}) \\ - \boldsymbol{k} \times (\boldsymbol{v} \times (\boldsymbol{k} \times \boldsymbol{v}))\big) + e \int \mathrm{d}^3 k |\boldsymbol{k}|^{-1} \widehat{\varphi}_r^{*\prime} \boldsymbol{k} \times \big((\boldsymbol{\omega} \times \nabla_k) \times \widehat{\boldsymbol{B}}_2\big) \\ = e \int \mathrm{d}^3 k |\boldsymbol{k}|^{-1} \widehat{\varphi}_r^{*\prime} \boldsymbol{k} \times (\nabla_k (\boldsymbol{\omega}\cdot\widehat{\boldsymbol{B}}_2) - \boldsymbol{\omega}\nabla_k \cdot \widehat{\boldsymbol{B}}_2) \\ = -e \int \mathrm{d}^3 k |\boldsymbol{k}|^{-1} \widehat{\varphi}_r^{*\prime} \times (\boldsymbol{k} \times \boldsymbol{\omega}) \nabla_k \cdot \widehat{\boldsymbol{B}}_2 \,. \end{aligned}$$

For the divergence of $\widehat{\boldsymbol{B}}_2$ we obtain

$$\nabla_k \cdot \widehat{\boldsymbol{B}}_2 = 2\big[\boldsymbol{k}^2 - (\boldsymbol{k}\cdot\boldsymbol{v})^2\big]^{-2} \boldsymbol{k}^2 (\boldsymbol{\omega}\cdot\nabla_k - (\boldsymbol{v}\cdot\boldsymbol{\omega})(\boldsymbol{v}\cdot\nabla_k)) e\widehat{\varphi} \quad (10.35)$$

and therefore zero Lorentz torque results in the condition

$$\int \mathrm{d}^3 k |\nabla_k \widehat{\varphi}|^2 2\big[\boldsymbol{k}^2 - (\boldsymbol{k}\cdot\boldsymbol{v})^2\big]^{-2} (\boldsymbol{k} \times \boldsymbol{\omega})(\boldsymbol{\omega}\cdot\boldsymbol{k} - (\boldsymbol{v}\cdot\boldsymbol{\omega})(\boldsymbol{v}\cdot\boldsymbol{k})) = 0 \,. \quad (10.36)$$

Taking into account that $\widehat{\varphi}$ is radial, the torque vanishes only if either $\boldsymbol{\omega} \parallel \boldsymbol{v}$ or $\boldsymbol{\omega} \perp \boldsymbol{v}$. If $\boldsymbol{v} = 0$, the torque always vanishes. For $\boldsymbol{\omega}$ oblique to $\boldsymbol{v}$ Eqs. (10.17)–(10.20) have no soliton-like solution.

Physically the charge distribution is rigid, but the electromagnetic fields are Lorentz contracted along $\boldsymbol{v}$. This mismatch yields a nonvanishing torque unless $\boldsymbol{\omega} \parallel \boldsymbol{v}$, respectively $\boldsymbol{\omega} \perp \boldsymbol{v}$. Clearly, the mismatch is an artifact of the semirelativistic Abraham model. As discussed in the previous section, for a relativistic extended charge distribution there is a charged soliton for every $\boldsymbol{v}$ and $\boldsymbol{\omega}$. Because in the Abraham model some charge solitons are "missing", an analysis of the adiabatic limit is hampered at an early stage and we do not really know what happens. Through radiation damping the spin could be forced to remain parallel to $\boldsymbol{v}(t)$. There could be an effective dynamics separately for the parallel and perpendicular components of $\boldsymbol{\omega}(t)$. Only one particular case lends itself to a more detailed analysis. We simply make sure that $\boldsymbol{q}(t) = 0$ for all $t$, e.g. by taking $\boldsymbol{E}_{\mathrm{ex}} = 0$, $\boldsymbol{B}_{\mathrm{ex}} = \varepsilon \boldsymbol{B}$ with $\boldsymbol{B}$ a spatially constant, possibly time-dependent vector, and suitable initial conditions for the Maxwell field. Then the Abraham model without external forces has a stationary solution for every $\boldsymbol{\omega}$ and the adiabatic limit is meaningful and of interest. We take up this problem in the following section.

In the quantized version of the Abraham model, the Pauli–Fierz Hamiltonian to be discussed in chapter 13, the spin couples differently and the Lorentz torque is not the quantization of the right-hand side of (10.20). The Pauli–Fierz model has a two-fold degenerate ground state for every fixed total momentum (smaller than some critical value $p_c$). Associated to this subspace there is an adiabatic evolution which admits an arbitrary spin orientation. Thus through quantization one regains some features of the relativistic model.

## 10.3 Adiabatic limit and the gyromagnetic ratio

We consider a spinning charge sitting forever at the origin and hence choose $\boldsymbol{E}_{\mathrm{ex}} = 0$, $\boldsymbol{B}_{\mathrm{ex}} = \varepsilon \boldsymbol{B}_0$ with a constant $\boldsymbol{B}_0$, the initial $\boldsymbol{E}$ field odd, and the initial $\boldsymbol{B}$ field even in $\boldsymbol{x}$. Then the equations of motion simplify. We recall them for completeness,

$$\partial_t \boldsymbol{B}(\boldsymbol{x},t) = -\nabla \times \boldsymbol{E}(\boldsymbol{x},t), \quad \partial_t \boldsymbol{E}(\boldsymbol{x},t) = \nabla \times \boldsymbol{B}(\boldsymbol{x},t) - (\boldsymbol{\omega}(t) \times \boldsymbol{x}) e\varphi(\boldsymbol{x}), \tag{10.37}$$

$$\nabla \cdot \boldsymbol{E}(\boldsymbol{x},t) = e\varphi(\boldsymbol{x}), \quad \nabla \cdot \boldsymbol{B}(\boldsymbol{x},t) = 0, \tag{10.38}$$

together with Newton's rotational equations of motion

$$I_{\mathrm{b}} \frac{\mathrm{d}}{\mathrm{d}t} \boldsymbol{\omega} = e \int \mathrm{d}^3 x \varphi(\boldsymbol{x}) \boldsymbol{x} \times \big( \boldsymbol{E}(\boldsymbol{x},t) + (\boldsymbol{\omega}(t) \times \boldsymbol{x}) \times (\varepsilon \boldsymbol{B}_0 + \boldsymbol{B}(\boldsymbol{x},t)) \big). \tag{10.39}$$

To obtain the effective dynamics let us first argue statically. The angular momentum, $\boldsymbol{s}$, of the charge soliton is the sum $\boldsymbol{s} = \boldsymbol{s}_{\mathrm{b}} + \boldsymbol{s}_{\mathrm{f}}$ with $\boldsymbol{s}_{\mathrm{b}} = I_{\mathrm{b}} \boldsymbol{\omega}$ and

$$\boldsymbol{s}_{\mathrm{f}} = \int \mathrm{d}^3 x \boldsymbol{x} \times (\boldsymbol{E} \times \boldsymbol{B}) \tag{10.40}$$

for $\boldsymbol{E}$, $\boldsymbol{B}$ the charge soliton field at $\boldsymbol{v} = 0$ and $\boldsymbol{\omega}$. Inserting from (10.28)–(10.32) we obtain

$$\boldsymbol{s}_{\mathrm{f}} = I_{\mathrm{f}} \boldsymbol{\omega} \quad \text{with} \quad I_{\mathrm{f}} = \frac{2}{3} e^2 \int \mathrm{d}^3 k |\nabla_k \widehat{\varphi}|^2 |\boldsymbol{k}|^{-2}. \tag{10.41}$$

Therefore

$$\boldsymbol{s} = (I_{\mathrm{b}} + I_{\mathrm{f}}) \boldsymbol{\omega}. \tag{10.42}$$

The external torque is $\boldsymbol{\mu} \times \boldsymbol{B}_{\mathrm{ex}}$ with the magnetic moment

$$\boldsymbol{\mu} = \mu \boldsymbol{\omega}, \quad \mu = \frac{1}{3} e \int \mathrm{d}^3 x \varphi(\boldsymbol{x}) \boldsymbol{x}^2, \tag{10.43}$$

and thus the spin precession reads

$$\frac{\mathrm{d}}{\mathrm{d}t} \boldsymbol{s} = \boldsymbol{\mu} \times \boldsymbol{B}_{\mathrm{ex}}, \quad (I_{\mathrm{b}} + I_{\mathrm{f}}) \frac{\mathrm{d}}{\mathrm{d}t} \boldsymbol{\omega} = \mu \boldsymbol{\omega} \times \boldsymbol{B}_{\mathrm{ex}}. \tag{10.44}$$

The conventional definition of the gyromagnetic ratio $g$ is through

$$\frac{\mathrm{d}}{\mathrm{d}t}\omega = g\frac{e}{2m}\omega \times \boldsymbol{B}_{\mathrm{ex}} , \tag{10.45}$$

where $m$ is the mass of the particle; compare with the BMT equation (10.7) for small velocities. Equating (10.44) and (10.45) we deduce the effective $g$-factor of the Abraham model as

$$g = \frac{(\mu/e)2m}{I_{\mathrm{b}} + I_{\mathrm{f}}} = \frac{1 + \frac{2}{3}(e^2/m_{\mathrm{b}}) \int \mathrm{d}^3k |\widehat{\varphi}|^2 |\boldsymbol{k}|^{-2}}{1 + (e^2/m_{\mathrm{b}}) \int \mathrm{d}^3k |\nabla_{\boldsymbol{k}}\widehat{\varphi}|^2 |\boldsymbol{k}|^{-2} / \int \mathrm{d}^3k |\nabla_{\boldsymbol{k}}\widehat{\varphi}|^2} . \tag{10.46}$$

For $e \to 0$ we obtain $g = 1$, as it has to be. In the opposite limit, $m_{\mathrm{b}} \to 0$, only the second summands survive. We did not discover any simple bounds, but for a uniformly charged sphere and ball the integrals have already been computed at the end of section 10.1. One obtains with $R = R_\varphi$ the radius of the sphere, respectively ball,

$$g_{\mathrm{sphere}} = \frac{1 + (e^2/4\pi Rm_{\mathrm{b}})(2/3)}{1 + (e^2/4\pi Rm_{\mathrm{b}})(1/3)} , \quad g_{\mathrm{ball}} = \frac{1 + (e^2/4\pi Rm_{\mathrm{b}})(4/5)}{1 + (e^2/4\pi Rm_{\mathrm{b}})(2/7)} . \tag{10.47}$$

Thus $g_{\mathrm{sphere}} \to 2$, respectively $g_{\mathrm{ball}} \to 14/5$, for $Rm_{\mathrm{b}} \to 0$. For $g = 2$ the spin and orbital precession are exactly in phase, whereas for $g = 1$ the spin turns once during two cyclotron revolutions.

To provide dynamical support we follow the scheme of chapter 7. One integrates (10.37), (10.38) and inserts in the Lorentz torque taking into account that the initial fields decay quickly. Then

$$I_{\mathrm{b}}\frac{\mathrm{d}}{\mathrm{d}t}\omega(t) = \varepsilon\mu\omega(t) \times \boldsymbol{B}_0 + \boldsymbol{N}_{\mathrm{self}}(t) , \tag{10.48}$$

where, after some rearrangement, the retarded torque simplifies to

$$\begin{aligned} \boldsymbol{N}_{\mathrm{self}}(t) = \int_0^t \mathrm{d}s \frac{2}{3} e^2 \int \mathrm{d}^3k |\nabla_{\boldsymbol{k}}\widehat{\varphi}|^2 \\ \times \Big( -(\cos|\boldsymbol{k}|(t-s))\omega(s) + \frac{1}{|\boldsymbol{k}|}(\sin|\boldsymbol{k}|(t-s))\omega(t) \times \omega(s) \Big) . \end{aligned} \tag{10.49}$$

Let us denote the solution to (10.48) by $\omega^\varepsilon(t) = \omega(\varepsilon t)$. We insert this ansatz in (10.49) and Taylor-expand. Then

$$\begin{aligned}
N^\varepsilon_{\text{self}}(\varepsilon^{-1}t) &= \int_0^{\varepsilon^{-1}t} \mathrm{d}s \frac{2}{3}e^2 \int \mathrm{d}^3k |\nabla_{\boldsymbol{k}}\widehat{\varphi}|^2 \Big( -(\cos|\boldsymbol{k}|(\varepsilon^{-1}t - s))\omega(\varepsilon s) \\
&\quad + \frac{1}{|\boldsymbol{k}|}(\sin|\boldsymbol{k}|(\varepsilon^{-1}t - s))\omega(\varepsilon t) \times \omega(\varepsilon s)\Big) \\
&\cong \int_0^{\varepsilon^{-1}t} \mathrm{d}s \frac{2}{3}e^2 \int \mathrm{d}^3k |\nabla_{\boldsymbol{k}}\widehat{\varphi}|^2 \Big( -(\cos|\boldsymbol{k}|s)\big(\omega(t) - \varepsilon s\dot{\omega}(t) \\
&\quad + \frac{1}{2}\varepsilon^2 s^2 \ddot{\omega}(t)\big) + \frac{1}{|\boldsymbol{k}|}(\sin|\boldsymbol{k}|s)\omega(t)\big(\omega(t) \\
&\quad - \varepsilon s\dot{\omega}(t) + \frac{1}{2}\varepsilon^2 s^2 \ddot{\omega}(t)\big)\Big). \qquad (10.50)
\end{aligned}$$

Let

$$I_p = \int_0^\infty \mathrm{d}t\, t^p \int \mathrm{d}^3k |\nabla_{\boldsymbol{k}}\widehat{\varphi}|^2 \frac{1}{|\boldsymbol{k}|} \sin|\boldsymbol{k}|t\,, \quad J_p = \int_0^\infty \mathrm{d}t\, t^p \int \mathrm{d}^3k |\nabla_{\boldsymbol{k}}\widehat{\varphi}|^2 \cos|\boldsymbol{k}|t\,. \qquad (10.51)$$

Then, using the fact that $\widehat{\varphi}$ is radial,

$$\begin{aligned}
I_0 &= \int \mathrm{d}^3k |\nabla_{\boldsymbol{k}}\widehat{\varphi}|^2 |\boldsymbol{k}|^{-2}\,, \quad I_1 = 0\,, \\
I_2 &= \frac{1}{4\pi}\int \mathrm{d}^3x \int \mathrm{d}^3x' \varphi(\boldsymbol{x})\varphi(\boldsymbol{x}')\boldsymbol{x}\cdot\boldsymbol{x}'|\boldsymbol{x}-\boldsymbol{x}'| = -\frac{1}{2\pi}\int \mathrm{d}^3k |\nabla_{\boldsymbol{k}}\widehat{\varphi}|^2 |\boldsymbol{k}|^{-4}\,,
\end{aligned} \qquad (10.52)$$

and

$$J_0 = 0, \quad J_p = -pI_{p-1}, \quad p = 1, 2, \ldots. \qquad (10.53)$$

Therefore to order $\varepsilon^2$

$$N^\varepsilon_{\text{self}}(t) = -\varepsilon \frac{2}{3}e^2 I_0 \dot{\omega}(t) + \varepsilon^2 \frac{1}{3}e^2 I_2 \omega(t) \times \ddot{\omega}(t)\,, \qquad (10.54)$$

and inserted in (10.48)

$$I_{\text{b}}\varepsilon\dot{\omega}(t) = \varepsilon\mu\omega(t) \times \boldsymbol{B}_0 - \varepsilon I_{\text{f}}\dot{\omega}(t) + \varepsilon^2 \frac{1}{3}e^2 I_2 \omega(t) \times \ddot{\omega}(t)\,, \qquad (10.55)$$

where $I_{\text{f}} = 2e^2 I_0/3$ in agreement with the static result (10.41).

Beyond the renormalization of $I_{\text{b}}$ we have also obtained the radiation reaction $\omega(t) \times \ddot{\omega}(t)$. As for the translational degrees of freedom only the solution on the center manifold is of physical relevance. To compute the effective dynamics we regard (10.44) as the unperturbed dynamics and reinsert in (10.55). To be somewhat more general let us take $\boldsymbol{B}_0$ to be time dependent and varying on the slow time scale. One obtains

$$(I_\mathrm{b} + I_\mathrm{f})\dot{\boldsymbol{\omega}} = \mu\boldsymbol{\omega} \times \boldsymbol{B}_0 + \varepsilon e^2\big(\mu I_2/3(I_\mathrm{b} + I_\mathrm{f})\big)\big(\dot{\boldsymbol{\omega}}(\boldsymbol{\omega} \cdot \boldsymbol{B}_0) + (\boldsymbol{\omega} \times (\dot{\boldsymbol{B}}_0 \times \boldsymbol{\omega}))\big) . \tag{10.56}$$

Since $\omega^2$ is conserved under (10.56), the radiation reaction only modifies the frequency of gyration to order $\varepsilon$. A second-order term like $\ddot{\boldsymbol{\omega}}$ would lead to friction in the effective equation. As can be seen from (10.53), its prefactor $J_2$ vanishes and radiation damping appears only at order $\varepsilon^4$ through $I_4\dddot{\boldsymbol{\omega}}$.

## Notes and references

### *Section 10.1*

BMT is an acronym for Bargmann, Michel and Telegdi (1959). The BMT equation is explained in Jackson (1999). Bailey and Picasso (1970) is an informative article on how the BMT equation is used in the analysis of the high-precision measurements of the electron and muon $g$-factor. The BMT equation with $g = 2$ is the semiclassical limit of the Dirac equation (Rubinow and Keller 1963; Bolte and Keppeler 1999; Spohn 2000b; Panati *et al.* 2002a). Appel and Kiessling (2001) compute the effective parameters for a charge distribution concentrated on a sphere.

Just as for translational degrees of freedom, one way to guess the effective spin dynamics is to impose Lorentz invariance. In addition, one could require that the equations of motion come from a Lagrangian action. In full generality, including an electric dipole moment, this program is carried out by Bhabha (1939), Bhabha and Corben (1941) with earlier work by Frenkel (1926). Alternative approaches are compared in Corben (1961) and Nyborg (1962). Concise summaries are Barut (1964), who discusses also how the BMT equation fits into the general scheme, Teitelbom *et al.* (1980), and Rohrlich (1990). A more microscopic approach would be to carry out the adiabatic limit for the Lorentz model of section 2.5. In Nodvik's version of the model such an expansion is pushed to the order where translational and rotational degrees of freedom couple (Nodvik 1964).

The Lorentz model simplifies if initial data are assumed such that the particle moves at constant velocity. Then translational and rotational degrees of freedom decouple. Appel and Kiessling (2002) study the existence of solutions and their long-time limit. In the adiabatic limit, compare with section 10.3, the angular momentum responds to an external torque through the effective gyromagnetic ratio of (10.9).

## Section 10.2

The nonrelativistic model of a rotating charge is introduced by Abraham (1903) and studied by Herglotz (1903), Schwarzschild (1903), and Thomas (1927). Schwarzschild (1903) notes that a stationary solution exists only if $\boldsymbol{\omega}$ is either parallel or orthogonal to $\boldsymbol{v}$. Kiessling (1999) remarks that the standard form of the total angular momentum is conserved only if the inner rotation of the charged particle is included.

## Section 10.3

Grandy and Aghazadeh (1982) compute the gyromagnetic ratio to order $e^2$. The validity of the equations of motion (10.56) is proved in Imaikin *et al.* (2004).

# 11
# Many charges

There is little effort in extending the Abraham model to several particles. We label their positions and velocities as $\boldsymbol{q}_j(t)$, $\boldsymbol{v}_j(t)$, $j = 1, \dots, N$. The $j$-th particle has bare mass $m_{\mathrm{b}j}$ and charge $e_j$, where for simplicity the form factor $\widehat{\varphi}$ is assumed to be the same for all particles. The motion of each particle is governed by the Lorentz force as before, and the current in the Maxwell equations now becomes the sum over the single-particle currents. Therefore the equations of motion read

$$
\begin{aligned}
c^{-1}\partial_t \boldsymbol{B}(\boldsymbol{x}, t) &= -\nabla \times \boldsymbol{E}(\boldsymbol{x}, t)\,, \\
c^{-1}\partial_t \boldsymbol{E}(\boldsymbol{x}, t) &= \nabla \times \boldsymbol{B}(\boldsymbol{x}, t) - \sum_{j=1}^{N} e_j \varphi(\boldsymbol{x} - \boldsymbol{q}_j(t)) c^{-1} \boldsymbol{v}_j(t)\,, \\
\nabla \cdot \boldsymbol{E}(\boldsymbol{x}, t) &= \sum_{j=1}^{N} e_j \varphi(\boldsymbol{x} - \boldsymbol{q}_j(t))\,, \quad \nabla \cdot \boldsymbol{B}(\boldsymbol{x}, t) = 0\,,
\end{aligned}
\tag{11.1}
$$

$$
\frac{\mathrm{d}}{\mathrm{d}t}\big(m_{\mathrm{b}i}\,\gamma_i \boldsymbol{v}_i(t)\big) = e_i \big(\boldsymbol{E}_\varphi(\boldsymbol{q}_i(t), t) + c^{-1}\boldsymbol{v}_i(t) \times \boldsymbol{B}_\varphi(\boldsymbol{q}_i(t), t)\big)\,, \tag{11.2}
$$

where $i = 1, \dots, N$ with $\gamma_i = (1 - (\boldsymbol{v}_i/c)^2)^{-1/2}$.

There are no external forces. The force acting on a given particle is due to the other particles, as mediated through the Maxwell field, and to the self-force, which we have discussed already at length. If two particles are at a distance of only a few times $R_\varphi$, then they interact strongly with forces which depend on the details of the phenomenological and unknown charge distribution. Thus physically we trust our model only if particles are far apart on the scale set by $R_\varphi$.

## 11.1 Retarded interaction

Let us take as a starting point the condition that initially particles are far apart, thus $|\boldsymbol{q}_i^0 - \boldsymbol{q}_j^0| = \mathcal{O}(\varepsilon^{-1} R_\varphi)$. The velocities are less than $c$, not necessarily small, and the initial fields are the linear superposition of $N$ charge soliton fields corresponding to the initial conditions $\boldsymbol{q}_i^0, \boldsymbol{v}_i^0$, $i = 1, \ldots, N$. To understand the scales involved it is convenient to switch to macroscopic coordinates, which simply amounts to replacing in (11.1), (11.2) $e_j$ by $\sqrt{\varepsilon} e_j$ and $\varphi$ by $\varphi_\varepsilon$ with $\varphi_\varepsilon(\boldsymbol{x}) = \varepsilon^{-3}\varphi(\varepsilon^{-1}\boldsymbol{x})$; compare with the second half of section 6.1. Then $|\boldsymbol{q}_i^0 - \boldsymbol{q}_j^0| = \mathcal{O}(1)$.

We insert the solution of the inhomogeneous Maxwell–Lorentz equations (11.1) into the Lorentz force of (11.2). The forces are additive and the force on particle $i$ naturally splits into a self-force ($i = j$) and a mutual force ($i \neq j$). For the self-force one uses the Taylor expansion of chapter 7. Thereby the mass is renormalized and the next order is the radiation reaction. For the mutual force we recall that in section 7.2 it was shown already that, to leading order, the field generated by charge $j$ is the Liénard–Wiechert field. Thus, one obtains as retarded equations of motion

$$
\begin{aligned}
m_i(\boldsymbol{v}_i)\dot{\boldsymbol{v}}_i = &\sum_{\substack{j=1\\ j\neq i}}^{N} \varepsilon e_i\big(\boldsymbol{E}_{\mathrm{ret}j}(\boldsymbol{q}_i, t) + \boldsymbol{v}_i \times \boldsymbol{B}_{\mathrm{ret}j}(\boldsymbol{q}_i, t)\big) \\
&+ \varepsilon(e_i^2/6\pi)\big[\gamma_i^4(\boldsymbol{v}_i\cdot\ddot{\boldsymbol{v}}_i)\boldsymbol{v}_i + 3\gamma_i^6(\boldsymbol{v}_i\cdot\dot{\boldsymbol{v}}_i)^2\boldsymbol{v}_i + 3\gamma_i^4(\boldsymbol{v}_i\cdot\dot{\boldsymbol{v}}_i)\dot{\boldsymbol{v}}_i + \gamma^2\ddot{\boldsymbol{v}}_i\big],
\end{aligned}
\tag{11.3}
$$

$t \geq 0$, which accounts for the effective mass $m_i$ and the radiation reaction of the $i$-th particle; compare with Eq. (8.1). $\boldsymbol{E}_{\mathrm{ret}j}(\boldsymbol{x}, t)$ equals (2.24) with $e$ replaced by $e_j$, $\boldsymbol{q}$ replaced by $\boldsymbol{q}_j$, and $t_{\mathrm{ret}}$ replaced by $t_{\mathrm{ret}j}$ which is implicitly defined through

$$
t_{\mathrm{ret}j} = t - |\boldsymbol{x} - \boldsymbol{q}_j(t_{\mathrm{ret}j})|\,. \tag{11.4}
$$

For $\boldsymbol{x} = \boldsymbol{q}_i$ the retarded time is of order 1. Similarly $\boldsymbol{B}_{\mathrm{ret}j}(\boldsymbol{x}, t)$ equals (2.25) with $\boldsymbol{q}$ replaced by $\boldsymbol{q}_j$ and $t_{\mathrm{ret}}$ replaced by $t_{\mathrm{ret}j}$. The strength $\varepsilon$ results from the charge, $\sqrt{\varepsilon} e_i$, and the scale factor $\sqrt{\varepsilon}$ in (8.47). Viewed differently, on the microscopic scale the force is of order (distance)$^{-2} = \varepsilon^2$ and thus of order $\varepsilon$ when accumulated over a time span $\varepsilon^{-1}$. To solve (11.3) one needs the trajectories for the whole past. Our assumption of no initial slip is equivalent to

$$
\boldsymbol{q}_i(t) + \boldsymbol{q}_i^0 + t\boldsymbol{v}_i^0\,,\ i = 1, \ldots, N\,,\ t \leq 0\,, \tag{11.5}
$$

which must be added to (11.3).

Using (11.3) one can estimate the size of the various contributions. The near fields of $\boldsymbol{E}_{\mathrm{ret}j}$ and $\boldsymbol{B}_{\mathrm{ret}j}$ are of order 1. Therefore the acceleration is of order $\varepsilon$, which implies that the far field of $\boldsymbol{E}_{\mathrm{ret}j}$ and $\boldsymbol{B}_{\mathrm{ret}j}$ is $\mathcal{O}(\varepsilon^2)$. The radiation reaction term involves $\ddot{v}_i$ and is therefore $\mathcal{O}(\varepsilon^3)$.

We see that the various contributions are well ordered in powers of $\varepsilon$. The forces are weak, however, and therefore over longer times the particles will move apart, which is of somewhat reduced interest. There are two limiting situations of physical relevance, which will be discussed in the following sections. One possibility is to take the initial velocity $|v_i/c| \ll 1$. Then to lowest order the particles interact through the static Coulomb potential and post-Coulombic corrections can be studied meaningfully. The other option is to let $N \to \infty$, which yields a kinetic description for charge densities as commonly used in plasma physics.

## 11.2 Limit of small velocities

We impose the condition that initially $|\boldsymbol{v}_j/c| \ll 1$. Then retardation effects should be negligible and the particles interact through the static Coulomb potential. According to the standard textbook recipe, $|\boldsymbol{v}_j/c| \ll 1$ is to be interpreted as $c \to \infty$. Indeed, from (11.1) one concludes $\boldsymbol{B} = 0$ and

$$\nabla \times \boldsymbol{E}(\boldsymbol{x}, t) = 0\,, \quad \nabla \cdot \boldsymbol{E}(\boldsymbol{x}, t) = \sum_{j=1}^{N} e_j \varphi(\boldsymbol{x} - \boldsymbol{q}_j(t))\,, \tag{11.6}$$

which together with Newton's equations of motion yields the desired result. Unfortunately, our argument fails on two counts. First, the interaction is obtained as the smeared Coulomb potential. More severely, in Newton's equations of motion only the bare mass of charge $i$ appears, whereas physically it should respond to forces with its renormalized mass. Of course, the reason is that $c \to \infty$ does not ensure charges to be far apart on the scale of $R_\varphi$.

To improve we require, as in the previous section, that the initial positions satisfy

$$|\boldsymbol{q}_i^0 - \boldsymbol{q}_j^0| = \mathcal{O}(\varepsilon^{-1} R_\varphi)\,, \quad i \neq j\,. \tag{11.7}$$

Then the force is of order $\varepsilon^2$. Under rescaling the dynamical variables should be of order 1 as $\varepsilon \to 0$. If in addition we demand the relation $\dot{\boldsymbol{q}} = \boldsymbol{v}$ to be preserved, the only choice remaining is

$$|\boldsymbol{v}_j| = \mathcal{O}(\sqrt{\varepsilon} c) \quad \text{and} \quad t = \varepsilon^{-3/2} R_\varphi / c\,. \tag{11.8}$$

Indeed, the accumulated force is of order $\sqrt{\varepsilon}$, which means that the magnitude of the velocity is preserved. We have arrived at the following scale transformation

$$\begin{aligned} &t=\varepsilon^{-3/2}t', \quad \boldsymbol{q}_j=\varepsilon^{-1}\boldsymbol{q}'_j, \quad v_j=\sqrt{\varepsilon}\boldsymbol{v}'_j, \\ &\boldsymbol{x}=\varepsilon^{-1}\boldsymbol{x}', \quad \boldsymbol{E}=\varepsilon^{3/2}\boldsymbol{E}', \quad \boldsymbol{B}=\varepsilon^{3/2}\boldsymbol{B}', \end{aligned} \tag{11.9}$$

where the primed quantities are considered to be of $\mathcal{O}(1)$. The field amplitudes are scaled by $\varepsilon^{3/2}$ so as to preserve the field energy. There is little risk in omitting the primes below. We set

$$\boldsymbol{q}_j^{\varepsilon}(t)=\varepsilon\boldsymbol{q}_j\big(\varepsilon^{-3/2}t\big), \quad \boldsymbol{v}_j^{\varepsilon}(t)=\varepsilon^{-1/2}\boldsymbol{v}_j(\varepsilon^{-3/2}t). \tag{11.10}$$

Then the rescaled Maxwell's and Newton's equations of motion are

$$\begin{aligned} &\sqrt{\varepsilon}\,\partial_t\boldsymbol{B}(\boldsymbol{x},t)=-\nabla\times\boldsymbol{E}(\boldsymbol{x},t), \\ &\sqrt{\varepsilon}\,\partial_t\boldsymbol{E}(\boldsymbol{x},t)=\nabla\times\boldsymbol{B}(\boldsymbol{x},t)-\sum_{j=1}^{N}\sqrt{\varepsilon}e_j\varphi_\varepsilon(\boldsymbol{x}-\boldsymbol{q}_j^{\varepsilon}(t))\sqrt{\varepsilon}\boldsymbol{v}_j^{\varepsilon}(t), \\ &\nabla\cdot\boldsymbol{E}(\boldsymbol{x},t)=\sum_{j=1}^{N}\sqrt{\varepsilon}e_j\varphi_\varepsilon(\boldsymbol{x}-\boldsymbol{q}_j^{\varepsilon}(t)), \quad \nabla\cdot\boldsymbol{B}(\boldsymbol{x},t)=0, \end{aligned} \tag{11.11}$$

$$\begin{aligned} \varepsilon\frac{\mathrm{d}}{\mathrm{d}t}\big(m_{\mathrm{b}i}(1-\varepsilon\boldsymbol{v}_i^{\varepsilon}(t)^2)^{-1/2}\boldsymbol{v}_i^{\varepsilon}(t)\big)=\sqrt{\varepsilon}e_i\big(&\boldsymbol{E}_{\varphi_\varepsilon}(\boldsymbol{q}_i^{\varepsilon}(t),t) \\ &+\sqrt{\varepsilon}\boldsymbol{v}_i^{\varepsilon}(t)\times\boldsymbol{B}_{\varphi_\varepsilon}(\boldsymbol{q}_i^{\varepsilon}(t),t)\big). \end{aligned} \tag{11.12}$$

On the new scale the velocity of light tends to infinity as $c/\sqrt{\varepsilon}$ and the charge distribution has total charge $\sqrt{\varepsilon}e_j$, finite electrostatic energy $m_{\mathrm{f}}$, and shrinks to a $\delta$-function as $\varphi_\varepsilon$. Recall that the scale parameter $\varepsilon$ is just a convenient way to order the magnitudes of the various contributions.

Before entering into more specific computations, it is useful first to sort out what should be expected. We follow our practice from before and denote positions and velocities of the comparison dynamics by $\boldsymbol{r}_j, \boldsymbol{u}_j$, $j=1,\ldots,N$, i.e. $\boldsymbol{q}_j^{\varepsilon}(t)\cong\boldsymbol{r}_j(t)$, $\boldsymbol{v}_j^{\varepsilon}(t)\cong\boldsymbol{u}_j(t)$. Since the velocities are small, the kinetic energy takes its nonrelativistic limit

$$T_0(\boldsymbol{u}_j)=\frac{1}{2}\Big(m_{\mathrm{b}j}+\frac{4}{3}m_{\mathrm{f}j}\Big)\boldsymbol{u}_j^2, \tag{11.13}$$

up to a constant; compare with (4.24). Note that the mass of the particle is renormalized through the interaction with the field. For small velocities, magnetic fields are small and retardation effects can be neglected. Thus the potential energy of the effective dynamics should be purely Coulombic and be given by

$$V_{\mathrm{coul}}(\boldsymbol{r}_1,\ldots,\boldsymbol{r}_N)=\frac{1}{2}\sum_{i\neq j=1}^{N}\frac{e_i e_j}{4\pi|\boldsymbol{r}_i-\boldsymbol{r}_j|}\,. \tag{11.14}$$

To obtain post-Coulombic corrections, one has to expand properly the self- and retarded forces, which we will carry to order $\varepsilon^{5/2}$ where the radiation reaction appears first. Since, as can be seen from (11.1), (11.2), the forces are additive, it suffices to consider two particles only. For initial conditions we choose the linear superposition of the two charge solitons corresponding to the initial data $\boldsymbol{q}_i^0$, $\boldsymbol{v}_i^0$, $i=1,2$. One solves the Maxwell equations and inserts them in the Lorentz force. As already explained, in the self-interaction the contribution from the initial fields vanishes for $t\geq\varepsilon\bar{t}_\varphi$. In the mutual interaction the initial fields take a time of order $\sqrt{\varepsilon}$ to reach the other particle and their contribution vanishes for $t\geq\sqrt{\varepsilon}|\boldsymbol{q}_1^0-\boldsymbol{q}_2^0|$. Thus for larger times one is allowed to insert in (11.2) the retarded fields only, which yields

$$\varepsilon\frac{\mathrm{d}}{\mathrm{d}t}\big(m_{\mathrm{b}1}\,\gamma_1\boldsymbol{v}_1^\varepsilon(t)\big)=\boldsymbol{F}_{\mathrm{ret},11}(t)+\boldsymbol{F}_{\mathrm{ret},12}(t)\,, \tag{11.15}$$

$$\varepsilon\frac{\mathrm{d}}{\mathrm{d}t}\big(m_{\mathrm{b}2}\,\gamma_2\boldsymbol{v}_2^\varepsilon(t)\big)=\boldsymbol{F}_{\mathrm{ret},21}(t)+\boldsymbol{F}_{\mathrm{ret},22}(t)\,, \tag{11.16}$$

where

$$\begin{aligned}\boldsymbol{F}_{\mathrm{ret},ij}(t)=e_ie_j&\int_0^t\mathrm{d}s\int\mathrm{d}^3k|\widehat{\varphi}(\varepsilon\boldsymbol{k})|^2\mathrm{e}^{\mathrm{i}\boldsymbol{k}\cdot(\boldsymbol{q}_i^\varepsilon(t)-\boldsymbol{q}_j^\varepsilon(s))}\\&\times\Big(-\varepsilon^{1/2}(|\boldsymbol{k}|^{-1}\sin(|\boldsymbol{k}|(t-s)/\sqrt{\varepsilon}))\mathrm{i}\boldsymbol{k}-\varepsilon(\cos(|\boldsymbol{k}|(t-s)/\sqrt{\varepsilon}))\boldsymbol{v}_j^\varepsilon(s)\\&+\varepsilon^{3/2}(|\boldsymbol{k}|^{-1}\sin(|\boldsymbol{k}|(t-s)/\sqrt{\varepsilon}))\boldsymbol{v}_i^\varepsilon(t)\times(\mathrm{i}\boldsymbol{k}\times\boldsymbol{v}_j^\varepsilon(s))\Big)\,,\end{aligned} \tag{11.17}$$

$i,j=1,2$.

For the self-interaction we set $\varepsilon\boldsymbol{k}=\boldsymbol{k}'$, $\varepsilon^{-3/2}t=t'$. Then

$$\begin{aligned}\boldsymbol{F}_{\mathrm{ret},11}(t)=\varepsilon^{-3/2}(e_1)^2&\int_0^\infty\mathrm{d}\tau\int\mathrm{d}^3k|\widehat{\varphi}(\boldsymbol{k})|^2\mathrm{e}^{\mathrm{i}\boldsymbol{k}\cdot(\boldsymbol{q}_1^\varepsilon(t)-\boldsymbol{q}_1^\varepsilon(t-\varepsilon^{3/2}\tau))/\varepsilon}\\&\times\Big(-\varepsilon^{1/2}(|\boldsymbol{k}|^{-1}\sin|\boldsymbol{k}|\tau)\mathrm{i}\boldsymbol{k}-\varepsilon(\cos|\boldsymbol{k}|\tau)\boldsymbol{v}_1^\varepsilon(t-\varepsilon^{3/2}\tau)\\&+\varepsilon^{3/2}(|\boldsymbol{k}|^{-1}\sin|\boldsymbol{k}|\tau)\boldsymbol{v}_1^\varepsilon(t)\times\big(\mathrm{i}\boldsymbol{k}\times\boldsymbol{v}_1^\varepsilon(t-\varepsilon^{3/2}\tau)\big)\Big)\,.\end{aligned} \tag{11.18}$$

One Taylor-expands as

$$\varepsilon^{-1}\big(\boldsymbol{q}_1^\varepsilon(t)-\boldsymbol{q}_1^\varepsilon(t-\varepsilon^{3/2}\tau)\big)=\varepsilon^{1/2}\tau\boldsymbol{v}-\frac{1}{2}\varepsilon^2\tau^2\dot{\boldsymbol{v}}+\frac{1}{6}\varepsilon^{7/2}\tau^3\ddot{\boldsymbol{v}},$$
$$\boldsymbol{v}_1^\varepsilon(t-\varepsilon^{3/2}\tau)=\boldsymbol{v}-\varepsilon^{3/2}\tau\dot{\boldsymbol{v}}+\frac{1}{2}\varepsilon^3\tau^2\ddot{\boldsymbol{v}}. \tag{11.19}$$

Then, up to errors of order $\varepsilon^3$,

$$\begin{aligned}\boldsymbol{F}_{\mathrm{ret},11}(t)=(e_1)^2\int_0^\infty\mathrm{d}\tau\int\mathrm{d}^3k|\widehat{\varphi}(\boldsymbol{k})|^2\Big(\varepsilon\big[-(|\boldsymbol{k}|^{-1}\sin|\boldsymbol{k}|\tau)\frac{1}{2}\tau^2(\boldsymbol{k}\cdot\dot{\boldsymbol{v}})\boldsymbol{k}\\
+(\cos|\boldsymbol{k}|\tau)\tau\dot{\boldsymbol{v}}\big]+\varepsilon^2\big[\big(-(|\boldsymbol{k}|^{-1}\sin|\boldsymbol{k}|\tau)\frac{1}{2}\tau^2(\boldsymbol{k}\cdot\dot{\boldsymbol{v}})\boldsymbol{k}\\
+(\cos|\boldsymbol{k}|\tau)\tau\dot{\boldsymbol{v}}\big)\Big(-\frac{1}{2}\tau^2(\boldsymbol{k}\cdot\boldsymbol{v})^2\Big)-(\cos|\boldsymbol{k}|\tau)\frac{1}{2}\tau^3(\boldsymbol{k}\cdot\dot{\boldsymbol{v}})(\boldsymbol{k}\cdot\boldsymbol{v})\boldsymbol{v}\\
+(|\boldsymbol{k}|^{-1}\sin|\boldsymbol{k}|\tau)\big(\tau^2(\boldsymbol{k}\cdot\boldsymbol{v})\boldsymbol{v}\times(\boldsymbol{k}\times\dot{\boldsymbol{v}})+\frac{1}{2}\tau^2(\boldsymbol{k}\cdot\dot{\boldsymbol{v}})(\boldsymbol{v}\times(\boldsymbol{k}\times\boldsymbol{v}))\big)\big]\\
+\varepsilon^{5/2}\big[(|\boldsymbol{k}|^{-1}\sin|\boldsymbol{k}|\tau)\frac{1}{6}\tau^3(\boldsymbol{k}\cdot\ddot{\boldsymbol{v}})\boldsymbol{k}-(\cos|\boldsymbol{k}|\tau)\frac{1}{2}\tau^2\ddot{\boldsymbol{v}}\big]\Big).\end{aligned} \tag{11.20}$$

For the mutual interaction we leave the $k$-integration and set $\varepsilon^{1/2}t=t'$. Then

$$\begin{aligned}\boldsymbol{F}_{\mathrm{ret},12}(t)=\sqrt{\varepsilon}e_1e_2\int_0^\infty\mathrm{d}\tau\int\mathrm{d}^3k|\widehat{\varphi}(\varepsilon\boldsymbol{k})|^2\mathrm{e}^{\mathrm{i}\boldsymbol{k}\cdot(\boldsymbol{q}_1^\varepsilon(t)-\boldsymbol{q}_2^\varepsilon(t-\sqrt{\varepsilon}\tau))}\\
\times\Big(-\varepsilon^{1/2}(|\boldsymbol{k}|^{-1}\sin|\boldsymbol{k}|\tau)\mathrm{i}\boldsymbol{k}-\varepsilon(\cos|\boldsymbol{k}|\tau)\boldsymbol{v}_2^\varepsilon(t-\sqrt{\varepsilon}\tau)\\
+\varepsilon^{3/2}(|\boldsymbol{k}|^{-1}\sin|\boldsymbol{k}|\tau)\boldsymbol{v}_1^\varepsilon(t)\times(\mathrm{i}\boldsymbol{k}\times\boldsymbol{v}_2^\varepsilon(t-\sqrt{\varepsilon}\tau))\Big).\end{aligned} \tag{11.21}$$

One Taylor-expands as

$$\boldsymbol{q}_1^\varepsilon(t)-\boldsymbol{q}_2^\varepsilon(t-\sqrt{\varepsilon}\tau)=\boldsymbol{r}+\varepsilon^{1/2}\tau\boldsymbol{v}_2-\frac{1}{2}\varepsilon\tau^2\dot{\boldsymbol{v}}_2+\frac{1}{6}\varepsilon^{3/2}\tau^3\ddot{\boldsymbol{v}}_2,$$
$$\boldsymbol{v}_1^\varepsilon(t)=\boldsymbol{v}_1,\quad\boldsymbol{v}_2^\varepsilon(t-\sqrt{\varepsilon}\tau)=\boldsymbol{v}_2-\varepsilon^{1/2}\tau\dot{\boldsymbol{v}}_2+\frac{1}{2}\varepsilon\tau^2\ddot{\boldsymbol{v}}_2 \tag{11.22}$$

with $\boldsymbol{r} = \boldsymbol{q}_1^\varepsilon(t) - \boldsymbol{q}_2^\varepsilon(t)$. Then, up to errors of order $\varepsilon^3$,

$$\begin{aligned}
\boldsymbol{F}_{\text{ret},12} &= e_1 e_2 \int_0^\infty \mathrm{d}\tau \int \mathrm{d}^3k |\widehat{\varphi}(\varepsilon \boldsymbol{k})|^2 \mathrm{e}^{\mathrm{i}\boldsymbol{k}\cdot\boldsymbol{r}} \Big( - \varepsilon(|\boldsymbol{k}|^{-1} \sin |\boldsymbol{k}|\tau)\, \mathrm{i}\boldsymbol{k} + \varepsilon^2 \big[ (|\boldsymbol{k}|^{-1} \sin |\boldsymbol{k}|\tau) \\
&\quad \times \big( - \frac{1}{2} \tau^2 (\boldsymbol{k} \cdot \dot{\boldsymbol{v}}_2)\boldsymbol{k} + \frac{1}{2} \tau^2 (\boldsymbol{k} \cdot \boldsymbol{v}_2)^2 \mathrm{i}\boldsymbol{k} + \boldsymbol{v}_1 \times (\mathrm{i}\boldsymbol{k} \times \boldsymbol{v}_2) \big) \\
&\quad + (\cos |\boldsymbol{k}|\tau)(\tau \dot{\boldsymbol{v}}_2 - \mathrm{i}\tau (\boldsymbol{k} \cdot \boldsymbol{v}_2)\boldsymbol{v}_2) \big] \\
&\quad + \varepsilon^{5/2} \big[ (|\boldsymbol{k}|^{-1} \sin |\boldsymbol{k}|\tau) \big( \frac{1}{6} \tau^3 (\boldsymbol{k} \cdot \ddot{\boldsymbol{v}}_2)\boldsymbol{k} - \frac{1}{2} \tau^3 (\boldsymbol{k} \cdot \boldsymbol{v}_2)(\boldsymbol{k} \cdot \dot{\boldsymbol{v}}_2)\mathrm{i}\boldsymbol{k} \\
&\quad - \frac{1}{6} \tau^3 (\boldsymbol{k} \cdot \boldsymbol{v}_2)^3 \boldsymbol{k} \big) - (\cos |\boldsymbol{k}|\tau) \frac{1}{2} \tau^2 \ddot{\boldsymbol{v}} \big] \Big) \\
&= (e_1 e_2 / 4\pi) \Big( - \varepsilon \nabla_{\boldsymbol{r}} |\boldsymbol{r}|^{-1} + \varepsilon^2 \big[ \big( \frac{1}{2} \nabla_{\boldsymbol{r}} (\dot{\boldsymbol{v}}_2 \cdot \nabla_{\boldsymbol{r}}) - \frac{1}{2} \nabla_{\boldsymbol{r}} (\boldsymbol{v}_2 \cdot \nabla_{\boldsymbol{r}})^2 \big) |\boldsymbol{r}| \\
&\qquad - (\dot{\boldsymbol{v}}_2 - \boldsymbol{v}_2 (\boldsymbol{v}_2 \cdot \nabla_{\boldsymbol{r}})) |\boldsymbol{r}|^{-1} + (\boldsymbol{v}_1 \times (\nabla_{\boldsymbol{r}} \times \boldsymbol{v}_2)) |\boldsymbol{r}|^{-1} \big] \\
&\qquad + \frac{2}{3} \frac{1}{4\pi} \varepsilon^{5/2} e_1 e_2 \ddot{\boldsymbol{v}}_2 \Big) .
\end{aligned} \tag{11.23}$$

We discuss each order separately, where we recall that in (11.15), (11.16) the acceleration is multiplied by $\varepsilon$. As anticipated, to order 1 one obtains the Coulomb dynamics with renormalized mass from $\boldsymbol{F}_{jj}(t)$. Let us define the Coulomb Lagrangian

$$L_{\text{coul}} = \sum_{j=1}^{N} \frac{1}{2} \Big( m_{\text{b}j} + \frac{4}{3} m_{\text{f}j} \Big) \boldsymbol{u}_j^2 - \frac{1}{2} \sum_{i \neq j=1}^{N} \frac{e_i e_j}{4\pi |\boldsymbol{r}_i - \boldsymbol{r}_j|} . \tag{11.24}$$

Then the comparison dynamics is

$$\frac{\mathrm{d}}{\mathrm{d}t} (\nabla_{\boldsymbol{u}_j} L_{\text{coul}}) - \nabla_{\boldsymbol{r}_j} L_{\text{coul}} = 0 , \quad j = 1, \dots , N , \tag{11.25}$$

with the error bounds

$$|\boldsymbol{q}_j^\varepsilon(t) - \boldsymbol{r}_j(t)| = \mathcal{O}(\varepsilon) , \quad |\boldsymbol{v}_j^\varepsilon(t) - \boldsymbol{u}_j(t)| = \mathcal{O}(\varepsilon) . \tag{11.26}$$

The first-order correction is $\mathcal{O}(\varepsilon)$. More conventionally the error is counted in powers of $|v/c|$ relative to the zeroth-order Coulomb dynamics. To convert, one only has to set $\varepsilon = 1$. The first correction is then of order $|v/c|^2$ ($= \mathcal{O}(\varepsilon)$, compare with (11.8)), and the next-order corrections $|v/c|^3$. The order $\varepsilon^2$ terms in (11.20), (11.23) combine in a simple fashion and yield the Darwin correction. Let us define the Darwin Lagrangian

$$L_{\text{darw}} = \sum_{j=1}^{N} \Big( \Big( m_{\text{b}j} + \frac{4}{3} m_{\text{f}j} \Big) \frac{1}{2} \boldsymbol{u}_j^2 + \varepsilon \Big( \frac{1}{8} m_{\text{b}j} + \frac{2}{15} m_{\text{f}j} \Big) c^{-2} \boldsymbol{u}_j^4 \Big)$$
$$- \frac{1}{2} \sum_{i \neq j=1}^{N} \frac{e_i e_j}{4\pi |\boldsymbol{r}_i - \boldsymbol{r}_j|} \Big[ 1 - \varepsilon \frac{1}{2c^2} \big( \boldsymbol{u}_i \cdot \boldsymbol{u}_j + (\boldsymbol{u}_i \cdot \widehat{\boldsymbol{r}}_{ij})(\boldsymbol{u}_j \cdot \widehat{\boldsymbol{r}}_{ij}) \big) \Big] \tag{11.27}$$

with $\widehat{\boldsymbol{r}}_{ij} = (\boldsymbol{r}_i - \boldsymbol{r}_j)/|\boldsymbol{r}_i - \boldsymbol{r}_j|$. In the first sum one recognizes the correction to the kinetic energy, while in the second-term corrections due to retardation and the magnetic field combine into a velocity-dependent potential. The comparison dynamics is governed by the improved Lagrangian,

$$\frac{\text{d}}{\text{d}t} \big( \nabla_{\boldsymbol{u}_j} L_{\text{darw}} \big) - \nabla_{\boldsymbol{r}_j} L_{\text{darw}} = 0 , \quad j = 1, \ldots, N , \tag{11.28}$$

with the error bounds

$$|\boldsymbol{q}_j^{\varepsilon}(t) - \boldsymbol{r}_j(t)| = \mathcal{O}(\varepsilon^{3/2}) , \quad |\boldsymbol{v}_j^{\varepsilon}(t) - \boldsymbol{u}_j(t)| = \mathcal{O}(\varepsilon^{3/2}) . \tag{11.29}$$

At order $|\boldsymbol{v}/c|^3$ one picks up terms proportional to $\ddot{\boldsymbol{v}}_j$. Remarkably, the prefactors in $\boldsymbol{F}_{\text{ret}jj}$ and $\boldsymbol{F}_{\text{ret}ji}$ are identical, and one obtains the comparison dynamics

$$\frac{\text{d}}{\text{d}t} \big( \nabla_{\boldsymbol{u}_j} L_{\text{darw}} \big) - \nabla_{\boldsymbol{r}_j} L_{\text{darw}} = \varepsilon^{3/2} \frac{e_j}{6\pi c^3} \sum_{i=1}^{N} e_i \ddot{\boldsymbol{u}}_i , \quad j = 1, \ldots, N . \tag{11.30}$$

The physical solutions have to be on the center manifold of (11.30). At the present level of precision it suffices to substitute the Lagrangian dynamics to lowest order, which yields

$$\frac{\text{d}}{\text{d}t} \big( \nabla_{\boldsymbol{u}_j} L_{\text{darw}} \big) - \nabla_{\boldsymbol{r}_j} L_{\text{darw}} = \varepsilon^{3/2} \frac{e_j}{6\pi c^3} \frac{1}{2} \sum_{\substack{i,i'=1 \\ i \neq i'}}^{N} \Big( \frac{e_i}{m_i} - \frac{e_{i'}}{m_{i'}} \Big) \frac{e_i e_{i'}}{4\pi |\boldsymbol{r}_i - \boldsymbol{r}_{i'}|^3}$$
$$\times \big( (\boldsymbol{u}_i - \boldsymbol{u}_{i'}) - 3 (\widehat{\boldsymbol{r}}_{ii'} \cdot (\boldsymbol{u}_i - \boldsymbol{u}_{i'})) \widehat{\boldsymbol{r}}_{ii'} \big) . \tag{11.31}$$

If the charge–mass ratio $e_j/m_j$ does not depend on $j$, the damping term is suppressed. The collection of charges has vanishing dipole moment. This can be seen also directly by considering the dipole moment $\boldsymbol{d} = \sum_{j=1}^{N} e_j \boldsymbol{q}_j = \sum_{j=1}^{N} (e_j/m_j) m_j \boldsymbol{q}_j$. If $(e_j/m_j) = \text{const.}$, then $\boldsymbol{d}$ equals the center of mass and $\ddot{\boldsymbol{d}} = 0$. Thus there is no dipole radiation. Only quadrupole radiation is allowed and radiation damping would appear at the scale $|\boldsymbol{v}/c|^5$.

We briefly return to the limit $c \to \infty$ from the beginning of this subsection. In fact, the expansion for computing the effective dynamics turns out to be not so drastically different as one might have anticipated. To lowest order the kinetic energy is $m_{\mathrm{b}j}\boldsymbol{u}_j^2/2$ and is modified to $(m_{\mathrm{b}j} + (4m_{\mathrm{f}j}/3c^2))\boldsymbol{u}_j^2/2$ at the Darwin order $|v/c|^2$. The correction to the quadratic behavior is visible only at order $|v/c|^4$. The friction term is identical to that of (11.30). Only in (11.27) must the Coulomb potential be smeared by the charge distribution $\varphi$.

## 11.3 The Vlasov–Maxwell equations

If $N$ is large, it is impractical to follow the trajectory of individual particles and, as widely used for example in plasma physics, a kinetic description is more appropriate. The basic object describing matter is now the distribution function $f_\alpha(\boldsymbol{x}, \boldsymbol{v}, t)$. For each component $\alpha$ it is a function on the one-particle phase space and defined through

$$f_\alpha(\boldsymbol{x}, \boldsymbol{v}, t)\mathrm{d}^3x\mathrm{d}^3v = \tfrac{1}{N}\big(\text{number of particles with charge } e_\alpha \text{ in the volume element } \mathrm{d}^3x\mathrm{d}^3v \text{ at time } t\big)\,.$$

The charge density of the $\alpha$-th component is then

$$\rho_\alpha(\boldsymbol{x}, t) = e_\alpha \int \mathrm{d}^3v f_\alpha(\boldsymbol{x}, \boldsymbol{v}, t) \tag{11.32}$$

and the total charge density

$$\rho(\boldsymbol{x}, t) = \sum_\alpha \rho_\alpha(\boldsymbol{x}, t)\,. \tag{11.33}$$

Similarly, the current density is

$$\boldsymbol{j}_\alpha(\boldsymbol{x}, t) = e_\alpha \int \mathrm{d}^3v \boldsymbol{v} f_\alpha(\boldsymbol{x}, \boldsymbol{v}, t), \quad \boldsymbol{j}(\boldsymbol{x}, t) = \sum_\alpha \boldsymbol{j}_\alpha(\boldsymbol{x}, t)\,. \tag{11.34}$$

The Maxwell field is governed by (2.2), (2.3) with $\rho$ from (11.33) and $\boldsymbol{j}$ from (11.34) as source terms. As densities on the one-particle phase space the distribution functions evolve according to

$$\partial_t f_\alpha(\boldsymbol{x}, \boldsymbol{v}, t) + \nabla_{\boldsymbol{x}} \cdot \big(\boldsymbol{v} f_\alpha(\boldsymbol{x}, \boldsymbol{v}, t)\big) + \big(\nabla_{\boldsymbol{v}} \cdot (m_\alpha \gamma)^{-1} \times (\boldsymbol{F}_\alpha - (\boldsymbol{v} \cdot \boldsymbol{F}_\alpha)\boldsymbol{v}) f_\alpha(\boldsymbol{x}, \boldsymbol{v}, t)\big) = 0 \tag{11.35}$$

with the Lorentz force

$$\boldsymbol{F}_\alpha = e_\alpha\big(\boldsymbol{E}(\boldsymbol{x}, t) + \boldsymbol{v} \times \boldsymbol{B}(\boldsymbol{x}, t)\big)\,. \tag{11.36}$$

The system of equations (2.2), (2.3), and (11.32)–(11.36) are called the Vlasov–Maxwell system. They were written down first by Vlasov in 1938 in the more conventional form where the velocity $v$ is replaced by the kinetic momentum $u = m_\alpha v/\sqrt{1-v^2}$. Then in (11.35), (11.36) $v$ is to be replaced by $u/\sqrt{m_\alpha^2 + u^2}$ and the Vlasov equation for the distribution function $f_\alpha(x, u, t)\mathrm{d}^3x\mathrm{d}^3u$ reads

$$\partial_t f_\alpha(x, u, t) + (m_\alpha^2 + u^2)^{-1/2} u \cdot \nabla_x f_\alpha(x, u, t) + F_\alpha \cdot \nabla_u f_\alpha(x, u, t) = 0 . \tag{11.37}$$

The static limit of the Vlasov–Maxwell system, namely $c \to \infty$ yielding $B = 0$, $\nabla \times E = 0$, $\nabla \cdot E = \rho$, is the Vlasov equation.

To establish the link to the Abraham model with $N$ charges it is convenient to start on the macroscopic scale, for simplicity for a single component, where

$$\partial_t B(x, t) = -\nabla \times E(x, t) ,$$

$$\partial_t E(x, t) = \nabla \times B(x, t) - \varepsilon \sum_{j=1}^{N} e\varphi_\varepsilon(x - q_j(t))v_j(t) ,$$

$$\nabla \cdot E(x, t) = \varepsilon \sum_{j=1}^{N} e\varphi_\varepsilon(x - q_j(t)) , \quad \nabla \cdot B(x, t) = 0 , \tag{11.38}$$

$$\frac{\mathrm{d}}{\mathrm{d}t}\big(m_\mathrm{b}\gamma_i v_i(t)\big) = e\big(E_{\varphi_\varepsilon}(q_i(t), t) + v_i(t) \times B_{\varphi_\varepsilon}(q_i(t), t)\big) . \tag{11.39}$$

We used here the freedom in the scale factor for the amplitude of the electromagnetic fields which accounts for an extra $\sqrt{\varepsilon}$ as compared to (6.11). On a formal level, the step to the Vlasov–Maxwell equation is immediate. We set $N = \varepsilon^{-1}$. The typical distance between particles is then $\varepsilon^{1/3} R_\varphi$ while the charge diameter is $\varepsilon R_\varphi \ll \varepsilon^{1/3} R_\varphi$. Thus particles are still very far apart. If we assume that at time $t$ the particle configuration is well approximated by a distribution function, then the source term of the Maxwell equations is of the form claimed in (11.33), (11.34). For (11.39) we have again to split into the self- and mutual parts. The self-part renormalizes the mass to $m(v)$ from (8.2) and the mutual part yields the force of (11.36) for the considered component. Put differently, in (11.39) the Maxwell fields $E$, $B$, smeared by $\varphi_\varepsilon$ and evaluated at $q_i(t)$, have a singular part which renormalizes the mass and a smooth part from all the other charges which is governed by (11.38). To carry out this program and to thereby derive the Vlasov–Maxwell equations along the lines indicated remains as a task for the future.

## 11.4 Statistical mechanics

For a system of many particles the first impetus is to investigate its equilibrium statistical mechanics. Although this means venturing into the domain of nonzero temperatures, let us see how much will be captured by our oversimplified model of matter. Statistical mechanics starts with a Hamiltonian defined on phase space.

Since this is also the starting point for canonical quantization, in Chapter 13, our discussion of the Pauli–Fierz model necessarily deals with the Lagrangian and Hamiltonian structure of the Abraham model. We preview the result (13.24) for a system of $N$ particles. The canonical coordinates for the particles are $(\boldsymbol{q}_j, \boldsymbol{p}_j)$, $j = 1, \ldots, N$. For the Maxwell field we adopt the Coulomb gauge, $\nabla \cdot \boldsymbol{A} = 0$. The canonical field variables are then $(\boldsymbol{A}(\boldsymbol{x}), -\boldsymbol{E}_\perp(\boldsymbol{x}))$, $\boldsymbol{x} \in \mathbb{R}^3$. Both fields are purely transverse, $\nabla \cdot \boldsymbol{A} = 0 = \nabla \cdot \boldsymbol{E}_\perp$. In terms of these variables the Hamiltonian for the Abraham model reads

$$H = \sum_{j=1}^{N} \frac{1}{2m_{\mathrm{b}j}} \big(\boldsymbol{p}_j - e_j \boldsymbol{A}_\varphi(\boldsymbol{q}_j)\big)^2 + \frac{1}{2} \int \mathrm{d}^3x \big(\boldsymbol{E}_\perp(\boldsymbol{x})^2 + (\nabla \times \boldsymbol{A}(\boldsymbol{x}))^2\big)$$

$$+ \frac{1}{2} \sum_{i,j=1}^{N} e_i e_j V_{\varphi\mathrm{coul}}(\boldsymbol{q}_i - \boldsymbol{q}_j)\,. \tag{11.40}$$

For simplicity we adopt the nonrelativistic kinetic energy, $\boldsymbol{p}^2/2m$. The potential $V_{\varphi\mathrm{coul}}$ originates from the longitudinal part of $\boldsymbol{E}$ and is defined through

$$V_{\varphi\mathrm{coul}}(\boldsymbol{q}) = \int \mathrm{d}^3k |\widehat{\varphi}(\boldsymbol{k})|^2 |\boldsymbol{k}|^{-2} \mathrm{e}^{\mathrm{i}\boldsymbol{k}\cdot\boldsymbol{q}}\,. \tag{11.41}$$

$V_{\varphi\mathrm{coul}}$ is the Coulomb potential smeared by the charge distribution $\varphi$, which appears twice, since both the $i$-th and the $j$-th particles carry a charge distribution.

The particles are confined to the box $\Lambda \subset \mathbb{R}^3$. We should also restrict the fields to the box $\Lambda$, but it will be somewhat simpler to regard them as filling all space. Then, formally, the equilibrium distribution at inverse temperature $\beta = 1/k_{\mathrm{B}}T$ is given by

$$\frac{1}{Z} \mathrm{e}^{-\beta H} \prod_{j=1}^{N} \chi_\Lambda(\boldsymbol{q}_j) \mathrm{d}^3 q_j \mathrm{d}^3 p_j \prod_{\boldsymbol{x}\in\mathbb{R}^3} \mathrm{d}^2 A(\boldsymbol{x}) \mathrm{d}^2 E_\perp(\boldsymbol{x})\,, \tag{11.42}$$

where $Z$ is the normalizing partition function and $\chi_\Lambda$ is the indicator function for the box $\Lambda$. Since the field energy is quadratic in $\boldsymbol{E}_\perp$ and $\boldsymbol{A}$, combined with the a priori measure and the normalization, it follows that $\boldsymbol{E}_\perp(\boldsymbol{x})$ and $\boldsymbol{A}(\boldsymbol{x})$ are Gaussian fields. We will only need $\boldsymbol{A}(\boldsymbol{x})$. It has mean zero and covariance

$$\langle \boldsymbol{A}(\boldsymbol{x}) \boldsymbol{A}(\boldsymbol{x}') \rangle_0 = \frac{1}{\beta} \int \mathrm{d}^3k |\boldsymbol{k}|^{-2} (\mathbb{1} - \widehat{\boldsymbol{k}} \otimes \widehat{\boldsymbol{k}}) \mathrm{e}^{\mathrm{i}\boldsymbol{k}\cdot(\boldsymbol{x}-\boldsymbol{x}')}\,. \tag{11.43}$$

From the experience with black-body radiation we have little trust in the statistics of the Maxwell field at large wave numbers and therefore concentrate on the particle degrees of freedom, only.

According to (11.40), (11.42) for fixed positions $\boldsymbol{q}_j$, $j = 1, \dots, N$, the momenta are Gaussian distributed with mean zero and covariance

$$\begin{aligned}\langle \boldsymbol{p}_i \boldsymbol{p}_j \rangle_{(q_1,\dots,q_N)} &= \langle (\boldsymbol{p}_i + e_i \boldsymbol{A}_\varphi(\boldsymbol{q}_i))(\boldsymbol{p}_j + e_j \boldsymbol{A}_\varphi(\boldsymbol{q}_j)) \rangle \\ &= \langle \boldsymbol{p}_i \boldsymbol{p}_j \rangle + e_i e_j \langle \boldsymbol{A}_\varphi(\boldsymbol{q}_i) \boldsymbol{A}_\varphi(\boldsymbol{q}_j) \rangle \\ &= \frac{1}{\beta}\Big( m_{\mathrm{b}i} \delta_{ij} \mathbb{1} + e_i e_j \int \mathrm{d}^3 k |\widehat{\varphi}(\boldsymbol{k})|^2 |\boldsymbol{k}|^{-2} (\mathbb{1} - \widehat{\boldsymbol{k}} \otimes \widehat{\boldsymbol{k}}) \mathrm{e}^{\mathrm{i}\boldsymbol{k}\cdot(\boldsymbol{q}_i - \boldsymbol{q}_j)} \Big) .\end{aligned} \tag{11.44}$$

Here in the first equality we shifted $\boldsymbol{p}_j$ by $e_j \boldsymbol{A}_\varphi(\boldsymbol{q}_j)$ which transforms $\langle\cdot\rangle$ to a Gaussian averaging factorized with respect to the $\boldsymbol{p}$'s and $\boldsymbol{A}$'s. For $i = j$ we recover the renormalized mass $m_{\mathrm{b}i} + m_{\mathrm{f}i}$. For $i \neq j$, there are momentum correlations which decay as $|\boldsymbol{q}_i - \boldsymbol{q}_j|^{-1}$ in the distance of the two particles.

For the distribution of the positions, we integrate first over $\boldsymbol{p}$ and then over $\boldsymbol{A}$ with the result

$$\frac{1}{Z} \mathrm{e}^{-\beta V} \prod_{j=1}^{N} \chi_\Lambda(\boldsymbol{q}_j) \mathrm{d}^3 q_j , \quad V = \frac{1}{2} \sum_{i,j=1}^{N} e_i e_j V_{\varphi\mathrm{coul}}(\boldsymbol{q}_i - \boldsymbol{q}_j) , \tag{11.45}$$

which is the standard Gibbs distribution for a Coulombic system of charges. The equilibrium statistics decouples into a positional part and, when conditioned on the positions, a Gaussian velocity part.

The equilibrium properties of Coulomb systems have been studied very extensively. To be specific, let us consider a two-component charge-symmetric plasma, which is neutral in the sense that both components have the same chemical potential. Since the system is very large, the natural quantities are the free energy and the correlation functions in the limit where the volume tends to infinity, $\Lambda \uparrow \mathbb{R}^3$. Indeed this limit has been established together with one major qualitative result, namely the validity of the Debye–Hückel theory at sufficiently low density. One inserts an extra charge at the origin into the system at thermal equilibrium. Then the charges of opposite sign screen in a statistical sense and the average charge density decays on the scale of the Debye length $l_{\mathrm{D}} = (4\pi e^2 \beta \rho)^{-1/2}$.

While we cannot enter into details, it might be useful to understand how the smearing of the charge distribution is needed even on the level of equilibrium statistical mechanics. Let us assume that the two components have equal charge of opposite sign, which means either $e_j = e$ or $e_j = -e$. Since $V_{\varphi\mathrm{coul}}$ is of positive type (the Fourier transform of a positive measure), one has

$$\begin{aligned}\frac{1}{2} \sum_{i \neq j = 1}^{N} e_i e_j V_{\varphi\mathrm{coul}}(\boldsymbol{q}_i - \boldsymbol{q}_j) &\geq -\frac{1}{2} \sum_{j=1}^{N} e_j^2 V_{\varphi\mathrm{coul}}(0) \\ &= -\Big( \frac{1}{2} e^2 \int \mathrm{d}^3 k |\widehat{\varphi}(\boldsymbol{k})|^2 |\boldsymbol{k}|^{-2} \Big) N .\end{aligned} \tag{11.46}$$

The energy is bounded from below by a constant proportional to $N$, which means that $V_{\varphi\mathrm{coul}}$ defines a thermodynamically stable interaction. To control the behavior for large $\Lambda$ one uses again the positive definiteness of $V_{\varphi\mathrm{coul}}$ and introduces the auxiliary Gaussian field $\phi(\boldsymbol{x})$, $\boldsymbol{x} \in \mathbb{R}^3$, with mean zero and covariance

$$\langle \phi(\boldsymbol{x})\phi(\boldsymbol{y})\rangle_G = V_{\varphi\mathrm{coul}}(\boldsymbol{x}-\boldsymbol{y}) , \tag{11.47}$$

which is well defined since $\widehat{V}_{\varphi\mathrm{coul}}(\boldsymbol{k}) \geq 0$. Then

$$\mathrm{e}^{-\beta V(\boldsymbol{q}_1,\dots,\boldsymbol{q}_N)} = \Big\langle \exp\Big[\mathrm{i}\sqrt{\beta}\sum_{j=1}^{N} e_j\phi(\boldsymbol{q}_j)\Big]\Big\rangle_G \tag{11.48}$$

and the grand canonical partition function becomes

$$\begin{aligned} Z_\Lambda &= \sum_{N=0}^{\infty}\frac{z^N}{N!}\sum_{\sigma_1,\dots,\sigma_N=\pm 1}\int_\Lambda \mathrm{d}^3 q_1 \dots \int_\Lambda \mathrm{d}^3 q_N \\ &\quad \times \exp\Big[-\beta e^2\frac{1}{2}\sum_{i,j=1}^{N}\sigma_i\sigma_j V_{\varphi\mathrm{coul}}(\boldsymbol{q}_i-\boldsymbol{q}_j)\Big] \\ &= \sum_{N=0}^{\infty}\frac{z^N}{N!}\Big\langle\Big(\int_\Lambda \mathrm{d}^3 q \sum_{\sigma=\pm 1}\exp\big[\mathrm{i}\sqrt{\beta}e\sigma\phi(\boldsymbol{q})\big]\Big)^N\Big\rangle_G \\ &= \Big\langle \exp\Big[2z\int_\Lambda \mathrm{d}^3 q \cos(\sqrt{\beta}e\phi(\boldsymbol{q}))\Big]\Big\rangle_G . \end{aligned} \tag{11.49}$$

Thus our system of charges has been converted into a field theory. The a priori measure $\langle\cdot\rangle_G$ is known as the Gaussian massless free field. In (11.49) it is perturbed by the interaction $\int_\Lambda \mathrm{d}^3 q \cos(\sqrt{\beta}e\phi(\boldsymbol{q}))$, which is clearly proportional to $|\Lambda|$. Thus we conclude that the pressure is extensive,

$$\log Z_\Lambda \cong |\Lambda| . \tag{11.50}$$

Despite the long-range forces, a neutral Coulomb system has extensive (volume-proportional) thermodynamics, provided the charges are somewhat smeared.

## Notes and references

### *Section 11.1*

On the quantized level the retarded interaction between neutral atoms shows as an attractive $R^{-7}$ decay of the interaction potential in contrast to the nonretarded, attractive van der Waals $R^{-6}$ law.

### Section 11.2

The Darwin Lagrangian is discussed in Jackson (1999). In Kunze and Spohn (2000c) the errors in (11.29) are estimated. Kunze and Spohn (2001) extend their analysis to include radiation reaction. The major novel difficulty is to properly match the initial conditions of the comparison dynamics (11.31). The next post-Coulombic correction, of order $|v/c|^4$, is computed formally by Landau and Lifshitz (1959), Barker and O'Connell (1980a, 1980b), and Damour and Schäfer (1991). It contains quadrupole corrections to the Coulomb interaction and terms proportional to $\dddot{v}$. It would be of interest to compare these results with the systematic expansion presented here.

A qualitatively rather similar problem arises in general relativity. The object of interest is a binary pulsar, like the famous Hulse–Taylor pulsar PSR 1913 + 16. It consists of two neutron stars, each with a mass of roughly 1.4 solar mass and a diameter of 10 km. They rotate around their common center of mass with a period of 7 h 45 min. The neutron stars move slowly with $|v/c| \cong 10^{-3}$. Since one of the neutron stars is rotating, it emits radio waves through which the orbit can be tracked with very high precision, in fact so precise that damping through the emission of gravitational waves can be verified quantitatively. I refer to Hulse (1994) and Taylor (1994). As in the case of charges, the theoretical challenge is to obtain the orbits of the two neutron stars in an expansion in $|v/c|$. For gravitation there is no dipole radiation and damping appears only at order $|v/c|^5$, with $|v/c|^0$ being the Newtonian orbit. Since experimental accuracy is expected to increase further (Will 1999) various groups have taken up the challenge with the present order at $|v/c|^7$ (Jaranowski and Schäfer 1998).

### Section 11.3

The relativistic Vlasov–Maxwell equations already appear in the original 1938 paper of Vlasov, see Vlasov (1961). The existence of solutions is studied at increasing level of generality in Glassey and Schaeffer (1991, 1997, 2000). In the nonretarded Vlasov–Poisson approximation the existence of solutions is now well understood (Pfaffelmoser 1992; Schaeffer 1991) and the link to the $N$-particle system has been established for a mollified potential (Neunzert 1975; Braun and Hepp 1977), a review being Spohn (1991). Physically the natural requirement is to have the charge diameter much smaller than the interparticle distance. Since this case is somewhat singular, a satisfactory derivation of the Vlasov–Poisson approximation is open, with a partial step towards its solution in Batt (2001).

As in the case of $N$ charges, the solution to the Vlasov–Maxwell system can be expanded in powers of $1/c$. The leading order is then Vlasov–Poisson, as established by Schaeffer (1986), which is corrected à la Darwin at order $c^{-2}$, as proved by Bauer and Kunze (2003). A one-component system can dissipate energy only through quadrupole radiation, which first appears at order $c^{-5}$. A two-component system emits dipole radiation at order $c^{-3}$. Properties of the formally derived Vlasov equation including radiative friction are studied by Kunze and Rendall (2001).

### *Section 11.4*

The statistical mechanics of charges plus Maxwell field is usually treated only on the level of thermodynamics (Alastuey and Appel 2000). Lebowitz and Lieb (1969) and Lieb and Lebowitz (1972) prove the existence of the, in fact shape-dependent, thermodynamic limit for Coulomb systems. A very readable review is Lieb and Lebowitz (1973). The existence of the infinite-volume limit of the correlation functions in the case of charge-symmetric systems is proved by Fröhlich and Park (1978). For the Debye–Hückel theory I recommend the excellent survey by Brydges and Martin (1999).

# 12

# Summary and preamble to the quantum theory

Within the framework of specific models for the coupling between charges and the electromagnetic field we have presented a fair amount of rather detailed arguments and computations. Thus before embarking on the quantized theory, it might be useful to summarize our main findings.

- *Extended charge.* To have a well-defined dynamics, a smeared charge distribution has to be used. This can be done either on the semirelativistic level of the Abraham model or in the form of a relativistically covariant theory, i.e. the Lorentz model. In the latter case internal rotation must be included by necessity.
- *Adiabatic regime.* Situations for which the classical electron theory can be experimentally tested fall in the adiabatic regime with a remarkable level of accuracy. Quantum mechanics must be used way before one leaves the domain of validity of the adiabatic approximation. A good example is the hydrogen atom in a bound state. Sufficiently far from the nucleus, which is certainly satisfied when at least a Bohr radius away from it, the assumptions for the adiabatic approximation are fulfilled and the dynamics of the electron is well governed by Eq. (9.14). On the other hand, it is known that the fluorescent spectrum of the hydrogen atom is accounted for only by quantum mechanics. To test the classical electron theory on the basis of this system is simply not feasible. Thus, in the range where the classical electron theory is applicable by necessity one is inside its adiabatic regime. In this regime the particle becomes point-like and is characterized by a charge, an effective mass, and, in the case of internal rotation, by an effective magnetic moment; compare with sections 4.2 and 10.1. From the full charge and mass distribution, which in principle constitute an infinite number of free parameters, only a few of their low-order moments are retained. They then enter in the Landau–Lifshitz equation (9.10), which governs the motion of the charge with great precision and properly accounts for friction through radiation. In addition,

the electromagnetic fields are determined from the Liénard–Wiechert potentials as generated by the motion of a point charge.

- *Point-charge limit.* As judged from the context of chapter 28 of the *Feynman Lectures*, "consistency" in the 1963 opinion of R. Feynman, compare with the citation at the end of section 3.3, refers to the point-charge limit $R_\varphi \to 0$. I agree, but as argued at length there is no need ever to take this limit. Letting the size $R_\varphi$ of the extended charge distribution shrink to zero yields objects of infinite mass. While the mere mathematical operation is admissible, it would result in a theory with very little physical content. The attempt to compensate through a proper adjustment of the bare mass fails, since the electromagnetic mass merely adds to the bare mass. Thus, the bare mass necessarily becomes negative which results in an unstable Hamiltonian.

The transition to the quantum theory of photons, electrons, and nuclei could hardly be less spectacular. I find it truly amazing that the simple rules of canonical quantization work so well for the Abraham model and thus open the gateway to a theory describing a vast territory of physical experience. Of course, just as the Abraham model, the theory is semirelativistic, and is thus also known as nonrelativistic quantum electrodynamics. No quantization of the Lorentz model seems to be available. In the relativistic domain one has to rely on conventional quantum electrodynamics.

We continue to adhere to the principle of restricting our attention to dynamical problems, for which the interaction between the charged particles and the photon field must be included. In particular, the emission and absorption of light by atoms and the scattering of photons from charges will play an important role. Adiabatic-type limits will be studied again. They show up in the limit of slow motion, where the photon field is approximated by the static Coulomb interaction, and in the derivation of the effective mass and the effective magnetic moment. To keep the topics within manageable size, many things had to be left out. In terms of applications the most serious omission is macroscopic electrodynamics, where the photon field is treated in the classical limit and matter is taken into account in a continuum description in terms of suitable electric and magnetic susceptibilities. Needless to say, they must be based on an atomistic quantum model of matter.

# Part II

## Quantum theory

# 13

# Quantizing the Abraham model

Classical theories must emerge from quantum mechanics and there is no reason to expect a simple recipe which would yield the physically correct quantum theory from the classical input. On the other hand, at least in the nonrelativistic domain, the rules of canonical quantization have served well and it is natural to apply them to the Abraham model. There is one immediate difficulty. Canonical quantization starts from identifying the canonical variables of the classical theory. Thus we first have to rewrite the equations of motion for the Abraham model in Hamiltonian form. For this purpose we adopt the Coulomb gauge, as usual, so as to eliminate the constraints. In the quantized version we thereby obtain the Pauli–Fierz Hamiltonian which has an obvious extension to include spin.

We have to ensure that the Pauli–Fierz Hamiltonian generates a unitary time evolution on the appropriate Hilbert space of physical states. Mathematically this means that we have to specify conditions under which the Pauli–Fierz Hamiltonian is a self-adjoint operator, an issue which can be satisfactorily resolved. Still, the true physical situation is more subtle and in fact not so well understood. It is related to the abundance of very low-energy photons, i.e the infrared problem, and to the arbitrariness of the cutoff at high energies, i.e. the ultraviolet problem. There are many items of interest before these, and it will take us a while to start discussing these subtleties.

Some words on our notation: In the beginning we keep $c$, $\hbar$, and later set them equal to one, mostly without notice. The vector notation, like $\boldsymbol{x}$, tends to be a little heavy, in particular since some of the objects become either operators or random variables. Therefore we stick with $x$, whose vector character has to be inferred from the context.

## 13.1* Lagrangian and Hamiltonian rewriting of the Abraham model

We consider $N$ charges coupled to the Maxwell field. Their motion is governed by (11.1), (11.2), which we repeat with the only difference that the relativistic kinetic energy is replaced by its Galilean cousin.

*(Inhomogeneous Maxwell–Lorentz equations)*

$$c^{-1}\partial_t B = -\nabla \times E\,, \quad c^{-1}\partial_t E = \nabla \times B - c^{-1} j\,, \tag{13.1}$$

$$\nabla \cdot E = \rho\,, \quad \nabla \cdot B = 0\,, \tag{13.2}$$

where the charge and current density are given by

$$\rho(x,t) = \sum_{j=1}^{N} e_j \varphi(x - q_j(t))\,, \quad j(x,t) = \sum_{j=1}^{N} e_j \varphi(x - q_j(t)) v_j(t) \tag{13.3}$$

satisfying charge conservation by fiat.

*(Newton's equations of motion)*

$$m_j \frac{\mathrm{d}}{\mathrm{d}t} v_j(t) = e_j\big(E_\varphi(q_j(t),t) + c^{-1} v_j(t) \times B_\varphi(q_j(t),t)\big)\,, \tag{13.4}$$

$j = 1, \ldots, N$. $\varphi$ is the charge distribution. It satisfies Condition $(C)$, Eq. (2.38).

The Lagrangian for a charge subject to external potentials is discussed in every text on classical mechanics. The Lagrangian of the coupled system, charges plus Maxwell field, can almost be guessed on that basis. We introduce the electromagnetic potentials through

$$E = -c^{-1}\partial_t A - \nabla\phi\,, \quad B = \nabla \times A\,, \tag{13.5}$$

hence guaranteeing $\nabla \cdot B = 0$ and the first half of (13.1), and regard as position-like variables $\{q_j, j = 1, \ldots, N, \phi(x), A(x), x \in \mathbb{R}^3\}$. Let us define the Lagrange density

$$\mathcal{L}_0(x) = \frac{1}{2}\big(E(x)^2 - B(x)^2\big) + c^{-1} j(x) \cdot A(x) - \rho(x)\phi(x)\,, \tag{13.6}$$

where, according to (13.3), $\rho, j$ depend on the positions and velocities of the charges. The Lagrangian of the Abraham model is then

$$L = \sum_{j=1}^{N} \frac{1}{2} m_j \dot{q}_j^2 + \int \mathrm{d}^3x\, \mathcal{L}_0(x)\,. \tag{13.7}$$

We only have to verify that the Euler–Lagrange equations for the action obtained from $L$ yield (13.1), (13.2), and (13.4). Indeed

$$\frac{\mathrm{d}}{\mathrm{d}t}\frac{\partial L}{\partial \dot{q}_j} - \frac{\partial L}{\partial q_j} = 0 \tag{13.8}$$

are Newton's equations of motion. Using '$\dot{}$' for $\partial_t$ as concise notation, variation with respect to $\phi$ yields

$$\frac{\mathrm{d}}{\mathrm{d}t}\frac{\delta L}{\delta \dot{\phi}} - \frac{\delta L}{\delta \phi} = 0\,, \tag{13.9}$$

which is equivalent to

$$-\nabla \cdot (c^{-1}\dot{A} + \nabla\phi) = \rho \tag{13.10}$$

and which we recognize as the first half of (13.2). Finally

$$\frac{\mathrm{d}}{\mathrm{d}t}\frac{\delta L}{\delta \dot{A}} - \frac{\delta L}{\delta A} = 0 \tag{13.11}$$

amounts to

$$c^{-1}(c^{-1}\ddot{A} + \nabla\dot{\phi}) = -\nabla \times (\nabla \times A) + c^{-1}j\,, \tag{13.12}$$

which is nothing but the second half of (13.1).

Since (13.9) represents only a constraint and is not an equation of motion, clearly we are using a redundant set of dynamical variables. Let us do the counting. We split the electromagnetic fields into longitudinal and transverse components,

$$E = E_{\parallel} + E_{\perp}\,, \quad B = B_{\parallel} + B_{\perp}\,. \tag{13.13}$$

Since $\nabla \cdot B = 0$, we have $B_{\parallel} = 0$. From $\nabla \cdot E = \rho$ we conclude

$$\widehat{E}_{\parallel} = -\mathrm{i}\widehat{\rho}k/k^2\,. \tag{13.14}$$

$E_{\perp}$ and $B_{\perp}$ satisfy a first-order evolution equation. Thus, in the sense of Lagrangian mechanics, there are two independent field degrees of freedom at every space point, while in (13.6) we employed four degrees of freedom.

We first eliminate $\phi$ through (13.10), i.e.

$$\widehat{\phi} = \frac{1}{k^2}\big(\mathrm{i}k \cdot c^{-1}\dot{\widehat{A}}_{\parallel} + \widehat{\rho}\big)\,. \tag{13.15}$$

Then, using Fourier transforms and Parseval's identity, (13.7) transforms to

$$L = \sum_{j=1}^{N} \frac{1}{2} m_j \dot{q}_j^2 + \frac{1}{2} \int \mathrm{d}^3k \big[ c^{-2} \dot{\widehat{A}}_\perp^* \cdot \dot{\widehat{A}}_\perp + k^{-2} \widehat{\rho}^* \widehat{\rho} - (k \times \widehat{A}_\perp^*) \cdot (k \times \widehat{A}_\perp) \big]$$
$$+ \frac{1}{2} \int \mathrm{d}^3k \big[ c^{-1} \widehat{j}^* \cdot \widehat{A} + c^{-1} \widehat{j} \cdot \widehat{A}^* - 2k^{-2} \widehat{\rho}^* \widehat{\rho}$$
$$- \mathrm{i} c^{-1} k^{-2} \big( \widehat{\rho}^* k \cdot \dot{\widehat{A}}_\| - \widehat{\rho} k \cdot \dot{\widehat{A}}_\|^* \big) \big] . \tag{13.16}$$

The term $\widehat{\rho}^* \widehat{\rho}$ depends only on the $q_j$'s and is recognized as the Coulomb potential,

$$\frac{1}{2} \int \mathrm{d}^3k k^{-2} \widehat{\rho}^* \widehat{\rho} = \frac{1}{2} \sum_{i,j=1}^{N} e_i e_j \int \mathrm{d}^3y \mathrm{d}^3y' \varphi(y) (4\pi |q_i - q_j - y + y'|)^{-1} \varphi(y')$$
$$= V_{\varphi\mathrm{coul}}(q_1, \ldots, q_N) . \tag{13.17}$$

The Coulomb potential is smeared by $\varphi$, which as before is indicated by the subscript. $\varphi$ appears twice, since both the $i$-th and the $j$-th particle carry a charge distribution. To simplify the last term of (13.16) we use the conservation law $\dot{\widehat{\rho}} + \mathrm{i}k \cdot \widehat{j} = 0$. Then

$$L = \sum_{j=1}^{N} \frac{1}{2} m_j \dot{q}_j^2 - V_{\varphi\mathrm{coul}} + \frac{1}{2} \int \mathrm{d}^3k \big[ c^{-2} \dot{\widehat{A}}_\perp^* \cdot \dot{\widehat{A}}_\perp - k^2 \widehat{A}_\perp^* \cdot \widehat{A}_\perp \big]$$
$$+ \frac{1}{2} \int \mathrm{d}^3k \big[ c^{-1} \widehat{j}^* \cdot \widehat{A}_\perp + c^{-1} \widehat{j} \cdot \widehat{A}_\perp^* \big]$$
$$+ c^{-1} \frac{\mathrm{d}}{\mathrm{d}t} \Big( \frac{1}{2} \int \mathrm{d}^3k |k|^{-1} \mathrm{i} \big[ \widehat{\rho} \widehat{A}_\|^* - \widehat{\rho}^* \widehat{A}_\| \big] \Big) . \tag{13.18}$$

Since $A_\|$ appears only inside a total time derivative, we have identified $A_\|$ as the second redundant field. To drop the redundant degrees of freedom, the simplest choice is to set $A_\| = 0$ by exploiting the gauge freedom, which means selecting the *Coulomb gauge* defined by

$$\nabla \cdot A = 0 . \tag{13.19}$$

The vector potential is purely transverse and we henceforth drop the subscript $\perp$. Transforming back to real space, the Lagrangian of the Abraham model reads

$$L = \sum_{j=1}^{N} \frac{1}{2} m_j \dot{q}_j^2 - V_{\varphi\mathrm{coul}} + \int \mathrm{d}^3x \mathcal{L}(x) \tag{13.20}$$

with the Lagrange density

$$\mathcal{L} = \frac{1}{2} \big[ (c^{-1} \dot{A})^2 - (\nabla \times A)^2 \big] + c^{-1} j \cdot A . \tag{13.21}$$

The transverse vector field $A(x)$, $x \in \mathbb{R}^3$, should be regarded as position-like variables.

The step from Lagrange to Hamilton is standard. One introduces the momentum $p_j$ canonically conjugate to $q_j$ by

$$p_j = m_j \dot{q}_j + c^{-1} e_j A_\varphi(q_j) . \tag{13.22}$$

For the momentum field canonically conjugate to $A$ we obtain

$$\frac{\delta L}{\delta \dot{A}} = c^{-2} \dot{A} = -c^{-1} E_\perp . \tag{13.23}$$

Then the Hamiltonian corresponding to $L$ reads

$$\begin{aligned} H = & \sum_{j=1}^{N} \frac{1}{2m_j} \big(p_j - c^{-1} e_j A_\varphi(q_j)\big)^2 + V_{\varphi\text{coul}} \\ & + \frac{1}{2} \int \mathrm{d}^3x \big[E_\perp(x)^2 + (\nabla \times A(x))^2\big] \end{aligned} \tag{13.24}$$

with the canonically conjugate pairs $q_j$, $p_j$ and $A(x)$, $-c^{-1}E_\perp(x)$.

## 13.2 The Pauli–Fierz Hamiltonian

In the form (13.24) we are ready to apply the rules of canonical quantization. The position and momentum of the $j$-th particle are elevated to algebraic objects (linear operators) which satisfy the commutation relations

$$[q_{i\alpha}, p_{j\beta}] = \mathrm{i}\hbar \delta_{\alpha\beta} \delta_{ij} , \tag{13.25}$$

$\alpha, \beta = 1, 2, 3$, $i, j = 1, \ldots, N$. In the Schrödinger representation, which will be used throughout, the Hilbert space of wave functions is

$$\mathcal{H}_\mathrm{p} = L^2(\mathbb{R}^{3N}) , \tag{13.26}$$

restricted to either the symmetric or antisymmetric subspace depending on whether the particles are bosons or fermions. Positions and momenta become

$$q_j = x_j , \quad p_j = -\mathrm{i}\hbar \nabla_{x_j} \tag{13.27}$$

as linear operators on $\mathcal{H}_\mathrm{p}$, i.e. if $\psi(x_1, \ldots, x_N) \in L^2(\mathbb{R}^{3N})$ is the wave function for the particles, then $q_j \psi(x_1, \ldots, x_N) = x_j \psi(x_1, \ldots, x_N)$, $p_j \psi(x_1, \ldots, x_N) = -\mathrm{i}\hbar \nabla_{x_j} \psi(x_1, \ldots, x_N)$.

For the fields $A(x)$, $-c^{-1}E_\perp(x)$ one is tempted to postulate commutation relations analogous to (13.25). The difficulty is that the quantization has to satisfy the transversality constraint (13.19) which is nonlocal. Fortunately it is linear and it becomes local in Fourier space as $k \cdot \widehat{A} = 0$. We thus introduce at each $k \in \mathbb{R}^3$ the standard dreibein

$$\widehat{k} = k/|k|\,, \quad e_1(k)\,, \quad e_2(k)\,, \tag{13.28}$$

which satisfies $\widehat{k} \cdot e_i(k) = 0$, $i = 1, 2$, $e_1(k) \cdot e_2(k) = 0$. There is some freedom of how to choose $e_1, e_2$, but the transverse projection $Q^{\perp}(k) = 1 - \widehat{k} \otimes \widehat{k} = 1 - |k|^{-2}|k\rangle\langle k|$ is unique. The two transverse components $e_1(k) \cdot \widehat{A}(k)$, $e_2(k) \cdot \widehat{A}(k)$ are regarded as independent variables, correspondingly for $-c^{-1}\widehat{E}_{\perp}$. Since $A$ is real, we have $\widehat{A}(k)^* = \widehat{A}(-k)$. Therefore one has to restrict $k$ to a half-space and take the real and imaginary parts of $\widehat{A}(k)$ as independent variables which are subject to the rules of canonical quantization. To achieve this goal it is helpful to introduce two standard Bose fields with creation and annihilation operators

$$a^*(k, \lambda)\,, \quad a(k, \lambda)\,, \quad k \in \mathbb{R}^3, \quad \lambda = 1, 2\,, \tag{13.29}$$

satisfying the canonical commutation relations

$$\begin{gathered}[a(k, \lambda), a^*(k', \lambda')] = \delta_{\lambda\lambda'}\delta(k - k')\,, \\ [a(k, \lambda), a(k', \lambda')] = 0\,, \quad [a^*(k, \lambda), a^*(k', \lambda')] = 0\,. \end{gathered} \tag{13.30}$$

◇ For a linear operator $A$, the adjoint operator is denoted by $A^*$. ◇

In terms of these Bose fields we set

$$\widehat{A}(k) = \sum_{\lambda=1,2} c\sqrt{\hbar/2\omega}\big(e_\lambda(k)a(k, \lambda) + e_\lambda(-k)a^*(-k, \lambda)\big)\,, \tag{13.31}$$

$$\widehat{E}_{\perp}(k) = \sum_{\lambda=1,2} \sqrt{\hbar\omega/2}\big(\mathrm{i}e_\lambda(k)a(k, \lambda) - \mathrm{i}e_\lambda(-k)a^*(-k, \lambda)\big) \tag{13.32}$$

with

$$\omega(k) = c|k|\,. \tag{13.33}$$

Then indeed $\widehat{A}$, $\widehat{E}_{\perp}$ are transverse, $\widehat{A}(k)^* = \widehat{A}(-k)$, $\widehat{E}(k)^* = \widehat{E}(-k)$, and

$$[e_\lambda(k) \cdot \widehat{A}(k), -c^{-1}e_{\lambda'}(k') \cdot \widehat{E}_{\perp}(k')^*] = \mathrm{i}\hbar\delta_{\lambda\lambda'}\delta(k - k') \tag{13.34}$$

which should be understood in analogy to (13.25).

In physical space (13.31), (13.32) become

$$A(x) = \sum_{\lambda=1,2} \int d^3k c\sqrt{\hbar/2\omega}\, e_\lambda(k)(2\pi)^{-3/2}\big(e^{ik\cdot x} a(k,\lambda) + e^{-ik\cdot x} a^*(k,\lambda)\big), \tag{13.35}$$

$$E_\perp(x) = \sum_{\lambda=1,2} \int d^3k \sqrt{\hbar\omega/2}\, e_\lambda(k)(2\pi)^{-3/2} i\big(e^{ik\cdot x} a(k,\lambda) - e^{-ik\cdot x} a^*(k,\lambda)\big). \tag{13.36}$$

Clearly $A^*(x) = A(x)$, $E^*_\perp(x) = E_\perp(x)$. The commutator (13.34) translates into

$$[A_\alpha(x), -c^{-1}E_{\perp\beta}(x')] = i\hbar\delta^\perp_{\alpha\beta}(x - x') \tag{13.37}$$

with the transverse delta function

$$\delta^\perp_{\alpha\beta}(x) = (2\pi)^{-3} \int d^3k e^{ik\cdot x}(\delta_{\alpha\beta} - \widehat{k}_\alpha \widehat{k}_\beta) = \frac{2}{3}\delta_{\alpha\beta}\delta(x) - \frac{1}{4\pi|x|^3}(\delta_{\alpha\beta} - 3\widehat{x}_\alpha\widehat{x}_\beta), \tag{13.38}$$

where $\widehat{x}_\alpha = x_\alpha/|x|$.

At this point we have left the classical world. $A(x)$, $E_\perp(x)$ and their Fourier transforms $\widehat{A}(k)$, $\widehat{E}_\perp(k)$ will now always stand for operator-valued fields. In the atomic and solid state physics literature by tradition one uses $a^\dagger$ as the boson creation operator adjoint to the annihilation operator $a$. We try to avoid such a profileration of symbols.

Next on the agenda should be the Fock representation of the Bose fields $a(k,\lambda)$ and the definition of $A(x)$, $E_\perp(x)$ as operator-valued fields acting on Fock space. But let us keep this for the beginning of the next section and proceed immediately to our goal, namely the Hamiltonian of the quantized Abraham model. All we have to do is to insert (13.35), (13.36) into the classical Hamiltonian. This results, after omitting the zero-point energy of photons, in the *(spinless) Pauli–Fierz Hamiltonian*

$$H = \sum_{j=1}^{N} \frac{1}{2m_j}\big(p_j - c^{-1}e_j A_\varphi(q_j)\big)^2 + V_{\varphi\mathrm{coul}} + H_\mathrm{f} \tag{13.39}$$

with the field Hamiltonian

$$H_\mathrm{f} = \sum_{\lambda=1,2} \int d^3k \hbar\omega(k) a^*(k,\lambda) a(k,\lambda). \tag{13.40}$$

There is no ambiguity in the operator ordering, since $p_j \cdot A_\varphi(q_j) = A_\varphi(q_j) \cdot p_j$ by the transversality condition (13.19). We recall that the spherically symmetric form factor $\widehat{\varphi}$ cuts couplings to the field, more explicitly

$$A_\varphi(q) = \sum_{\lambda=1,2} \int \mathrm{d}^3k c\sqrt{\hbar/2\omega}\, e_\lambda(k)\big(\widehat{\varphi}(k)\mathrm{e}^{\mathrm{i}k\cdot q}a(k,\lambda) + \widehat{\varphi}^*(k)\mathrm{e}^{-\mathrm{i}k\cdot q}a^*(k,\lambda)\big)\,. \tag{13.41}$$

To simplify notation, $\widehat{\varphi}$ will be assumed to be real, which can always be achieved through a suitable canonical transformation of the form $a(k,\lambda) \to \mathrm{e}^{\mathrm{i}\theta(k)}a(k,\lambda)$.

Two immediate generalizations are noted. First of all it is convenient to add external potentials $\phi_{\mathrm{ex}}$, $A_{\mathrm{ex}}$, where the abbreviation $e\phi_{\mathrm{ex}}(x) = V(x)$ will be employed frequently. This should be thought of as a limiting case of (13.39): we imagine that some charges are nailed down by letting their masses $m \to \infty$; then their kinetic term in (13.39) disappears and $V_{\varphi\mathrm{coul}}$ splits into an external potential plus an interaction potential for the movable charges. Similarly one can produce an external current which then generates $A_{\mathrm{ex}}$. Thus the external potentials are not quantized and are added into the Hamiltonian as in the classical theory which yields

$$H = \sum_{j=1}^{N} \frac{1}{2m_j}\big(p_j - c^{-1}e_jA_\varphi(q_j) - c^{-1}e_jA_{\mathrm{ex}}(q_j)\big)^2 + V_{\varphi\mathrm{coul}} + \sum_{j=1}^{N} e_j\phi_{\mathrm{ex}}(q_j) + H_{\mathrm{f}}\,. \tag{13.42}$$

Secondly, particles have spin. Of course, an electron has spin $\frac{1}{2}$. In our approximation nuclei are modeled as structureless particles carrying a nuclear spin, ranging from 0 to 9/2 according to experimental evidence. The classical theory is now of little help. The natural guess is to include spin as in the nonrelativistic one-particle Schrödinger theory. For a single electron in infinite space, no external potentials, the Hamiltonian then becomes

$$H = \frac{1}{2m}\big(\sigma\cdot(p - c^{-1}e_jA_\varphi(q))\big)^2 + H_{\mathrm{f}}\,, \tag{13.43}$$

where $\sigma = (\sigma_1, \sigma_2, \sigma_3)$ is the vector of Pauli spin-$\frac{1}{2}$ matrices. If necessary, one could include higher terms in (13.43) as they emerge from the Foldy–Wouthuysen expansion of the Dirac equation.

Having introduced the Pauli–Fierz Hamiltonian as the major player of the quantum part of the treatise, we pause for a while with a few general remarks.

*Zero-point energy*. In the Pauli–Fierz Hamiltonian we have omitted the zero-point energy $\int \mathrm{d}^3k\hbar\omega$, which is infinite. The Heisenberg equations of motion remain unaltered by this reset in the zero of energy. However, one has to be careful. If one wants to compute the change in energy of the quantized Maxwell field through the insertion of a pair of perfectly conducting plates, then in this energy difference the zero-point energy has to be properly handled; compare with section 13.6. A further change in the zero of the energy scale comes from the Coulomb self-interaction, namely the diagonal part

$$\frac{1}{2}\sum_{j=1}^{N} e_j^2 \int \mathrm{d}^3k |\widehat{\varphi}(k)|^2 k^{-2} \tag{13.44}$$

in the sum (13.17), which is finite only because the form factor cuts off the high-frequency modes.

*Range of validity, limiting cases.* The claimed range of validity of the Pauli–Fierz Hamiltonian is flabbergasting. To be sure, on the high-energy side, nuclear physics and high-energy physics are omitted. On the long-distance side, we could phenomenologically include gravity on the Newtonian level, but anything beyond that is ignored. As the bold claim goes, any physical phenomenon in between, including life on Earth, is accurately described through the Pauli–Fierz Hamiltonian (13.39) (and a suitably chosen initial wave function). There have been speculations that quantum mechanics is modified roughly at the $10^{-5}$ m scale. But so far there seems to be no evidence in this direction. On the contrary, whenever a detailed comparison with the theory can be made, it reassuringly seems to work well. Of course, our trust is not based on strict mathematical deductions from the Pauli–Fierz Hamiltonian. This is too difficult a program. Our confidence comes from well-studied limit cases. In the static limit we imagine turning off the interaction to the quantized part of the Maxwell field. This clearly results in Schrödinger particles interacting through a purely Coulombic potential, for which many predictions are accessible to experimental verification. But beware, even there apparently simple questions remain to be better understood. For example, the size of atoms as we see them in nature remains mysterious if only the Coulomb interaction and the Pauli exclusion principle are allowed. Another limiting case is a region completely free of charges. At standard field strengths there are sufficiently many photons per unit volume for the predictions from the quantized Maxwell field to match with the ones of the classical Maxwell field. As will be discussed, radiation phenomena are well grasped by the Pauli–Fierz Hamiltonian. These and many other limiting cases are the reason for regarding (13.39) as an accurate description of low-energy phenomena.

*Model parameters, renormalization.* If we focus our attention on (13.43), there are four model parameters: the mass $m$, the charge $e$, the gyromagnetic ratio $g = 2$, and the form factor $\widehat{\varphi}$. $c$ and $\hbar$, which also appear, are constants of nature. As discussed at length for the classical theory, what is observed experimentally is always the compound object consisting of the particle and its photon cloud. Thus $m$, $e$, $g$

have to be regarded as bare parameters and their observed value must be computed from the theory. The bare values are renormalized through the interaction with the Maxwell field. As will be shown below, the charge $e$ is not renormalized, since there is no vacuum polarization. One way to argue is to imagine two charged particles with a very large mass separated by a distance $R$. According to (13.39) their mutual force is then $e_1 e_2/4\pi R^2$ with the *bare* charges $e_1$, $e_2$. Further support is the response of a particle to slowly varying external potentials. In this adiabatic limit, $e$ enters in the effective equation with its bare value while $m$ and $g$ are renormalized. The Pauli–Fierz model is not in a position to predict the experimental value of the mass, since the bare mass is unaccessible, in principle. The renormalized (effective) mass has to be given as an empirical input, to which the bare mass is correspondingly adjusted. On the other hand, the dimensionless gyromagnetic ratio $g$ is a definite (though empirically slightly inaccurate) prediction of the theory; compare with sections 16.6 and 19.3.5. Perhaps the most unwanted feature of the Pauli–Fierz Hamiltonian is the form factor $\widehat{\varphi}$. The pragmatic attitude is to choose $\varphi$ with some taste. On the classical level we concluded that the form factor cannot be removed. In the limit $\varphi(x) \to \delta(x)$ the particle-like objects become infinitely heavy. The simple structure of the energy–momentum relation (4.11) does not allow for compensation, since in a stable theory the bare mass has to be positive. The quantum theory has a richer structure and it seems that one can carry out the limit $\varphi(x) \to \delta(x)$ and at the same time take $m \to 0$ such that the observed mass remains fixed. We will come back to this point in due course.

*The quest for a closed physical theory.* We have commented on this point already. But let us expand on it in the present context. The static limit of the Pauli–Fierz Hamiltonian, i.e. Schrödinger particles interacting through the static Coulomb potential, is a closed theory for electrons and nuclei. The Hamiltonian is a self-adjoint linear operator and generates a unitary time evolution. This is also the case for the quantized Maxwell field without charges. Of course, this does not mean that we have solved any physical problem. It just assures us of a definite mathematical framework within which consequences can be explored. One would hope to have such a secure foundation also for the Pauli–Fierz model and it remains to be seen how much of this program can be realized.

We still have to complete the story of the Pauli–Fierz model. One defines the time-evolved linear operator $A(t)$ through

$$A(t) = \mathrm{e}^{\mathrm{i}Ht/\hbar} A \mathrm{e}^{-\mathrm{i}Ht/\hbar} \tag{13.45}$$

in the Heisenberg picture. Then

$$\frac{\mathrm{d}}{\mathrm{d}t} A(t) = \frac{\mathrm{i}}{\hbar}[H, \mathrm{e}^{\mathrm{i}Ht/\hbar} A \mathrm{e}^{-\mathrm{i}Ht/\hbar}] = \frac{\mathrm{i}}{\hbar}[H, A](t) \,. \tag{13.46}$$

On this level, so to speak, as a control of the quantization prescription, we use the commutation relations (13.25) and (13.37) to verify that the operator-valued fields indeed satisfy the Maxwell equations and that the particles satisfy Newton's equations of motion. Computing the commutators as in (13.46) one obtains

$$\begin{aligned} &c^{-1}\partial_t B = -\nabla \times E\,, \quad c^{-1}\partial_t E = \nabla \times B - c^{-1} j\,, \\ &\nabla \cdot E = \rho\,, \quad \nabla \cdot B = 0\,, \end{aligned} \tag{13.47}$$

where now

$$\begin{aligned} \rho(x,t) &= \sum_{j=1}^{N} e_j \varphi(x - q_j(t))\,, \\ j(x,t) &= \sum_{j=1}^{N} \frac{1}{2} e_j \big(v_j(t)\varphi(x - q_j(t)) + \varphi(x - q_j(t)) v_j(t)\big) \end{aligned} \tag{13.48}$$

with the velocity operator

$$v_j = \big(p_j - c^{-1} e_j A_\varphi(q_j)\big)/m_j\,. \tag{13.49}$$

Similarly, one obtains the symmetrized Lorentz force as

$$m_j \dot{v}_j(t) = e_j\Big(E_\varphi(q_j(t),t) + \frac{1}{2c}\big(v_j(t) \times B_\varphi(q_j(t),t) - B_\varphi(q_j(t),t) \times v_j(t)\big)\Big)\,. \tag{13.50}$$

If there are external fields, $E_\varphi$, $B_\varphi$ is to be replaced by $E_\varphi + E_{\mathrm{ex}}$, $B_\varphi + B_{\mathrm{ex}}$. In (13.47)–(13.49), $q_j(t)$, $p_j(t)$, respectively $A(t)$, $-c^{-1}E_\perp(t)$, are operators satisfying the commutation relations (13.25), respectively (13.37), at all times.

Also of interest is to record the Heisenberg equations of motion for the Pauli–Fierz Hamiltonian (13.43) including spin. The Maxwell equations are as before. However, in the case of a single charged particle, the current density is now

$$j(x) = \frac{1}{2} e\big(v\varphi(x-q) + \varphi(x-q)v\big) + \frac{e\hbar}{2m}\sigma \times \nabla_q \varphi(x-q) \tag{13.51}$$

with the velocity operator $v = (p - c^{-1} e A_\varphi(q))/m$. The Schrödinger equation reads

$$m\ddot{q} = e\big(E_\varphi(q,t) + \frac{1}{2c}(v \times B_\varphi(q,t) - B_\varphi(q,t) \times v)\big) + \frac{e\hbar}{2mc}\sigma \cdot \nabla_q B_\varphi(q,t)\,, \tag{13.52}$$

consistent with the general rule that the magnetic force equals $c^{-1}\int d^3x j(x) \times B(x)$, and the Pauli equation for the spin reads

$$\dot{\sigma} = -\frac{e}{mc} B_\varphi(q,t) \times \sigma \,. \tag{13.53}$$

If one compares (13.52), (13.53) with the classical equations of motion of a spinning charge, cf. section 10.2, then one observes that in quantum mechanics the spin degrees of freedom couple somewhat differently to the Maxwell field than the classical internal angular momentum. Since $\varphi$ is radial, in fact $\nabla_x \varphi_r(|x|) = \varphi_r'(|x|)\widehat{x}$ and the spin part of the current (13.51) has the effective charge distribution $\varphi_r'(|x|)/|x|$. However, the evolution equation for $\sigma$ has only superficial similarity with Eq. (10.20) for $\omega$.

## 13.3 Fock space, self-adjointness

To define the Pauli–Fierz Hamiltonian as a linear operator, one has to introduce a suitable Hilbert space of wave functions. Provisionally we assume that the number of photons, either virtual or real in the usual parlance, is finite, though necessarily arbitrary, since $H$ does not conserve the number of photons. This means that we will have to work in the Fock representation of the Bose fields $a(k,\lambda)$. We introduce the one-particle Hilbert space

$$\mathfrak{h} = L^2(\mathbb{R}^3) \otimes \mathbb{C}^2 \,. \tag{13.54}$$

$\mathfrak{h}$ consists of wave functions $\psi(k,\lambda)$, with the photon wave number $k \in \mathbb{R}^3$ and the helicity $\lambda = 1, 2$. The inner product in $\mathfrak{h}$ is $\langle \varphi, \psi \rangle_{\mathfrak{h}} = \sum_{\lambda=1,2} \int d^3k \varphi^*(k,\lambda)\psi(k,\lambda)$. Out of $\mathfrak{h}$ we construct the Fock space $\mathcal{F}$ in the usual way

$$\mathcal{F} = \bigoplus_{n=0}^{\infty} \left(\mathfrak{h}^{\otimes n}\right)_{\text{sym}} \,, \tag{13.55}$$

where $\mathfrak{h}^{\otimes n}$ denotes the $n$-fold tensor product and where "sym" means that we restrict to the subspace of wave functions symmetric under interchange of labels, i.e.

$$\psi_n(k_1, \lambda_1, \ldots, k_n, \lambda_n) = \psi_n(k_{\pi(1)}, \lambda_{\pi(1)}, \ldots, k_{\pi(n)}, \lambda_{\pi(n)}) \tag{13.56}$$

for an arbitrary permutation $\pi$. By definition an element $\psi \in \mathcal{F}$ is of the form $(\psi_0, \psi_1, \ldots)$ and

$$\langle \varphi, \psi \rangle_{\mathcal{F}} = \sum_{n=0}^{\infty} \langle \varphi_n, \psi_n \rangle_{\mathfrak{h}^{\otimes n}} \,. \tag{13.57}$$

The Fock vacuum, $\psi_0$, will be denoted by $\Omega$.

◇ As the reader will have noticed, for the inner product in a Hilbert space we use the notation $\langle \varphi, \psi \rangle$, which is linear in the second and antilinear in the first argument. The standard physics notation would be the Dirac bracket $\langle \varphi | \psi \rangle$, which is also linear in the second argument. (A further widespread convention is a scalar product linear in the first argument.) The subscript in $\langle \varphi, \psi \rangle_{\mathcal{F}}$ is used to indicate the Hilbert space under consideration: it will be omitted if it is obvious from the context. The length of a vector is $\|\psi\| = \langle \psi, \psi \rangle^{1/2}$. ◇

For $f \in \mathfrak{h}$ one defines the smeared creation and annihilation operators

$$a(f) = \sum_{\lambda=1,2} \int \mathrm{d}^3k f^*(k, \lambda) a(k, \lambda), \quad a^*(f) = \sum_{\lambda=1,2} \int \mathrm{d}^3k f(k, \lambda) a^*(k, \lambda). \tag{13.58}$$

As operators in $\mathcal{F}$ they act through

$$(a(f)\psi)_n(k_1, \lambda_1, \ldots, k_n, \lambda_n) = \sqrt{n+1} \sum_{\lambda=1,2} \int \mathrm{d}^3k f^*(k, \lambda) \times \psi_{n+1}(k_1, \lambda_1, \ldots, k_n, \lambda_n, k, \lambda), \tag{13.59}$$

$$(a^*(f)\psi)_n(k_1, \lambda_1, \ldots, k_n, \lambda_n) = \frac{1}{\sqrt{n}} \sum_{j=1}^{n} f(k_j, \lambda_j) \times \psi_{n-1}(k_1, \lambda_1, \ldots, \widetilde{k}_j, \widetilde{\lambda}_j, \ldots, k_n, \lambda_n), \tag{13.60}$$

where $\tilde{}$ means that this variable is to be omitted. The field Hamiltonian

$$H_{\mathrm{f}} = \sum_{\lambda=1,2} \int \mathrm{d}^3k \hbar\omega(k) a^*(k, \lambda) a(k, \lambda) \tag{13.61}$$

acts as multiplication by $\sum_{j=1}^{n} \hbar\omega(k_j)$ on the $n$-particle subspace $\mathfrak{h}^{\otimes n}$. With all these definitions we see that the Pauli–Fierz Hamiltonian operates on the Hilbert space

$$\mathcal{H} = \mathcal{H}_{\mathrm{p}} \otimes \mathcal{H}_{\mathrm{f}} \tag{13.62}$$

with $\mathcal{H}_{\mathrm{p}} = L^2(\mathbb{R}^{3N})$ and $\mathcal{H}_{\mathrm{f}} = \mathcal{F}$. Physically the particle Hilbert space $\mathcal{H}_{\mathrm{p}}$ is too large, since in nature only symmetric, respectively antisymmetric, wave functions are realized. Still mathematically it is convenient to work with all of $L^2(\mathbb{R}^{3N})$.

In any dynamical theory, usually the first step is to establish the existence of solutions of the evolution equations. In our case this means to prove that $H$ is a self-adjoint operator on a suitable domain of functions, where for concreteness we consider the Pauli–Fierz operator of (13.43) for a single electron. If not even the self-adjointness question can be resolved, there is little hope of rigorously handling qualitative properties of interest.

We observe that $\langle\psi, H\psi\rangle_{\mathcal{H}} \geq 0$, clearly. This means that $H$ has equal defect indices and therefore at least one self-adjoint extension. Amongst those there is a distinguished extension, called the Friedrichs extension, which is obtained through the closure of the quadratic form $\langle\psi, H\psi\rangle_{\mathcal{H}}$ with smooth wave functions of a finite number of photons. The Friedrichs extension gives no information on the domain of self-adjointness and, in principle, there could be other extensions. A more concrete approach is to prove that, for the purpose of the existence of dynamics, the interaction can be regarded as small. We decompose $H$ as

$$
\begin{aligned}
H &= H_0 + H_1 \\
&= \frac{1}{2m}p^2 + H_{\mathrm{f}} - \frac{e}{2mc}\big(p \cdot A_\varphi(x) + A_\varphi(x) \cdot p\big) \\
&\quad + \frac{e^2}{2mc^2}A_\varphi(x)^2 - \frac{e\hbar}{2mc}\sigma \cdot B_\varphi(x)\,,
\end{aligned}
\tag{13.63}
$$

$B_\varphi(x) = \nabla \times A_\varphi(x)$, $p_1 = p$ for the momentum, and $q_1 = x$ for the position of the particle, and want to prove that $H_1$ is small compared to $H_0$, $H_0 = (p^2/2m) + H_{\mathrm{f}}$.

Abstractly one uses the Kato–Rellich theorem. We consider the densely defined linear operators $A$, $B$ on a Hilbert space $\mathcal{H}$ with inner product $\langle\cdot,\cdot\rangle$ and suppose that

(i) for the domains $D(B) \supset D(A)$,
(ii) for some constants $a, b$ and all $\psi \in D(A)$

$$
\|B\psi\| \leq a\|A\psi\| + b\|\psi\|\,.
\tag{13.64}
$$

Then $B$ is said to be $A$-bounded. The smallest $a$ is called the relative bound. Usually $a$ can be made smaller at the expense of $b$.

**Theorem 13.1** (Kato–Rellich theorem). *Suppose $A$ is self-adjoint, $B$ is symmetric, and $B$ is $A$-bounded with relative bound $a < 1$. Then $A + B$ is self-adjoint on $D(A)$ and essentially self-adjoint on any core of $A$.*

For multiparticle Schrödinger operators of the form $-\frac{1}{2}\Delta + V$ the Kato–Rellich theorem is a standard technique and yields the existence of dynamics for a very large class of potentials $V$ including the Coulomb potential. For the Pauli–Fierz operator the form version of Theorem 13.1 is more convenient.

**Theorem 13.2** (KLMN theorem). *Let $A$ be a positive self-adjoint operator. Let $\beta(\psi,\varphi) = \langle\psi, B\varphi\rangle$ be a symmetric quadratic form defined for all $\psi, \varphi \in D(A^{1/2})$ such that for some constants $a < 1$, $b < \infty$*

$$
|\langle\psi, B\psi\rangle| \leq a\langle\psi, A\psi\rangle + b\langle\psi,\psi\rangle
\tag{13.65}
$$

*for all $\psi \in D(A^{1/2})$. Then there exists a unique self-adjoint operator $C$ with*

*$D(C) \subset D(A^{1/2})$ such that*

$$\langle \psi, C\psi \rangle = \langle \psi, A\psi \rangle + \langle \psi, B\psi \rangle. \tag{13.66}$$

*Moreover, $C$ is bounded from below by $-b$.*

Let us see how the KLMN theorem works in the case of the Pauli–Fierz Hamiltonian $H$, which means that one has to establish

$$|\langle \psi, H_1\psi \rangle_{\mathcal{H}}| \leq a\langle \psi, H_0\psi \rangle_{\mathcal{H}} + b\langle \psi, \psi \rangle_{\mathcal{H}} \tag{13.67}$$

with $a < 1$. We set $\hbar = c = m = 1$ and, following the convention (13.58), put

$$A_\varphi(x) = a(f_x) + a^*(f_x) \tag{13.68}$$

with

$$f_x(k, \lambda) = \widehat{\varphi}(k)\sqrt{1/2\omega}\, e_\lambda(k) \mathrm{e}^{-ik\cdot x}\,. \tag{13.69}$$

The creation and annihilation operators are bounded through $(H_{\mathrm{f}})^{1/2}$ as

$$\begin{aligned} \|a^*(f)\psi\|_{\mathcal{F}} &\leq \|f/\sqrt{\omega}\|_{\mathfrak{h}} \|(H_{\mathrm{f}})^{1/2}\psi\|_{\mathcal{F}} + \|f\|_{\mathfrak{h}}\|\psi\|_{\mathcal{F}}\,, \\ \|a(f)\psi\|_{\mathcal{F}} &\leq \|f/\sqrt{\omega}\|_{\mathfrak{h}} \|(H_{\mathrm{f}})^{1/2}\psi\|_{\mathcal{F}} \end{aligned} \tag{13.70}$$

and by the Schwarz inequality

$$\begin{aligned} |\langle \psi, (a(f) + a^*(f))^2\psi \rangle_{\mathcal{F}}| &\leq 2\langle \psi, a^*(f)a(f)\psi \rangle_{\mathcal{F}} + \|f\|_{\mathfrak{h}}^2 \|\psi\|_{\mathcal{F}}^2 \\ &\quad + 2|\langle \psi, a^*(f)a^*(f)\psi \rangle_{\mathcal{F}}| \\ &\leq 5\|f/\sqrt{\omega}\|_{\mathfrak{h}}^2 \langle \psi, H_{\mathrm{f}}\psi \rangle_{\mathcal{F}} + 3\|f\|_{\mathfrak{h}}^2\|\psi\|_{\mathcal{F}}^2\,. \end{aligned} \tag{13.71}$$

Therefore the $A_\varphi^2$-term has a relative bound less than 1 only if $e$ is sufficiently small.

We do not attempt to optimize the constants and thus write

$$|\langle \psi, p \cdot A_\varphi(x)\psi \rangle_{\mathcal{H}}| \leq \frac{1}{2}\langle \psi, p^2\psi \rangle_{\mathcal{H}} + \frac{1}{2}\langle \psi, A_\varphi(x)^2\psi \rangle_{\mathcal{H}}\,, \tag{13.72}$$

$$|\langle \psi, \sigma \cdot B_\varphi(x)\psi \rangle_{\mathcal{H}}| \leq \frac{1}{2}\langle \psi, B_\varphi(x)^2\psi \rangle_{\mathcal{H}} + \frac{3}{2}\|\psi\|_{\mathcal{H}}^2\,. \tag{13.73}$$

Also, by using (13.69), (13.71),

$$\langle \psi, A_\varphi(x)^2\psi \rangle_{\mathcal{H}} \leq 5\|\widehat{\varphi}/\omega\|_{\mathfrak{h}}^2 \langle \psi, H_{\mathrm{f}}\psi \rangle_{\mathcal{H}} + 3\|\widehat{\varphi}/\sqrt{\omega}\|_{\mathfrak{h}}^2\|\psi\|_{\mathcal{H}}^2\,, \tag{13.74}$$

$$\langle \psi, B_\varphi(x)^2\psi \rangle_{\mathcal{H}} \leq 5\||k|\widehat{\varphi}/\omega\|_{\mathfrak{h}}^2 \langle \psi, H_{\mathrm{f}}\psi \rangle_{\mathcal{H}} + 3\||k|\widehat{\varphi}/\sqrt{\omega}\|_{\mathfrak{h}}^2\|\psi\|_{\mathcal{H}}^2\,. \tag{13.75}$$

Thus if

$$\int d^3k|\widehat{\varphi}(k)|^2(\omega^{-2}+\omega) < \infty\,, \tag{13.76}$$

one can find a constant $e_0$ such that for $|e| \le e_0$ the operator $H_1$ is $H_0$ form-bounded with a bound less than 1. By a similar reasoning form-bounded can be replaced by bounded. From Theorem 13.2 we conclude

**Theorem 13.3** (Self-adjointness, Kato–Rellich). *If* $|e| \le e_0$ *with suitable* $e_0$ *and if the form factor* $\widehat{\varphi}$ *satisfies the condition (13.76), then the Pauli–Fierz operator* $H$ *of (13.63) is self-adjoint on the domain* $D(\frac{1}{2m}p^2 + H_{\mathrm{f}})$.

Since $\widehat{\varphi}(0) = (2\pi)^{-3/2}$, the condition (13.76) is satisfied if, as assumed, $\widehat{\varphi}$ cuts off ultraviolet wave numbers.

◇ We denote constants by $c_0, c_1, \ldots, e_0$, etc., depending on the context. The numerical value of these constants may change from equation to equation. Since we always work with computable bounds, in principle these constants can be expressed through the parameters of the Pauli–Fierz Hamiltonian. To do so actually would overburden the notation. ◇

The restriction on $e$ is intrinsic to the method, since only then is $e^2 A_\varphi(x)^2$ small compared to $H_{\mathrm{f}}$. To go beyond one needs a completely different technique which is based on functional integration, as will be explained in chapter 14.

**Theorem 13.4** (Self-adjointness, functional integration). *If (13.76) holds, then the* $N$*-particle Pauli–Fierz Hamiltonian* $H$ *of (13.39) is self-adjoint on the domain* $D(\sum_{j=1}^{N}(p_j^2/2m_j) + H_{\mathrm{f}})$. *Furthermore* $H$ *is bounded from below.*

*Proof*: Hiroshima (2002).

Theorem 13.4 remains valid under the inclusion of spin and the addition of external potentials with very mild conditions on their regularity.

In summary, the Pauli–Fierz Hamiltonian uniquely generates the unitary time evolution $\mathrm{e}^{-\mathrm{i}Ht/\hbar}$ on $\mathcal{H}$ provided the condition (13.76) holds. Under a suitable ultraviolet cutoff the quantum dynamics of charges and photons is well defined.

## 13.4 Energy and length scales

The characteristic energy and length scales will depend on the physical situation. In our context two distinct cases are of particular importance. For the point-charge (= ultraviolet) limit relativistic units are used, which means that lengths are measured in units of the Compton wavelength

$$\lambda_{\mathrm{c}} = \hbar/m_{\mathrm{e}}c \tag{13.77}$$

and energies in units of the rest energy $m_e c^2$ of the electron. For applications in atomic physics and quantum optics, atomic units are more appropriate, where the size of an atom is set by the Bohr radius

$$r_B = \frac{4\pi\hbar^2}{m_e e^2} = \alpha^{-1}\frac{\hbar}{m_e c} \tag{13.78}$$

and the energy scale is set by the ionization energy

$$\frac{e^2}{4\pi r_B} = \frac{e^4 m_e}{(4\pi)^2\hbar^2} = \alpha^2(m_e c^2) \tag{13.79}$$

with

$$\alpha = \frac{e^2}{4\pi\hbar c}, \tag{13.80}$$

the Sommerfeld fine-structure constant written in Heaviside–Lorentz units. The ionization energy corresponds to the length $\alpha^{-1} r_B$ which approximately equals the wavelength of the Lyman alpha line. The scales compare as

| | | | | |
|---|---|---|---|---|
| | *ionization* | | *rest* | |
| | $\alpha^2$ | $\alpha$ | $1$ | energy $[m_e c^2]$ |
| length $[\hbar/m_e c]$ | $\alpha^{-2}$ | $\alpha^{-1}$ | $1$ | |
| | *Lyman alpha* | *Bohr radius* | *Compton* | |

Since $\alpha \simeq 1/137$ in nature, the scales are well separated.

These scales necessarily reappear in the Pauli–Fierz Hamiltonian with the crucial difference that the physical mass $m_e$ of the electron is replaced by its bare mass $m$. To have a concrete example let us discuss the hydrogen atom as the simplest two-particle case. We assume that the nucleus is infinitely heavy. Then the Pauli–Fierz Hamiltonian reads

$$H = \frac{1}{2m}\left(\sigma \cdot (p + c^{-1} e A_\varphi(x))\right)^2 + H_f - e^2 V_{\varphi\text{coul}}(x), \tag{13.81}$$

where $-e$ is the charge of the electron and for the purpose of this subsection only we set

$$V_{\varphi\text{coul}}(x) = \int d^3k |\widehat{\varphi}(k)|^2 |k|^{-2} e^{-ik\cdot x}. \tag{13.82}$$

We transform to dimensionless form, such that the energy unit is $\alpha^2(mc^2)$, the length scale for the electron is $r_B = \hbar/\alpha mc$, and that for the photons is $\alpha^{-1}r_B$. This is achieved through the canonical transformation $U$ defined through

$$U^*a(k,\lambda)U = (\alpha^{-2}\lambda_c)^{3/2}a(\alpha^{-2}\lambda_c k,\lambda)\,,$$
$$U^*xU = \alpha^{-1}\lambda_c x\,,\quad U^*pU = \alpha\lambda_c^{-1}p\,, \tag{13.83}$$

where now the Compton wavelength

$$\lambda_c = \hbar/mc \tag{13.84}$$

depends on the bare mass rather than the physical mass as in (13.77). Then

$$U^*HU = \alpha^2 mc^2\Big(\frac{1}{2}(\sigma\cdot(-\mathrm{i}\nabla_x - \sqrt{4\pi\alpha}\alpha A_{\tilde\varphi}(\alpha x))\big)^2 + \sum_{\lambda=1,2}\int \mathrm{d}^3k|k|a^*(k,\lambda)a(k,\lambda) - 4\pi V_{\varphi_\alpha\mathrm{coul}}(x)\Big) \tag{13.85}$$

with

$$A_{\tilde\varphi}(x) = \sum_{\lambda=1,2}\int \mathrm{d}^3k\widehat{\varphi}(\alpha^2\lambda_c^{-1}k)\frac{1}{\sqrt{2|k|}}e_\lambda(k)\big(\mathrm{e}^{\mathrm{i}k\cdot x}a(k,\lambda) + \mathrm{e}^{-\mathrm{i}k\cdot x}a^*(k,\lambda)\big) \tag{13.86}$$

and $\widehat{\varphi}_\alpha(k) = \widehat{\varphi}(\alpha k/\lambda_c)$. We infer from (13.85) that the Maxwell field is weakly coupled to the electron. Thus, *cum grano salis*, perturbation theory around $\alpha = 0$ should provide a qualitatively correct picture. In particular, spectral lines should be rather sharp. In addition, since $A_{\tilde\varphi}$ varies only on the scale $\alpha^{-1}r_B$, the dipole approximation $A_{\tilde\varphi}(\alpha x) \cong A_{\tilde\varphi}(0)$ will suffice as long as the electron remains bound to the nucleus.

The dimensionless form (13.85) teaches us also how to choose the wave number cutoff $\widehat{\varphi}$. Thus, if $\widehat{\varphi} = (2\pi)^{-3/2}$ for $|k| < \Lambda$, $\widehat{\varphi} = 0$ for $|k| \geq \Lambda$, then $\Lambda \gg 1/r_B$ to have a negligible smearing of the Coulomb potential. On the other hand, at the scale of the rest energy of the electron, the Pauli–Fierz model cannot be expected to describe the physics correctly. Thus the cutoff should satisfy

$$1 \ll \Lambda r_B \ll \alpha^{-1}\,. \tag{13.87}$$

It is instructive to compare the atomic units with relativistic units. In the latter case the scale transformation $U$ reads

$$U^*a(k,\lambda)U = \lambda_c^{3/2}a(\lambda_c k,\lambda)\,,\quad U^*xU = \lambda_c x\,,\quad U^*pU = \lambda_c^{-1}p\,. \tag{13.88}$$

Then

$$U^*HU = mc^2\Big(\frac{1}{2}\big(\sigma\cdot(-\mathrm{i}\nabla_x - \sqrt{4\pi\alpha}A_{\varphi_{\lambda_c}}(x))\big)^2 + \sum_{\lambda=1,2}\int d^3k|k|a^*(k,\lambda)a(k,\lambda) - 4\pi\alpha V_{\varphi_{\lambda_c}\mathrm{coul}}(x)\Big) \quad (13.89)$$

with the form factor in units of the Compton wavelength, $\widehat{\varphi}_{\lambda_c} = \widehat{\varphi}(k/\lambda_c)$. Note that the cutoff depends through the Compton wavelength on the bare electron mass.

## 13.5 Conservation laws

The Pauli–Fierz Hamiltonian (13.42) is invariant under translations and rotations. Therefore the total momentum and the total angular momentum will be conserved. One only has to identify the generators of these symmetries. The generator for the translations of the $j$-th particle is its momentum $p_j$, which means

$$\mathrm{e}^{\mathrm{i}a\cdot p_j/\hbar}q_j\mathrm{e}^{-\mathrm{i}a\cdot p_j/\hbar} = q_j + a\,. \quad (13.90)$$

Similarly the field translations are generated by the momentum of the Maxwell field

$$P_\mathrm{f} = \sum_{\lambda=1,2}\int d^3k\hbar k a^*(k,\lambda)a(k,\lambda) \quad (13.91)$$

with the property that

$$\mathrm{e}^{\mathrm{i}a\cdot P_\mathrm{f}/\hbar}a(k,\lambda)\mathrm{e}^{-\mathrm{i}a\cdot P_\mathrm{f}/\hbar} = \mathrm{e}^{-\mathrm{i}a\cdot k}a(k,\lambda)\,. \quad (13.92)$$

Thus the total momentum

$$P = \sum_{j=1}^{N} p_j + P_\mathrm{f} \quad (13.93)$$

must be conserved and indeed

$$[H, P] = 0\,. \quad (13.94)$$

Next we consider a rotation $R$ by an angle $\theta$ relative to the axis of rotation $\widehat{n}$ through the origin. For position and momentum we have

$$\mathrm{e}^{\mathrm{i}\theta\widehat{n}\cdot(q_j\times p_j)/\hbar}q_j\mathrm{e}^{-\mathrm{i}\theta\widehat{n}\cdot(q_j\times p_j)/\hbar} = Rq_j\,,\quad \mathrm{e}^{\mathrm{i}\theta\widehat{n}\cdot(q_j\times p_j)/\hbar}p_j\mathrm{e}^{-\mathrm{i}\theta\widehat{n}\cdot(q_j\times p_j)/\hbar} = Rp_j\,. \quad (13.95)$$

For the Maxwell field we define the angular momentum relative to the origin

$$J_\mathrm{f} = -\sum_{\lambda=1,2}\int d^3k a^*(k,\lambda)(k\times \mathrm{i}\hbar\nabla_k)a(k,\lambda) \quad (13.96)$$

and the helicity

$$S_{\mathrm{f}} = c^{-1} \int \mathrm{d}^3x E(x) \times A(x) = \mathrm{i}\hbar \int \mathrm{d}^3k \widehat{k}\big(a^*(k,2)a(k,1) - a^*(k,1)a(k,2)\big) \tag{13.97}$$

with $\widehat{k} = k/|k|$. Their sum rotates the vector potential as

$$\mathrm{e}^{\mathrm{i}\theta\widehat{n}\cdot(J_{\mathrm{f}}+S_{\mathrm{f}})/\hbar} A(x) \mathrm{e}^{-\mathrm{i}\theta\widehat{n}\cdot(J_{\mathrm{f}}+S_{\mathrm{f}})/\hbar} = RA(R^{-1}x) \tag{13.98}$$

and correspondingly for the transverse electric field $E_\perp(x)$. We conclude that the total angular momentum

$$J = \sum_{j=1}^{N} (q_j \times p_j) + J_{\mathrm{f}} + S_{\mathrm{f}} \tag{13.99}$$

is conserved and indeed

$$[H, J] = 0\,. \tag{13.100}$$

If the $j$-th particle carries spin $\sigma_j$, then

$$J = \sum_{j=1}^{N} (q_j \times p_j) + \sum_{j=1}^{N} \frac{1}{2}\hbar\sigma_j + J_{\mathrm{f}} + S_{\mathrm{f}} \tag{13.101}$$

is the conserved total angular momentum.

The helicity $S_{\mathrm{f}}$ is diagonalized through transforming to circularly polarized photons. We define the left-circularly and right-circularly polarized annihilation operators

$$a_+(k) = \frac{1}{\sqrt{2}}\big(a(k,1) - \mathrm{i}a(k,2)\big)\,, \quad a_-(k) = \frac{1}{\sqrt{2}}\big(a(k,1) + \mathrm{i}a(k,2)\big)\,. \tag{13.102}$$

Then

$$S_{\mathrm{f}} = \int \mathrm{d}^3k \widehat{k}\big(a_+^*(k)a_+(k) - a_-^*(k)a_-(k)\big)\,, \tag{13.103}$$

which establishes that the photon has spin 1. However, only two helicity states are admissible, +1 for left and −1 for right polarization. The corresponding one-photon states are

$$e_\pm(k)(2\pi)^{-3/2}\mathrm{e}^{\mathrm{i}(k\cdot x \mp \omega t)}, \quad e_\pm(k) = \frac{1}{\sqrt{2}}\big(e_1(k) \pm \mathrm{i}e_2(k)\big)\,. \tag{13.104}$$

For the + index the photon state represents a plane wave whose polarization vector rotates in a right-handed sense about $k$ and thus appears to an observer facing the incoming wave as left polarized.

## 13.6 Boundary conditions and the Casimir effect

So far we took for granted that the Maxwell field lives in infinite space. In many applications one has a macroscopically finite geometry, like a cavity or a wave guide, and it is necessary to include it as a boundary condition into the Hamiltonian. For concreteness, let us assume then some bounded region $\Lambda$ whose surface $\partial\Lambda$ is defined through a perfect, grounded conductor. Momentarily there are no charges inside $\Lambda$. Then the Maxwell equations are

$$c^{-1}\partial_t B = -\nabla \times E\,, \quad c^{-1}\partial_t E = \nabla \times B\,, \quad \nabla \cdot E = 0\,, \quad \nabla \cdot B = 0\,. \tag{13.105}$$

If $\widehat{n}(x)$ denotes the outward normal at $x \in \partial\Lambda$, the boundary conditions for a perfect conductor are

$$\widehat{n} \cdot B(x) = 0\,, \quad \widehat{n} \times E(x) = 0 \quad \text{at} \quad x \in \partial\Lambda\,. \tag{13.106}$$

The rules of canonical quantization apply as before, only the final expressions are less explicit, since (13.105) together with the boundary conditions (13.106) cannot be solved through simple Fourier transformation. Let $L^2(\Lambda, \mathbb{R}^3)$ be the space of (complex valued) vector fields on $\Lambda$. $A \in L^2$ is divergence free if $\nabla \cdot A = 0$ and we denote by $Q_\Lambda^\perp$ the projection onto all such fields. The quantum mechanical Fock space is built up from $Q_\Lambda^\perp L^2$ as one-particle Hilbert space. Notationally it is slightly more convenient to start from $L^2(\Lambda, \mathbb{R}^3)$ and incorporate the projection into the definition of the quantized fields. We introduce then the three-component Bose field $a(x), a^*(x)$ satisfying

$$[a_\alpha(x), a^*_{\alpha'}(x')] = \delta_{\alpha\alpha'}\delta(x - x') \tag{13.107}$$

with all other commutators vanishing. The quantized Maxwell field will depend only on $Q_\Lambda^\perp a$ and $Q_\Lambda^\perp a^*$.

As before the vector potential $A$ satisfies the Coulomb gauge, which implies

$$c^{-2}\partial_t^2 A = \Delta A\,, \quad \nabla \cdot A = 0 \tag{13.108}$$

with boundary conditions

$$\widehat{n} \cdot (\nabla \times A) = 0\,, \quad \widehat{n} \times A = 0 \quad \text{at} \quad x \in \partial\Lambda\,. \tag{13.109}$$

Since $E_\perp = -c^{-1}\partial_t A$, one can write the solution to (13.108), (13.109) on $Q_\Lambda^\perp L^2$ as in (13.108)

$$\begin{pmatrix} A(t) \\ E_\perp(t) \end{pmatrix} = \begin{pmatrix} \cos \Omega t & -c\Omega^{-1} \sin \Omega t \\ c^{-1}\Omega \sin \Omega t & \cos \Omega t \end{pmatrix} \begin{pmatrix} A \\ E_\perp \end{pmatrix}, \tag{13.110}$$

where, as a linear operator, $\Omega = c(-\Delta \otimes \mathbb{1})^{1/2}$ restricted to $Q_\Lambda^\perp L^2$ and with the mixed Dirichlet–Neumann boundary condition (13.109). $\Omega$ is a positive self-adjoint operator. In analogy to (13.35), (13.36) the canonically quantized fields are obtained as

$$A(x) = c\sqrt{\hbar/2}\,\Omega^{-1/2} Q_\Lambda^\perp\big(a(x) + a^*(x)\big)\,, \tag{13.111}$$

$$E_\perp(x) = \sqrt{\hbar/2}\,\Omega^{1/2} Q_\Lambda^\perp \mathrm{i}\big(a(x) - a^*(x)\big)\,. \tag{13.112}$$

Clearly their commutation relations are

$$[A_\alpha(x), -c^{-1}E_{\perp\alpha'}(x')] = \mathrm{i}\hbar (Q_\Lambda^\perp)_{\alpha\alpha'}(x, x') \tag{13.113}$$

with the right-hand side denoting the integral kernel of $Q_\Lambda^\perp$ in $L^2(\Lambda, \mathbb{R}^3)$. The field energy is a sum over the energy in each mode, which in position space becomes

$$H_\mathrm{f} = \hbar \int_\Lambda \mathrm{d}^3x a^*(x) \cdot \Omega Q_\Lambda^\perp a(x)\,. \tag{13.114}$$

In case there are charges enclosed in the cavity, their mutual Coulomb interaction has to respect the perfect conductor boundary condition (13.106). For example, since $E_\parallel$ is not quantized, for a single charge at $q$ the potential $\phi_\Lambda$ satisfies the Poisson equation

$$\Delta\phi_\Lambda(x) = e\varphi(x - q)\,, \quad \phi_\Lambda(x) = 0 \quad \text{for} \quad x \in \partial\Lambda \tag{13.115}$$

and the potential acting on the particle is given by

$$e\phi_{\Lambda\varphi}(q) = e \int \mathrm{d}^3x \varphi(q - x)\phi_\Lambda(x)\,. \tag{13.116}$$

Close to the surface $\phi_\Lambda(x)$ is determined by the image charge and looks like an attractive Coulomb potential. Thus we have to add phenomenologically to the Hamiltonian a surface potential $V_\mathrm{sur}$ which keeps the particle confined to the cavity $\Lambda$. Altogether the Pauli–Fierz Hamiltonian for a single charge enclosed in a cavity is

$$H = \frac{1}{2m}\big(p - c^{-1}eA_\varphi(q)\big)^2 + e\phi_{\Lambda\varphi}(q) + V_\mathrm{sur}(q) + H_\mathrm{f}\,. \tag{13.117}$$

To return to the charge-free situation, according to (13.114) we calibrated the ground state energy of the cavity at zero, which is an acceptable choice for a closed cavity. If, however, the cavity is open, as for example two plane parallel, grounded metal plates, then the natural zero of energy refers to the energy of the field vacuum in infinite space. In the presence of the plates this vacuum energy is lowered by an amount which depends on the separation of the plates. Therefore there is an effective attractive force between the plates – the famous Casimir effect. Together with the spectrum of the black-body radiation it provides the most direct evidence for the quantum nature of the Maxwell field.

If one adopts the boundary conditions as in (13.109), the energy difference – with and without plates – diverges because of high-frequency modes, which reflects the fact that the metal plates cannot be perfect conductors up to arbitrarily high frequencies. We therefore choose a cutoff function $g$ with $g(\omega) = 1$ for small $\omega$ and rapidly decreasing at infinity. The plates are parallel to each other, have a distance $d$, and an area $\ell^2$ which is taken to be very large. Then the energy difference per unit area is given by

$$\frac{1}{\ell^2}\Delta E(d) = \frac{\pi^2\hbar c}{4d^3}\Big(\frac{1}{2}G(0) + \sum_{n=1}^{\infty} G(n) - \int_0^\infty \mathrm{d}\kappa\, G(\kappa)\Big), \qquad (13.118)$$

where

$$G(\kappa) = 2\int_\kappa^\infty \mathrm{d}u\, u^2 g(\pi u/d). \qquad (13.119)$$

For analytic $g$ one can use in (13.118) the Euler–MacLaurin summation formula,

$$\frac{1}{2}F(0) + \sum_{n=1}^{\infty} F(n) - \int_0^\infty \mathrm{d}\kappa\, F(\kappa) = -\frac{1}{12}F'(0) + \frac{1}{720}F'''(0) + \text{higher derivatives}, \qquad (13.120)$$

and note that $F'(0) = 0$, $F'''(0) = -4$, since $g(0) = 1$, whereas every extra derivative carries a factor $1/d$. Thus to leading order

$$\frac{1}{\ell^2}\Delta E(d) = -\frac{\pi^2\hbar c}{720 d^3} + \mathcal{O}(d^{-4}), \qquad (13.121)$$

independently of the choice of the cutoff function $g$, and the force per unit area between the conducting plates is given by

$$\frac{1}{\ell^2}F(d) = -\frac{\pi^2\hbar c}{240 d^4} + \mathcal{O}(d^{-5}). \qquad (13.122)$$

## 13.7 Dipole and single-photon approximation

Even for a single charge the Pauli–Fierz Hamiltonian resists exact diagonalization and one has to rely on approximations. As suggested by (13.85), since the coupling to the photon field is weak, an obvious strategy is to expand in $\alpha$. Such a perturbative treatment is covered extensively in standard texts and there is no need to repeat it here. Since one of our aims is to explain why perturbation theory works so well, we will make contact with the conventional results later on. Another strategy is to truncate the Hamiltonian to taste, so as not to throw out the physics. In essence there are only two such schemes, the dipole approximation and the single-photon approximation.

*(i) Dipole approximation*

We consider a single charge confined by an external potential $e\phi_{\mathrm{ex}}$, centered at the origin. Since the potential inhibits large excursions, one loses little by evaluating the vector potential at the origin instead of at $q$, the true position of the charged particle. This leads to the dipole Hamiltonian

$$H = \frac{1}{2m}\big(p - c^{-1}eA_\varphi(0)\big)^2 + e\phi_{\mathrm{ex}}(q) + H_{\mathrm{f}}\,. \tag{13.123}$$

The interaction $p \cdot A_\varphi(0)$ couples $p$ to the fluctuating vector potential at the origin. We can transform it to a fluctuating electric field coupled to the position $q$ through the unitary operator

$$U = \exp[\mathrm{i}c^{-1}eq \cdot A_\varphi(0)/\hbar]\,. \tag{13.124}$$

Then

$$\begin{aligned} U^*pU &= p + c^{-1}eA_\varphi(0)\,, \quad U^*qU = q\,,\\ U^*a(k,\lambda)U &= a(k,\lambda) + \mathrm{i}q \cdot e_\lambda(k)e\widehat{\varphi}(k)\sqrt{1/2\hbar\omega}\,, \end{aligned} \tag{13.125}$$

which imply

$$U^*HU = \frac{1}{2m}p^2 + e\phi_{\mathrm{ex}}(q) + H_{\mathrm{f}} - eq \cdot E_{\perp\varphi}(0) + \frac{1}{2}\Big(\frac{2}{3}\int \mathrm{d}^3k e^2|\widehat{\varphi}(k)|^2\Big)q^2\,. \tag{13.126}$$

The extra harmonic potential balances $q \cdot E_{\perp\varphi}$ so as to make the sum of the last three terms positive.

Even in the form (13.123), respectively (13.126), $H$ is not tractable and in a second approximation one assumes the external potential to be harmonic. Then the dipole Hamiltonian reads

$$H = \frac{1}{2m}\big(p - c^{-1}eA_\varphi(0)\big)^2 + \frac{1}{2}m\omega_0^2q^2 + H_{\mathrm{f}}\,. \tag{13.127}$$

Clearly, the Hamiltonian is quadratic in the dynamical variables and consequently the Heisenberg equations of motion are linear,

$$\begin{aligned} \dot q(t) &= \frac{1}{m}\big(p(t) - c^{-1}eA_\varphi(0,t)\big)\,, \quad \dot p(t) = -m\omega_0^2q(t)\,,\\ c^{-2}\partial_t^2A(x,t) &= \Delta A(x,t) + (e/c)\delta_\varphi^\perp(x)\dot q(t) \end{aligned} \tag{13.128}$$

with $\delta^\perp$ the transverse $\delta$-function of (13.38). As before the index $\varphi$ denotes convolution with the form factor $\varphi$. At this point (13.128) can be solved as classical equations of motion. One obtains the exact line shape, the Lamb shift, and the Rayleigh scattering of light from a bound charge. It should be noted that, since the energy levels of the harmonic oscillator are equidistant, several emitted photons will interfere, which makes the emission spectrum distinct from, say, the hydrogen atom; compare with section 17.4.

Even though the equations of motion (13.128) are linear, their solution is not a back of the envelope computation and one often resorts to yet another approximation, the rotating wave approximation. One starts from (13.126) with the harmonic potential $\frac{1}{2}m\omega_0^2 q^2$, which already includes the last summand in (13.126), and rewrites the harmonic oscillator in terms of its creation and annihilation operator $b, b^*$. Then

$$H = \hbar\omega_0 b^* b - \mathrm{i}\sqrt{\hbar/2m\omega_0}(b - b^*) \cdot eE_{\perp\varphi}(0) + H_{\mathrm{f}}\,. \qquad (13.129)$$

In the coupling, one ignores the counter-rotating terms $ba$ and $b^*a^*$, which results in

$$H_{\mathrm{rw}} = (b, a) \cdot \mathsf{h}(b, a)^t\,. \qquad (13.130)$$

Our notation emphasizes that the rotating wave Hamiltonian $H_{\mathrm{rw}}$ is quadratic in $(b, a)$ and should be regarded as the second quantization of the one-particle Hamiltonian $\mathsf{h}$. The one-particle space is $\mathcal{K} = \mathbb{C}^3 \oplus (L^2(\mathbb{R}^3) \otimes \mathbb{C}^2)$, the $\mathbb{C}^3$ subspace corresponding to $b, b^*$. A wave function in $\mathcal{K}$ is of the form $(\chi, \psi(k, \lambda))$, $\chi$ the one-particle amplitude for the oscillator and $\psi(k, \lambda)$ the one-particle photon amplitude. $\mathsf{h}$ acting on a pair $(\chi, \psi)$ is defined by

$$\mathsf{h}\begin{pmatrix} \chi \\ \psi(k, \lambda) \end{pmatrix} = \begin{pmatrix} \hbar\omega_0\chi - \frac{1}{2}\sum_{\lambda=1,2}\int \mathrm{d}^3k e\widehat{\varphi}^* \hbar\sqrt{\omega/m\omega_0}e_\lambda(k)\psi(k, \lambda) \\ -\frac{1}{2}e\widehat{\varphi}\hbar\sqrt{\omega/m\omega_0}e_\lambda(k) \cdot \chi + \hbar\omega\psi(k, \lambda) \end{pmatrix}, \qquad (13.131)$$

$\widehat{\varphi} = \widehat{\varphi}(k)$, $\omega = \omega(k)$. $\mathsf{h}$ will reappear as the Friedrichs–Lee Hamiltonian. For $e = 0$, the eigenvalue $\hbar\omega_0$ is embedded in the continuous spectrum $[0, \infty)$. The coupling turns this eigenvalue into a resonance; compare with section 17.3.

A further popular variant is to set $\omega_0 = 0$ in (13.127) and to regard the Hamiltonian as describing a freely propagating charge. One finds that the mass of the particle is increased due to the coupling with the field. However, quantitatively such a result cannot be trusted, since the dipole approximation is based on the assumption that the electron remains close to the origin. There is no such mechanism for a free particle.

*(ii) Single-photon approximation*

We restrict the Fock space to $\mathbb{C} \oplus \mathfrak{h}$. Then the wave functions $\psi$ are pairs $(\psi_0(x), \psi_1(x, k, \lambda))$. $\psi_0(x)$ is the wave function for an electron and no photon present, while $\psi_1(x,k,\lambda)$ is the wave function for the electron plus one photon with momentum $\hbar k$ and helicity $\lambda$. The correspondingly restricted Pauli–Fierz Hamiltonian is denoted by $H_1$. From (13.39), setting $N=1$, $\hbar=1=c$, one infers

$$(H_1\psi)_0(x) = \frac{1}{2m}p^2\psi_0(x) + \sum_{\lambda=1,2}\int \mathrm{d}^3k e\widehat{\varphi}\frac{1}{\sqrt{2\omega}}\mathrm{e}^{ik\cdot x}\frac{1}{m}e_\lambda\cdot p\psi_1(x,k,\lambda)\,,$$
$$(H_1\psi)_1(x,k,\lambda) = \Big(\frac{1}{2m}p^2+\omega\Big)\psi_1(x,k,\lambda) + e\widehat{\varphi}\frac{1}{\sqrt{2\omega}}\mathrm{e}^{-ik\cdot x}\frac{1}{m}e_\lambda\cdot p\psi_0(x)\,, \tag{13.132}$$

where the $A_\varphi^2$ contribution has been neglected. $H_1$ is a two-particle problem with a translation-invariant interaction. The electron has kinetic energy $\frac{1}{2m}p^2$. The photon can be either "dead" ($\psi_0$) or "alive" ($\psi_1$). The kinetic energy is zero in the dead state and $\hbar\omega$ in the alive state. Through the interaction a photon is either created or annihilated, which corresponds to a transition between dead and alive. Because of the $1/\sqrt{\omega}$-factor, this interaction has a long range and decays only as $r^{-3/2}$ in the relative distance between the electron and the photon.

## Notes and references

### *Section 13.1*

The Hamiltonian form of the Abraham model in the Coulomb gauge is standard and explained in Cohen-Tannoudji *et al.* (1989) and Sakurai (1986), for example.

### *Section 13.2*

The name "Pauli–Fierz" is not accurate historically. The Hamiltonian (13.42) appears at the beginning of paragraph two of Pauli and Fierz (1938) as a matter of fact, without citation. Pauli and Fierz study the generation of infrared photons in Compton scattering. Cohen-Tannoudij *et al.* (1989) call (13.42) "of basic importance" and Milonni (1994) refers to (13.42) simply as "the Hamiltonian". Thus despite its fundamental nature the Hamiltonian (13.42) carries no specific name in the literature. Lately, "nonrelativistic quantum electrodynamics" and "Pauli–Fierz" have become common usage in some quarters. We stick to the latter convention, which is certainly better than to be speechless.

The quantization of the electromagnetic field as a system of harmonic oscillators was common knowledge right after the advent of quantum mechanics through the work of Dirac (1927), Landau (1927), Jordan and Pauli (1928), Fermi (1930), and Landau and Peierls (1930), and was immediately applied to atomic radiation by many quantum theorists. The systematic derivation of the Hamiltonian (13.42) is not so well documented and was presumably regarded as more or less obvious, although the advantage of the Coulomb gauge was only slowly realized. The review articles by Breit (1932) and by Fermi (1932) and the research monograph by Heitler (1936, 1958) explain the quantization in its modern form, in essence. Since "one cannot comb the hair on a sphere", the polarization vectors $e_\lambda(k)$ are necessarily discontinuous in $k$, which causes poor decay in their Fourier transform. We refer to Lieb and Loss (2004) for a formulation using only the transverse projection.

The size of atoms as based exclusively on the Coulomb Hamiltonian is a longstanding open problem. We refer to Lieb (1990, 2001).

Textbooks on nonrelativistic quantum electrodynamics are listed in Notes and References to section 3.2.

### *Section 13.3*

Criteria for self-adjointness are given in Reed and Simon (1980, 1975). In our context the Kato–Rellich theorem has been applied by Nelson (1964b) and Fröhlich (1974), amongst others. Self-adjointness without restriction on the magnitude of the charge is proved by Hiroshima (2000b, 2002). A review is Hiroshima (2001).

### *Section 13.5*

A more detailed treatment of conservation laws is Huang (1998).

### *Section 13.6*

Casimir (1948) discovered the attraction of two conducting plates through vacuum fluctuations. Casimir and Polder (1948) compute the attractive force between two atoms, the retarded van der Waals force, and the force between an atom and a wall. The forces are minute and direct experimental evidence had to wait for a while. We refer to Sparnaay (1958) and Lamoreaux (1997). On the theoretical side a complete coverage is Milloni (1994), Huang (1998), with the finite-temperature corrections discussed by Schwinger *et al.* (1978), Bordag *et al.* (2000), and Feinberg *et al.* (2000).

### *Section 13.7*

Apparently the first systematic study of the dipole approximation with a harmonic external potential is Kramers (1948) and van Kampen (1951). Various aspects are covered by Senitzky (1960), Schwabl and Thirring (1964), Ford, Kac and Mazur

(1965), Ullersma (1966), Ford, Lewis and O'Connell (1988a, 1988b), Grabert, Schramm and Ingold (1988), Unruh and Zurek (1989). A mathematical study is the series by Arai (1981, 1983a, 1983b, 1990, 1991). Since the dipole approximation provides a reasonable description of radiation processes, one might regard the harmonic potential as the lowest-order approximation and expand in the anharmonicity. This program has been carried through in Maassen (1984), Spohn (1997), Maassen, Guţă and Botvich (1999), and Fidaleo and Liverani (1999). If the anharmonicity is small, in fact so small that the external potential remains convex and grows as $\frac{1}{2}m\omega_0^2 q^2$ for large $q$, then the convergence of the time-dependent Dyson series can be controlled uniformly in $t$. With such a strong estimate one can show that qualitatively the properties of the damped harmonic oscillator persist into the nonlinear regime.

The dipole approximation is not restricted to a single particle. For example one may consider two harmonically bound charges with their center of charge at $r_1$ and $r_2$. Then the kinetic energies are approximated by $(p_j - c^{-1}e_j A_\varphi(r_j))^2/2m_j$, $j = 1, 2$. Denoting $R = |r_1 - r_2|$, one is interested in the ground state energy, $E(R)$, as a function of the separation. Because of retardation $E(R) \cong -R^{-7}$ for large $R$ and $E(R) \cong -R^{-6}$ in an intermediate regime.

If $\phi_{\mathrm{ex}} = 0$, then the Hamiltonian (13.123) can be unitarily transformed to $H' = (p^2/2m_{\mathrm{eff}}) + H_{\mathrm{f}}$. $m_{\mathrm{eff}}$ agrees with the effective mass of the Abraham model to lowest order in $|v|/c$; compare with section 4.1.

The single-photon approximation was already used in disguise by Dirac (1927) and Weisskopf and Wigner (1930). It is instructive to extend this approximation by cutting Fock space at $N$ photons (Hübner and Spohn, unpublished manuscript; Skibsted 1998). If one artificially adds to the space of single-photon wave functions a one-dimensional subspace for a "dead" photon, then the theory has a structure very similar to an $(N+1)$-particle Schrödinger equation. The photons interact only indirectly through the atom. The cluster decomposition consists of $n$ free photons and $N - n$ photons bound by the atom, $n = 0, 1, \ldots, N$.

# 14

# The statistical mechanics connection

Models from quantum mechanics can be converted into statistical mechanics systems through the Wick rotation $t \rightsquigarrow -it$. Within quantum field theory this technique has been very powerful, both in proving qualitative properties and as a computational tool. Besides these more practical aspects, the statistical mechanics formulation is an additional source of intuition which cannot so easily be extracted from the Schrödinger differential equation. The price to pay is that, in essence only ground state properties can be handled. Truly time-dependent problems must be treated in physical time. For charges interacting with the Maxwell field the Wick rotation is equally attractive. There is one additional bonus: since the field Hamiltonian is quadratic and since the coupling to the field is linear, as first observed by Feynman, the Gaussian integration over the Maxwell field can be done explicitly. This results in a fairly concise statistical-mechanical description for the particles.

In Euclidean language the possible paths of the charge and the fields become fluctuating quantities. To distinguish in notation we use $t \mapsto q_t$ for a random path of the charge and $t \mapsto A_t(x)$ for a random history of the transverse vector field. $\mathbb{E}(\cdot)$ refers to expectation with respect to the measure of integration, either specified through the context or indicated by a subscript. Sometimes we also use the statistical mechanics shorthand $\langle \cdot \rangle$ for averages.

## 14.1 Functional integral representation

For a single particle, subject to the potential $V(x)$, the imaginary time Schrödinger equation is, setting $\hbar = 1 = m$,

$$\partial_t \psi = -H_\mathrm{p} \psi \,, \quad H_\mathrm{p} = -\frac{1}{2}\Delta + V \tag{14.1}$$

and its solution for $t \geq 0$ is constructed through the Trotter product formula as

$$(\mathrm{e}^{-tH_\mathrm{p}}\psi)(x) = \lim_{n\to\infty} (\mathrm{e}^{t\Delta/2n}\mathrm{e}^{-tV/n})^n \psi(x)\,. \tag{14.2}$$

We recognize $\exp[\frac{1}{2}t\Delta]$ as the transition probability for a Brownian motion, whose paths will be denoted here by $t \mapsto q_t$. Brownian motion is a Gaussian process and therefore defined through the mean and covariance. Explicitly,

$$\mathbb{E}(q_t) = 0\,, \quad \mathbb{E}(q_{s\alpha}q_{t\beta}) = \delta_{\alpha\beta}\min(s,t)\,. \tag{14.3}$$

If the Brownian motion starts at $x$, we indicate the start point as a subscript in the expectation and have $\mathbb{E}_x(q_t) = x$, $\mathbb{E}_x((q_s - x)_\alpha(q_t - x)_\beta) = \delta_{\alpha\beta}\min(s,t)$, $\alpha, \beta = 1, 2, 3$. In particular the transition probability is obtained as

$$\mathbb{P}_x(\{q_t \in \mathrm{d}^3y\}) = (2\pi t)^{-3/2}\exp[-(y-x)^2/2t]\mathrm{d}^3y = (\mathrm{e}^{t\Delta/2})(x,y)\mathrm{d}^3y\,. \tag{14.4}$$

Writing out (14.2) in position space representation, one infers the Feynman–Kac formula

$$(\mathrm{e}^{-tH_\mathrm{p}}\psi)(x) = \mathbb{E}_x\Big(\exp\Big[-\int_0^t \mathrm{d}s\, V(q_s)\Big]\psi(q_t)\Big)\,. \tag{14.5}$$

The Brownian motion path has acquired a non-Gaussian weight, which is the exponential of the potential energy integrated along the path $q_t$.

The statistical mechanics connection becomes more obvious upon discretizing time in units of $\tau$. We set $\phi_n = q_{n\tau}$, $\phi_n \in \mathbb{R}^3$. Then, in approximation, (14.5) reads

$$\begin{aligned}&\frac{1}{Z}\int \mathrm{d}^3\phi_0 \ldots \mathrm{d}^3\phi_N\, \delta(\phi_0 - x)\exp\Big[-\frac{1}{2\tau}\sum_{j=0}^{N-1}(\phi_{j+1}-\phi_j)^2\Big]\\ &\quad\times\exp\Big[-\tau\sum_{j=1}^{N}V(\phi_j)\Big]\psi(\phi_N)\,,\end{aligned} \tag{14.6}$$

$N\tau = t$. The statistical mechanics model lives on a one-dimensional lattice and has at each site a continuous "spin" with three components. The first exponential is a quadratic nearest-neighbor interaction and, except for the normalization, represents a discrete-time Gaussian random walk in $\mathbb{R}^3$. The potential can be combined with the Lebesgue measure as $\exp[-\tau V(\phi_j)]\mathrm{d}^3\phi_j$ and thus provides a non-Gaussian single-site measure.

For (14.5) to make sense one needs some minimal conditions on $V$ to ensure that the expectation is defined. An obvious sufficient condition is to have $V \geq c_0 > -\infty$. Equation (14.5) indicates that very roughly there are three families of potentials: (i) *Binding*, $V$ increases at infinity. Under the measure in (14.5) $q_t$ has in essence bounded fluctuations. For $t \to \infty$ the path measure for $q_t$ becomes a stationary diffusion process. (ii) *No binding*, e.g. a repulsive potential decaying to zero at infinity or a bounded periodic potential. A typical path $q_t$ fluctuates and diffuses to infinity as a Brownian motion with some effective diffusion coefficient. (iii) *Local binding*, like an attractive square-well potential. For the purpose of discussion let us set the potential as $\lambda V$ with $V$ attractive near the origin and decaying to zero at infinity. For large $\lambda$ the potential dominates and $q_t$ is confined as a stationary diffusion process. As $\lambda$ decreases, $q_t$ makes longer and longer excursions until it unbinds at some critical $\lambda_c$. For $\lambda < \lambda_c$ the Brownian motion dominates. Since Brownian motion is recurrent in dimension $d = 1, 2$, one has $\lambda_c = 0$, whereas for $d = 3$ generically $\lambda_c > 0$.

It is of use to translate the path properties of the particle to spectral properties of the particle Hamiltonian $H_p = -\frac{1}{2}\Delta + V$. We denote by $\Sigma$ the continuum edge of $H_p$ and, if it exists, by $\psi_0$ the unique ground state of $H_p$, i.e. $H_p\psi_0 = E_0\psi_0$. In case (i) the spectrum of $H_p$ is purely discrete, formally $\Sigma = \infty$. In the second case $H_p$ has a purely continuous spectrum and no eigenvalues. For a locally binding potential which decays to zero at infinity, case (iii), the continuum edge is $\Sigma = 0$. For sufficient attraction there are bound states with an energy below $\Sigma$, in particular $E_0 < 0$. In dimension $d = 1, 2$ an arbitrarily weak attraction results in a bound state, whereas for $d \geq 3$ a minimal strength is required. As is well understood, there is more complicated spectral behavior around with various borderline cases. For our purposes the schematic classification above will suffice.

Our goal is to extend the Feynman–Kac formula (14.5) to $\mathrm{e}^{-tH}$ with $H$ the Pauli–Fierz Hamiltonian. This will be done in two steps. Firstly we study a one-particle Hamiltonian including an external vector potential, and secondly we write $\mathrm{e}^{-tH_f}$ in terms of a suitable Gaussian measure. Combining both elements yields the desired generalization.

Let us assume then that the quantum particle is subject to a magnetic field and denote the corresponding vector potential by $a(x)$, to distinguish from the fluctuating vector potential $A_t$ used later on. The imaginary time Schrödinger equation becomes

$$\partial_t\psi = -H_p\psi\,, \quad H_p = \frac{1}{2}(-\mathrm{i}\nabla - a)^2 + V\,. \tag{14.7}$$

Then, as before, we represent $\mathrm{e}^{-tH_p}$ through the Trotter product formula. The vector potential yields a term proportional to $\dot{q}$, as can be guessed from the corresponding classical action. More precisely one obtains

$$(\mathrm{e}^{-tH_p}\psi)(x) = \mathbb{E}_x\Big(\exp\Big[-\mathrm{i}\int_0^t \mathrm{d}q_s\cdot a(q_s) - \frac{\mathrm{i}}{2}\int_0^t \mathrm{d}s\,\nabla\cdot a(q_s) - \int_0^t \mathrm{d}s\,V(q_s)\Big]\psi(q_t)\Big)\,. \tag{14.8}$$

The stochastic integral appearing in (14.8) is defined as Ito integral, which means that the discretization of $a(x)$ is evaluated at the left end point,

$$\int_0^t \mathrm{d}q_s\cdot a(q_s) = \lim_{n\to\infty}\sum_{m=1}^{nt} a(q_{(m-1)/n})\cdot(q_{m/n} - q_{(m-1)/n})\,. \tag{14.9}$$

This limit exists almost surely with respect to Brownian motion. Through the Ito convention one picks up in (14.8) the additional term containing $\nabla \cdot a$. It disappears, if in (14.9) we were to use the, in our context perhaps more natural, Feynman–Stratonovich midpoint rule where $a(q_{(m-1)/n})$ is replaced by $\frac{1}{2}(a(q_{m/n}) + a(q_{(m-1)/n}))$. Note that in the Coulomb gauge the stochastic integral does not depend on the particular choice of the rule for the discretization, since $\nabla \cdot a = 0$.

On a purely formal level, following Feynman, the quantum propagator is written as a sum over all paths from $x'$ to $x$ in the time span $t$ "weighted" by the exponential of the classical action,

$$(\mathrm{e}^{-\mathrm{i}H_{\mathrm{p}}t})(x, x') = \int \prod_{0 \le s \le t} \mathrm{d}^3 q_s \delta(q_0 - x')\delta(q_t - x) \exp\Big[\mathrm{i}\int_0^t \mathrm{d}s L(q_s, \dot{q}_s)\Big] \tag{14.10}$$

with the classical Lagrangian $L(q, \dot{q}) = \frac{1}{2}\dot{q}^2 - V(q) + \dot{q} \cdot a(q)$. Note that compared to the right side of (14.8) the role of $x$ and $x'$ has been interchanged. Upon Wick rotation $t \rightsquigarrow -\mathrm{i}t$ and time reversal $q_s \rightsquigarrow q_{t-s}$ (14.10) becomes

$$(\mathrm{e}^{-tH_{\mathrm{p}}})(x, x') = \int \prod_{0 \le s \le t} \mathrm{d}^3 q_s \delta(q_0 - x)\delta(q_t - x') \times \exp\Big[-\int_0^t \mathrm{d}s\Big(\frac{1}{2}\dot{q}_s^2 + V(q_s) + \mathrm{i}\dot{q}_s \cdot a(q_s)\Big)\Big]. \tag{14.11}$$

One recognizes the potential term and the stochastic integral $-\mathrm{i}\int_0^t \mathrm{d}s\dot{q}_s \cdot a(q_s)$ with the mid point rule. The exponential of the kinetic term combines with the infinite-product Lebesgue measure to Brownian motion, denoted by $\mathbb{E}_x$ in (14.8), which starts at $x$ according to the factor $\delta(q_0 - x)$.

We turn to the functional integral for the Maxwell field, which we can think of as an infinite collection of harmonic oscillators. Let us first recall the single harmonic oscillator with Hamiltonian

$$H = \frac{1}{2}\big(-\partial_x^2 + \omega^2 x^2 - \omega\big) \tag{14.12}$$

as a differential operator acting on $L^2(\mathbb{R}, \mathrm{d}x)$. It has the normalized eigenvectors $|n\rangle$, $n = 0, 1, \ldots$, i.e.

$$H|n\rangle = \varepsilon_n |n\rangle\,, \quad \varepsilon_n = \omega n\,. \tag{14.13}$$

$|0\rangle$ is the ground state of $H$. In the position representation $H\psi_0 = 0$ with $\psi_0(x)^2 = \sqrt{\omega/\pi}\mathrm{e}^{-\omega x^2}$. Thus, alternatively we can use the linear span of the $|n\rangle$'s as the Hilbert space of states. This corresponds to the Fock space $\mathcal{F}$ over the one-particle space $\mathbb{C}$, which means $\psi \in \mathcal{F}$ is of the form $\psi = (\psi_0, \psi_1, \dots)$, $\psi = \sum_{n=0}^{\infty} \psi_n |n\rangle$. A further, as it will turn out natural, choice is the Hilbert space $\mathcal{H}_0 = L^2(\mathbb{R}, \psi_0(x)^2 \mathrm{d}x)$ with weight given by the square of the ground state wave function.

Of course, these Hilbert spaces are unitarily equivalent. Of interest is the unitary map from $\mathcal{F}$ to $\mathcal{H}_0$ which is achieved through the Wick ordering of polynomials. We regard $x$ as a random variable on $\mathbb{R}$ equipped with the normalized Gaussian measure $\psi_0(x)^2 \mathrm{d}x$. Then, denoting expectation by $\langle \cdot \rangle$, the Wick order of $x$ is defined recursively through $:x^0: = 1$, $\partial_x :x^n: = n :x^{n-1}:$, and $\langle :x^n: \rangle = 0$, $n = 1, 2, \dots$. Note that the Wick order depends both on the random variable and on the underlying measure. Thus $:1: = 1, :x: = x - \langle x \rangle = x, :x^2: = x^2 - 2\langle x \rangle x - \langle x^2 \rangle + 2\langle x \rangle^2 = x^2 - (1/2\omega)$, etc., in our case. Let $P_n$ denote the $n$-th Hermite polynomial,

$$P_n(x) = \sum_{j=0}^{[n/2]} \frac{n!}{(n-2j)!\,j!} (-\tfrac{1}{2})^j x^{n-2j} , \tag{14.14}$$

with $[n]$ the integer part. Then the Wick-ordered mononomial of order $n$ is given by

$$:x^n: = (2\omega)^{-n/2} P_n(\sqrt{2\omega}x) . \tag{14.15}$$

One has

$$\langle :x^n: :x^m: \rangle = \langle :x^n:, :x^m: \rangle_{\mathcal{H}_0} = (2\omega)^{-n} n! \delta_{mn} . \tag{14.16}$$

By linearity Wick order extends to all finite polynomials. Let us also introduce

$$a^* = \frac{1}{\sqrt{2\omega}} (\omega x - \partial_x) , \quad a = \frac{1}{\sqrt{2\omega}} (\omega x + \partial_x) \tag{14.17}$$

as creation and annihilation operators of the harmonic oscillator. Their Wick order means that all annihilation operators are moved to the right, e.g. $:aa^*: = a^* a$. Then

$$:\Big(\frac{1}{\sqrt{2\omega}} (a^* + a)\Big)^n: |0\rangle = (2\omega)^{-n/2} \sqrt{n!} |n\rangle . \tag{14.18}$$

Comparing with (14.16) the map $U$ from $\mathcal{F}$ to $\mathcal{H}_0$ should be defined through

$$:\Big(\frac{1}{\sqrt{2\omega}} (a^* + a)\Big)^n: |0\rangle \mapsto :x^n: \tag{14.19}$$

and extended by linearity. By the very construction the closure of $U$ as a linear map $\mathcal{F} \to \mathcal{H}_0$ is then unitary. Note that $\mathrm{e}^{-tH}$ is implemented as

$$U\mathrm{e}^{-tH}U^{-1}{:}x^n{:} = {:}(\mathrm{e}^{-\omega t}x)^n{:} = \mathrm{e}^{-n\omega t}{:}x^n{:}\,. \tag{14.20}$$

Through the Feynman–Kac formula (14.5) we boost (14.12) to a Gaussian stochastic process denoted by $x_t$. It takes real values, is stationary in time, has mean zero, and covariance

$$\mathbb{E}(x_t x_s) = \frac{1}{2\omega}\mathrm{e}^{-\omega|t-s|}\,. \tag{14.21}$$

We recognize $x_t$ as the stationary Ornstein–Uhlenbeck process governed by the stochastic differential equation

$$\mathrm{d}x_t = -\omega x_t \mathrm{d}t + \mathrm{d}b_t\,, \tag{14.22}$$

where $b_t$ is standard one-dimensional Brownian motion. Note that

$$\mathbb{E}(f(x_t)) = \mathbb{E}(f(x_0)) = \langle \psi_0, f\psi_0\rangle = \langle 1, f\rangle_{\mathcal{H}_0} = \int \mathrm{d}x\,\psi_0(x)^2 f(x)\,, \tag{14.23}$$

$$\mathbb{E}(f(x_t)g(x_s)) = \langle \psi_0, f\mathrm{e}^{-|t-s|H}g\psi_0\rangle = \langle f, \mathrm{e}^{|t-s|L}g\rangle_{\mathcal{H}_0}\,, \tag{14.24}$$

where we used the similarity transformation

$$\psi_0^{-1}\mathrm{e}^{-tH}\psi_0 = \mathrm{e}^{tL}\,,\quad t \geq 0\,, \tag{14.25}$$

with $L$ the generator of the Ornstein–Uhlenbeck process $x_t$,

$$L = -\omega x\partial_x + \tfrac{1}{2}\partial_x^2\,. \tag{14.26}$$

According to (14.23) the Ornstein–Uhlenbeck process $x_t$ has $\psi_0(x)^2$ as stationary measure. With probability one $t \mapsto x_t$ is continuous and we may choose $C(\mathbb{R},\mathbb{R})$, the space of all continuous functions over $\mathbb{R}$, as path space. In fact, $x_t$ has in essence bounded fluctuations and increases at most logarithmically for large $t$.

The point of our exercise is that it carries over essentially verbatim to the infinite-dimensional setting, except for the flat Hilbert space $L^2(\mathbb{R}, \mathrm{d}x)$. $H_\mathrm{f}$ plays the role of the harmonic oscillator. The boson Fock space over the transverse vector fields $L^2_\perp(\mathbb{R}^3, \mathbb{R}^3)$ plays the role of the Fock space over $\mathbb{C}$. The Ornstein–Uhlenbeck process $x_t$ is replaced by the infinite-dimensional Ornstein–Uhlenbeck process $A_t(x)$. Let us start with the latter. $A_t(x)$ is a Gaussian process with mean zero and covariance

$$\mathbb{E}\big(A_{t\alpha}(x)A_{t'\alpha'}(x')\big) = (2\pi)^{-3}\int \mathrm{d}^3k\,\mathrm{e}^{\mathrm{i}k\cdot(x-x')}Q^\perp_{\alpha\alpha'}(k)\frac{1}{2\omega}\mathrm{e}^{-\omega|t-t'|}\,, \tag{14.27}$$

$\alpha, \alpha' = 1, 2, 3$. Because of the transverse projection $Q^{\perp}_{\alpha\alpha'}(k) = \delta_{\alpha\alpha'} - \widehat{k}_\alpha \widehat{k}_{\alpha'}$ the covariance (14.27) implies that

$$\nabla \cdot A_t = 0 \tag{14.28}$$

almost surely. $A_t(x)$ becomes a proper Gaussian random variable once it is integrated over the real test function $f$,

$$A_t(f) = \sum_{\alpha=1}^{3} \int d^3x f_\alpha(x) A_{t\alpha}(x) \,. \tag{14.29}$$

From (14.27) we conclude that

$$\mathbb{E}\big(A_t(f)^2\big) = \int d^3k (2\omega)^{-1} \widehat{f}^* \cdot Q^{\perp} \widehat{f} \,. \tag{14.30}$$

Thus $A_t(f)$ has a bounded variance provided $\| f/\sqrt{\omega} \|_{\mathfrak{h}} < \infty$.

In quantum field theory Lorentz invariance is of central importance; this becomes more evident by treating time and space on an equal footing. We thus Fourier transform in (14.27) also with respect to $t$ and obtain

$$\mathbb{E}\big(\widehat{A}_\alpha(k_0, k)^* \widehat{A}_{\alpha'}(k_0', k')\big) = \delta(k - k')\delta(k_0 - k_0') Q^{\perp}_{\alpha\alpha'}(k)(k^2 + k_0^2)^{-1} \,, \tag{14.31}$$

which is more symmetric. However, fixing the Coulomb gauge spoils full rotation invariance in $\mathbb{R}^4$.

In our context time is singled out and we prefer to think of $t \mapsto A_t$ as a stochastic process with values in the transverse vector fields. Most conveniently, we regard $A_t$ as the element of a Hilbert space $\mathcal{K}'$, which is chosen such that $t \mapsto A_t$ is continuous in $t$. $A_t(x)$ is somewhat singular in $x$, which has to be balanced by defining the norm of the Hilbert space $\mathcal{K}'$ through the inner product

$$\langle f, g \rangle_{\mathcal{K}'} = \sum_{\lambda=1,2} \int d^3k \widehat{f}(k, \lambda)^* \omega^{1/2} (-\Delta_k + k^2)^{-\kappa} \omega^{1/2} \widehat{g}(k, \lambda) \tag{14.32}$$

with some $\kappa \geq 0$. The predual Hilbert space is denoted by $\mathcal{K}$. It has the inner product

$$\langle f, g \rangle_{\mathcal{K}} = \sum_{\lambda=1,2} \int d^3k \widehat{f}(k, \lambda)^* \omega^{-1/2} (-\Delta_k + k^2)^{\kappa} \omega^{-1/2} \widehat{g}(k, \lambda) \,. \tag{14.33}$$

**Lemma 14.1** (Regularity properties for sample paths of the Ornstein–Uhlenbeck process). *We regard the Ornstein–Uhlenbeck process $A_t(x)$ with covariance* (14.27) *as taking values in the Hilbert space $\mathcal{K}'$ with $\kappa > \frac{7}{2}$. Then $t \mapsto A_t \in \mathcal{K}'$ is almost surely (norm) continuous. The path space of the Ornstein–Uhlenbeck* *process can be taken as $C(\mathbb{R}, \mathcal{K}')$, the space of continuous functions with values in $\mathcal{K}'$.*

*Proof*: The Ornstein–Uhlenbeck process $A_t$ is Markov and time reversible. A general estimate for such processes gives

$$\mathbb{E}\big(\sup_{0\le t\le T} A_t(f)^2\big) \le 3\mathbb{E}\big(A_0(f)^2\big) + 72T\mathcal{D}\big(A_0(f), A_0(f)\big), \tag{14.34}$$

where $\mathcal{D}$ is the Dirichlet form defined through

$$\mathcal{D}\big(A_0(f), A_0(f)\big) = \lim_{t\to 0}\frac{1}{t}\big(\mathbb{E}\big(A_t(f)A_0(f)\big) - \mathbb{E}\big(A_0(f)^2\big)\big). \tag{14.35}$$

Therefore

$$\mathbb{E}\big(\sup_{0\le t\le T} A_t(f)^2\big) \le c_0 \sum_{\lambda=1,2} \int d^3k |\widehat{f}(k,\lambda)|^2(1+\omega^{-1}). \tag{14.36}$$

The eigenfunctions of $(-\Delta_k + k^2)$ are the Hermite functions $h_n$, $n \in \mathbb{N}^3$, with eigenvalue $\lambda_n = 1 + 2\sum_{\alpha=1}^{3} n_\alpha$. Therefore

$$\begin{aligned}\mathbb{E}\big(\sup_{0\le t\le T} \|A_t\|^2_{\mathcal{K}'}\big) &= \mathbb{E}\big(\sup_{0\le t\le T} \sum_{n\in\mathbb{N}^3} (\lambda_n)^{-\kappa} A_t(\sqrt{\omega}h_n)^2\big) \\ &\le c_0 \sum_{n\in\mathbb{N}^3} (\lambda_n)^{-\kappa} \int d^3k |\widehat{h}_n(k)|^2(1+\omega).\end{aligned} \tag{14.37}$$

Using operator monotonicity as $(k^2)^{1/2} \le (-\Delta_k + k^2)^{1/2}$ yields the bound

$$c_0 \sum_{n\in\mathbb{N}^3} (\lambda_n)^{-\kappa+\frac{1}{2}}, \tag{14.38}$$

which is finite provided $\kappa > \frac{7}{2}$.

The inequality (14.37) establishes that $A_t$ lies in $\mathcal{K}'$ with probability one. Continuity is proved by a similar argument. The complete details can be found, e.g., in Giacomin *et al.* (2001), Lemma 5.5. □

The path measure for $A_t(x)$, as a probability measure on $C(\mathbb{R}, \mathcal{K}')$, is denoted by $\mathrm{dP}$. The time-zero field is $A_0(x)$. $A_0(x)$ has the distribution $\mathrm{dP}^0$ as a probability measure on $\mathcal{K}'$. According to (14.30) $\mathrm{dP}^0$ is Gaussian with mean zero and covariance

$$\mathbb{E}_{\mathrm{dP}^0}\big(A_0(f)A_0(g)\big) = \int d^3k (2\omega)^{-1} \widehat{f}^* \cdot Q^\perp \widehat{g}. \tag{14.39}$$

As in the case of a single oscillator, there is a natural unitary map $U$ from Fock space $\mathcal{F}$ to $L^2(\mathcal{K}', \mathrm{dP}^0)$ which is achieved through Wick order. The Wick order for operators on $\mathcal{F}$ is defined by moving all creation operators to the left. The Wick-ordered polynomials on $\mathcal{K}'$ are defined through a multilinear extension of the orthogonalization scheme for a single oscillator. Let $X_1, \dots, X_k$ be $k$ random variables. Their Wick order, relative to $\langle\cdot\rangle$, is defined recursively by $:(X_1)^0 \dots (X_k)^0: = 1$, $\langle :(X_1)^{n_1} \dots (X_k)^{n_k}: \rangle = 0$, and $\partial/\partial X_j :(X_1)^{n_1} \dots (X_k)^{n_k}: = n_j :(X_1)^{n_1} \dots (X_j)^{n_j - 1} \dots (X_k)^{n_k}:$. Clearly, for a single degree of freedom, i.e. $\mathcal{K}' = \mathbb{R}$, $\mathrm{dP}^0 = \sqrt{\omega/\pi}\mathrm{e}^{-\omega^2 x^2}\mathrm{d}x$, the Wick order agrees with the construction in (14.15). The unitary map $U : \mathcal{F} \to L^2(\mathcal{K}', \mathrm{dP}^0)$ is then given by

$$U\Omega = 1\,, \quad U{:}A(f_1) \dots A(f_n){:}\,\Omega = {:}A_0(f_1) \dots A_0(f_n){:}\ . \tag{14.40}$$

Here $\Omega$ denotes the Fock vacuum of $\mathcal{F}$. $A(f_j)$ is the quantized vector potential (13.35) smeared by $f_j$ as $A(f_j) = \int \mathrm{d}^3x f_j(x) \cdot A(x)$, whereas to the right stands the Wick order of polynomials as functions on $\mathcal{K}'$. We note that the dynamics is implemented as

$$U\mathrm{e}^{-tH_\mathrm{f}}U^{-1}{:}A_0(f_1) \dots A_0(f_n){:} = {:}A_0(\mathrm{e}^{-\omega t} f_1) \dots A_0(\mathrm{e}^{-\omega t} f_n){:} \tag{14.41}$$

for $t \geq 0$. $UH_\mathrm{f}U^{-1}$, a linear operator acting on $L^2(\mathcal{K}', \mathrm{dP}^0)$, is referred to as the Schrödinger representation of $H_\mathrm{f}$.

Next we couple the charge and the Maxwell field. According to (14.39) the natural Hilbert space is

$$\mathcal{H}_\mathrm{s} = L^2(\mathbb{R}^3, \mathrm{d}^3x) \otimes L^2(\mathcal{K}', \mathrm{dP}^0)\,, \tag{14.42}$$

the subscript 's' standing for Schrödinger. The particle Hamiltonian reads $H_\mathrm{p} = -\frac{1}{2}\Delta + V$, with the shorthand $V(q) = e\phi_\mathrm{ex}(q)$, and the field Hamiltonian $UH_\mathrm{f}U^{-1}$ is defined through (14.41). Let us denote by $\mathbb{E}_{\mathrm{dW}\times\mathrm{dP}}$ expectation with respect to the path measure $\mathrm{dW} \times \mathrm{dP}$, where $\mathrm{dP}$ is the path measure for the Ornstein–Uhlenbeck process $A_t(x)$ and $\mathrm{dW}$ the Wiener measure for $q_t$, i.e. the path measure of Brownian motion with starting distribution $\mathrm{d}^3x$. Let $F, G \in \mathcal{H}_\mathrm{s}$. Then, combining (14.5) and the infinite-dimensional analog of (14.24), we conclude that for the uncoupled system

$$\begin{aligned} &\mathbb{E}_{\mathrm{dW}\times\mathrm{dP}}\Big(F(q_0, A_0)^* \exp\Big[-\int_0^t \mathrm{d}s\, V(q_s)\Big] G(q_t, A_t)\Big) \\ &\quad = \langle 1 \otimes U^{-1}F, \mathrm{e}^{-t(H_\mathrm{p}\otimes 1 + 1 \otimes H_\mathrm{f})} 1 \otimes U^{-1}G\rangle_{\mathcal{H}_\mathrm{s}}\,, \end{aligned} \tag{14.43}$$

$t \geq 0$. In the following, the somewhat pedantic $1\otimes$ will be omitted, in particular $U$ acts on $L^2(\mathbb{R}^3, \mathrm{d}^3x) \otimes \mathcal{F}$ as 1 on the first and as (14.40) on the second factor.

The missing step is to include the minimal coupling to the field through the vector potential. For this purpose we note that in the Hilbert space $L^2(\mathcal{K}', \mathrm{dP}^0)$ of the Schrödinger representation the transverse vector potential $A(x)$ acts as a multiplication operator, compare with (14.40), and in the functional integral the operator $A(x)$ becomes a fluctuating vector potential $A_t(x)$, which is to be inserted in the minimal coupling as $\frac{1}{2}(p - eA_{t\varphi}(q))^2$. Thus one can use (14.7) and (14.8), properly adapted to time-dependent vector potentials respecting the Coulomb gauge $\nabla \cdot A_t = 0$. For later convenience let us reintroduce the mass of the quantum particle, which amounts to replacing $(p - eA_{t\varphi}(q))^2/2$ by $(p - eA_{t\varphi}(q))^2/2m$ and hence taking the Wiener process dW with diffusion coefficient $1/m$ instead of 1, i.e. $\mathbb{E}_0(q_{s\alpha}q_{t\beta}) = m^{-1}\delta_{\alpha\beta}\min(s,t)$. As a result we obtain the functional integral representation for the semigroup $\mathrm{e}^{-tH}, t \geq 0$, of the spinless Pauli–Fierz Hamiltonian (13.39) for a single particle as

$$\langle F, U\mathrm{e}^{-tH}U^{-1}G\rangle_{\mathcal{H}_\mathrm{s}}$$
$$= \mathbb{E}_{\mathrm{dW}\times\mathrm{dP}}\Big(F(q_0, A_0)^* \exp\Big[-\int_0^t \mathrm{d}s\, V(q_s) - \mathrm{i}e\int_0^t \mathrm{d}q_s \cdot A_{s\varphi}(q_s)\Big]G(q_t, A_t)\Big). \tag{14.44}$$

Recall that $A_{t\varphi}(q) = \int \mathrm{d}^3x\varphi(q-x)A_t(x)$. Equation (14.44) is the basic result of this section. It says that the measure on paths is weighted by the exponential of the classical action. The quadratic terms yield dW × dP and constitute the Gaussian a priori measure of the uncoupled system. The external potential and the minimal coupling to the quantized transverse vector potential are displayed explicitly.

We still have to check that the random variable in the exponential of (14.44) remains finite almost surely. The function $q, s \mapsto A_{s\varphi}(q)$ is (almost surely) continuous in both variables, which makes the stochastic integral well defined. To compute the variance, one notes

$$\mathbb{E}_{\mathrm{dP}}\Big(\Big(\int_0^t \mathrm{d}q_s \cdot A_{s\varphi}(q_s)\Big)^2\Big) = \int_0^t\int_0^t \mathrm{d}q_s \cdot W(q_s - q_{s'}, s - s')\mathrm{d}q_{s'}\,. \tag{14.45}$$

$W$ is the transverse photon propagator,

$$W_{\alpha\beta}(x,t) = \int \mathrm{d}^3k|\widehat{\varphi}(k)|^2 Q^{\perp}_{\alpha\beta}(k)\frac{1}{2\omega}\mathrm{e}^{-\omega|t|}\mathrm{e}^{\mathrm{i}k\cdot x}\,, \tag{14.46}$$

which is bounded by our assumption on $\widehat{\varphi}$. The average of (14.45) with respect to Brownian motion yields

$$\mathbb{E}_{\mathrm{dW}\times\mathrm{dP}}\Big(\delta(q_0)\Big(\int_0^t \mathrm{d}q_s \cdot A_{s\varphi}(q_s)\Big)^2\Big) = t\frac{2}{3m}\int \mathrm{d}^3k|\widehat{\varphi}|^2/2\omega\,, \tag{14.47}$$

since one of the two stochastic differentials points in the future except at the diagonal where $\mathrm{d}q_{t\alpha}\mathrm{d}q_{t\beta} = m^{-1}\delta_{\alpha\beta}\mathrm{d}t$. Thus the action appearing in the exponential of (14.44) has a bounded variance.

## 14.2 Integrating out the Maxwell field

We return to the basic formula (14.44) and assume that $F, G$ are of the special form $F(q, A) = G(q, A) = \psi(q)$ with $\psi \geq 0$ and of rapid decrease. The Gaussian integration over $\mathrm{dP}$ can then be carried out with the result

$$\langle \psi \otimes \Omega, \mathrm{e}^{-tH} \psi \otimes \Omega \rangle_{\mathcal{H}} = \mathbb{E}_{\mathrm{dW}}\big(\psi(q_0) \exp\Big[-\int_0^t \mathrm{d}s\, V(q_s) - \frac{1}{2}e^2 \int_0^t \int_0^t \mathrm{d}q_s \cdot W(q_s - q_{s'}, s - s')\mathrm{d}q_{s'}\Big] \psi(q_t)\big) . \tag{14.48}$$

Since $\mathrm{d}q_{t\alpha}\mathrm{d}q_{t\beta} = m^{-1}\delta_{\alpha\beta}\mathrm{d}t$ almost surely, we may remove the diagonal cut in the double stochastic integral at the expense of the factor $t(2/3m) \int \mathrm{d}^3k|\widehat{\varphi}|^2/2\omega$. $W$ is the transverse photon propagator (14.46), written more traditionally

$$W(x, t) = \frac{1}{2\pi} \int \mathrm{d}^3k\mathrm{d}k_0 |\widehat{\varphi}(k)|^2 (k^2 + k_0^2)^{-1} \mathrm{e}^{\mathrm{i}(k \cdot x - k_0 t)} Q^{\perp}(k) \tag{14.49}$$

as a $3 \times 3$ matrix. If one removes the ultraviolet cutoff by replacing $\widehat{\varphi}(k)$ by $(2\pi)^{-3/2}$, then (14.49) can be computed explicitly. For our purpose it suffices that qualitatively

$$W(x, t) \cong (x^2 + t^2)^{-1} \tag{14.50}$$

with some modifications due to the transverse projection. Reintroducing $\widehat{\varphi}$ smooths this function at $(x, t) = 0$, but keeps the slow $t^{-2}$ decay. For massive photons, $\omega(k) = (k^2 + m_{\mathrm{ph}}^2)^{1/2}$, this decay would switch to an exponential.

Equation (14.48) looks like the partition function of an equilibrium statistical mechanics system. We regard $\mathrm{dW}$ as the a priori measure on continuous paths in three-dimensional space. The time interval $[0, t]$ corresponds to the volume. From the point of view of statistical mechanics it is more natural to place it symmetric relative to the origin, i.e. as $[-t, t]$. Configurations are paths $q_s$, $|s| \leq t$. The factors $\psi(q_{-t})$, $\psi(q_t)$ constrain their end points to be most likely close to the origin. The paths have a Boltzmann weight consisting of two contributions, a single time integral from the external potential and a double time integral induced through the Maxwell field. Our observation suggests that the basic object must be the Gibbs measure for paths $q_s$, $|s| \leq t$, as given through

$$Z(2t)^{-1}\psi(q_{-t})\psi(q_t) \times \exp\Big[-\int_{-t}^t \mathrm{d}s\, V(q_s) - \frac{1}{2}e^2 \int_{-t}^t \int_{-t}^t \mathrm{d}q_s \cdot W(q_s - q_{s'}, s - s')\mathrm{d}q_{s'}\Big]\mathrm{dW} \tag{14.51}$$

relative to the Wiener measure dW with $Z(2t)$ the normalizing constant (14.48). The average with respect to the probability measure (14.51) is denoted below by $\langle\cdot\rangle_t$ and by $\langle\cdot\rangle_t^0$ for $e = 0$.

The relationship to usual spin systems becomes even more evident upon discretizing time in steps of $\tau$; compare with (14.6). Then, setting $q_{n\tau} = \phi_n$, $\phi_n \in \mathbb{R}^3$, $N\tau = t$, (14.51) becomes

$$\frac{1}{Z}\prod_{n=-N}^{N} \mathrm{d}^3\phi_n \psi(\phi_{-N})\psi(\phi_N)\exp\Big[-\frac{1}{2}\frac{m}{\tau}\sum_{j=-N}^{N-1}(\phi_{j+1}-\phi_j)^2 - \tau\sum_{j=-N}^{N}V(\phi_j) - \frac{1}{2}e^2\sum_{i,j=-N}^{N-1}(\phi_{j+1}-\phi_j)\cdot W(\phi_i-\phi_j, i-j)(\phi_{i+1}-\phi_i)\Big], \tag{14.52}$$

which is the Gibbs measure for a three-component continuous spin system with external potential $V$, a quadratic nearest-neighbor interaction, and a long-range interaction $W$. The spin configurations are over a one-dimensional lattice. Alternatively, we may interpret $\phi_j$ as the position of the $j$-th monomer of an elastic string (polymer) curling in three-dimensional space. The term $(\phi_{j+1} - \phi_j)^2$ is the usual nearest-neighbor elastic energy. Integrating over the Maxwell field results in an additional long-range elastic interaction between the monomers.

In the picture of an elastic string, cf. figure 14.1, it is natural to distinguish between the case $V = 0$ and a confining potential. Let us first discuss $V = 0$ and for definiteness pin the polymer at both end points, i.e. $q_{-t} = 0 = q_t$. If $e = 0$, then the mean square displacement at the midpoint, given by

$$\langle (q_0)^2\rangle_t^0 = 3t/2m\,, \tag{14.53}$$

reflects the stiffness of the free string. We expect that the interaction renormalizes the stiffness as

$$\langle (q_0)^2\rangle_t \cong 3t/2\sigma \tag{14.54}$$

for large $t$, which defines the (effective) stiffness $\sigma$. The expectation in (14.54) is with respect to the interacting measure (14.51). The long-range interaction should make the polymer stiffer as compared to the free case $e = 0$, which means that the effective stiffness should be increasing with increasing coupling $e^2$.

To gain a crude idea whether such a picture is at least qualitatively correct we replace $W(q, t)$ by $W(0, t)$ in (14.51). Going back to (14.44) this is equivalent to replacing $A_{s\varphi}(q_s)$ by $A_{s\varphi}(0)$ which is the dipole approximation. By rotation

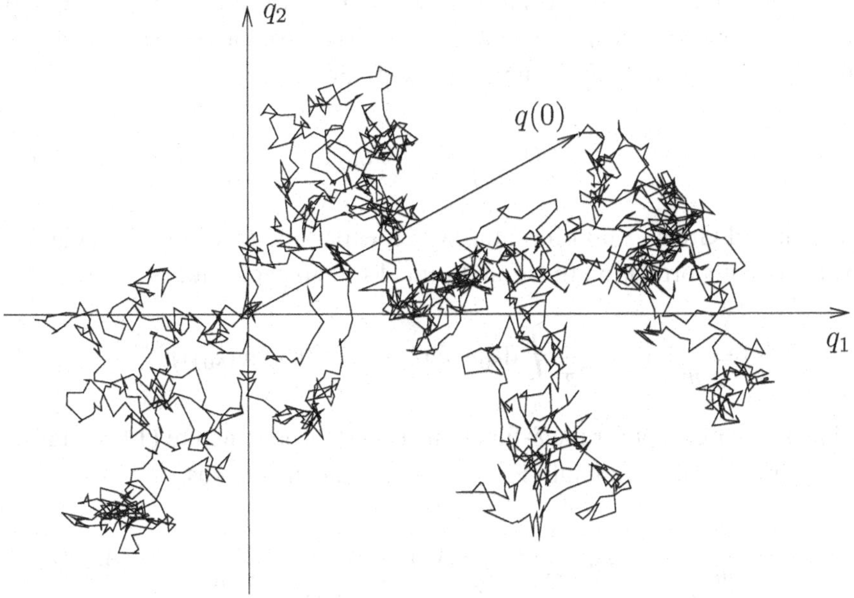


Figure 14.1: Elastic string with end points pinned at the origin.

invariance

$$W_{\alpha\beta}(0,t) = \delta_{\alpha\beta} w(t) \tag{14.55}$$

and we recall that $w(t) \cong 1/t^2$ for large $t$. In the dipole approximation the Gibbs measure (14.51) is Gaussian and (14.54) can be computed explicitly. One obtains

$$\frac{1}{\sigma} = \int \langle \mathrm{d}q_t \cdot \mathrm{d}q_0 \rangle \, , \tag{14.56}$$

where $\langle \cdot \rangle$ is the infinite-time limit in the dipole approximation, which is Gaussian and has the covariance

$$\langle \mathrm{d}q_t \cdot \mathrm{d}q_0 \rangle = \mathrm{d}t \frac{1}{2\pi} \int \mathrm{d}k_0 (m + e^2 \widehat{w}(k_0))^{-1} \mathrm{e}^{\mathrm{i}k_0 t} \, . \tag{14.57}$$

Therefore,

$$\sigma = m + e^2 \widehat{w}(0) = m + \frac{2}{3} e^2 \int \mathrm{d}^3 k |\widehat{\varphi}|^2 \frac{1}{\omega^2} \, , \tag{14.58}$$

which as anticipated is increasing, in fact linearly in $e^2$. We remark that if $w(t)$ decays like $1/t$ or even slower, the interaction is so strong that the stiffness is infinite, in the sense that the typical fluctuations of $q_0$ are no longer of the order $\sqrt{t}$ but grow more slowly with $t$.

If one pins only the left end point, $q_{-t} = 0$, one may think of $q_t$ as a random walk with mean square displacement $\langle q_t^2 \rangle = 3D(2t)$ for large $t$. $D$ is the diffusion coefficient and $D = \sigma^{-1}$ in our units. Thus (14.56), written as

$$D = \int \mathrm{d}t \, \langle \dot{q}_t \cdot \dot{q}_0 \rangle \,, \tag{14.59}$$

is the standard Green–Kubo formula, which expresses $D$ as a time integral over the velocity autocorrelation function. From (14.57) one concludes

$$\langle \dot{q}_t \cdot \dot{q}_0 \rangle = \frac{1}{m}\delta(t) - \frac{1}{2\pi} \int \mathrm{d}k_0 e^2 \widehat{w}(k_0) \big( m(m + e^2 \widehat{w}(k_0)) \big)^{-1} \mathrm{e}^{\mathrm{i}k_0 t} \,, \tag{14.60}$$

which is regular except for the $\delta$-function at $t = 0$. The structure (14.60) turns out to be general. For the full Pauli–Fierz Hamiltonian one obtains

$$\langle \dot{q}_t \cdot \dot{q}_0 \rangle = \frac{1}{m}\delta(t) - \langle \psi_0 \,, \frac{1}{m}(P_\mathrm{f} + eA_\varphi) \cdot \mathrm{e}^{-|t|(H_0 - E(0))} \frac{1}{m}(P_\mathrm{f} + eA_\varphi) \psi_0 \rangle_{\mathcal{F}} \tag{14.61}$$

with a notation which will be explained in section 15.2. Here we just state that with the Definition 15.3 of the effective mass one has the identity

$$\frac{1}{m_\mathrm{eff}} = \int \mathrm{d}t \, \langle \dot{q}_t \cdot \dot{q}_0 \rangle = D = \frac{1}{\sigma} \,. \tag{14.62}$$

Thus the stiffness of the polymer in the Euclidean framework equals the effective mass of the charge coupled to the Maxwell field. Note that the regular part of (14.61) is negative, which means that the stiffness is increased as compared to the bare value $m$. With this background the result (14.58) looks familiar. It is the effective mass of the Abraham model in the nonrelativistic limit; compare with (4.24). The true effective mass of the Pauli–Fierz model has a more complicated dependence on the bare parameters $e$ and $m$, however.

The second case of interest is a confining potential. For large $t$ the partition function is dominated by the ground state of $H$, provided it exists at all. In fact, as we will see, ground state expectations can be computed through the limit $t \to \infty$. Thus, as for thermodynamic systems, the infinite-volume limit is of direct physical interest. If the ground state exists, it should be unique and independent of the particular limit procedure. Translated to (14.51) uniqueness means that the limit $t \to \infty$ exists and is independent of the boundary conditions $q_{-t}$ and $q_t$, at least if they are not allowed to increase too fast. Since $t$ is one-dimensional, such a property will hold, if the energy across the origin is bounded uniformly in the volume, i.e. if

$$\int_{-\infty}^{0}\int_{0}^{\infty} \mathrm{d}q_s \cdot W(q_s - q_{s'}, s - s')\mathrm{d}q_{s'} \leq c_0 \,. \tag{14.63}$$

Because of the stochastic integration, (14.63) cannot be true literally, but only in the sense that there is a small probability for the interaction across the origin to take large values. Stochastic integrals like (14.63) are not easily estimated, but if we set $q_s - q_{s'} = 0$, which is reasonable since $V$ is supposed to be confining, then the interaction energy is

$$\int_{-\infty}^{0} \mathrm{d}s \int_{0}^{\infty} \mathrm{d}s'(q_s \cdot q_{s'})^2 w''(s - s') \,. \tag{14.64}$$

Note that from the stochastic integration we obtain two extra derivatives, which means that $w''(t) \cong t^{-4}$ for large $t$. If the path $q_s$ does not make too wild excursions, the interaction energy in (14.63) is essentially bounded, which implies uniqueness of the Gibbs measure in (14.51). To have a phase transition for a Gibbs measure in one dimension the interaction has to decay as $t^{-2}$ or slower, which is avoided by two powers in our context.

The statistical mechanics intuition applied to (14.51) suggests that if $H_\mathrm{p}$ has a ground state $\psi_0(x)$, i.e. if the ground state for the uncoupled system is $\psi_0 \otimes \Omega$, then, as the coupling is turned on, the ground state will persist and remain unique at any coupling strength. For large $e^2$ fluctuations are suppressed and the ground state must be essentially classical.

## 14.3 Some applications

*(i) Positivity improvement*

Let us consider a general measure space $(\mathcal{M}, \mu)$ and the corresponding Hilbert space $L^2(\mathcal{M}, \mu)$ of square integrable functions on $\mathcal{M}$. In addition, we have the semigroup $\mathrm{e}^{-tH}$, $t \geq 0$, acting on $L^2(\mathcal{M}, \mu)$ with $(\mathrm{e}^{-tH})^* = \mathrm{e}^{-tH}$ and $\inf \sigma(H) = 0$, i.e. $\|\mathrm{e}^{-tH}\| = 1$ for $t \geq 0$. We say that $\mathrm{e}^{-tH}$ is *positivity preserving*, if for $f \geq 0$ we have $\mathrm{e}^{-tH} f \geq 0$. $\mathrm{e}^{-tH}$ is *positivity improving* if $f \geq 0$ implies $\mathrm{e}^{-tH} f > 0$ for $t > 0$. We remark that positivity is not a Hilbert space notion, it depends on the choice of $\mathcal{M}$. Positivity means that, up to normalization, $\mathrm{e}^{-tH}$ is a Markov semigroup and some sort of stochastic model is lurking behind. Our interest in the notion of positivity improvement comes from the fact that it implies uniqueness of the ground state. In essence, positivity improvement is the only general criterion available. The reason for uniqueness is simple. Let $\psi$ be an eigenfunction of $H$ with eigenvalue 0. Then by positivity $|\mathrm{e}^{-tH}\psi| \leq \mathrm{e}^{-tH}|\psi|$ and thus

$$\langle|\psi|, \mathrm{e}^{-tH}|\psi|\rangle \geq \langle|\psi|, |\mathrm{e}^{-tH}\psi|\rangle \geq \langle\psi, \mathrm{e}^{-tH}\psi\rangle = \langle\psi, \psi\rangle . \qquad (14.65)$$

As a consequence, since $\mathrm{e}^{-tH}$ is a contraction, one has $\mathrm{e}^{-tH}|\psi| = |\psi|$ and, since $\mathrm{e}^{-tH}$ is positivity improving, $\mathrm{e}^{-tH}|\psi| = |\psi| > 0$. But then also $\mathrm{e}^{-tH}(|\psi| - \psi) = |\psi| - \psi$. Either $|\psi| - \psi = 0$ in which case $\psi > 0$ or else $|\psi| - \psi > 0$ in which case $\psi < 0$. We conclude that a second eigenvector with eigenvalue zero could not be orthogonal to $\psi$.

In view of this technique, it is desirable to prove that $U\mathrm{e}^{-tH}U^{-1}$, where $H$ is the spinless Pauli–Fierz Hamiltonian (13.39) with $A_{\mathrm{ex}} = 0$, is positivity improving on $\mathbb{R}^3 \times \mathcal{K}'$ with measure $\mathrm{d}^3x \times \mathrm{dP}^0$. A look at (14.44) makes positivity an unlikely fact because of the fluctuating phase. The trick to achieve the desired property is to interchange the role of $A$ and $E_\perp$ through the unitary transformation $\mathrm{e}^{-\mathrm{i}\pi N_{\mathrm{f}}/2}$ with

$$N_{\mathrm{f}} = \sum_{\lambda=1,2} \int \mathrm{d}^3k a^*(k, \lambda) a(k, \lambda) \qquad (14.66)$$

the total number of photons.

**Theorem 14.2** (Positivity improving). *Let* $H = \frac{1}{2}(p - eA_\varphi(x))^2 + H_{\mathrm{f}} + V(x)$ *be the spinless Pauli–Fierz Hamiltonian with external potential* $V$*. Then the semigroup* $U\mathrm{e}^{\mathrm{i}\pi N_{\mathrm{f}}/2}\mathrm{e}^{-tH}\mathrm{e}^{-\mathrm{i}\pi N_{\mathrm{f}}/2}U^{-1}$ *is positivity improving on* $\mathbb{R}^3 \times \mathcal{K}'$ *with measure* $\mathrm{d}^3x \times \mathrm{dP}^0$.

*Proof*: Hiroshima (2000a).

**Corollary 14.3** (Uniqueness of the ground state). *If the spinless Pauli–Fierz Hamiltonian has a ground state, then the ground state is necessarily unique.*

The actual proof of Theorem 14.2 is somewhat technical. But there is a simple heuristic reason to see that it should be correct. We have

$$\mathrm{e}^{\mathrm{i}\pi N_{\mathrm{f}}/2} H \mathrm{e}^{-\mathrm{i}\pi N_{\mathrm{f}}/2} = \frac{1}{2}(p - eE_{\perp\widetilde{\varphi}}(x))^2 + H_{\mathrm{f}} + V(x) , \qquad (14.67)$$

where the smoothing function $\varphi$ is replaced by $\widetilde{\varphi}$ with $\widehat{\widetilde{\varphi}} = \widehat{\varphi}/\omega$. We formally discretize the Maxwell field in (14.67) as

$$\frac{1}{2}\Big(p - e\sum_j \widetilde{\varphi}(j - x)p_j\Big)^2 + \frac{1}{2}\sum_j p_j^2 + \frac{1}{2}\sum_{|i-j|=1}(q_i - q_j)^2 + V(x) \qquad (14.68)$$

up to a constant. Here $(q_j, p_j)$ are a canonical pair of position and momentum operators and the sum is over a discrete lattice in position space. We employ the usual Feynman–Kac formula. The first two terms define a multidimensional Brownian motion. It has a position-dependent diffusion matrix, which by inspection is strictly positive. Thus the "free" measure is positivity improving, a property which is preserved when adding the potential.

*(ii) Diamagnetic inequality*

In (14.44) the fluctuating magnetic field appears as a phase, which leads immediately to the *diamagnetic inequality*

$$|\langle F, U\mathrm{e}^{-tH}U^{-1}G\rangle_{\mathcal{H}_\mathrm{s}}| \le \langle |F|, U\mathrm{e}^{-t(H_\mathrm{p}+H_\mathrm{f})}U^{-1}|G|\rangle_{\mathcal{H}_\mathrm{s}} . \tag{14.69}$$

As one application we derive a bound on the electronic charge density in the ground state. We assume the existence of a ground state, $H\psi_\mathrm{g} = E_\mathrm{g}\psi_\mathrm{g}$, with ground state energy $E_\mathrm{g}$. Then the electronic charge density is

$$\rho_\mathrm{g}(x) = \|\psi_\mathrm{g}(x,\cdot)\|_{\mathcal{F}}^2 = \sum_{n=0}^{\infty}\sum_{\lambda}\int \mathrm{d}^{3n}k|\psi_{\mathrm{g}n}(x,k_1,\lambda_1,\dots,k_n,\lambda_n)|^2 . \tag{14.70}$$

We choose $F = f(x)U\psi_\mathrm{g}$, $f \ge 0$ and bounded, $G = U\psi_\mathrm{g}$. Since $\mathrm{e}^{-tH}\psi_\mathrm{g} = \mathrm{e}^{-tE_\mathrm{g}}\psi_\mathrm{g}$ and since $\mathrm{e}^{-tH_\mathrm{f}}$ is a contraction, one concludes from the diamagnetic inequality that

$$\begin{aligned}\mathrm{e}^{-(t+\tau)E_\mathrm{g}}\int \mathrm{d}^3x f(x)\rho_\mathrm{g}(x) &\le \langle f\rho_\mathrm{g}^{1/2}, \mathrm{e}^{-(t+\tau)H_\mathrm{p}}\rho_\mathrm{g}^{1/2}\rangle_{L^2}\\ &= \langle \mathrm{e}^{-tH_\mathrm{p}}f\rho_\mathrm{g}^{1/2}, \mathrm{e}^{-\tau H_\mathrm{p}}\rho_\mathrm{g}^{1/2}\rangle_{L^2} .\end{aligned} \tag{14.71}$$

From the Feynman–Kac formula (14.5) it follows that $(\mathrm{e}^{-\tau H_\mathrm{p}}\rho_\mathrm{g}^{1/2})(x) \le c_1$. Using this bound in (14.71) and letting $f$ shrink to a $\delta$-function at $x$ we obtain

$$\rho_\mathrm{g}(x) \le [c_1\mathrm{e}^{tE_\mathrm{g}}(\mathrm{e}^{-tH_\mathrm{p}}1)(x)]^2 \tag{14.72}$$

with 1 the constant function. Inequality (14.72) is the desired bound on the electronic charge density.

To make this bound explicit we rewrite

$$(\mathrm{e}^{-tH_\mathrm{p}}1)(x) = \mathbb{E}_x\big(\mathrm{e}^{-\int_0^t \mathrm{d}s V(q_s)}\big) \tag{14.73}$$

according to (14.5) for the particular choice $\psi(x) = 1$. If the potential has a lower bound as $V(x) \ge c_0 + c_1|x|^\gamma$, $c_1 > 0$, $\gamma > 1$, then for fixed $t$ the weight in (14.73) is dominated by the potential and one has $\rho_\mathrm{g}(x) \le c\mathrm{e}^{-V(x)}$. On the other hand if $V(x) \to 0$ as $|x| \to \infty$, then the expression in (14.73) tends to 1 as $x \to \infty$. Thus we should optimize in $t$ for fixed $x$. Very crudely this means minimizing the action $\int_0^t \mathrm{d}s\big(\frac{1}{2}\dot{q}_s^2 + V(q_s)\big)$ for a fixed initial condition $q_0 = x$ and then to optimize in $t$. In this variation one has to include the contribution from the exponentially growing factor $e^{E_g t}$. To have a bound state at all, $V_{\min} = \min_x V(x) < 0$. For sufficiently small $e$ also $V_{\min} - E_g < 0$ and the variational bound decays exponentially in $x$, i.e. $\rho_g(x) \le c e^{-\gamma|x|}$. For larger $e$ one can no longer balance $E_g$ and the bound (14.72) becomes vacuous.

The diamagnetic inequality suggests that the decay of the electronic charge density in the ground state does not worsen by the coupling to the Maxwell field. If one imagines, rather crudely, the effective mass of the electron to be increased through the interaction with the field, then the electron density should become even better localized for larger $e$ and point-like as $e \to \infty$.

*(iii) Photon expectations*

We discovered in section 14.2 that, through integrating over the Maxwell field, one obtains a path integral (functional measure) for the electron paths which has the structure of an equilibrium measure relative to an a priori weight given by the Wiener measure. Here we expand on this observation by computing averages for the photon field in the ground state. Let $\psi_0$ be the ground state of $H_p$ and let us introduce the approximate ground state

$$\psi_T = e^{-TH}\psi_0 \otimes \Omega / \| e^{-TH}\psi_0 \otimes \Omega \| \tag{14.74}$$

of $H$, which is normalized to one and, if the limit does not vanish, converges as $T \to \infty$ to the unique ground state $\psi_g$ of $H$. For observables of the form $f(x)$ the same argument as for (14.48) leads to

$$\begin{aligned} &\langle \psi_T, f(x)\psi_T \rangle_{\mathcal{H}} \qquad\qquad (14.75) \\ &= \langle \psi_0 \otimes \Omega, e^{-2TH}\psi_0 \otimes \Omega \rangle_{\mathcal{H}}^{-1} \langle \psi_0 \otimes \Omega, e^{-TH} f(x) e^{-TH}\psi_0 \otimes \Omega \rangle_{\mathcal{H}} \\ &= \mathbb{E}^{G}_{[-T,T]}\big(f(q_0)\big)\,. \end{aligned}$$

The "volume" $[-T, T]$ is arranged symmetrically relative to the origin. $\mathbb{E}^{G}_{[-T,T]}$ refers to the normalized expectation

$$\mathbb{E}^{G}_{[-T,T]}(\,\circ\,) = Z(2T)^{-1} \mathbb{E}_{\mathrm{dW}}\big(\psi_0(q_{-T})\psi_0(q_T) e^{-\int_{-T}^{T} dt\, V(q_t)} e^{-S_{[-T,T]}}\,\circ\,\big) \tag{14.76}$$

with the normalizing partition function

$$Z(2T) = \mathbb{E}_{\mathrm{dW}}\Big(\psi_0(q_{-T})\psi_0(q_T) \exp\Big[-\int_{-T}^{T} dt\, V(q_t) - S_{[-T,T]}\Big]\Big) \tag{14.77}$$

and with the effective action

$$S_{[-T,T]} = \frac{1}{2} e^2 \int_{-T}^{T}\int_{-T}^{T} dq_t \cdot W(q_t - q_s, t - s) dq_s\,. \tag{14.78}$$

Clearly, in the infinite-volume limit

$$\lim_{T\to\infty}\langle\psi_T, f(x)\psi_T\rangle_{\mathcal{H}} = \langle\psi_{\rm g}, f(x)\psi_{\rm g}\rangle_{\mathcal{H}} = \int \mathrm{d}^3x f(x)\rho_{\rm g}(x) = \langle f(q_0)\rangle\,. \tag{14.79}$$

Thus the electronic charge density is the distribution of $q_0$, the position of the path at time $t = 0$, under the infinite-volume Gibbs measure $\langle\cdot\rangle$, i.e. under the probability measure obtained in the limit $T \to \infty$ in (14.76) which we denote by $\langle\cdot\rangle$.

With (14.79) we have opened the first page in the dictionary for the translation from Fock space expectations to Gibbs averages. We plan to expand the dictionary by considering a bounded operator $1 \otimes B$ referring only to the photons and want to compute the expectation

$$\langle\psi_T, 1\otimes B\psi_T\rangle_{\mathcal{H}}\,, \tag{14.80}$$

which for large $T$ goes over to the ground state expectation $\langle\psi_{\rm g}, 1\otimes B\psi_{\rm g}\rangle_{\mathcal{H}}$.

Using the basic identity (14.44) one can write

$$\begin{aligned}&(U\mathrm{e}^{-TH}\psi_0\otimes\Omega)(q,A)\\ &\quad=\mathbb{E}_q\mathbb{E}_A\Big(\psi_0(q_t)\exp\Big[-\int_0^T \mathrm{d}t\,V(q_t) - \mathrm{ie}\int_0^T \mathrm{d}q_t\cdot A_{t\varphi}(q_t)\Big]\Big),\end{aligned} \tag{14.81}$$

where $\mathbb{E}_A$ refers to the Ornstein–Uhlenbeck process $A_t(x)$ with fixed initial field $A_0 = A$. The Gaussian expectation $\mathbb{E}_A$ can be carried out with the result

$$\begin{aligned}\mathbb{E}_A\big(\mathrm{e}^{-\mathrm{ie}\int_0^T \mathrm{d}q_t\cdot A_{t\varphi}(q_t)}\big) &= \mathrm{e}^{-\mathrm{ie}A(f_+)}\exp\Big[-\frac{1}{2}e^2\int_0^T\int_0^T\int \mathrm{d}^3k|\widehat{\varphi}(k)|^2\\ &\quad\times \mathrm{d}q_t\cdot Q^{\perp}(k)\mathrm{d}q_s\mathrm{e}^{\mathrm{i}k\cdot(q_t-q_s)}\frac{1}{2\omega}\big(\mathrm{e}^{-\omega|t-s|}-\mathrm{e}^{-\omega t}\mathrm{e}^{-\omega s}\big)\Big],\end{aligned} \tag{14.82}$$

where

$$\widehat{f}_+(k) = \int_0^T \mathrm{d}q_t\widehat{\varphi}\mathrm{e}^{-\mathrm{i}k\cdot q_t}\mathrm{e}^{-\omega t}\,, \tag{14.83}$$

which depends on the path $q_t$, $0 \le t \le T$.

We have to take the expectation of $1 \otimes B$ with respect to the wave function (14.81), for which it is convenient to regard the adjoint wave function as coming from an integration relative to a Brownian motion running from 0 to $-T$. For this purpose one time-reverses the Brownian motion, which starts then at $q_{-T}$ and ends at $q$. Upon integrating over $\mathrm{d}q_0$ one obtains the Wiener measure for Brownian paths $t \mapsto q_t$, $|t| \le T$. The expectation $\mathbb{E}_A$ for the adjoint wave function yields an expression as in (14.82) where $f_+$ is replaced by $f_-$ and

$$\widehat{f}_-(k) = -\int_{-T}^{0} \mathrm{d}q_t \widehat{\varphi} \mathrm{e}^{-\mathrm{i}k\cdot q_t} \mathrm{e}^{-\omega|t|} \tag{14.84}$$

with a minus sign, since $\mathrm{d}q_t$ is odd under time-reversal. The expectation for $B$ is most easily written in Fock space. Then

$$\begin{aligned} &\langle \psi_0 \otimes \Omega, \mathrm{e}^{-TH}(1 \otimes B)\mathrm{e}^{-TH} \psi_0 \otimes \Omega \rangle_{\mathcal{H}} \\ &= \mathbb{E}_{\mathrm{dW}}\big(\psi_0(q_{-T})\psi_0(q_T)\mathrm{e}^{-\int_{-T}^{T} V(q_t)\mathrm{d}t} \langle \Omega, \mathrm{e}^{\mathrm{i}eA(f_-)} B \mathrm{e}^{-\mathrm{i}eA(f_+)} \Omega \rangle_{\mathcal{F}} \\ &\times \exp\Big[ -\frac{1}{2} e^2 \Big( \int_{-T}^{0}\int_{-T}^{0} + \int_{0}^{T}\int_{0}^{T} \Big) \int \mathrm{d}^3k |\widehat{\varphi}|^2 \mathrm{d}q_t \cdot Q^{\perp} \mathrm{d}q_s \mathrm{e}^{\mathrm{i}k\cdot(q_t - q_s)} \\ &\times \frac{1}{2\omega} \big( \mathrm{e}^{-\omega|t-s|} - \mathrm{e}^{-\omega|t|}\mathrm{e}^{-\omega|s|} \big) \Big] \big). \end{aligned} \tag{14.85}$$

To make further progress we have to choose particular observables. One example is the generating function for the photon number density in momentum space, i.e.

$$B = \exp\Big[ -\sum_{\lambda=1,2} \int \mathrm{d}^3k \mu(k) a^*(k,\lambda) a(k,\lambda) \Big] \tag{14.86}$$

with $\mu \ge 0$. Then

$$\begin{aligned} \langle \Omega, \mathrm{e}^{\mathrm{i}eA(f_-)} B \mathrm{e}^{-\mathrm{i}eA(f_+)} \Omega \rangle_{\mathcal{F}} &= \exp\Big[ -\frac{1}{4} \langle \widehat{f}_-, \omega^{-1/2} Q^{\perp} \omega^{-1/2} \widehat{f}_- \rangle_{\mathfrak{h}} \\ &- \frac{1}{4} \langle \widehat{f}_+, \omega^{-1/2} Q^{\perp} \omega^{-1/2} \widehat{f}_+ \rangle_{\mathfrak{h}} - \frac{1}{2} \langle \widehat{f}_-, \omega^{-1/2} Q^{\perp} \mathrm{e}^{-\mu} \omega^{-1/2} \widehat{f}_+ \rangle_{\mathfrak{h}} \Big]. \end{aligned} \tag{14.87}$$

Collecting all terms yields

$$\begin{aligned} &\langle \psi_T, \exp\Big[ -\sum_{\lambda=1,2} \int \mathrm{d}^3k \mu(k) a^*(k,\lambda) a(k,\lambda) \Big] \psi_T \rangle_{\mathcal{H}} \\ &= \mathbb{E}^{\mathrm{G}}_{[-T,T]}\Big( \exp\Big[ -e^2 \int_{-T}^{0}\int_{0}^{T} \int \mathrm{d}^3k |\widehat{\varphi}|^2 \mathrm{d}q_t \cdot Q^{\perp} \mathrm{d}q_s \mathrm{e}^{\mathrm{i}k\cdot(q_t - q_s)} \\ &\times \frac{1}{2\omega} \mathrm{e}^{-\omega|t|} \mathrm{e}^{-\omega|s|} (\mathrm{e}^{-\mu} - 1) \Big] \Big). \end{aligned} \tag{14.88}$$

We differentiate with respect to $\mu(k)$ and obtain the ground state photon number density in momentum space,

$$\sum_{\lambda=1,2} \langle \psi_{\rm g}, a^*(k,\lambda)a(k,\lambda)\psi_{\rm g}\rangle_{\mathcal{H}}$$
$$= -e^2 \frac{1}{2\omega}|\widehat{\varphi}|^2 \int_{-\infty}^{0}\int_{0}^{\infty} \langle {\rm d}q_t \cdot Q^{\perp}{\rm d}q_s {\rm e}^{-\omega|t|}{\rm e}^{-\omega|s|}{\rm e}^{{\rm i}k\cdot(q_t-q_s)}\rangle\,, \quad (14.89)$$

the average with respect to the infinite-volume Gibbs measure. In particular one has the remarkable identity that $\langle \psi_{\rm g}, N_{\rm f}\psi_{\rm g}\rangle_{\mathcal{H}}$ equals the average interaction energy between the right and left half-line in the statistical mechanics system. By the same technique, the photon density in physical space is given by

$$\sum_{\lambda=1,2} \langle \psi_{\rm g}, a^*(x,\lambda)a(x,\lambda)\psi_{\rm g}\rangle_{\mathcal{H}}$$
$$= -e^2 \sum_{\lambda=1,2}\int_{0}^{\infty}\int_{-\infty}^{0} \langle {\rm d}q_t \cdot f_\lambda(x-q_t,t){\rm d}q_s \cdot f_\lambda(x-q_s,s)\rangle\,, \quad (14.90)$$

where

$$\widehat{f_\lambda}(k,t) = \widehat{\varphi}\frac{1}{\sqrt{2\omega}}{\rm e}^{-\omega|t|}e_\lambda(k)\,. \quad (14.91)$$

Equations (14.89) and (14.90) are only partially useful, since there is too little information on the ${\rm d}q_t \cdot Q^{\perp}{\rm d}q_s$ correlations, except for the soft photon bound (15.9), (15.14) from which one concludes that

$$-c_0(1+|k|^3) \le \int_{-\infty}^{0}\int_{0}^{\infty} \langle {\rm d}q_t \cdot Q^{\perp}{\rm d}q_s {\rm e}^{-\omega|t|}{\rm e}^{-\omega|s|}{\rm e}^{{\rm i}k\cdot(q_t-q_s)}\rangle \le 0 \quad (14.92)$$

with some positive constant $c_0$. The interaction energy between right and left is bounded and negative on the average. One would expect also its exponential moments to be bounded. If so, $\langle \psi_{\rm g}, {\rm e}^{-\lambda N_{\rm f}}\psi_{\rm g}\rangle_{\mathcal{H}} < \infty$ for all $\lambda$ by (14.88), which implies that in the ground state the number of photons has a super-exponential decay.

Our method may be applied to other observables of interest. For example the ground state expectation and variance of the vector potential is given by

$$\langle \psi_{\rm g}, A(x)\psi_{\rm g}\rangle_{\mathcal{H}} = 0\,, \quad (14.93)$$

$$\langle \psi_{\rm g}, A(x)^2\psi_{\rm g}\rangle_{\mathcal{H}} = \langle \Omega, A(x)^2\Omega\rangle_{\mathcal{F}} - \sum_{\lambda=1,2}\Big\langle\Big(\int_{-\infty}^{\infty} {\rm d}q_t \cdot f_{{\rm A}\lambda}(x-q_t,t)\Big)^2\Big\rangle, \quad (14.94)$$

where

$$\widehat{f}_{A\lambda} = e_\lambda \widehat{\varphi} \frac{1}{2\omega} e^{-\omega|t|}. \tag{14.95}$$

Similarly for the transverse electric field one has

$$\langle \psi_g, E_\perp(x)\psi_g \rangle_{\mathcal{H}} = 0, \tag{14.96}$$

$$\langle \psi_g, E_\perp(x)^2 \psi_g \rangle_{\mathcal{H}} = \langle \Omega, E_\perp(x)^2 \Omega \rangle_{\mathcal{F}} + \sum_{\lambda=1,2} \Big\langle \Big( \int_{-\infty}^{\infty} dq_t \cdot f_{E\lambda}(x - q_t, t) \Big)^2 \Big\rangle \tag{14.97}$$

with $\widehat{f}_{E\lambda} = \partial_t \widehat{f}_{A\lambda}$. In fact, the vacuum variances are infinite but become finite when the fields are slightly smeared out. Through the presence of a bound electron the electric field fluctuations are increased whereas the vector field fluctuations are suppressed. Their product remains constant, as required by the uncertainty relation.

We recall that $\langle E \rangle = \langle E_\| \rangle + \langle E_\perp \rangle$, the second term being zero by (14.96). From the equations of motion, $\nabla \cdot \langle \psi_g, E(x)\psi_g \rangle_{\mathcal{H}} = e\langle \varphi(x - q_0) \rangle = e\varphi * \rho_g(x)$, $\rho_g$ being the electron ground state density of (14.70). Thus, at large distances the average electric field generated by a charge bound in the ground state is the Coulomb field with a strength determined through the bare charge $e$, from which we conclude that in the Pauli–Fierz model there is *no* charge renormalization.

## Notes and references

### *Section 14.1*

Gentle introductions to path integrals are Schulman (1981) and Kleinert (1995) emphasizing statistical mechanics aspects. Roepstorff (1994) treats in detail the quantized Maxwell field. Path integrals with a focus on relativistic quantum field theory are explained in the advanced textbook of Huang (1998). Simon (1979) is a beautiful discussion on the connection between functional integration and the Schrödinger equation. In particular, he explains the Feynman–Kac–Ito formula used in (14.8). Gaussian processes, Wick ordering, and the Schrödinger representation are exhaustively covered in Simon (1974) and Glimm and Jaffe (1987). The functional measure for the Pauli–Fierz Hamiltonian is discussed by Hiroshima (1997b). A standard reference on infinite-dimensional Ornstein–Uhlenbeck processes is Holley and Stroock (1978). In Giacomin *et al.* (2001) martingale-type estimates are explained.

Functional integration has two historical roots which developed apparently completely independently. Feynman (1948), cf. also the textbook by Feynman and

Hibbs (1965), uses space-time histories to visualize quantum processes. This led to quantum propagators as a "sum over histories". On the other hand, Wiener, Levy, and many other probabilists developed the theory of probability measures on function space (= the space of trajectories) to have a mathematical framework for Brownian motion and diffusion processes. Kac (1950) realized that the two approaches are related through the Wick rotation. The extension to models of quantum fields is achieved by Nelson (1966, 1973). With his insights functional integration became the "secret weapon" and is at the heart of the technical development in constructive quantum field theory through the hands of Glimm, Jaffe, Spencer, Simon, and many, many others. I refer to Glimm and Jaffe (1987).

### *Section 14.2*

The integration over field degrees of freedom is discussed in Feynman and Hibbs (1965) and in Feynman (1948). He tackled a variety of physical problems with this technique. The most widely known is the ground state energy of the polaron (Feynman 1955) for which the analog of (14.48) is estimated through a variational method with a result which covered both the intermediate and strong coupling regime for the first time. To view the effective mass as the stiffness of a polymer is proposed in Spohn (1987). If the Maxwell field is replaced by a scalar field, cf. section 19.2, the double stochastic integral becomes a double Riemann integral, which is much easier to handle. In particular, one obtains reasonable bounds on the effective stiffness with a technique borrowed from Brascamp, Lieb and Lebowitz (1976). To view the path measure (14.51) as a Gibbs measure relative to Brownian motion is stressed in Lőrinczi and Minlos (2001), Betz *et al.* (2002), and Lőrinczi *et al.* (2002a, 2002b).

### *Section 14.3*

Positivity-improving semigroups are treated in Reed and Simon (1978), Chapter XIII.12. For the existence of the ground state we refer to section 15.1. Whenever magnetic fields are involved, the diamagnetic inequality is very helpful; compare for example with Cycon, Froese, Kirsch and Simon (1987). Carmona (1978) uses Brownian motion to estimate ground state properties of $-\Delta + V$. His techniques extend to a charge coupled to a scalar field as discussed in Betz *et al.* (2002). There is also a functional analytic proof of exponential localization, which is patterned after Agmon (1982) in the case of the Schrödinger equation, see Theorem 20.1.

# 15

# States of lowest energy: statics

Quantizing the Abraham model results in the Pauli–Fierz Hamiltonian which is a self-adjoint operator under rather general conditions. Thus the dynamics is well defined and we can start to investigate some of its properties. The most basic item is the states of lowest energy. They really come in two varieties: (i) If the electron is bound by a strong external electrostatic potential, like the Coulomb potential of a nailed-down nucleus, then the lowest energy state is the ground state, where the electron is at rest modulo quantum fluctuations. (ii) If there are no external potentials, then the total momentum is conserved and the state of lowest energy must be determined for every fixed total momentum, which then describes the electron together with its surrounding photon cloud traveling at constant velocity. Physically the most important information is the energy–momentum relation which gives the lowest energy $E$ at given total momentum $P$. Both item (i) and item (ii) are discussed in this chapter. In case (i) one expects to have always a ground state provided the external potential is binding. In case (ii) the infrared divergence of the Pauli–Fierz model becomes visible. As will be explained in more detail in section 19.1, for total momentum $P \neq 0$ the state of lowest energy is not in Fock space. An electron traveling at nonzero velocity binds an infinite number of photons. To avoid such a subtlety, for item (ii) we proceed as if the photon had a tiny mass.

The external fields manufactured with macroscopic devices under laboratory conditions are weak and have a slow variation when measured in units of the effective size of the charge, roughly given through the inverse size of the form factor $\widehat{\varphi}$. Such external fields thus constitute a small perturbation in item (ii) and, as for the Abraham model, an important dynamical issue is to understand the motion of the charge in terms of an effective one-particle Hamiltonian. The energy–momentum relation must play an important role, but there will be additional pieces accounting for the spin precession. Our discussion of this topic is postponed to section 16, to keep the lengths of the chapters in reasonable proportion.

## 15.1 Bound charge

The hydrogen atom has a stable ground state and thus makes the size of atoms of the order of a few ångströms. The problem under discussion is whether this ground state persists as the quantized transverse modes of the Maxwell field are taken into consideration. Since the electron now has the opportunity to bind photons, one would expect it to have effectively a larger mass. This intuition is confirmed through the path integral of chapter 14, which suggests that the fluctuations in the stochastic trajectories are reduced due to the additional interaction energy $W$ from the integration over the Maxwell field. Thus the coupling to the photons should enhance binding.

To put such reasoning on more solid grounds, we recall that for a Schrödinger operator $H_S = -(1/2m)\Delta + V$ with a Coulomb-like potential, i.e. a potential $V$ such that $\lim_{|x|\to\infty} V(x) = 0$, it is rather straightforward to ensure a stable ground state. Let us assume that $V$ is infinitesimally bounded with respect to $-\Delta$. Then the bottom of the continuous spectrum, denoted by $E_c$, satisfies $E_c = 0$ and one only has to make sure that the energy is lowered when the electron is moved from infinity to the potential region. This means that one has to find a trial wave function such that $\langle \psi , H_S\psi\rangle < 0$. By the Kato–Rellich theorem $H_S$ is bounded from below. Thus $H_S$ must have an eigenvalue at the bottom of its spectrum. The ground state wave function $\psi_g$ is nodeless, since $e^{-tH_S}$ is positivity improving; compare with section 14.3(i). Hence the ground state is unique. To adapt such reasoning to the Pauli–Fierz Hamiltonian

$$H = \frac{1}{2m}(p - eA_\varphi(x))^2 + H_f + V(x) = H^0 + V, \tag{15.1}$$

one faces the difficulty that there are photon excitations of arbitrarily small energies. Thus $H$ has no spectral gap and a variational bound will not do. The convential approach is to first assume an infrared cutoff in the form factor $\widehat{\varphi}$ by setting $\widehat{\varphi}(k) = 0$ for $|k| \le \sigma$ and to adopt the construction explained in property (vi) of section 15.2.1. This yields the existence of a ground state $\psi_{g,\sigma}$ for the cutoff Hamiltonian $H_\sigma$. One is then left to show that as $\sigma \to 0$ the sequence of ground states $\psi_{g,\sigma}$ has a limit $\psi_g$ which is the desired ground state for $H$. The difficulty is that as $\sigma \to 0$ the number of bound photons could increase without limit resulting in the physical ground state lying outside of Fock space. This is one aspect of the infrared problem to be discussed in more detail in section 19.1. Thus one has to establish a bound on the number of low-energy (soft) photons in the ground state. We explain some parts of the argument which allow us to illustrate the pull-through formula that will also be handy later on.

**Theorem 15.1** (Soft photon bound). *Let $\psi_g$ be a ground state of the Pauli–Fierz Hamiltonian $H$ of (15.1), $H\psi_g = E\psi_g$. Then the average number of photons is bounded as*

$$\langle \psi_g, N_f \psi_g \rangle \leq c_0 \langle \psi_g, x^2 \psi_g \rangle. \tag{15.2}$$

*Proof*: Clearly

$$\langle \psi_g, N_f \psi_g \rangle = \sum_{\lambda=1,2} \int \mathrm{d}^3k \|a(k,\lambda)\psi_g\|^2. \tag{15.3}$$

Through a virial-type argument we plan to make use of the fact that $\psi_g$ is an eigenfunction, and start with the pull-through formula

$$[H, a(k,\lambda)] = -\omega(k)a(k,\lambda) + e\widehat{\varphi}\frac{1}{\sqrt{2\omega}}\mathrm{e}^{-\mathrm{i}k\cdot x}\frac{1}{m}e_\lambda(k)\cdot(p - eA_\varphi(x)). \tag{15.4}$$

Note that

$$\frac{1}{m}(p - eA_\varphi(x)) = \mathrm{i}[H, x]. \tag{15.5}$$

Therefore

$$\begin{aligned}(H+\omega)a(k,\lambda) - a(k,\lambda)H = e\widehat{\varphi}\frac{1}{\sqrt{2\omega}}\Big(&\mathrm{i}[H, \mathrm{e}^{-\mathrm{i}k\cdot x}e_\lambda(k)\cdot x] \\ &-\mathrm{i}[H, \mathrm{e}^{-\mathrm{i}k\cdot x}]e_\lambda(k)\cdot x\Big).\end{aligned} \tag{15.6}$$

The commutator with $\mathrm{e}^{-\mathrm{i}k\cdot x}$ is

$$[H, \mathrm{e}^{-\mathrm{i}k\cdot x}] = -\frac{1}{m}k\cdot(p - eA_\varphi(x))\mathrm{e}^{-\mathrm{i}k\cdot x} - \frac{1}{2m}k^2\mathrm{e}^{-\mathrm{i}k\cdot x} \tag{15.7}$$

and applied to $\psi_g$,

$$\begin{aligned}a(k,\lambda)\psi_g &= \mathrm{i}e\widehat{\varphi}\frac{1}{\sqrt{2\omega}}(H - E + \omega)^{-1} \\ &\Big((H-E) + \frac{1}{2m}k^2 + \frac{1}{m}k\cdot(p - eA_\varphi(x))\Big)\mathrm{e}^{-\mathrm{i}k\cdot x}e_\lambda(k)\cdot x\psi_g \\ &= \mathrm{i}e\widehat{\varphi}\frac{1}{\sqrt{2\omega}}(\phi_1 + \phi_2 + \phi_3).\end{aligned} \tag{15.8}$$

Thus, choosing $e > 0$ for notational convenience,

$$\|a(k,\lambda)\psi_g\| \leq e|\widehat{\varphi}|\frac{1}{\sqrt{2\omega}}(\|\phi_1\| + \|\phi_2\| + \|\phi_3\|) \tag{15.9}$$

and

$$\|\phi_1\| \le \|x\psi_g\|, \quad \|\phi_2\| \le \frac{1}{2m}k^2\frac{1}{\omega}\|x\psi_g\|,$$
$$\|\phi_3\| \le |k|\|(H - E + \omega)^{-1}\frac{1}{m}\widehat{k}\cdot(p - eA_\varphi(x))\mathrm{e}^{-ik\cdot x}e_\lambda(k)\cdot x\psi_g\|. \tag{15.10}$$

To estimate the norm of $\phi_3$ we use

$$\|(H - E + \omega)^{-1}\widehat{k}\cdot(p - eA_\varphi(x))\| = \|\widehat{k}\cdot(p - eA_\varphi(x))(H - E + \omega)^{-1}\| \tag{15.11}$$

and

$$\begin{aligned}
&\langle\psi, (H - E + \omega)^{-1}(\widehat{k}\cdot(p - eA_\varphi(x)))^2(H - E + \omega)^{-1}\psi\rangle \\
&\quad\le \langle\psi, (H - E + \omega)^{-1}(p - eA_\varphi(x))^2(H - E + \omega)^{-1}\psi\rangle \\
&\quad\le \langle\psi, (H - E + \omega)^{-1}(c_1H + c_2)(H - E + \omega)^{-1}\psi\rangle \\
&\quad\le \langle\psi, \psi\rangle\Big[\sup_{\lambda\ge 0}\big(c_1(\lambda + E) + c_2\big)(\lambda + \omega)^{-2}\Big],
\end{aligned} \tag{15.12}$$

provided $V_-$ is $H$ bounded. Inserting in (15.10)

$$\|\phi_3\| \le |k|\Big(c_1\frac{1}{\omega} + c_2\sqrt{\omega}\Big)\|x\psi_g\| \tag{15.13}$$

is obtained. With these estimates we return to (15.3) to get

$$\langle\psi_g, N_f\psi_g\rangle \le c_0\int \mathrm{d}^3k|\widehat{\varphi}(k)|^2\frac{1}{\omega}(1 + \omega^{-2}k^4 + \omega^{-2}k^2 + \omega k^2)\|x\psi_g\|^2, \tag{15.14}$$

which proves (15.2). □

Bounds on $\|x\psi_g\|$ are available from the diamagnetic inequality combined with functional integration, see section 14.3(i), and from yet another pull-through-type argument, see section 20.1.

Note that in (15.14) we can still afford the two extra powers $\omega^{-2}$ close to $k = 0$. This is consistent with a decay as $|t|^{-4}$ in the effective action given at the end of section 14.2.

The modern variant for the existence of a ground state relies on having an energy gain when the electron is moved from infinity to the potential region. Thereby, as discussed at length in section 20.1, the existence of a ground state for atoms and molecules is also ensured. To be complete we now state

**Theorem 15.2** (Unique ground state). *Let $V = V_+ - V_-$ be the decomposition of the external potential $V$ into positive and negative parts. It is assumed that $V_-$ is*

*infinitesimally bounded relative to* $p^2$, *i.e.* $|\langle\psi, V_-\psi\rangle| \le \varepsilon\langle\psi, p^2\psi\rangle + b(\varepsilon)\langle\psi,\psi\rangle$ *for every* $\varepsilon > 0$, *and that* $\frac{1}{2m}p^2 + V$ *has a ground state with isolated ground state energy. Then the Hamiltonian* $H$ *of (15.1) has a unique ground state* $\psi_\mathrm{g} \in \mathcal{H}$, *i.e.* $H\psi_\mathrm{g} = E\psi_\mathrm{g}$ *and* $E$ *is the lowest energy.*

*Proof*: The existence is proved by Griesemer, Lieb and Loss (2001). The uniqueness relies on the fact that the semigroup $\mathrm{e}^{-tH}$ is positivity improving in a suitable basis, see Hiroshima (2000a) and section 14.3. □

Note that in Theorem 15.2 there is no restriction on the magnitude of the charge.

## 15.2 Energy–momentum relation, effective mass

For the Abraham model the motion of the charge subject to slowly varying external potentials is determined by the energy–momentum relation $E(P)$. There is good reason to expect the same scenario quantum mechanically, which poses two problems. First of all one has to study $E(P)$, which makes a two-line computation classically but turns out to be much harder in quantum theory. Secondly, given $E(P)$, we have to explain how it governs the effective one-particle theory. This topic is deferred to chapter 16.

Since there are no external forces acting on the electron, the Pauli–Fierz Hamiltonian reads

$$H = \frac{1}{2m}(p - eA_\varphi(x))^2 + H_\mathrm{f}. \tag{15.15}$$

As shown already, the total momentum

$$P = p + \sum_{\lambda=1,2}\int \mathrm{d}^3k k a^*(k,\lambda)a(k,\lambda) = p + P_\mathrm{f} \tag{15.16}$$

is conserved, $[H, P] = 0$. Therefore $H$ can be decomposed according to the subspaces of constant $P$. This is achieved through the unitary transformation

$$U = \mathrm{e}^{\mathrm{i}x\cdot P_\mathrm{f}}, \tag{15.17}$$

which more explicitly is given by

$$(U\psi)_n(k, k_1, \lambda_1, \ldots, k_n, \lambda_n) = \psi_n\Big(k - \sum_{j=1}^{n} k_j, k_1, \lambda_1, \ldots, k_n, \lambda_n\Big), \tag{15.18}$$

using the momentum representation $p = k$, $x = \mathrm{i}\nabla_k$. Then

$$UHU^{-1} = \frac{1}{2m}(P - P_\mathrm{f} - eA_\varphi)^2 + H_\mathrm{f} \tag{15.19}$$

with the shorthand

$$A_\varphi = A_\varphi(0). \tag{15.20}$$

Not to overload notation we return to $p$ instead of $P$, remembering that $p$ is still canonically conjugate to $x$ but now stands for the total momentum. The Hamiltonian under study is then

$$H_p = \frac{1}{2m}(p - P_{\mathrm{f}} - eA_\varphi)^2 + H_{\mathrm{f}}. \tag{15.21}$$

For each fixed $p$, $H_p$ acts on Fock space $\mathcal{F}$. Thus we may think of the unitary $U$ as a map from $L^2(\mathbb{R}^3) \otimes \mathcal{F}$ to the direct integral $\int^\oplus \mathrm{d}^3 p \mathcal{F}_p$ such that $UHU^{-1} = \int^\oplus \mathrm{d}^3 p H_p$. For the remainder of this section we will regard $p$ simply as a parameter. The scalar product $\langle \cdot, \cdot \rangle$ is in Fock space throughout.

**Definition 15.3** *The energy–momentum relation, $E(p)$, of the Pauli–Fierz Hamiltonian is given by*

$$E(p) = \inf_{\psi, \|\psi\|=1} \langle \psi, H_p \psi \rangle. \tag{15.22}$$

*The effective mass $m_{\mathrm{eff}}$ is the inverse curvature of $E(p)$ at $p = 0$. Since $E(p)$ is rotation-invariant,*

$$(m_{\mathrm{eff}})^{-1} \delta_{\alpha\beta} = \partial_{p_\alpha} \partial_{p_\beta} E(p)|_{p=0}. \tag{15.23}$$

There is no simple scheme to compute $E(p)$ and $m_{\mathrm{eff}}$, but we will establish some qualitative properties of $E(p)$ which point in the right direction. In order not to lose sight of the goal we state

**Claim 15.4** (Energy–momentum relation). *Let $\omega(k) = \sqrt{m_{\mathrm{ph}}^2 + k^2}$ with $m_{\mathrm{ph}} > 0$. There exists a threshold value, $p_{\mathrm{c}}$, of the total momentum such that for all $|p| < p_{\mathrm{c}}$, $H_p$ has a unique ground state $\psi_p \in \mathcal{F}$,*

$$H_p \psi_p = E(p) \psi_p. \tag{15.24}$$

*$E(p)$ is separated by a gap from the continuous spectrum, i.e. if $E_{\mathrm{c}}(p)$ denotes the bottom of the continuous spectrum, then*

$$E_{\mathrm{c}}(p) - E(p) = \Delta(p) > 0. \tag{15.25}$$

In Claim 15.4 we assumed a small photon mass $m_{\mathrm{ph}}$. Thus at $p = 0$ excitations require at least an energy $m_{\mathrm{ph}}$. For physical photons $m_{\mathrm{ph}} = 0$, however. Arbitrarily small-energy excitations are possible and the spectral gap closes, which is one part of the infrared behavior of the Pauli–Fierz model. The assumption $m_{\mathrm{ph}} > 0$ introduces a spectral gap, so to speak, by hand. An alternative scheme to separate the ground state band from the continuum is to decouple all modes with $|k| \leq \sigma$ by replacing the true $\widehat{\varphi}$ by $\widehat{\varphi}_\sigma$, where $\widehat{\varphi}_\sigma = \widehat{\varphi}$ for $|k| \geq \sigma$ and $\widehat{\varphi}_\sigma = 0$ for $|k| < \sigma$.

We made the proviso that the ground state band ceases to exist beyond the threshold $p_c$, where we allow for $p_c = \infty$. If $p_c < \infty$, then the electron cannot be accelerated beyond the maximal momentum $p_c$. For $|p| > p_c$, $H_p$ has no ground state. States with $|p| > p_c$ decay into lower-momentum states through the emission of Čerenkov radiation. In fact the same phenomenon occurs classically if in the given medium the speed of light propagation is less than the maximal speed of the charge.

To investigate $E(p)$, let us first have a look at the uncoupled system, $e = 0$. Then the eigenstate in (15.24) is the Fock vacuum $\Omega$ with eigenvalue $p^2/2m$. The energies in the one-photon subspace are $\omega(k) + (p - k)^2/2m$, which is already part of the continuous spectrum. The energy in the $n$-photon subspace is $(2m)^{-1}(p - \sum_{j=1}^n k_j)^2 + \sum_{j=1}^n \omega(k_j) \geq (2m)^{-1}(p - \sum_{j=1}^n k_j)^2 + \omega(\sum_{j=1}^n k_j)$ and for low energies it suffices to take the one-photon part of the continuous spectrum into account. If $p$ is small, $|p| < m\ (= mc)$, the lowest energy is $p^2/2m$ separated by a gap of order $\omega(0) = m_{\mathrm{ph}}$ from the continuum. On the other hand, for $|p| > m$, the eigenvalue $p^2/2m$ is embedded in the continuum and expected to turn into a resonance, once $e$ is different from zero. In some model systems it is found that $p_c < \infty$ for $e = 0$, but $p_c = \infty$ at any $e \neq 0$. Whether $p_c = \infty$ depends also on the form of the kinetic energy of the electron. If instead of $\frac{1}{2m}p^2$ as kinetic energy one repeats the argument just given for the relativistic cousin $\sqrt{p^2 + m^2}$, then $p_c = \infty$ at $e = 0$ and it remains so for $e > 0$. For the Pauli–Fierz model (in three dimensions) the accepted opinion is that the electron cannot be accelerated beyond $p_c \cong \mathcal{O}(mc)$.

Perturbation theory assures us that the isolated ground state energy band for $|p| < p_c$ at $e = 0$ will persist for small nonzero $e$. The range of validity of perturbation theory is set by $\omega(0) = m_{\mathrm{ph}}$ and is therefore very narrow. To improve and to be able to let $m_{\mathrm{ph}} \to 0$ we have to employ nonperturbative techniques, for which we follow Fröhlich (1974). Only the core of each argument is explained; the shorter ones are given immediately in the text and the longer ones are shifted to an appendix. Here is our list.

**Property (i)**: *$E(p)$ is rotation invariant.*

According to section 13.5 there is a unitary operator $U_R$ such that $U_R^* H_p U_R = H_{Rp}$ with $R$ an arbitrary rotation. Therefore $E(p) = E(Rp)$.

**Property (ii)**: *The bound*

$$E(0) \leq E(p) \tag{15.26}$$

*holds.*

From the functional integral representation, compare with chapter 14 and the further explanations in the appendix, it will become clear that

$$|\langle F, U \mathrm{e}^{\mathrm{i}\pi N_\mathrm{f}/2} \mathrm{e}^{-tH_p} \mathrm{e}^{-\mathrm{i}\pi N_\mathrm{f}/2} U^{-1} F\rangle| \leq \langle |F|, U \mathrm{e}^{-tH_0} U^{-1} |F|\rangle \tag{15.27}$$

for $t \geq 0$. We choose $\mathrm{e}^{-\mathrm{i}\pi N_\mathrm{f}/2} U^{-1} F = \psi_p$, or else an approximate ground state if $\psi_p$ does not exist. Let $\mu(\mathrm{d}\lambda)$ be the spectral measure for $U^{-1}|F|$ under $H_0$ and $\lambda_{\min}$ be the left edge of its support. Taking the limit $t \to \infty$ in (15.27), we obtain

$$E(p) \geq \lambda_{\min} \geq E(0). \tag{15.28}$$

One would expect $E(p)$ to be increasing in $|p|$, but no conclusive argument seems to be available.

**Property (iii)**: *As a bound we have*

$$E(p) - E(0) \leq \frac{1}{2m} p^2. \tag{15.29}$$

The inequality (15.29) follows from a variational argument. One has

$$\begin{aligned} E(p) \leq \langle \psi_0, H_p \psi_0\rangle &= \langle \psi_0, \Big(\frac{1}{2m} p^2 + H_0 - \frac{1}{m} p \cdot (P_\mathrm{f} + eA_\varphi)\Big)\psi_0\rangle \\ &= E(0) + \frac{1}{2m} p^2 - \frac{1}{m} p \cdot \langle \psi_0, (P_\mathrm{f} + eA_\varphi)\psi_0\rangle \\ &= E(0) + \frac{1}{2m} p^2, \end{aligned} \tag{15.30}$$

since $H_0 \psi_0 = E(0)\psi_0$ and $\frac{1}{m}\langle \psi_0, (P_\mathrm{f} + eA_\varphi)\psi_0\rangle = \nabla E(0) = 0$ by rotation invariance.

**Property (iv):** *As a bound we have*

$$E(p) \leq E(p - k) + \omega(k). \tag{15.31}$$

*In particular,* $E(p) - E(0) \leq \omega(p)$.

The proof is given in the appendix. There is also a corresponding lower bound.

**Property (v)**: *There are constants* $c_1 > 0$, $c_2$ *such that* $E(p) \geq c_1|p| + c_2$. The proof is given in the appendix.

The next property expresses the stability against one-photon excitations. Define

$$\Delta(p) = \inf_k \{E(p - k) - E(p) + \omega(k)\}. \tag{15.32}$$

Then by property (iv) $\Delta(p) \geq 0$.

**Property (vi)**: *For the bottom of the continuous spectrum we have*

$$E_{\mathrm{c}}(p) = E(p) + \Delta(p). \tag{15.33}$$

*If $\Delta(p) > 0$, then $H_p$ has a ground state at $E(p)$.*

The proof is given in the appendix. We want to infer from the bounds on $E(p)$ that $\Delta(p) > 0$, at least for small $|p|$. As a substitute for the missing proof of the monotonicity of $E(p)$, note that from second-order perturbation in $p$

$$\begin{aligned}\partial_{p_\alpha}\partial_{p_\beta}E(p) = \frac{1}{m}\delta_{\alpha\beta} &- 2\langle\psi_p, \big(m^{-1}(p - P_{\mathrm{f}} - eA_\varphi) - \nabla E(p)\big)_\alpha (H_p - E(p))^{-1} \\ &\times \big(m^{-1}(p - P_{\mathrm{f}} - eA_\varphi) - \nabla E(p)\big)_\beta \psi_p\rangle. \end{aligned}\tag{15.34}$$

This leads to

**Property (vii)**: $E(p) = \frac{1}{2m}p^2 + t(p)$. *$t$ is convex down.*

From property (ii) we conclude that $t(p) - t(0) \geq -\frac{1}{2m}p^2$, which means that $t(p) - t(0)$ cannot bend down too fast. This allows us to establish

**Property (viii)**: *If $|p| \leq (\sqrt{3} - 1)m$, then $\Delta(p) > 0$ and $H_p$ has a ground state separated by the gap $\Delta(p)$ from the continuum.*

Finally, the uniqueness follows from the overlap with the Fock vacuum.

**Property (ix)**: *If $|p| < p_{\mathrm{c}}$ and if*

$$\frac{2e^2}{m}\int \mathrm{d}^3k|\widehat{\varphi}|^2\omega^{-1}E(p)(E(p-k) - E(p) + \omega)^{-2} < \frac{1}{2}, \tag{15.35}$$

*then $H_p$ has a unique ground state.*

Again the proof is given in the appendix. If $|p| < (\sqrt{3} - 1)m \leq p_{\mathrm{c}}$ and (15.35) holds, then $E(p, e)$ is analytic jointly in $p$ and $e$ as a standard consequence of perturbation theory.

In summary, properties (i)–(ix) lend support to the qualitative behavior of the energy–momentum relation as schematically presented in figure 15.1. The bold line indicates the ground state. $E(0)$ increases with the coupling. The gap of size $m_{\mathrm{ph}}$ is not shown. As $m_{\mathrm{ph}} \to 0$ the gap closes. To understand what really happens in this limit, one has to study the infrared scaling of the Pauli–Fierz Hamiltonian with care. Explicit expressions for $E(p)$ do not seem to be available. Computationally only perturbation in $e$ is accessible. To second order one obtains

$$\begin{aligned}E(p) = &\frac{e^2}{2m}\frac{2}{3}\int \mathrm{d}^3k|\widehat{\varphi}|^2(2\omega)^{-1} \\ &+ \frac{1}{2m}p^2\Big(1 - \frac{e^2}{m}\frac{1}{3}\int \mathrm{d}^3k|\widehat{\varphi}|^2\Big(2\omega(\omega + \frac{1}{2m}k^2)\Big)^{-1}\Big) + \mathcal{O}(e^4), \end{aligned}\tag{15.36}$$

which can be trusted only for sufficiently small $p$. $E(0)$ increases in $e$ and in the

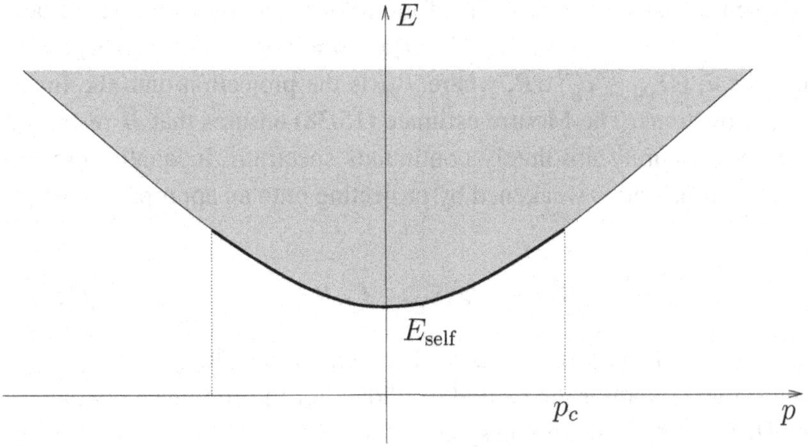


Figure 15.1: The energy–momentum relation for the Pauli–Fierz Hamiltonian.

ultraviolet cutoff, as does $m_{\text{eff}}$. Equation (15.36) confirms the physical intuition that the coupling to the Maxwell field effectively increases the mass of the electron. Note that already to order $e^2$ the effective mass differs from that obtained in the dipole approximation, compare with (14.58), and thus from the effective mass of the classical cousin, the Abraham model.

The nature of the excited states, even close to the ground state band, is left untouched by the present considerations. Physically one expects, as we have indeed established for the Abraham model, a dynamically transient stage when by radiating photons the electron adjusts to the long-time freely propagating state of the form $\mathrm{e}^{-\mathrm{i}tE(p)} f(p)\psi_{\mathrm{g}}(p)$. Here the amplitudes $f(p)$ vanish for $|p| > p_{\mathrm{c}}$ and are determined through the initial conditions. In spectral terms, this implies that $H_p$ has a purely absolutely continuous spectrum except for the possible eigenvalue at $E(p)$. The only powerful technique available for establishing such a property is the method of positive commutators and, as its sisters, Mourre estimates and complex dilations, cf. chapter 17. Let us see how this method applies to the Pauli–Fierz Hamiltonian $H_p$.

In the abstract setting one starts from a self-adjoint operator $H$ on some Hilbert space $\mathcal{H}$ and searches for another self-adjoint operator, the conjugate operator $D$, such that

$$[H, \mathrm{i}D] \geq c_0 > 0. \tag{15.37}$$

Then $H$ has a purely absolutely continuous spectrum. The example to keep in mind here is $H = x$ and $\mathrm{i}D = -\partial_x$. In our context, clearly, (15.37) is too strong. The appropriate modification reads

$$[H, \mathrm{i}D] \geq c_0 - R \tag{15.38}$$

with $R$ a positive trace class operator. This form allows one to count eigenvalues. If $H\psi_n = E_n\psi_n$, $\|\psi_n\| = 1$, then $\langle\psi_n, [H, \mathrm{i}D]\psi_n\rangle_{\mathcal{H}} = 0 \geq c_0 - \langle\psi_n, R\psi_n\rangle$ and, by summing over $n$, $\mathrm{tr}P_{\mathrm{pp}} \leq c_0^{-1}\mathrm{tr}R$, where $P_{\mathrm{pp}}$ is the projection onto the linear span of all eigenfunctions. The Mourre estimate (15.38) ensures that $H$ restricted to $(1 - P_{\mathrm{pp}})\mathcal{H}$ has a purely absolutely continuous spectrum. Inequality (15.38) could be still too strong and is weakened by projecting onto an appropriate energy interval $\Delta$ as

$$E_\Delta[H, \mathrm{i}D]E_\Delta \geq c_0 E_\Delta - E_\Delta R E_\Delta, \tag{15.39}$$

where $E_\Delta$ is the spectral projection of $H$ for the interval $\Delta \subset \mathbb{R}$.

For the Pauli–Fierz operator the natural candidate for the conjugate operator is the generator $D_1$ of dilations in photon space, i.e. $(\mathrm{e}^{-\mathrm{i}D_1 t} f)(k) = t^{3/2} f(tk)$. Then

$$\mathrm{i}D_1 = -\frac{1}{2}\big(\widehat{k} \cdot \partial_k + \partial_k \cdot \widehat{k}\big) \tag{15.40}$$

as operator on $L^2(\mathbb{R}^3, \mathrm{d}^3k)$. We denote the second quantization of $D_1$ by

$$D = \sum_{\lambda=1,2} \int \mathrm{d}^3k a^*(k,\lambda) D_1 a(k,\lambda). \tag{15.41}$$

With these preparations

$$[H_p, \mathrm{i}D] = N_{\mathrm{f}} - \frac{1}{m}\mathrm{d}\Gamma(\widehat{k}) \cdot (p - P_{\mathrm{f}} - eA_\varphi) + \frac{1}{m} eA_{\varphi_1} \cdot (p - P_{\mathrm{f}} - eA_\varphi), \tag{15.42}$$

where $\mathrm{d}\Gamma(\widehat{k}) = \sum_{\lambda=1,2} \int \mathrm{d}^3k \widehat{k} a^*(k,\lambda)a(k,\lambda)$ and $\widehat{\varphi}_1 = \sqrt{\omega}\mathrm{i}D_1 \frac{1}{\sqrt{\omega}}\widehat{\varphi}$.

Let us abbreviate $B = p - P_{\mathrm{f}} - eA_\varphi$. By the Kato–Rellich theorem

$$\begin{aligned}\frac{e}{m}(A_{\varphi_1} \cdot B) &\leq \frac{e}{2m}\big((A_{\varphi_1})^2 + B^2\big) \\ &\leq \frac{e}{2m}(c_1 H_p + c_2) + eH_p \leq e(c_1 H_p + c_2)\end{aligned} \tag{15.43}$$

with coefficients $c_1, c_2$ independent of $p$ and $e$ and whose value may change from line to line. Similarly, using the fact that $[N_{\mathrm{f}}, B]$ is $H_p$-bounded and $\mathcal{O}(e)$,

$$\begin{aligned}\frac{1}{m}\mathrm{d}\Gamma(\widehat{k}) \cdot B &= \frac{1}{m} B N_{\mathrm{f}}^{1/2} \cdot N_{\mathrm{f}}^{-1/2}\mathrm{d}\Gamma(\widehat{k}) \\ &\leq \frac{1}{m}\frac{1}{2m}\big(BN_{\mathrm{f}}^{1/2}\big)^2 + \frac{1}{2}N_{\mathrm{f}} \\ &\leq \frac{1}{m}N_{\mathrm{f}}^{1/2} H_p N_{\mathrm{f}}^{1/2} + \frac{1}{2}N_{\mathrm{f}} + e(c_1 H_p + c_2).\end{aligned} \tag{15.44}$$

Let $E_\Sigma$ be the spectral projection of $H_p$ onto the interval $(-\infty, \Sigma]$. Combining (15.42), (15.43), and (15.44) and using the property that $N_\mathrm{f} \geq 1 - P_\Omega$, the final result reads

$$E_\Sigma[H_p, \mathrm{i}D]E_\Sigma \geq E_\Sigma(1 - P_\Omega)E_\Sigma\Big(\frac{1}{2} - \frac{1}{m}\Sigma\Big) - e(c_1\Sigma + c_2)E_\Sigma. \quad (15.45)$$

Inequality (15.45) has the structure anticipated in (15.39) with $\Delta = (-\infty, \Sigma]$ and $R$ the one-dimensional projection $P_\Omega$. Thus we count the number of eigenvalues in $(-\infty, \Sigma]$ as

$$\mathrm{tr}[P_{\mathrm{pp}}E_\Sigma] \leq \Big(1 - e(c_1\Sigma + c_2)\Big(\frac{1}{2} - \frac{1}{m}\Sigma\Big)^{-1}\Big)^{-1} \quad (15.46)$$

which can be made strictly less than 2 by adjusting $e$. We have not tried to optimize the constants. But the net result is that, upon fixing $e_0$, $p_\mathrm{c}$ sufficiently small and $\Sigma = p_\mathrm{c}^2/4m$, say, in the interval $(-\infty, \Sigma]$ the operator $H_p$ has a purely absolutely continuous spectrum and a single, nondegenerate eigenvalue located at $E(p)$, provided $|e| < e_0$ and $|p| \leq p_\mathrm{c}$. To study the high-energy/high-momentum part of the spectrum other methods will have to be developed.

### 15.2.1 Appendix: Properties of $E(p)$

We prove properties (iv), (v), (vi), (viii), and (ix).

*Property (vi)*: Fix $p$ and choose the momentum lattice $(\delta\mathbb{Z})^3$ with lattice spacing $\delta > 0$. The 3-axis of the lattice is parallel to $p$. Correspondingly, $\mathbb{R}^3$ is partitioned into cubes $C_\delta(n) = \{k|(n_\alpha - \frac{1}{2})\delta \leq k_\alpha < (n_\alpha + \frac{1}{2})\delta, \alpha = 1, 2, 3\}$ with integer $n_\alpha$. The one-particle space $L^2(\mathbb{R}^3) \otimes \mathbb{C}^2 = \mathfrak{h}$ is decomposed into a discrete and a fluctuating part,

$$\mathfrak{h} = \mathfrak{h}_\mathrm{d} \oplus \mathfrak{h}_\mathrm{f}. \quad (15.47)$$

$\psi \in \mathfrak{h}_\mathrm{d}$ is constant over each cube and $\psi \in \mathfrak{h}_\mathrm{f}$ satisfies $\int_{C_\delta(n)} \mathrm{d}^3k\psi(k, \lambda) = 0$ for all $n \in \mathbb{Z}^3$. Such an orthogonal decomposition of the one-particle space factorizes the Fock space as

$$\mathcal{F} = \mathcal{F}_\mathrm{d} \otimes \mathcal{F}_\mathrm{f}. \quad (15.48)$$

If $\Omega_\mathrm{f}$ is the Fock vacuum of $\mathcal{F}_\mathrm{f}$, we set $\mathcal{F}_\delta = \mathcal{F}_\mathrm{d} \otimes \Omega_\mathrm{f}$ and $\mathcal{F} = \mathcal{F}_\delta \oplus \mathcal{F}_\delta^\perp$.

We want $H_p$ to respect the factorization (15.48). This is achieved by replacing $k, \widehat{\varphi}/\sqrt{2\omega}$, and $\omega$ by their lattice approximation $k_\delta$, $(\widehat{\varphi}/\sqrt{2\omega})_\delta$, and $\omega_\delta$, where we set $f_\delta(k) = \delta^{-3} \int_{C_\delta(n)} \mathrm{d}^3kf(k)$ for $k \in C_\delta(n)$. Then $H_p$ is approximated by $H_p(\delta) = \frac{1}{2m}(p - P_{\mathrm{f}}(\delta) - eA_\varphi(\delta))^2 + H_{\mathrm{f}}(\delta)$, which factorizes according to (15.48) as

$$H_p(\delta) = \frac{1}{2m}\big(p - P_{\mathrm{f,d}} \otimes 1 - 1 \otimes P_{\mathrm{f,f}} - eA_{\varphi,\mathrm{d}} \otimes 1\big)^2 + H_{\mathrm{f,d}} \otimes 1 + 1 \otimes H_{\mathrm{f,f}}. \tag{15.49}$$

The fluctuating part of $A_\varphi(\delta)$ vanishes, since $\int \mathrm{d}^3k(\widehat{\varphi}/\sqrt{2\omega})_\delta \psi = 0$ for each $\psi \in \mathfrak{h}_{\mathrm{f}}$. Note that $[H_p(\delta), 1 \otimes P_{\Omega_{\mathrm{f}}}] = 0$, with $P_{\Omega_{\mathrm{f}}}$ the projection onto $\Omega_{\mathrm{f}}$, and therefore $H_p(\delta)$ is reduced by the subspaces $\mathcal{F}_\delta, \mathcal{F}_\delta^\perp$. The bottom of the spectrum of $H_p(\delta)$ is denoted by $E(p, \delta)$.

We want to establish a lower bound on $H_p(\delta) \upharpoonright \mathcal{F}_\delta^\perp$. We choose $\psi \in \mathcal{F}_{\mathrm{d}}$ and $\theta \in \mathcal{F}_{\mathrm{f}}$ with fixed $n$, i.e. $\theta(\underline{k}, \underline{\lambda}) = \theta(k_1, \lambda_1, \ldots, k_n, \lambda_n)$, $n \geq 1$. Then, with $\varphi = \psi \otimes \theta$,

$$\begin{aligned}
\langle \varphi, H_p(\delta)\varphi\rangle_{\mathcal{F}} &= \langle \psi \otimes \theta, H_p(\delta)\psi \otimes \theta\rangle_{\mathcal{F}} \\
&= \sum_{\underline{\lambda}} \int \mathrm{d}^{3n}\underline{k} |\theta(\underline{k}, \underline{\lambda})|^2 \langle \psi, \frac{1}{2m}\Big(p - P_{\mathrm{f,d}} - \sum_{j=1}^n k_{j\delta} - eA_{\varphi,\mathrm{d}}\Big)^2 \psi\rangle_{\mathcal{F}_{\mathrm{d}}} \\
&\quad + \langle \psi, H_{\mathrm{f,d}}\psi\rangle_{\mathcal{F}_{\mathrm{d}}} \langle \theta, \theta\rangle_{\mathcal{F}_{\mathrm{f}}} + \langle \psi, \psi\rangle_{\mathcal{F}_{\mathrm{d}}} \langle \theta, H_{\mathrm{f,f}}\theta\rangle_{\mathcal{F}_{\mathrm{f}}} \\
&= \sum_{\underline{\lambda}} \int \mathrm{d}^{3n}\underline{k} |\theta(\underline{k}, \underline{\lambda})|^2 \langle \psi, H_{p - \sum_{j=1}^n k_{j\delta},\mathrm{d}}\psi\rangle_{\mathcal{F}_{\mathrm{d}}} \\
&\quad + \langle \psi, \psi\rangle_{\mathcal{F}_{\mathrm{d}}} \langle \theta, H_{\mathrm{f,f}}\theta\rangle_{\mathcal{F}_{\mathrm{f}}} \\
&\geq \inf_{\underline{k}}\{E(p - \sum_{j=1}^n k_{j\delta}, \delta) + \sum_{j=1}^n \omega_\delta(k_j)\} \langle \psi, \psi\rangle_{\mathcal{F}_{\mathrm{d}}} \langle \theta, \theta\rangle_{\mathcal{F}_{\mathrm{f}}} \\
&\geq \inf_k\{E(p - k, \delta) + \omega_\delta(k)\} \langle \varphi, \varphi\rangle_{\mathcal{F}}.
\end{aligned} \tag{15.50}$$

By finite linear combinations this bound extends to a dense set: if $\varphi = \psi_1 \otimes \theta_1 + \psi_2 \otimes \theta_2$ with both $\theta_1$ and $\theta_2$ in the $n$-photon subspace, one only has to repeat the computation in (15.50). If they belong to different photon numbers, we use $\theta_1 \perp \theta_2$. If $E^\perp(p, \delta)$ denotes the bottom of the spectrum of $H_p(\delta) \upharpoonright \mathcal{F}_\delta^\perp$, we conclude that

$$E^\perp(p, \delta) \geq \inf_k\{E(p - k, \delta) + \omega_\delta(k)\}. \tag{15.51}$$

$H_p(\delta) \upharpoonright \mathcal{F}_\delta$ consists of a large, but finite number of oscillators with strictly positive frequencies. Therefore $H_p(\delta) \upharpoonright \mathcal{F}_\delta$ has a discrete spectrum. Let

$$\Delta(p, \delta) = \inf_k\{E(p - k, \delta) - E(p, \delta) + \omega_\delta(k)\}. \tag{15.52}$$

If $\Delta(p, \delta) \geq \Delta_0 > 0$ independently of $\delta$, then $E^\perp(p, \delta) - E(p, \delta) \geq \Delta(p, \delta) \geq \Delta_0$ by (15.51) and the ground state of $H_p(\delta)$ is in $\mathcal{F}_\delta$. The spectral projection $\chi_{[E(p,\delta), E(p,\delta)+\Delta_0]}(H_p(\delta))$ is a nonzero compact operator.

The next step is to show that $H_p(\delta)$ converges to $H_p$ as $\delta \to 0$. Technically one proves that for the difference of resolvents the limit

$$\lim_{\delta\to 0} \|(H_p(\delta) - z)^{-1} - (H_p - z)^{-1}\| = 0 \tag{15.53}$$

holds provided $z$ is sufficiently negative. The argument uses the first-order expansion for the resolvent and Kato–Rellich bounds of the type used in the proof of Theorem 13.3. The norm resolvent convergence (15.53) ensures that $\chi_{[E(p,\delta),E(p,\delta)+\Delta_0]}(H_p(\delta))$ converges in norm to $\chi_{[E(p),E(p)+\Delta_0]}(H_p)$ and that this operator is compact as a norm limit of compact operators. Since the limit operator is nonzero by construction, $H_p$ has a ground state at $E(p)$.

To confirm that $E_{\mathrm{c}}(p) = E(p) + \Delta(p)$ with $\Delta(p) = \inf_k\{E(p-k) - E(p) + \omega(k)\}$ the first part of (15.50) is repeated with a one-photon wave function $\theta(k_1, \lambda_1)$ well concentrated at $k_0$ with $k_0$ such that $\Delta(p) = E(p-k_0) - E(p) + \omega(k_0)$. There is an infinite number of orthogonal states, which by construction have an energy arbitrarily close to $E(p) + \Delta(p)$. This proves (vi).

*Property (iv)*: From the pull-through formula for $a^*$ we obtain

$$H_p a^*(k,\lambda) = a^*(k,\lambda)(H_{p-k} + \omega(k)) - \frac{e}{m\sqrt{2\omega(k)}}\widehat{\varphi}(k) e_\lambda \cdot (p - P_{\mathrm{f}} - eA_\varphi). \tag{15.54}$$

Let $\psi_{p-k,\delta}$ be an approximate ground state for $H_{p-k}$ with energies in the interval $[E(p-k), E(p-k)+\delta]$ (or let $\psi_{p-k}$ be equal to the ground state if it exists), and let us consider the one-photon excitation $\varphi_\delta = a^*(f_\delta)\psi_{p-k,\delta}$ with $f_\delta$ sharply centered at $k$. From (15.54) one infers

$$\begin{aligned}
E(p)\langle\varphi_\delta, \varphi_\delta\rangle &\le \langle\varphi_\delta, H_p\varphi_\delta\rangle \\
&= \langle\varphi_\delta, H_p a^*(f_\delta)\psi_{p-k,\delta}\rangle \\
&= \omega(k)\langle\varphi_\delta, \varphi_\delta\rangle + \langle\varphi_\delta, a^*(f_\delta) H_{p-k}\psi_{p-k,\delta}\rangle \\
&\quad - \sum_{\lambda'}\int \mathrm{d}^3k' \frac{e}{m\sqrt{2\omega(k')}}\widehat{\varphi}(k') f_\delta(k',\lambda') \\
&\quad \times \langle\varphi_\delta, e_{\lambda'}(k')\cdot(p - P_{\mathrm{f}} - eA_\varphi)\psi_{p-k,\delta}\rangle \\
&\le \langle\varphi_\delta, \varphi_\delta\rangle\big(\omega(k) + E(p-k) + \mathcal{O}(\delta)\big) \\
&\quad + \frac{1}{\sqrt{m}}\langle f_\delta, \frac{1}{\sqrt{\omega}}\widehat{\varphi}\rangle_{\mathfrak{h}} \langle\varphi_\delta, \varphi_\delta\rangle^{1/2} \langle\psi_{p-k,\delta}, H_p\psi_{p-k,\delta}\rangle^{1/2}.
\end{aligned} \tag{15.55}$$

We can now choose $f_\delta$ such that the last term multiplied by $\langle\varphi_\delta, \varphi_\delta\rangle^{-1}$ vanishes in the limit $\delta \to 0$. Thereby the bound of property (iv) results.

*Property (v)*: We have for $0 < a_1 < 1, a_2 > 0, (1 - a_1)(1 + a_2) = 1$,

$$\begin{aligned} 2m\langle\psi, H_p\psi\rangle &\geq a_1\langle\psi, (p - P_{\mathrm{f}})^2\psi\rangle - a_2\langle\psi, e^2 A_\varphi^2\psi\rangle + 2m\langle\psi, H_{\mathrm{f}}\psi\rangle \\ &\geq a_1\langle\psi, (p - P_{\mathrm{f}})^2\psi\rangle + (2m - a_2 a)\langle\psi, H_{\mathrm{f}}\psi\rangle - a_2 b\langle\psi, \psi\rangle, \end{aligned} \tag{15.56}$$

where the relative bound $\langle\psi, e^2 A_\varphi^2\psi\rangle \leq a\langle\psi, H_{\mathrm{f}}\psi\rangle + b\langle\psi, \psi\rangle$ is used. We choose $a_2$ such that $2m - a_2 a > 0$. Since for $\alpha > 0$

$$\frac{1}{2}\alpha(p - P_{\mathrm{f}})^2 + H_{\mathrm{f}} \geq |p| - \frac{1}{2}\alpha^{-1}, \tag{15.57}$$

the constants in (15.57) and (15.56) can be adjusted so as to give the desired bound.

*Property (viii)*: By rotational invariance it suffices to consider (15.32) along a line passing through the origin. We will denote these functions by the same symbol as before. Using properties (ii) and (iv) we obtain

$$\begin{aligned} E(p-k) - E(p) + \omega(k) &= E(p-k) - E(0) - E(p) + E(0) + \omega(k) \\ &\geq -\omega(p) + \omega(k) \end{aligned} \tag{15.58}$$

and it suffices to take the minimum over the interval $|k| \leq |p|$. By reflection symmetry, one may pick $p \geq 0$. We use the decomposition of $E$ from property (vii) and will show that

$$\Delta(p) = \min_{|k|\leq p}\left\{\frac{1}{2m}(p-k)^2 - \frac{1}{2m}p^2 + t(p-k) - t(p) + \omega(k)\right\} > 0 \tag{15.59}$$

provided $p < m/2$. This will come about by

**Lemma**: *Let $f : \mathbb{R} \to \mathbb{R}$ be convex, even, with $f(0) = 0$ and $f(x) \leq \frac{1}{2m}x^2$. Then the bounds*

$$-1 + \frac{1}{m}x \leq f'(x) \leq 1 + \frac{1}{m}x \tag{15.60}$$

*hold for $|x| \leq m$.*

*Proof*: If (15.60) holds for $x \geq 0$, by reflection symmetry it also holds for $x \leq 0$. So let us take $x \geq 0$. $f'(0) = 0$ and $f'$ is increasing. Therefore we only have to check the upper bound. Let $x_0$ be the smallest $x$ such that $f'(x_0) = 1 + \frac{1}{m}x_0$. Then, since $f'$ is increasing,

$$\begin{aligned}0 &\le -f(x) + \frac{1}{2m}x^2 \\ &= \frac{1}{2m}x_0^2 - f(x_0) - \int_{x_0}^{x} \mathrm{d}y\Big(f'(y) - \frac{1}{m}y\Big) \\ &\le \frac{1}{2m}x_0^2 - \int_{x_0}^{x} \mathrm{d}y\Big(f'(x_0) - \frac{1}{m}y\Big) \\ &= \frac{1}{2m}x^2 + \Big(1 + \frac{1}{m}x_0\Big)(x_0 - x) \end{aligned} \tag{15.61}$$

for all $x \ge x_0$, which can be satisfied only if $x_0 \ge m$. □

The lemma is used in (15.59) with $f(k) = -t(k) + t(0)$, which by properties (ii) and (vii) satisfies the assumptions, and we set

$$t(p-k) - t(p) = \int_p^{p-k} \mathrm{d}x t'(x). \tag{15.62}$$

If $k > 0$, the lower bound in (15.60) is applicable provided $p < m$, $0 \le k \le 2p$. If $k < 0$, the upper bound in (15.60) is applicable provided $p - k \le m$ and thus $p \le m/2$. The bounds put together yield

$$\frac{1}{2m}(p-k)^2 - \frac{1}{2m}p^2 + t(p-k) - t(p) + \omega(k) \ge -|k| + \omega(k) > 0 \tag{15.63}$$

for $p \le m/2$ and $|k| \le p$. Refining the last step of the argument the bound can be improved to $p \le (\sqrt{3}-1)m$, which implies $p_\mathrm{c} \ge (\sqrt{3}-1)m$.

*Property (ix)*: As in the second part of the proof of Theorem 15.5 below, one estimates the overlap of the ground state vector with $\Omega$ by using the analog of the pull-through formula (15.76). (15.69) is replaced then by (15.35).

Finally we have to show (15.27), for which purpose we Trotterize $H_p$ as the sum of $\frac{1}{2m}(p - P_\mathrm{f} - eA_\varphi)^2$ and $H_\mathrm{f}$ in the function space representation. We have

$$|U\mathrm{e}^{\mathrm{i}\pi N_\mathrm{f}/2}\mathrm{e}^{-tH_\mathrm{f}}\mathrm{e}^{-\mathrm{i}\pi N_\mathrm{f}/2}U^{-1}F| \le U\mathrm{e}^{\mathrm{i}\pi N_\mathrm{f}/2}\mathrm{e}^{-tH_\mathrm{f}}\mathrm{e}^{-\mathrm{i}\pi N_\mathrm{f}/2}U^{-1}|F|, \tag{15.64}$$

since $[H_\mathrm{f}, N_\mathrm{f}] = 0$ and $\mathrm{e}^{-tH_\mathrm{f}}$ has a positive kernel in function space. Recall the transformation (14.67). Linearizing the square with the Gaussian measure $\mu_G$ of mean zero and variance $t/m$, one obtains

$$\begin{aligned}
&(U\mathrm{e}^{\mathrm{i}\pi N_\mathrm{f}/2}\mathrm{e}^{-t(p-P_\mathrm{f}-eA_\varphi)^2/2m}\mathrm{e}^{-\mathrm{i}\pi N_\mathrm{f}/2}U^{-1}F)(A(\cdot))\\
&= (U\mathrm{e}^{-t(p-P_\mathrm{f}-eE_{\perp\tilde{\varphi}})^2/2m}U^{-1}F)(A(\cdot))\\
&= \int \mu_G(\mathrm{d}\lambda)\mathrm{e}^{\mathrm{i}\lambda\cdot p}U\mathrm{e}^{-\mathrm{i}\lambda\cdot(P_\mathrm{f}+eE_{\perp\tilde{\varphi}})}U^{-1}F(A(\cdot))\\
&= \int \mu_G(\mathrm{d}\lambda)\mathrm{e}^{\mathrm{i}\lambda\cdot p}F(A(\cdot+\lambda)+\lambda e\tilde{\varphi}(\cdot)), \qquad (15.65)
\end{aligned}$$

since $P_\mathrm{f}$ shifts and $E_{\perp\tilde{\varphi}}$ translates the field. In fact the components of $(p - P_\mathrm{f} - eE_{\perp\tilde{\varphi}})$ do not commute and in (15.65) there are errors of order $t^2$ which vanish as the Trotter spacing tends to zero. Taking absolute values on both sides of (15.65) yields

$$|\cdot| \le \int \mu_G(\mathrm{d}\lambda)|F(A(x+\lambda)+\lambda e\tilde{\varphi}(x))| \qquad (15.66)$$

and similarly for functionals of a finite number of fields. Therefore

$$\begin{aligned}
&|U\mathrm{e}^{\mathrm{i}\pi N_\mathrm{f}/2}\mathrm{e}^{-t(p-P_\mathrm{f}-eA_\varphi)^2/2m}\mathrm{e}^{-\mathrm{i}\pi N_\mathrm{f}/2}U^{-1}F|\\
&\le U\mathrm{e}^{\mathrm{i}\pi N_\mathrm{f}/2}\mathrm{e}^{-t(P_\mathrm{f}+eA_\varphi)^2/2m}\mathrm{e}^{-\mathrm{i}\pi N_\mathrm{f}/2}U^{-1}|F|. \qquad (15.67)
\end{aligned}$$

Iterating the bounds (15.64) and (15.67) results in (15.27).

## 15.3 Two-fold degeneracy in the case of spin

For the effective spin dynamics a crucial input is the two-fold degeneracy of the ground state of the Pauli–Fierz operator with spin, which will be established here for sufficiently small $e$. The restriction on $e$ is presumably an artifact of the method.

The Hamiltonian under consideration is

$$H_p = \frac{1}{2m}(p - P_\mathrm{f} - eA_\varphi)^2 - \frac{e}{2m}\sigma\cdot B_\varphi + H_\mathrm{f} \qquad (15.68)$$

acting on $\mathbb{C}^2 \otimes \mathcal{F}$, where $A_\varphi = A_\varphi(0)$, $B_\varphi = B_\varphi(0)$; compare with Eq. (15.21). We require $m_\mathrm{ph} > 0$. Let $P_\mathrm{g}$ be the projection onto the ground state subspace and $P_0$ be the projection onto the subspace spanned by $\chi\otimes\Omega$, $\chi\in\mathbb{C}^2$, $\mathrm{tr}P_0 = 2$. We assume $|p| < p_\mathrm{c}$. Then $\mathrm{tr}P_\mathrm{g} \ge 1$ by the arguments for the proof of property (vi).

**Theorem 15.5** (Two-fold degeneracy of the ground state band). *If $\Delta(p) > 0$ and whenever*

$$\frac{2e^2}{m}\int \mathrm{d}^3k|\widehat{\varphi}|^2\omega^{-1}\Big(E(p)+\frac{1}{2m}k^2\Big)\big(E(p-k)-E(p)+\omega\big)^{-2} < \frac{1}{3}, \qquad (15.69)$$

*then* $\mathrm{tr}P_\mathrm{g} = 2$.

For the Pauli–Fierz model with spin a proof of property (ii) is missing. If, very reasonably, it is assumed, then $\Delta(p) > 0$ for $|p| \leq (\sqrt{3} - 1)m$.

*Proof*: We assume $\widehat{\varphi}$ to be real which can always be achieved through a suitable canonical transformation.

Let $z$ be real and sufficiently negative. We claim that

$$P_0(z - H_p)^{-1} P_0 = a(z) P_0 \tag{15.70}$$

with real coefficient $a(z)$.

In (15.68) we set $H_0 = \frac{1}{2m}(p - P_\mathrm{f})^2 + H_\mathrm{f}$ and $H_p = H_0 + H_1$. $H_0$ does not depend on spin and when restricted to the $n$-photon subspace it is multiplication by a real function. By the Kato–Rellich theorem the resolvent expansion

$$\langle \chi \otimes \Omega, (z - H_p)^{-1} \chi \otimes \Omega \rangle = \sum_{n=0}^{\infty} \langle \chi \otimes \Omega, (z - H_0)^{-1} \big(H_1 (z - H_0)^{-1}\big)^n \chi \otimes \Omega \rangle \tag{15.71}$$

is convergent. Expanding the product yields as generic term

$$\prod_{j=1}^{m} (a_j + \mathrm{i} b_j \cdot \sigma) \tag{15.72}$$

with real coefficients $a_j, b_j$, depending on $k_1, \lambda_1, \ldots, k_m, \lambda_m$. Using the equality

$$(a_1 + \mathrm{i} b_1 \cdot \sigma)(a_2 + \mathrm{i} b_2 \cdot \sigma) = a_1 a_2 - b_1 \cdot b_2 + \mathrm{i} \sigma \cdot (a_1 b_2 + a_2 b_1 - b_1 \times b_2) \tag{15.73}$$

it follows that

$$\langle \chi \otimes \Omega, (z - H_p)^{-1} \chi \otimes \Omega \rangle = a(z) \langle \chi, \chi \rangle + \mathrm{i} b(z) \cdot \langle \chi, \sigma \chi \rangle \tag{15.74}$$

with real coefficients $a(z), b(z)$. Since the left-hand side is real, $b(z) = 0$ which proves (15.70).

Equation (15.70) holds on the negative real axis and therefore extends by analyticity to the full resolvent set. In particular, one can integrate (15.70) over a small contour encircling $E(p)$, the ground state energy of $H_p$. Then

$$P_0 P_\mathrm{g} P_0 = c_1 P_0. \tag{15.75}$$

By the pull-through argument

$$\begin{aligned}[a(k, \lambda), H_p] &= (H_{p-k} - H_p + \omega(k)) a(k, \lambda) \\ &\quad - \frac{e}{m} \frac{\widehat{\varphi}}{\sqrt{2\omega}} \Big( e_\lambda \cdot (p - P_\mathrm{f} - e A_\varphi) - \frac{1}{2} (e_\lambda \times \mathrm{i} k) \cdot \sigma \Big).\end{aligned} \tag{15.76}$$

Let now $\psi \in P_{\mathrm{g}}\mathcal{H}$. Then

$$\begin{aligned}\langle\psi, N_{\mathrm{f}}\psi\rangle &= \sum_{\lambda=1,2}\int \mathrm{d}^3k\|a(k,\lambda)\psi\|^2 \\ &\le \frac{2e^2}{m}\int \mathrm{d}^3k|\widehat{\varphi}|^2\omega^{-1}\Big(E(p)+\frac{1}{2m}k^2\Big) \\ &\quad\times\big(E(p-k)-E(p)+\omega\big)^{-2} = c_0.\end{aligned} \tag{15.77}$$

Since $\mathrm{tr}[P_{\mathrm{g}}(1-P_0)] \le \mathrm{tr}[P_{\mathrm{g}}N_{\mathrm{f}}] \le c_0\mathrm{tr}P_{\mathrm{g}}$, one concludes

$$(1-c_0)\mathrm{tr}P_{\mathrm{g}} \le \mathrm{tr}[P_{\mathrm{g}}P_0] \le 2. \tag{15.78}$$

If $c_0 < 1$, then $c_1 > 0$, with $c_1$ the constant in (15.75). Suppose $\mathrm{tr}P_{\mathrm{g}} = 1$. Then $P_{\mathrm{g}}$ projects along $\psi$ and $P_0P_{\mathrm{g}}P_0$ along $P_0\psi$ which contradicts (15.75). Thus $\mathrm{tr}P_{\mathrm{g}} \ge 2$. On the other hand if $c_0 < \frac{1}{3}$, then $\mathrm{tr}P_{\mathrm{g}} < 3$. In conjunction, $\mathrm{tr}P_{\mathrm{g}} = 2$ as was to be shown. □

An alternative approach would be to use the positive commutator technique as explained at the end of section 15.2. It says that, provided $|e| < e_0$, $|p| < p_{\mathrm{c}}$, the ground state of $H_p$ is exactly two-fold degenerate and that in a band above the ground state energy there is only an absolutely continuous spectrum.

## Notes and references

### *Section 15.1*

Our discussion of the soft photon bound is taken from Bach (private lecture notes) and Bach, Fröhlich and Sigal (1998a). If the potential $V$ is attractive, but so weak that $H_{\mathrm{at}}$ has no ground state, then a sufficiently strong coupling to the radiation field will generate a ground state, since the mass of the particle is effectively increased (Hiroshima and Spohn 2001; Hainzl 2002; Hainzl *et al.* 2003; Chen *et al.* 2003). The property of $\mathrm{e}^{-tH}$ to be positivity improving is not known to hold under additional terms, for instance including an external vector potential or spin. As explained to us by V. Bach, a soft photon bound as in Theorem 15.1 automatically estimates the overlap with the Fock vacuum. If $|e|$ is sufficiently small, this overlap is larger than $1/2$ and uniqueness is guaranteed.

With the Maxwell field replaced by a scalar field, compare with section 19.2, ground state properties are investigated in Gérard (2000) and Betz *et al.* (2002), where references to earlier work are given.

### Section 15.2

The key properties of the energy–momentum relation are established in Fröhlich (1974), where also the missing points of rigor are supplied. In fact Fröhlich discusses Nelson's model of a particle coupled to a scalar field, compare with section 19.2. In that case, as for example explained in Spohn (1988), $e^{-tH(p)}$, $t > 0$, is positivity improving in Fock space, within an exponentially small error $e^{-t}$. From this property uniqueness of the ground state $\psi_p$ is deduced by the argument explained in section 14.3. The overlap argument of property (ix) is a substitute which works only for small $e$. In his recent PhD thesis Chen (2001) establishes that $E(p)$ has a limit as $m_{\text{ph}} \to 0$. The limit function $E(p)$ is twice continuously differentiable for $|p|$ sufficiently small. Thus the effective mass of the electron remains well defined even in the physical case $m_{\text{ph}} = 0$, under the restriction of small $e$ and, of course, an ultraviolet cutoff. An example where $p_{\text{c}} = 1$ for $e = 0$ and $p_{\text{c}} = \infty$ for $e > 0$ is the Fröhlich polaron in two dimensions (Spohn 1988). Positive commutator methods at fixed total momentum are developed in the highly recommended paper by Fröhlich, Griesemer and Schlein (2003), where the complete proof for the Nelson model, see section 19.2 for its definition, can be found. Positive-commutator methods and the related Mourre estimates are most useful also in cases where the electron is confined by an external potential. We refer to Skibsted (1998), Bach, Fröhlich and Sigal (1998b), Dereziński and Gérard (1999), Bach, Fröhlich, Sigal and Soffer (1999), and Georgescu, Gérard and Møller (2004). A precusor is Hübner and Spohn (1995b).

### Section 15.3

The material is taken from Hiroshima and Spohn (2002).

# 16

# States of lowest energy: dynamics

As for classical dynamics, in many applications the external potentials have a slow variation in space-time. The standard procedure is then to ignore the quantized Maxwell field and to proceed with an effective one-particle Hamiltonian. This is justified since the photons very rapidly adjust to the motion of the electron. To put it differently, if a classical trajectory of the electron is prescribed, then the photons are governed by a Hamiltonian of slow time-dependence and essentially remain in their momentarily lowest state of energy. We propose first to study slow time variation, which abstractly falls under the auspices of the time-adiabatic theorem. However, the real issue is how, from the slow variation in space, to extract, rather than assume, the slow variation in time. It seems appropriate to call such a situation *space*-adiabatic.

We will work for a start with time-dependent perturbation theory using the insights gained from the time-adiabatic theorem. It turns out that these methods lead us astray in the case of slowly varying external vector potentials. Thus we are forced to develop more powerful techniques. They come from the area of pseudo-differential operators. In fact this theory provides a much sharper picture of adiabatic decoupling and a systematic scheme for computing effective Hamiltonians. To avoid technical complications we restrict ourselves to matrix-valued symbols. Transcribing these results formally to the Pauli–Fierz Hamiltonian we will compute the effective Hamiltonian governing the motion of the electron in the band of lowest energy, including spin precession. The effective Hamiltonian can be analysed through semiclassical methods which eventually leads to the nonperturbative definition of the gyromagnetic ratio.

There are other properties of the Pauli–Fierz Hamiltonian which can be handled semiclassically. Most notably we may consider a physical situation, where classical currents are prescribed. Then the Pauli–Fierz operator reduces to a time-dependent operator on Fock space quadratic in the bosonic annihilation/creation operators. Such quasi-free theories can be studied in great detail. In particular, coherent states of the photon field evolve in time according to the classical inhomogeneous Maxwell equations. Under standard macroscopic conditions field fluctuations are small and the classical Maxwell theory can be used safely. For example, a city radio station with a power of 100 kW at a wavelength of 100 m emits $10^{30}$ photons per second, and at a distance of 100 km a flux of $10^{15}$ photons $\mathrm{s}^{-1}$ $\mathrm{cm}^{-2}$ is still observed. On the other hand, experimentally even the smallest field intensities can be controlled and quantum features are of importance, as for example in photon counting statistics. For the Maxwell field an amazingly wide span of scales can be probed, from the classical deterministic behavior down to single-photon randomness.

## 16.1 The time-adiabatic theorem

In the case where no external forces are present, the total momentum is conserved; compare with sections 13.5 and 15.2. Thus under slowly varying external potentials the total momentum can be expected to change slowly, and the appropriate starting point is the Pauli–Fierz Hamiltonian in the representation diagonal with respect to the total momentum, i.e.

$$H = \frac{1}{2m}\big(\sigma \cdot (p - P_{\mathrm{f}} - eA_{\varphi} - eA_{\mathrm{ex}}(\varepsilon x))\big)^2 + e\phi_{\mathrm{ex}}(\varepsilon x) + H_{\mathrm{f}}. \tag{16.1}$$

Here $p$ refers to the total momentum and $\varepsilon$ is a dimensionless parameter regulating the variation of the external potentials $\phi_{\mathrm{ex}}, A_{\mathrm{ex}}$. Let us assume for the moment that a classical trajectory of the electron is given. Because of the slow variation of $\phi_{\mathrm{ex}}, A_{\mathrm{ex}}$ it has to be of the form $(q_{\varepsilon t}, p_{\varepsilon t})$, $0 \le t \le \varepsilon^{-1}\tau$ with $(\varepsilon q_{\varepsilon t}, p_{\varepsilon t})$ of order 1. Inserting in (16.1), the time-dependent Hamiltonian can be written as

$$\begin{aligned} H(\varepsilon t) = {} & \frac{1}{2m}\big(p_{\varepsilon t} - P_{\mathrm{f}} - eA_{\varphi} - eA_{\mathrm{ex}}(\varepsilon q_{\varepsilon t})\big)^2 - \frac{e}{2m}\sigma \cdot (B_{\varphi} + \varepsilon B_{\mathrm{ex}}(\varepsilon q_{\varepsilon t})) \\ & + e\phi_{\mathrm{ex}}(\varepsilon q_{\varepsilon t}), \end{aligned} \tag{16.2}$$

which governs the motion of photons and acts on $\mathcal{F}$. $t$ is measured in atomic units. $B_{\varphi} = B_{\varphi}(0)$ is the quantized magnetic field. We have already studied the spectrum of $H(t)$ for fixed $t$. The term proportional to $B_{\mathrm{ex}}$ is of order $\varepsilon$ and can be neglected. Provided $|p_t| < p_{\mathrm{c}}$, $H(t)$ has a two-fold degenerate ground state with energy

$$E(t) = E(p_t - eA_{\mathrm{ex}}(\varepsilon q_t)) + e\phi_{\mathrm{ex}}(\varepsilon q_t). \tag{16.3}$$

Physically it is expected that through radiation the photons approach very rapidly a state of lowest energy. Subsequently only very few photons escape, since the time variation is slow and $E(t)$ is separated by a gap from the continuous spectrum.

The time-adiabatic theorem of quantum mechanics makes an abstraction of the particular situation and simply postulates the time-dependent Hamiltonian $H(t)$ as given and acting on the Hilbert space $\mathcal{H}$. The role of the ground state subspace is played by a physically distinguished, "relevant" subspace with corresponding instantaneous spectral projection $P(t)$ and energy $E(t)$, i.e. $H(t)P(t) = E(t)P(t)$. It is assumed that for every $t$ the energy $E(t)$ is isolated by a finite gap from the rest of the spectrum of $H(t)$. The slow variation in time is introduced through $H(\varepsilon t)$ with $\varepsilon \ll 1$ as a dimensionless adiabaticity parameter and one is interested in the solution of the Schrödinger equation

$$\mathrm{i}\partial_t \psi(t) = H(\varepsilon t)\psi(t), \tag{16.4}$$

where the initial wave function $\psi(0)$ is assumed to lie already in the relevant subspace, $P(0)\psi(0) = \psi(0)$. $t$ is chosen to be so long that $P(t)$ rotates by some finite amount, implying that

$$0 \le t \le \varepsilon^{-1}\tau, \quad \tau = \mathcal{O}(1). \tag{16.5}$$

Sometimes it is convenient to switch to the slow time scale

$$t' = \varepsilon t. \tag{16.6}$$

Then our problem becomes

$$\mathrm{i}\varepsilon\partial_{t'}\psi(t') = H(t')\psi(t'), \quad P(0)\psi(0) = \psi(0), \quad 0 \le t' \le \tau. \tag{16.7}$$

To stress the similarity with the space-adiabatic situation, however, we stick to the fast time scale of (16.4).

As one of the basic results it is established that the subspace $P(\varepsilon t)$ is adiabatically protected in the sense that

$$\|(1 - P(\varepsilon t))\psi(t)\| \le c_0\varepsilon \quad \text{for} \quad 0 \le t \le \varepsilon^{-1}\tau \tag{16.8}$$

with some suitable constant $c_0$. Up to an error of order $\varepsilon$ the solution to the Schrödinger equation (16.4) clings to the relevant subspace $P(\varepsilon t)\mathcal{H}$.

It is of interest briefly to recall the proof of (16.8), since some central elements will reappear later. We denote the unitary propagator for (16.4) by $U^\varepsilon(t, s)$. The idea is to define a "diagonal" propagator $U^\varepsilon_{\mathrm{dg}}(t, s)$ such that it preserves $P(t)$ exactly, i.e.

$$P(\varepsilon t)U^\varepsilon_{\mathrm{dg}}(t, s) = U^\varepsilon_{\mathrm{dg}}(t, s)P(\varepsilon s). \tag{16.9}$$

The unitary propagator $U^\varepsilon_{\mathrm{dg}}(t, s)$ is generated by the Hamiltonian $H_{\mathrm{dg}}(\varepsilon t)$. From (16.9) it follows that

$$[H_{\mathrm{dg}}(\varepsilon t), P(\varepsilon t)] = \mathrm{i}\varepsilon\dot{P}(\varepsilon t). \tag{16.10}$$

We look for a solution which is $\varepsilon$-close to $H(\varepsilon t)$. Using the identities $P(t)\dot{P}(t)P(t) = 0$, $(1 - P(t))\dot{P}(t)(1 - P(t)) = 0$, we obtain

$$H_{\mathrm{dg}}(\varepsilon t) = H(\varepsilon t) + \mathrm{i}\varepsilon[\dot{P}(\varepsilon t), P(\varepsilon t)]. \tag{16.11}$$

To prove (16.8) one thus has to estimate the difference

$$U^{\varepsilon}(t,0) - U^{\varepsilon}_{\mathrm{dg}}(t,0) = -\varepsilon \int_0^t \mathrm{d}s U^{\varepsilon}(t,s)[\dot{P}(\varepsilon s), P(\varepsilon s)]U^{\varepsilon}_{\mathrm{dg}}(s,0). \tag{16.12}$$

While $H(\varepsilon t) - H_{\mathrm{dg}}(\varepsilon t)$ is of order $\varepsilon$, this is not good enough, since errors might add up over the long times $\varepsilon^{-1}\tau$. To make progress we note that $P[\dot{P}, P]P = 0 = (1 - P)[\dot{P}, P](1 - P)$, whereas $P[\dot{P}, P](1 - P) \neq 0$. Thus to improve on (16.12) one has to exploit the time averaging, which is most easily achieved by writing $[\dot{P}, P]$ as a time derivative. Let us assume for a moment that the commutator equation

$$[H(t), X(t)] = [\dot{P}(t), P(t)] \tag{16.13}$$

has a bounded solution $X(t)$. Then, using again (16.11),

$$\begin{aligned}
&U^{\varepsilon}(\varepsilon^{-1}\tau, 0) - U^{\varepsilon}_{\mathrm{dg}}(\varepsilon^{-1}\tau, 0) \qquad (16.14)\\
&= -\varepsilon \int_0^{\varepsilon^{-1}\tau} \mathrm{d}s U^{\varepsilon}(\varepsilon^{-1}\tau, s)\big(H(\varepsilon s)X(\varepsilon s) - X(\varepsilon s)H_{\mathrm{dg}}(\varepsilon s)\big)U^{\varepsilon}_{\mathrm{dg}}(s,0) + \mathcal{O}(\varepsilon)\\
&= \mathrm{i}\varepsilon \int_0^{\varepsilon^{-1}\tau} \mathrm{d}s \Big(\frac{\mathrm{d}}{\mathrm{d}s} U^{\varepsilon}(\varepsilon^{-1}\tau, s)X(\varepsilon s)U^{\varepsilon}_{\mathrm{dg}}(s,0) - U^{\varepsilon}(\varepsilon^{-1}\tau, s)X(\varepsilon s)\frac{\mathrm{d}}{\mathrm{d}s}U^{\varepsilon}_{\mathrm{dg}}(s,0)\Big)\\
&\quad + \mathcal{O}(\varepsilon)\\
&= \mathrm{i}\varepsilon \int_0^{\varepsilon^{-1}\tau} \mathrm{d}s \Big(\frac{\mathrm{d}}{\mathrm{d}s}\big(U^{\varepsilon}(\varepsilon^{-1}\tau, s)X(\varepsilon s)U^{\varepsilon}_{\mathrm{dg}}(s,0)\big) - U^{\varepsilon}(\varepsilon^{-1}\tau, 0)\varepsilon\dot{X}(\varepsilon s)U^{\varepsilon}_{\mathrm{dg}}(s,0)\Big)\\
&\quad + \mathcal{O}(\varepsilon),
\end{aligned}$$

which implies

$$\|U^{\varepsilon}(\varepsilon^{-1}\tau, 0) - U^{\varepsilon}_{\mathrm{dg}}(\varepsilon^{-1}\tau, 0)\| \leq c_0(1 + \tau)\varepsilon. \tag{16.15}$$

The adiabatic theorem (16.8) follows from

$$\begin{aligned}
\|(1 - P(\tau))U^{\varepsilon}(\varepsilon^{-1}\tau, 0)P(0)\psi\| &= \|(1 - P(\tau))U^{\varepsilon}_{\mathrm{dg}}(\varepsilon^{-1}\tau, 0)P(0)\psi\| + \mathcal{O}(\varepsilon)\\
&= \|(1 - P(\tau))P(\tau)U^{\varepsilon}_{\mathrm{dg}}(\varepsilon^{-1}\tau, 0)\psi\| + \mathcal{O}(\varepsilon)\\
&= \mathcal{O}(\varepsilon),
\end{aligned} \tag{16.16}$$

where (16.9) has been used.

It remains to see whether the commutator equation (16.13) has a solution. Because of the spectral gap we may set

$$X = P\dot{P}(1-P)(H-E)^{-1} + (H-E)^{-1}(1-P)\dot{P}P \tag{16.17}$$

and verify (16.13) directly. In particular, $\|X(t)\| \leq g^{-1}\|\dot{P}(t)\|$, with $g$ the width of the gap and $\|\dot{X}(t)\| \leq 3g^{-2}\|\dot{H}(t)\|\|\dot{P}(t)\|$.

While undoubtedly correct the estimate (16.8) does not specify the origin of the error. As we will explain below the order $\varepsilon$ is not due to dispersion into all of $\mathcal{H}$. Rather the true solution $\psi(t)$ is slightly tilted out of the subspace $P(\varepsilon t)\mathcal{H}$. If this effect is properly taken into account, the error in (16.8) can be made smaller than any given power $\varepsilon^n$ at the expense of adjusting the projection $P(t)$ to the slightly tilted projection $P^\varepsilon(t)$. The second missing aspect is more of a computational nature. Since $(1-P^\varepsilon(t))\mathcal{H}$ is in essence decoupled from the relevant subspace, one would like to have an, in our case time-dependent, effective Hamiltonian governing the solution in the subspace $P^\varepsilon(t)\mathcal{H}$, at least approximately. We will return to this point below.

## 16.2 The space-adiabatic limit

With these preparations done we return to the Pauli–Fierz model with the slowly varying electrostatic potential $V(\varepsilon x) = e\phi_{\mathrm{ex}}(\varepsilon x)$,

$$H = \frac{1}{2m}(p - P_{\mathrm{f}} - eA_\varphi)^2 + H_{\mathrm{f}} + V(\varepsilon x) = H_0 + V(\varepsilon x). \tag{16.18}$$

The case of a slowly varying vector potential will be discussed in section 16.6. Spin is omitted only for notational simplicity. $H$ acts on $L^2(\mathbb{R}^3, \mathrm{d}^3x) \otimes \mathcal{F}$. For the wave functions it is convenient to use the momentum representation $\psi(k, \underline{k})$, also for the electron, with the shorthand $\underline{k} = (k_1, \lambda_1, \ldots, k_n, \lambda_n)$, $n$ arbitrary, $\psi(k, \emptyset) = \psi(k) \otimes \Omega$. $H_0$ then has the direct integral decomposition

$$H_0 = \int^{\oplus} \mathrm{d}^3k\, H_0(k). \tag{16.19}$$

We assume a small photon mass and the validity of claim 15.4. Then, for every $k$, $|k| < p_{\mathrm{c}}$, $H_0(k)$ has a unique ground state $\psi_{\mathrm{g}}$, $H_0(k)\psi_{\mathrm{g}}(k, \underline{k}) = E(k)\psi_{\mathrm{g}}(k, \underline{k})$. Since in the momentum representation $H_0(k)$ is a real operator, the phase of $\psi_{\mathrm{g}}(k)$ can be chosen such that the wave function is real. In particular, using $\langle\psi_{\mathrm{g}}(k), \psi_{\mathrm{g}}(k)\rangle_{\mathcal{F}} = 1$, this implies $\langle\psi_{\mathrm{g}}(k), \nabla_k\psi_{\mathrm{g}}(k)\rangle_{\mathcal{F}} = 0$. $E(k)$ is separated by a finite gap from the continuum edge $E_{\mathrm{c}}(k)$. Since our aim is to demonstrate the basic principle, we deliberately ignore the fact that the ground state band exists only up to $p_{\mathrm{c}}$ and continue as if $p_{\mathrm{c}} = \infty$. At the cost of a suitable restriction on the initial state, the assumption $p_c = \infty$ can be avoided. We refer to the Notes at the end of the chapter for further explanations.

The ground state band is the subspace of wave functions of the form $\widehat{f}(k)\psi_g(k, \underline{k})$, and the corresponding projection is denoted by $P_g$. $P_g\mathcal{H}$ is invariant under $e^{-iH_0 t}$, $[H_0, P_g] = 0$, and

$$(e^{-iH_0 t}\widehat{f}\psi_g)(k, \underline{k}) = \big(e^{-iE(k)t}\widehat{f}(k)\big)\psi_g(k, \underline{k}). \tag{16.20}$$

Thus wave functions in the ground state band propagate according to a free quantum evolution with the effective energy–momentum relation $E(k)$.

If the slowly varying potential is turned on, the subspace $P_g\mathcal{H}$ is no longer invariant. The in-band dynamics is modified and there are transitions to excited states. For times which are not too long their effect remains negligible and one expects that

$$(e^{-iHt}\widehat{f}\psi_g)(k, \underline{k}) = \big(e^{-iH_{\mathrm{eff}}t}\widehat{f}(k)\big)\psi_g(k, \underline{k}) + \mathcal{O}(\varepsilon), \quad 0 \le t \le \varepsilon^{-1}\tau, \tag{16.21}$$

where, as for the time-adiabatic theorem, the time scale is determined by the condition that the electron should feel the presence of the potential $V$. The effective one-particle Hamiltonian, $H_{\mathrm{eff}}$, is defined through the *Peierls substitution*

$$H_{\mathrm{eff}} = E(p) + V(\varepsilon x). \tag{16.22}$$

Coupling to the Maxwell field renormalizes the kinetic energy of the quantum particle. In particular, for small velocities we have

$$H_{\mathrm{eff}} = \frac{1}{2m_{\mathrm{eff}}}p^2 + V(\varepsilon x). \tag{16.23}$$

The mass is renormalized, but the coupling to the electrostatic potential is still given by the bare charge $e$.

Let us now argue with some care that the Peierls substitution gives the correct time evolution in the ground state band. The Hamiltonian is the one specified in (16.18) and the relevant subspace is the ground state band $P_g\mathcal{H}$. In particular, initially $P_g\psi(0) = \psi(0)$. By construction, $[H_0, P_g] = 0$ and one has to understand the transitions between $P_g\mathcal{H}$ and $(1 - P_g)\mathcal{H} = Q_g\mathcal{H}$ induced by $V(\varepsilon x)$. For this purpose we decompose into a diagonal and an off-diagonal piece as

$$\begin{aligned} &V = V_{\mathrm{dg}} + V_{\mathrm{od}}, \\ &V_{\mathrm{dg}} = P_g V P_g + Q_g V Q_g, \quad V_{\mathrm{od}} = P_g V Q_g + Q_g V P_g. \end{aligned} \tag{16.24}$$

It should be recalled that the time evolution must be controlled over the time span $\varepsilon^{-1}\tau$, $\tau = \mathcal{O}(1)$. Thus only terms of order $\varepsilon^2$ can be ignored safely.

We consider first $P_g V P_g$, which in the ground state band acts as

$$(P_g V P_g \psi)(k, \underline{k}) = \int d^3k' \widehat{V}(k') \widehat{f}(k - \varepsilon k') \langle \psi_g(k), \psi_g(k - \varepsilon k') \rangle_{\mathcal{F}} \psi_g(k, \underline{k}) \tag{16.25}$$

with $\widehat{f}(k) = \langle \psi_g(k), \psi(k) \rangle_{\mathcal{F}}$. The Peierls substitution amounts to $V(\varepsilon P_g x P_g)$, since

$$P_g x P_g \psi(k, \underline{k}) = i \nabla_k \widehat{f}(k) \psi_g(k, \underline{k}) + i \widehat{f}(k) \langle \psi_g(k), \nabla_k \psi_g(k) \rangle_{\mathcal{F}} \psi_g(k, \underline{k}) \tag{16.26}$$

with the second term vanishing by the argument given above. The difference is estimated as

$$\begin{aligned} &\big(P_g V(\varepsilon x) P_g - V(\varepsilon P_g x P_g)\big) \psi(k, \underline{k}) \\ &= \int d^3k' \widehat{V}(k') \widehat{f}(k - \varepsilon k') \big( \langle \psi_g(k), \psi_g(k - \varepsilon k') \rangle_{\mathcal{F}} - 1 \big) \psi_g(k, \underline{k}). \end{aligned} \tag{16.27}$$

In the Taylor expansion, the first order vanishes, since $\langle \psi_g(k), \nabla_k \psi_g(k) \rangle_{\mathcal{F}} = 0$ as before, and the error is $\mathcal{O}(\varepsilon^2)$. Thus we are left with showing that $V_{od}$ acts as a small perturbation only.

Since $H_0(k) = \frac{1}{2m}(k - P_f - eA_\varphi)^2 + H_f$, one has $\nabla_k H_0(k) = \frac{1}{m}(k - P_f - eA_\varphi)$ and, with $P(k)$ denoting the projection onto $\psi_g(k)$, then $P_g = \int^\oplus d^3k P(k)$, $Q(k) = 1 - P(k)$, and $Q_g = 1 - P_g$. If clear from the context, the variable "$\underline{k}$" will be dropped. With these conventions

$$\begin{aligned} V(\varepsilon x) P_g \psi(k) &= \int d^3k' \widehat{V}(k') P(k - \varepsilon k') \psi(k - \varepsilon k') \\ &= \int d^3k' \widehat{V}(k') P(k) \psi(k - \varepsilon k') \\ &\quad - \varepsilon \int d^3k' \widehat{V}(k') k' \cdot \nabla_k P(k) \psi(k - \varepsilon k') + \mathcal{O}(\varepsilon^2). \end{aligned} \tag{16.28}$$

By first-order perturbation theory

$$\nabla_k P(k) = -Q(k)(H_0(k) - E(k))^{-1} \nabla_k H_0(k) P(k) + \text{h.c.}, \tag{16.29}$$

h.c. denoting the Hermitian conjugate. Therefore

$$Q_g V(\varepsilon x) P_g = -i\varepsilon Q_g \nabla P_g \cdot F(\varepsilon x) + \mathcal{O}(\varepsilon^2) \tag{16.30}$$

with the shorthand $\nabla P_g = \int^\oplus d^3k \nabla_k P(k)$ and the force $F(x) = -\nabla V(x)$.

The approximate time evolution is generated by

$$H_{dg} = H_0 + V_{dg}, \quad U_{dg}(t) = e^{-iH_{dg}t}, \tag{16.31}$$

and our goal is to compare it with the full time evolution $e^{-iHt} = U(t)$ over times of order $\varepsilon^{-1}$, i.e. to estimate the difference

$$U(\varepsilon^{-1}t) - U_{dg}(\varepsilon^{-1}t) = -i\int_0^{t/\varepsilon} ds\, U(\varepsilon^{-1}t - s)V_{od}U_{dg}(s) \tag{16.32}$$

with $t = \mathcal{O}(1)$.

At this point we have arrived at a structure very similar to the time-adiabatic difference (16.14). $V_{od} = \mathcal{O}(\varepsilon)$ and time averaging must be used. As before, the trick is to write $V_{od}$ as a time derivate, i.e. as a commutator with $H_0$, up to unavoidable errors of order $\varepsilon^2$. We set

$$B(k) = -Q(k)(H_0(k) - E(k))^{-2}\nabla_k H_0(k)P(k). \tag{16.33}$$

Then

$$Q(k)\nabla_k P(k) = -[H_0(k), B(k)]. \tag{16.34}$$

With the shorthand $B = \int^{\oplus} d^3k B(k)$ one has

$$Q_g\nabla P_g \cdot F = [H_0, B]\cdot F = [H_0, B\cdot F] - B\cdot[H_0, F] = [H_0, B\cdot F] + \mathcal{O}(\varepsilon), \tag{16.35}$$

since $[H_0, F] = \frac{1}{m}(p - P_f - eA_\varphi)\cdot[p, F] +$ h.c. and $[p, F] = -i\varepsilon\nabla_x F(\varepsilon x)$. It remains to substitute $H_{dg}$ for $H_0$. One has $[V_{dg}, B] = Q_g[V, B]P_g$. Since $B = \int^{\oplus} d^3k B(k)$ and $V = V(i\varepsilon\nabla_k)$, the commutator is of order $\varepsilon$, hence

$$Q_g\nabla P_g \cdot F = [H_{dg}, B\cdot F] + \mathcal{O}(\varepsilon). \tag{16.36}$$

On inserting in (16.32), we get

$$\begin{aligned}
U(\varepsilon^{-1}t) - U_{dg}(\varepsilon^{-1}t) &= -\varepsilon\int_0^{t/\varepsilon} ds\, U(\varepsilon^{-1}t - s)[H_{dg}, B\cdot F + F\cdot B^*]U_{dg}(s) \\
&\quad + \mathcal{O}(\varepsilon) \\
&= -\varepsilon\int_0^{t/\varepsilon} ds\, U(\varepsilon^{-1}t - s)U_{dg}(s)U_{dg}(-s)[H_{dg}, B\cdot F + F\cdot B^*]U_{dg}(s) + \mathcal{O}(\varepsilon) \\
&= i\varepsilon\int_0^{t/\varepsilon} ds\, U(\varepsilon^{-1}t - s)U_{dg}(s)\frac{d}{ds}(B\cdot F + F\cdot B^*)(s) + \mathcal{O}(\varepsilon) \\
&= i\varepsilon(B\cdot F + F\cdot B^*)U_{dg}(\varepsilon^{-1}t) - i\varepsilon U(\varepsilon^{-1}t)(B\cdot F + F\cdot B^*) \\
&\quad - i\varepsilon\int_0^{t/\varepsilon} ds\Big(\frac{d}{ds}U(\varepsilon^{-1}t - s)U_{dg}(s)\Big)(B\cdot F + F\cdot B^*)(s) + \mathcal{O}(\varepsilon) \\
&= \mathcal{O}(\varepsilon),
\end{aligned} \tag{16.37}$$

since $\frac{\mathrm{d}}{\mathrm{d}s}U(-s)U_{\mathrm{dg}}(s) = \mathrm{i}U(-s)V_{\mathrm{od}}U_{\mathrm{dg}}(s) = \mathcal{O}(\varepsilon)$ by (16.30). As in the time-adiabatic setting the leakage out of the ground state subspace $P_{\mathrm{g}}\mathcal{H}$ is $\mathcal{O}(\varepsilon)$ for times of order $\varepsilon^{-1}$. In addition we have identified the effective Hamiltonian (16.22) which approximately governs the time evolution inside $P_{\mathrm{g}}\mathcal{H}$.

## 16.3 Matrix-valued symbols

If in (16.18) a slowly varying vector potential is added through minimal coupling, then even on a formal level the argument of the previous section breaks down. The reason is that the ground state subspace $P_{\mathrm{g}}\mathcal{H}$ is no longer even approximately invariant under the time evolution. There is another subspace to take its role, but it has to be computed rather than guessed. We immediately consider the general case (16.1) and switch to the macroscopic space scale through the substitution $x$ for $\varepsilon x$. Then the Hamiltonian under study is

$$H = \frac{1}{2m}\big( - \mathrm{i}\varepsilon\nabla_x - P_{\mathrm{f}} - eA_\varphi - eA_{\mathrm{ex}}(x)\big)^2 - \frac{e}{2m}\sigma\cdot(B_\varphi + \varepsilon B_{\mathrm{ex}}(x)) \\ + e\phi_{\mathrm{ex}}(x) + H_{\mathrm{f}}. \tag{16.38}$$

As before, $-\mathrm{i}\nabla_x$ refers to the total momentum, $A_\varphi = A_\varphi(0)$, $B_\varphi = B_\varphi(0)$.

The first step is to mold (16.38) into the canonical space-adiabatic form. For this purpose we have to distinguish between the classical phase space variable $(q, p)$ and the corresponding operators, which exclusively for the purpose of sections 16.3–16.5 are denoted by $\widehat{q} = x$, $\widehat{p} = -\mathrm{i}\varepsilon\nabla_x$. To the Hamiltonian (16.38) in the obvious way we associate the operator-valued function (= symbol)

$$\begin{aligned} H(q, p) &= H_0(q, p) + \varepsilon H_1(q, p),\\ H_0(q, p) &= \frac{1}{2m}\big(p - P_{\mathrm{f}} - eA_\varphi - eA_{\mathrm{ex}}(q)\big)^2 - \frac{e}{2m}\sigma\cdot B_\varphi + e\phi_{\mathrm{ex}}(q) + H_{\mathrm{f}},\\ H_1(q, p) &= -\frac{e}{2m}\sigma\cdot B_{\mathrm{ex}}(q). \end{aligned} \tag{16.39}$$

For fixed $(q, p)$, $H(q, p)$ acts as an operator on $\mathbb{C}^2 \otimes \mathcal{F}$, $\mathbb{C}^2$ standing for the spin degrees of freedom. $H_0$ is called the leading symbol and $H_1$ the subleading symbol for $H$ because of the extra prefactor of $\varepsilon$ in the first line of (16.39). To a symbol one associates an operator through the Weyl quantization, which can be thought of as a specific prescription for ordering $x$ and $-\mathrm{i}\varepsilon\nabla_x$. To be general, let $A(q, p)$ be an operator-valued function with Fourier transform $\widetilde{A}(\eta, \xi)$,

$$A(q, p) = (2\pi)^{-3}\int \mathrm{d}^3\eta\mathrm{d}^3\xi\,\widetilde{A}(\eta, \xi)\mathrm{e}^{\mathrm{i}(\eta\cdot q + \xi\cdot p)}. \tag{16.40}$$

The Weyl quantization of $A$ is then simply

$$\mathcal{W}_\varepsilon(A) = (2\pi)^{-3} \int \mathrm{d}^3\eta \mathrm{d}^3\xi \widetilde{A}(\eta, \xi) \mathrm{e}^{\mathrm{i}(\eta\cdot\widehat{q}+\xi\cdot\widehat{p})}. \tag{16.41}$$

$A(q, p)$ is an operator-valued function and $\mathcal{W}_\varepsilon(A)$ is an operator on the large Hilbert space $\mathcal{H} = L^2(\mathbb{R}^3) \otimes \mathbb{C}^2 \otimes \mathcal{F}$. We will also use the notation

$$\mathcal{W}_\varepsilon(A) = A(\widehat{q}, \widehat{p}) = \widehat{A} \tag{16.42}$$

as a shorthand. Using the inverse Fourier transform in (16.40), $\mathcal{W}_\varepsilon(A)$ can be written in the form of an integral operator as

$$\mathcal{W}_\varepsilon(A)\psi(x) = (2\pi)^{-3} \int \mathrm{d}^3\xi \mathrm{d}^3 y A(\tfrac{1}{2}(x+y), \varepsilon\xi) \mathrm{e}^{\mathrm{i}\xi\cdot(x-y)}\psi(y). \tag{16.43}$$

Here $A$ acts on $\psi(x)$ which is a $\mathbb{C}^2 \otimes \mathcal{F}$-valued wave function, $\psi \in L^2(\mathbb{R}^3, \mathbb{C}^2 \otimes \mathcal{F}) = L^2(\mathbb{R}^3) \otimes (\mathbb{C}^2 \otimes \mathcal{F}) = \mathcal{H}$, and $\mathcal{W}_\varepsilon(A)$ is an operator acting on $\mathcal{H}$. Note that $f(\widehat{q}) = f(x)$, $f(\widehat{p}) = f(-\mathrm{i}\varepsilon\nabla_x)$ as operators. Also $\mathcal{W}_\varepsilon(A)$ being Hermitian is equivalent to $A(q, p) = A(q, p)^*$ for all $(q, p)$. For the Weyl quantization of $H(q, p)$ from (16.39) one obtains simply

$$H(\widehat{q}, \widehat{p}) = H, \tag{16.44}$$

as it should be. Thus the adiabatic evolution problem associated with (16.38) can be written as

$$\mathrm{i}\varepsilon\frac{\partial}{\partial t}\psi(x, t) = H(x, -\mathrm{i}\varepsilon\nabla_x)\psi(x, t) \tag{16.45}$$

with the Weyl rule for the ordering of operators. Consistent with the macroscopic space scale we switched also to macroscopic times through the substitution of $t$ for $\varepsilon t$. Equation (16.45) looks like a standard Schrödinger equation, apart from the fact that $\psi(x, t)$ takes values in $\mathbb{C}^2 \otimes \mathcal{F}$ and $H(q, p)$ acts as an operator on $\mathbb{C}^2 \otimes \mathcal{F}$.

$H_0(q, p)$ has a subspace of lowest energy with the corresponding projection denoted by $P(q, p)$. Deliberately ignoring $p_\mathrm{c} < \infty$, from section 15.3 we know already that $\mathrm{tr}[P(q, p)] = 2$ and

$$H_0(q, p)P(q, p) = E(q, p)P(q, p) \tag{16.46}$$

with the eigenvalue

$$E(q, p) = E(p - eA_{\mathrm{ex}}(q)) + e\phi_{\mathrm{ex}}(q). \tag{16.47}$$

One would expect that the Peierls substitution $E(\widehat{q}, \widehat{p})$ somehow plays the role of the effective one-particle Hamiltonian. Note that this would leave spin precession still hidden and, in fact, it will appear as the $\varepsilon$-order correction to the Peierls substitution $E(\widehat{q}, \widehat{p})$.

At this stage, as for the time-adiabatic theorem, it is convenient to abstract from the specific origin of the space-adiabatic evolution (16.45). Thereby the general structure of space-adiabatic problems becomes visible with the bonus of wide applicability. For simplicity $\mathbb{C}^2 \otimes \mathcal{F}$ is replaced by $\mathbb{C}^n$ with $n$ arbitrary. In fact, a finite-dimensional internal Hilbert space is not essential and only allows us to remain in familiar territory. We record that the Hamiltonian $H(q, p) = H_0(q, p) + \varepsilon H_1(q, p)$ is a matrix-valued function, assumed to be smooth in $q$, $p$. There is a relevant subspace of physical interest with energy band $E(q, p)$ of constant multiplicity $\ell$. This means that $H_0(q, p)$ has the eigenprojection $P(q, p)$, $[H_0(q, p), P(q, p)] = 0$, with $\mathrm{tr}[P(q, p)] = \ell$, $1 \le \ell < n$, such that

$$H_0(q, p)P(q, p) = E(q, p)P(q, p). \tag{16.48}$$

Most importantly, $H_0$ is assumed to have a spectral gap in the sense that

$$|E(q, p) - E_j(q, p)| \ge g > 0 \tag{16.49}$$

for all $(q, p)$ and all other eigenvalues $E_j(q, p)$ of $H_0(q, p)$. As before, the space-adiabatic evolution is governed by

$$\mathrm{i}\varepsilon \frac{\partial}{\partial t}\psi(x, t) = H(\widehat{q}, \widehat{p})\psi(x, t) \tag{16.50}$$

with $\psi(x, t)$ an $n$-spinor, i.e. the Hilbert space for the Schrödinger equation (16.50) is $L^2(\mathbb{R}^3) \otimes \mathbb{C}^n = \mathcal{H}$. Note that, if in (16.50) $H(\widehat{q}, \widehat{p})$ is replaced by $H(t)$, then (16.50) turns into its time-adiabatic cousin (16.7) where the role of the relevant projection $P(q, p)$ is taken over by $P(t)$.

The analysis of (16.50) will be carried out in such a way as to make use only of (16.48) and (16.49) with no further assumptions at all on the spectrum of $H_0(q, p)$ in the subspace orthogonal to $P(q, p)\mathbb{C}^n$. For this reason we are confident that the final result will apply also to the Pauli–Fierz Hamiltonian.

With the more general perspective gained, one can understand why the case $A_{\mathrm{ex}} = 0$ can be handled by more elementary means. In that case $H_0(q, p) = \frac{1}{2m}(p - P_{\mathrm{f}} - eA_{\varphi})^2 + e\phi_{\mathrm{ex}}(q)$. Thus $P(q, p)$ depends only on $p$ and $P(\widehat{q}, \widehat{p}) = P_{\mathrm{g}}$, the projection onto the ground state subspace. This suggests that also in the general case $P(\widehat{q}, \widehat{p})\mathcal{H}$ is the adiabatically decoupled subspace. Unfortunately $P(\widehat{q}, \widehat{p})^2 \neq P(\widehat{q}, \widehat{p})$, in general, although $P(q, p)^2 = P(q, p)$. On the other hand, as will be shown, $P(\widehat{q}, \widehat{p})(1 - P(\widehat{q}, \widehat{p})) = \mathcal{O}(\varepsilon)$. Since $P(\widehat{q}, \widehat{p})$ is Hermitian, its spectrum is of order $\varepsilon$ concentrated near 0 and 1. Thus, at the expense of an error of order $\varepsilon$, we can associate to $P(\widehat{q}, \widehat{p})$ a true projection operator $\widetilde{P}(\widehat{q}, \widehat{p})$, and $\widetilde{P}(\widehat{q}, \widehat{p})\mathcal{H}$ is the adiabatically protected subspace in lowest-order approximation.

From the example of $P(\widehat{q}, \widehat{p})$ just discussed, it is clear that for a study of the Schrödinger equation (16.50) in the limit of small $\varepsilon$ one has to understand the relationship between the multiplication of symbols and the multiplication of their Weyl quantization, which is taken up next. Let $A$, $B$ be two matrix-valued functions. One defines their Moyal product $A\#B$ implicitly through the condition

$$\mathcal{W}_\varepsilon(A)\mathcal{W}_\varepsilon(B) = \mathcal{W}_\varepsilon(A\#B). \tag{16.51}$$

The Moyal product is best grasped in the case where the symbols are given as formal power series,

$$A(q, p) = \sum_{j\geq 0} \varepsilon^j A_j(q, p), \quad B(q, p) = \sum_{j\geq 0} \varepsilon^j B_j(q, p), \tag{16.52}$$

where the expansion coefficients $A_j$, $B_j$ do not depend on $\varepsilon$. The equality is understood as $|A - \sum_{j\geq 0}^{n-1} \varepsilon^j A_j| \leq c_n \varepsilon^n$ with constants $c_n$ possibly growing so fast in $n$ that the partial sums in (16.52) do not converge. Then $A\#B$ also has a formal power series, which is written as

$$A\#B = \sum_{j\geq 0} \varepsilon^j (A\#B)_j. \tag{16.53}$$

Equating power by power in (16.51) one finds

$$(A\#B)_j(q, p) = \sum_{|\alpha|+|\beta|+l+m=j} (2i)^{-(|\alpha|+|\beta|)} \frac{(-1)^{|\beta|}}{|\alpha|!|\beta|!} \partial_q^\alpha \partial_p^\beta A_l(q, p) \partial_p^\alpha \partial_q^\beta B_m(q, p), \tag{16.54}$$

where it is understood that $j, l, m \in \mathbb{N}$ and $\alpha, \beta$ are multi-indices, $\alpha, \beta \in \mathbb{N}^3$. To lowest order

$$(A\#B)_0 = A_0 B_0, \quad (A\#B)_1 = A_0 B_1 + A_1 B_0 - \frac{\mathrm{i}}{2}\{A_0, B_0\}. \tag{16.55}$$

We introduced here the Poisson bracket $\{\cdot, \cdot\}$ for matrix-valued functions. It is defined by

$$\{A, B\} = \nabla_p A \cdot \nabla_q B - \nabla_q A \cdot \nabla_p B, \tag{16.56}$$

the dot referring to the scalar product of the two gradients. Thus even if the formal power series for $A$, $B$ consists only of the leading term, $A = A_0$, $B = B_0$, as is the case for $P(q, p)$, their Moyal product is a formal power series starting with

$$A\#B = AB - \varepsilon\frac{\mathrm{i}}{2}\{A, B\} + \mathcal{O}(\varepsilon^2) \tag{16.57}$$

and, by definition, the lowest-order product becomes

$$\mathcal{W}_\varepsilon(A)\mathcal{W}_\varepsilon(B) = \mathcal{W}_\varepsilon(AB - \varepsilon\frac{\mathrm{i}}{2}\{A, B\}) + \mathcal{O}(\varepsilon^2). \tag{16.58}$$

Note that in (16.56) the order of matrices must be respected. In general, it is *not* true that $\{A, A\} = 0$, or $\{A, B\} = -\{B, A\}$, as one is used to from the standard calculus of Poisson brackets.

In the sequel, very roughly the idea is to use (16.51) as a link between functions of operators, like the time-evolved position operator $\widehat{q}(t) = \mathrm{e}^{\mathrm{i}Ht/\varepsilon}\widehat{q}\mathrm{e}^{-\mathrm{i}Ht/\varepsilon}$, and matrix-valued symbols. In particular, one can regard the matrix-valued function $P(q, p)$ as the lowest-order symbol for the true Hilbert space projection onto the adiabatically decoupled relevant subspace.

## 16.4 Adiabatic decoupling, effective Hamiltonians

As noticed already, in general $P(\widehat{q}, \widehat{p})$ is not a projection, due to errors of order $\varepsilon$. This suggests to successively correct $P(q, p)$ with the goal in Weyl quantization to get a projection up to precision $\varepsilon^n$, $n$ arbitrary, a situation denoted by the symbol $\mathcal{O}(\varepsilon^\infty)$. We make the ansatz

$$\pi(q, p) = \sum_{j\geq 0} \varepsilon^j \pi_j(q, p), \quad \pi_0(q, p) = P(q, p) \tag{16.59}$$

and recall that in general

$$H(q, p) = \sum_{j\geq 0} \varepsilon^j H_j(q, p), \tag{16.60}$$

where in our specific application $H_j = 0$ for $j \geq 2$. The Weyl quantization for $\pi$ should be a projection and commute with $H(\widehat{q}, \widehat{p})$ up to errors $\mathcal{O}(\varepsilon^\infty)$. $\pi$ has then to satisfy the conditions

$$\pi^* = \pi, \quad \pi \# \pi = \pi, \quad \pi \# H = H \# \pi. \tag{16.61}$$

Through an iterative procedure it can be shown that the symbol $\pi$ is in fact uniquely determined by (16.61). By construction $\mathcal{W}_\varepsilon(\pi)^2 = \mathcal{W}_\varepsilon(\pi) + \mathcal{O}(\varepsilon^\infty)$ and there is a projection operator $\Pi$ on $\mathcal{H}$ naturally associated to $\mathcal{W}_\varepsilon(\pi)$. If we assume the initial wave function $\psi$ to lie in $\Pi\mathcal{H}$, $\Pi\psi = \psi$, then for the true solution $\psi(t) = \mathrm{e}^{-\mathrm{i}Ht/\varepsilon}\psi$ one has

$$(1 - \Pi)\psi(t) = \mathcal{O}(\varepsilon^\infty). \tag{16.62}$$

For this reason $\Pi\mathcal{H}$ is called an almost invariant subspace, associated to the relevant projection $P(q, q)$. On the adiabatic scale transitions out of $\Pi\mathcal{H}$ are exponentially suppressed as $e^{-(1/\varepsilon)}$ and the dynamics on $\Pi\mathcal{H}$ is governed by the diagonal Hamiltonian $H_{\mathrm{dg}} = \Pi\widehat{H}\Pi$.

Equation (16.62) solves the adiabatic problem only in principle. To have a workable scheme it is required to have a basis in $\Pi\mathcal{H}$ which is in some sense naturally adapted to the slow degrees of freedom and in which $H_{\mathrm{dg}}$ can be computed perturbatively. Of course, the hope is that low-order perturbation will suffice. For this purpose we pick a fixed $(q, p)$-independent basis $|\chi_\alpha\rangle$, $\alpha = 1, \ldots, n$, in $\mathbb{C}^n$ and define the $\ell$-dimensional *reference projection*

$$\pi_{\mathrm{r}} = \sum_{\alpha=1}^{\ell} |\chi_\alpha\rangle\langle\chi_\alpha|. \tag{16.63}$$

Since $|\chi_\alpha\rangle$ does not depend on $(q, p)$, $1 \otimes \pi_{\mathrm{r}} = \widehat{\pi}_{\mathrm{r}} = \mathcal{W}_\varepsilon(\pi_{\mathrm{r}})$ is a projection and its range defines the *reference Hilbert space* $L^2(\mathbb{R}^3) \otimes \pi_{\mathrm{r}}\mathbb{C}^n = \mathcal{H}_{\mathrm{r}}$ as a subspace of $\mathcal{H}$. Of course, at this stage the reference subspace is fairly arbitrary and a convenient choice must be made in concrete applications. The projection $P(q, p)$ is spanned by the eigenvectors $\psi_\alpha(q, p)$, $\alpha = 1, \ldots, \ell$, of $H_0(q, p)$, $\langle\psi_\alpha(q, p), \psi_\beta(q, p)\rangle_{\mathbb{C}^n} = \delta_{\alpha\beta}$. The unitary map from $P(q, p)\mathbb{C}^n$ to the reference subspace is then

$$u_0(q, p) = \sum_{\alpha=1}^{\ell} |\chi_\alpha\rangle\langle\psi_\alpha(q, p)|. \tag{16.64}$$

If $u_0$ were completed to a unitary operator $\widetilde{u}_0$ on $\mathbb{C}^n$, then for every $q, p$ the $n \times n$ matrix $\widetilde{u}_0 H_0 \widetilde{u}_0^*$ is block diagonal, with block sizes $\ell$ and $n - \ell$, and has in the $\ell \times \ell$ left upper block only the diagonal entries $E(q, p)$.

As in the case of the projection $P(q, p)$, $\mathcal{W}_\varepsilon(u_0)$ is in general not unitary with an error of order $\varepsilon$. Thus we iteratively correct so as to obtain a proper unitary operator from $\Pi\mathcal{H}$ to the reference subspace $\mathcal{H}_{\mathrm{r}}$. The ansatz is

$$u(q, p) = \sum_{j \geq 0} \varepsilon^j u_j(q, p), \tag{16.65}$$

with $u_0$ as in (16.64). Unitarity and transformation of $\pi$ to $\pi_{\mathrm{r}}$ translates into

$$u^* \# u = 1, \quad u \# u^* = 1, \quad u \# \pi \# u^* = \pi_{\mathrm{r}}. \tag{16.66}$$

One can show that such a symbol $u$ exists. Since $u_0$ is already not unique, neither is $u$. As with $\pi(q, p)$, one associates with $u$ a unitary operator $U : \Pi\mathcal{H} \to \mathcal{H}_{\mathrm{r}}$. On $\mathcal{H}_{\mathrm{r}}$ the motion is governed by $U\Pi\widehat{H}\Pi U^*$ and it agrees with the true solution up to $\mathcal{O}(\varepsilon^\infty)$. $U\Pi\widehat{H}\Pi U^*$ has a symbol determined through

$$h = u \# H \# u^*. \tag{16.67}$$

We call $h$ the *effective Hamiltonian*, associated to the almost invariant subspace $\Pi\mathcal{H}$. The crux of the construction is that $h$ can be represented by a formal power series,

$$h = \sum_{j\geq 0} \varepsilon^j h_j \tag{16.68}$$

and the effective Hamiltonian is successively approximated through the Weyl quantization

$$\mathcal{W}_\varepsilon(h) = \mathcal{W}_\varepsilon(\pi_\mathrm{r} h_0 \pi_\mathrm{r}) + \varepsilon \mathcal{W}_\varepsilon(\pi_\mathrm{r} h_1 \pi_\mathrm{r}) + \cdots . \tag{16.69}$$

Let us work out the two lowest orders. Clearly

$$\pi_\mathrm{r} h_0 \pi_\mathrm{r} = \pi_\mathrm{r} u_0^* H_0 u_0 \pi_\mathrm{r} = E(q, p)\pi_\mathrm{r}. \tag{16.70}$$

Its Weyl quantization is $E(\widehat{q}, \widehat{p})\pi_\mathrm{r}$ which is the anticipated Peierls substitution. In spinor space $E(q, p)\pi_\mathrm{r}$ is diagonal, see (16.63), and there is no internal motion at this order yet. For $h_1$ it is easier to rewrite (16.67) as $H\#u = u\#h$ and therefore $(H_0 + \varepsilon H_1)\#(u_0 + \varepsilon u_1) = (u_0 + \varepsilon u_1)\#(h_0 + \varepsilon h_1)$. Using (16.57) one thus obtains

$$h_1 = \Big(u_1 H_0 + u_0 H_1 - h_0 u_1 - \frac{\mathrm{i}}{2}\{u_0, H_0\} + \frac{\mathrm{i}}{2}\{h_0, u_0\}\Big) u_0^*. \tag{16.71}$$

Projecting onto $\pi_\mathrm{r}$, the terms $H_0 u_1$ and $u_1 h_0$ cancel and $h_1$ simplifies to

$$\pi_\mathrm{r} h_1 \pi_\mathrm{r} = \pi_\mathrm{r}\Big(u_0 H_1 u_0^* - \frac{\mathrm{i}}{2}\{u_0, H_0\}u_0^* + \frac{\mathrm{i}}{2}\{E, u_0\}u_0^*\Big)\pi_\mathrm{r}. \tag{16.72}$$

$u_0$ is inserted from Eq. (16.64). In the basis of the reference Hilbert space one then obtains to first order

$$\begin{aligned}\langle \chi_\alpha, (h_0 + \varepsilon h_1)\chi_\beta\rangle_{\mathbb{C}^n} = E\delta_{\alpha\beta} + \varepsilon\langle\psi_\alpha, H_1\psi_\beta\rangle_{\mathbb{C}^n} - \varepsilon\frac{\mathrm{i}}{2}\langle\psi_\alpha, \{H_0 + E, \psi_\beta\}\rangle_{\mathbb{C}^n} \\ + \mathcal{O}(\varepsilon^2),\end{aligned} \tag{16.73}$$

where $\alpha, \beta = 1, \ldots, \ell$, and where the Poisson bracket is understood as

$$\{H_0, \psi_\alpha\} = \nabla_p H_0 \cdot \nabla_q \psi_\alpha - \nabla_q H_0 \cdot \nabla_p \psi_\alpha \tag{16.74}$$

with $H_0$ acting on $\psi_\alpha$ as a matrix. The Weyl quantization of $h_0 + \varepsilon h_1$ is the effective Hamiltonian in $L^2(\mathbb{R}^3) \otimes \mathbb{C}^\ell$ to that order.

In principle, our scheme can be pushed up to arbitrary order. Formulas for $h_2$ are available, but they are already so involved that $h_3$ is out of reach. Physically the dominant effects are in $h_0$, $h_1$, and to some extent in $h_2$. Further terms will add only a minute correction. Of course, the adiabatic decoupling relies on the gap assumption (16.49). In the case where the energy bands of $H_0(q,p)$ cross, or almost cross, transition between bands become possible and the qualitative picture developed so far breaks down. Away from crossings the description through the effective Hamiltonian is still accurate, but close to nearly avoided crossings new techniques come into play.

The formula (16.73) looks unfamiliar. To get acquainted, a simple but instructive way is to return to the time-adiabatic setting of section 16.1, where $H(t)$ is a time-dependent $n \times n$ matrix and the relevant subspace has a constant multiplicity $\ell$. It is spanned by the instantaneous eigenvectors $\varphi_\alpha(t)$, $H(t)\varphi_\alpha(t) = E(t)\varphi_\alpha(t)$, $\alpha = 1,\dots,\ell$, and the projection onto the relevant subspace is given by $P(t) = \sum_{\alpha=1}^{\ell} |\varphi_\alpha(t)\rangle\langle\varphi_\alpha(t)|$. As before, one needs a reference subspace of dimension $\ell$ with time-*independent* basis $|\chi_\alpha\rangle$, $\alpha = 1,\dots,\ell$. We do not spell out the details of the computation, but state the final result. Including order $\varepsilon$, the unitary $U^\varepsilon(t)^*$ from the reference space $\mathbb{C}^\ell$ into $\mathbb{C}^n = \mathcal{H}_\mathrm{f}$ is given by

$$U^\varepsilon(t)^* = \sum_{\alpha=1}^{\ell}(|\varphi_\alpha(t)\rangle + |\mathrm{i}\varepsilon(H(t)-E(t))^{-1}(1-P(t))\dot{\varphi}_\alpha(t)\rangle)\langle\chi_\alpha| + \mathcal{O}(\varepsilon^2). \tag{16.75}$$

$U^\varepsilon(t)^*$ should be thought of as a kinematical component. It says, for each $t$, how the adiabatically protected subspace lies in $\mathbb{C}^n$. To order 1 the subspace is just $P(t)\mathbb{C}^n$ and (16.75) provides the first-order correction. The dynamical piece provides the information of how the solution vector rotates inside the almost invariant subspace. It is governed by the effective Hamiltonian acting in $\mathbb{C}^\ell$, which to order $\varepsilon^2$ has the form

$$\begin{aligned} h_{\alpha\beta}(t) = {} & \delta_{\alpha\beta}E(t) - \mathrm{i}\varepsilon\langle\varphi_\alpha(t), \dot{\varphi}_\beta(t)\rangle_{\mathbb{C}^n} \\ & + \frac{1}{2}\varepsilon^2\langle\dot{\varphi}_\alpha(t), (H(t)-E(t))^{-1}(1-P(t))\dot{\varphi}_\beta(t)\rangle_{\mathbb{C}^n} + \mathcal{O}(\varepsilon^3), \end{aligned} \tag{16.76}$$

$\alpha, \beta = 1,\dots,\ell$. The second term of $h(t)$ is the Berry phase. The approximate solution to (16.7) is obtained by first solving the time-dependent Schrödinger equation with $h_\mathrm{eff}(t)$ in the reference subspace $\mathbb{C}^\ell$ and then mapping into $\mathcal{H}$ through the unitary (16.75). Thereby the error in (16.8) is improved to order $\varepsilon^2$. In addition we know how the vector $\psi(t)$ rotates inside the relevant subspace. With some effort the precision could be improved to $\mathcal{O}(\varepsilon^3)$. Abstractly, an error $\mathcal{O}(\varepsilon^\infty)$ is guaranteed.

Matrix-valued symbols are a very powerful tool in the analysis of the space-adiabatic limit. But, in the end, one would like to have a result on the Schrödinger equation (16.45). This is always possible because the two frames of description are linked through Weyl quantization. To order $\varepsilon$ the result is

$$e^{-iH(\widehat{q},\widehat{p})t/\varepsilon}\psi = u_0(\widehat{q},\widehat{p})^* e^{-i(h_0(\widehat{q},\widehat{p})+\varepsilon h_1(\widehat{q},\widehat{p}))t/\varepsilon} u_0(\widehat{q},\widehat{p})\psi + (1+|t|)\mathcal{O}(\varepsilon) \tag{16.77}$$

provided the initial wave function lies in the relevant subspace, i.e. $\pi_0(\widehat{q},\widehat{p})\psi = \psi$. On the right, one has the effective dynamics in the reference subspace $L^2(\mathbb{R}^3) \otimes \mathbb{C}^\ell$ as generated by $\mathcal{W}_\varepsilon(h_0 + \varepsilon h_1)$. Then $\mathcal{W}_\varepsilon(u_0)$ which, up to error $\varepsilon$, is unitary turns the effective evolution into the physical Hilbert space $L^2(\mathbb{R}^3) \otimes \mathbb{C}^n$. The error $(1+|t|)\mathcal{O}(\varepsilon)$ comes from the correction of $\pi_0$ to $\pi_0 + \varepsilon\pi_1$, of $u_0$ to $u_0 + \varepsilon u_1$, and from the correction of $h_0 + \varepsilon h_1$ to $h_0 + \varepsilon h_1 + \varepsilon^2 h_2$. Equation (16.77) agrees with our findings for the particular case studied in section 16.2. There $h_0(q,p) = E(p) + V(q)$ and $h_1(p) = -\mathrm{i}\langle\psi_{\mathrm{g}}(p), \nabla_p\psi_{\mathrm{g}}(p)\rangle_{\mathcal{F}} = 0$ by our choice of the phase for $\psi_{\mathrm{g}}(p)$. Once the spin is included, $h_1$ no longer vanishes, see section 16.6.

At the risk of repeating the obvious: expectations of physical observables have the form $\langle\psi_t, A\psi_t\rangle$. Thus if $\psi_t$ is unitarily transformed so must be the observable $A$. When using the effective Hamiltonian of (16.67) one has to properly transform the observables of physical interest. To lowest order $x$ and $-\mathrm{i}\nabla_x$ transform into themselves. But, in general, to first order there will be corrections. Also, the basis $\psi_\alpha(q,p), \alpha = 1,\dots,\ell$, of the relevant subspace must be selected judiciously such that in the $|\chi_\alpha\rangle$-basis observables of interest have a simple representation. We will come back to this point in the context of the Pauli–Fierz operator; see section 16.6 below. The Weyl quantization of the effective Hamiltonian (16.67) still carries the small parameter $\varepsilon$ which suggests using semiclassical methods, a subject to be taken up in the following section. For general $E(q,p)$, the semiclassical regime is limited by the Ehrenfest time which in our units is of order $\log\varepsilon^{-1}$. We stress that the adiabatic limit has no such restrictions, as can be seen from (16.77): if one had included the term $h_2$, the approximation with the given precision would be valid for macroscopic times of order $\varepsilon^{-1}$.

## 16.5 Semiclassical limit

According to Eq. (16.73) the effective Hamiltonian has the form

$$H = H(\widehat{q},\widehat{p}) = E(\widehat{q},\widehat{p})\mathbb{1} + \varepsilon H_{\mathrm{sp}}(\widehat{q},\widehat{p}) \tag{16.78}$$

acting on $L^2(\mathbb{R}^3) \otimes \mathbb{C}^\ell$, where for clarity $\mathbb{1}$ denotes the $\ell \times \ell$ unit matrix. The last two terms in (16.73) have been renamed as $H_{\mathrm{sp}}$ anticipating that they are responsible for the precession of the $\ell$-spinor.

The semiclassical limit can be guessed most directly by considering the Heisenberg evolution of the semiclassical observable $\widehat{a} = a(\widehat{q}, \widehat{p})$ as

$$\widehat{a}(t) = e^{iHt/\varepsilon}\widehat{a}e^{-iHt/\varepsilon}. \tag{16.79}$$

$\widehat{a}(t)$ has a semiclassical representation through $a(q, p, t) = \sum_{j\geq 0} \varepsilon^j a_j(q, p, t)$. From the equations of motion

$$\varepsilon\frac{d}{dt}\widehat{a}(t) = i[H, \widehat{a}(t)], \tag{16.80}$$

using $[E\mathbb{1}, a_j(t)] = 0$, one finds to lowest order

$$\frac{d}{dt}a_0(t) = \{E, a_0(t)\} + i[H_{sp}, a_0(t)] + \mathcal{O}(\varepsilon) \tag{16.81}$$

with initial conditions $a_0(0) = a$.

Ignoring the error $\mathcal{O}(\varepsilon)$, the solution to (16.81) is easily constructed. First one defines the classical flow $\Phi_t$ on phase space through

$$\dot{q}_t = \nabla_p E(q_t, p_t), \quad \dot{p}_t = -\nabla_q E(q_t, p_t). \tag{16.82}$$

Secondly, given the initial condition $(q, p)$ with corresponding trajectory $(q_t, p_t)$ one obtains the time-dependent spin Hamiltonian $H_{sp}(t) = H_{sp}(q_t, p_t)$. It determines the spinor evolution as

$$i\frac{d}{dt}\chi(t) = H_{sp}(t)\chi(t), \quad \chi(t) \in \mathbb{C}^{\ell}. \tag{16.83}$$

The unitary propagator for (16.83) from $s$ to $t$ is denoted by $U(t, s|q, p)$, recalling that it depends on the trajectory through its initial conditions. Then

$$a_0(q, p, t) = U(t, 0|q, p)^* a(\Phi_t(q, p))U(t, 0|q, p), \tag{16.84}$$

as can be verified by inserting in (16.81).

In the semiclassical limit there is no back-reaction of the spin on the orbit. Such an effect could be seen in corrections to the semiclassical limit and in the next-order correction, $h_2$, to the effective Hamiltonian.

The predictions of the semiclassical limit move more sharply into focus through considering the dual Schrödinger picture. One picks a possibly $\varepsilon$-dependent initial wave function such that for expectations of semiclassical observables the limit

$$\lim_{\varepsilon\to 0}\langle\psi^{\varepsilon}, a(\widehat{q}, \widehat{p})\psi^{\varepsilon}\rangle = \int \operatorname{tr}[\rho_{cl}(d^3q d^3p)a(q, p)] \tag{16.85}$$

holds, examples being listed below. Here tr is over $\mathbb{C}^{\ell}$. $\rho_{cl}(d^3q d^3p)$ is a matrix-valued classical probability measure on phase space, $\rho_{cl}(d^3q d^3p) \geq 0$ as a matrix and $\int \mathrm{tr}[\rho_{\mathrm{cl}}(\mathrm{d}^3q\mathrm{d}^3p)] = 1$. Then at later times, from (16.82) and (16.84),

$$\lim_{\varepsilon\to 0}\langle \mathrm{e}^{-\mathrm{i}Ht/\varepsilon}\psi^\varepsilon, \widehat{a}\mathrm{e}^{-\mathrm{i}Ht/\varepsilon}\psi^\varepsilon\rangle = \lim_{\varepsilon\to 0}\langle\psi^\varepsilon, \widehat{a}(t)\psi^\varepsilon\rangle$$
$$= \int \mathrm{tr}[\rho_{\mathrm{cl}}(\mathrm{d}^3q\mathrm{d}^3p)U(t,0|q,p)^* a(\Phi_t(q,p))U(t,0|q,p)]$$
$$= \int \mathrm{tr}[U(t,0|q,p)\rho_{\mathrm{cl}}\circ\Phi_{-t}(\mathrm{d}^3q\mathrm{d}^3p)U(t,0|q,p)^* a(q,p)]. \quad (16.86)$$

The classical part of the measure is transported through the classical flow, while the spinor part evolves through the spin Hamiltonian $H_{\mathrm{sp}}(q_t, p_t)$. In this sense the quantum expectation on the left of (16.86) is approximated by the classical average on the right, keeping in mind that the internal spinor motion remains of full quantum nature.

We list a few conventional choices, where the position variable refers to the macroscopic scale. In *wave packet dynamics* one assumes a sharp concentration as $\rho_{\mathrm{cl}}(\mathrm{d}^3q\mathrm{d}^3p) = |\chi\rangle\langle\chi|\delta(q-q_0)\delta(p-p_0)\mathrm{d}^3q\mathrm{d}^3p$. Then at later times the wave packet is concentrated at $(q_t, p_t)$ and the spin $\chi_t$ precesses according to (16.83). A particular choice would be an initially Gaussian wave packet, which depends on $\varepsilon$ such that $\langle x\rangle_\varepsilon = q_0$, $\langle -\mathrm{i}\varepsilon\nabla_x\rangle = p_0$, $\langle (x-q_0)^2\rangle_\varepsilon \to 0$, and $\langle(-\mathrm{i}\varepsilon\nabla_x - p_0)^2\rangle_\varepsilon \to 0$ as $\varepsilon\to 0$. Note that to achieve the concentration in momentum the position is necessarily broadly distributed on the atomic scale. A *WKB wave function* is of the form $\psi^\varepsilon(x) = \chi(x)\mathrm{e}^{\mathrm{i}S(x)/\varepsilon}$. In the limit $\varepsilon\to 0$ it defines the initial distribution $\rho_{\mathrm{cl}}(\mathrm{d}^3q\mathrm{d}^3p) = |\chi(q)\rangle\langle\chi(q)|\delta(p-\nabla S(q))\mathrm{d}^3q\mathrm{d}^3p$. As a measure on the six-dimensional phase space it is concentrated on a three-dimensional hypersurface, a property which is retained by the flow $\Phi_t$. Since this surface may in general fold up in the course of time, it cannot be represented as the graph of a function. For fixed $q$ there could be several values of $p$. The wave function $U^\varepsilon(t)\psi^\varepsilon$ has the standard WKB form only locally in phase space. A further choice is a *microscopic wave packet* which in our units reads as $\psi^\varepsilon(x) = \chi\varepsilon^{-3/2}\psi(x/\varepsilon)$ with some given wave function $\psi$ on the microscopic scale. Then $\rho_{\mathrm{cl}}(\mathrm{d}^3q\mathrm{d}^3p) = |\chi\rangle\langle\chi|\delta(q)|\widehat{\psi}(p)|^2\mathrm{d}^3q\mathrm{d}^3p$. The wave packet is spatially localized, necessarily with a spread in momentum. $\rho_{\mathrm{cl}}$ is concentrated on the three-dimensional surface $\{(q,p)|q=0\}$ in phase space. Thus at a later time it will be of WKB form locally.

If we look back at our starting point, an electron subject to slowly varying external potentials governed by the Hamiltonian (16.1), it may appear that we have lost sight of our goal. To improve, we summarize our main findings on a qualitative level. First, slow variation is satisfied for all laboratory fields including those employed in the big accelerator machines. The translational degrees of freedom of the electron are thus governed in an excellent approximation by an effective Hamiltonian obtained from the Peierls substitution, $H_{\mathrm{eff}} = E(\widehat{p} - eA_{\mathrm{ex}}(\widehat{q})) + e\phi_{\mathrm{ex}}(\widehat{q})$.

In particular for small velocities, relying on the results from chapter 15,

$$H_{\text{eff}} = \frac{1}{2m_{\text{eff}}}(\widehat{p} - eA_{\text{ex}}(\widehat{q}))^2 + e\phi_{\text{ex}}(\widehat{q}). \tag{16.87}$$

To understand the spin precession, one has to compute the first-order correction $h_1$ to the effective Hamiltonian, which is the topic of the section to follow.

## 16.6 Spin precession and the gyromagnetic ratio

The time of pleasant harvest has come. The Hamiltonian is (16.38) with principal symbol

$$H_0(q, p) = H(p - eA_{\text{ex}}(q)) + e\phi_{\text{ex}}(q); \tag{16.88}$$

compare with (16.39). $H(p)$ acts on $\mathbb{C}^2 \otimes \mathcal{F}$ and is defined in (15.68), where for notational convenience we use $H(p)$ instead of $H_p$. From section 15.3 we know that $H(p)$ has a two-fold degenerate ground state with energy $E(p)$ and projector $P(p)$, $\text{tr}[P(p)] = 2$ provided $|p| \le p_{\text{c}}$ $(\cong m)$. Therefore $P(q, p) = P(p - eA_{\text{ex}}(q))$ as a projection operator on $\mathbb{C}^2 \otimes \mathcal{F}$ defines the relevant subspace for $H_0(q, p)$ with corresponding eigenvalue $E(q, p) = E(p - eA_{\text{ex}}(q)) + e\phi_{\text{ex}}(q)$. To lowest order the symbol of the effective Hamiltonian is then

$$h_0(q, p) = E(q, p)\mathbb{1} = \big(E(p - eA_{\text{ex}}(q)) + e\phi_{\text{ex}}(q)\big)\mathbb{1}, \tag{16.89}$$

with $\mathbb{1}$ the $2 \times 2$ unit matrix, and the orbital motion is approximately governed by

$$h_0(\widehat{q}, \widehat{p}) = \big(E(-\mathrm{i}\varepsilon\nabla_x - eA_{\text{ex}}(x)) + e\phi_{\text{ex}}(x)\big)\mathbb{1}. \tag{16.90}$$

The spin precession requires more attention. First of all one has to specify a basis in $P(p)\mathbb{C}^2 \otimes \mathcal{F}$. The singled-out choice is the eigenvectors of the total angular momentum component parallel to $p$, which we denote by $\psi_{\text{g}\pm}(p, \underline{k})$, $\langle\psi_{\text{g}-}(p), \psi_{\text{g}+}(p)\rangle_{\mathbb{C}^2\otimes\mathcal{F}} = 0$. To define them properly, we follow section 13.5 and introduce the total angular momentum

$$J = \frac{1}{2}\sigma + J_{\text{f}} + S_{\text{f}}, \tag{16.91}$$

see (13.96), (13.97). If $R$ is a rotation by angle $\theta$ relative to the axis of rotation $\widehat{n}$ through the origin, then

$$\mathrm{e}^{\mathrm{i}\theta\widehat{n}\cdot J} e_\lambda(k)a(k, \lambda)\mathrm{e}^{-\mathrm{i}\theta\widehat{n}\cdot J} = Re_\lambda(R^{-1}k)a(R^{-1}k, \lambda) \tag{16.92}$$

and therefore

$$\mathrm{e}^{\mathrm{i}\theta\widehat{n}\cdot J} A_\varphi \mathrm{e}^{-\mathrm{i}\theta\widehat{n}\cdot J} = RA_\varphi, \quad \mathrm{e}^{\mathrm{i}\theta\widehat{n}\cdot J} B_\varphi \mathrm{e}^{-\mathrm{i}\theta\widehat{n}\cdot J} = RB_\varphi, \quad \mathrm{e}^{\mathrm{i}\theta\widehat{n}\cdot J} \sigma \mathrm{e}^{-\mathrm{i}\theta\widehat{n}\cdot J} = R\sigma. \tag{16.93}$$

If $\widehat{n}$ is parallel to $p$, $\widehat{n} = p/|p|$, these relations imply that the component of $J$ along $p$ is conserved,

$$[H(p), p \cdot J] = 0. \tag{16.94}$$

$|p|^{-1} p \cdot J$ has the eigenvalues $\pm\frac{1}{2}, \pm\frac{3}{2}, \ldots$. For $e = 0$, $|p|^{-1} p \cdot J$ has eigenvalues $\pm\frac{1}{2}$ in the ground state subspace of $H(p)$. By continuity, for $e \neq 0$, the eigenvalue equations $H(p)\psi_{\mathrm{g}\pm}(p) = E(p)\psi_{\mathrm{g}\pm}(p)$, $|p|^{-1} p \cdot J \psi_{\mathrm{g}\pm}(p) = \pm\frac{1}{2}\psi_{\mathrm{g}\pm}(p)$ uniquely determine the basis vectors $\psi_{\mathrm{g}\pm}(p)$, up to phase factors $\mathrm{e}^{-\mathrm{i}\theta_\pm(p)}$. We interpret these states as having spin pointing parallel, eigenvalue $\frac{1}{2}$, and anti-parallel, eigenvalue $-\frac{1}{2}$, to $p$. On the other hand, except for $p = 0$, one has $[H(p), p' \cdot J] \neq 0$ unless $|p|^{-1}p = \pm|p'|^{-1}p'$.

The effective spin Hamiltonian in the $p \cdot J$-basis is derived with the help of (16.73), recalling the subprincipal symbol $H_1(q, p)$ from (16.39). Setting $\psi_{\mathrm{g}\pm}(q, p) = \psi_{\mathrm{g}\pm}(p - eA_{\mathrm{ex}}(q))$ one obtains

$$\begin{aligned} \langle\alpha|H_{\mathrm{sp}}(q, p)|\beta\rangle = & -\frac{e}{2m} B_{\mathrm{ex}}(q) \cdot \langle\psi_{\mathrm{g}\alpha}(q, p), \sigma\psi_{\mathrm{g}\beta}(q, p)\rangle_{\mathbb{C}^2\otimes\mathcal{F}} \\ & - \frac{\mathrm{i}}{2}\langle\psi_{\mathrm{g}\alpha}(q, p), \{H_0(q, p) + E(q, p), \psi_{\mathrm{g}\beta}(q, p)\}\rangle_{\mathbb{C}^2\otimes\mathcal{F}}, \end{aligned} \tag{16.95}$$

$\alpha, \beta = \pm$. Working out the Poisson bracket yields

$$\begin{aligned} \langle\alpha|H_{\mathrm{sp}}(q, p)|\beta\rangle = & - B_{\mathrm{ex}}(q) \cdot \Big(\frac{e}{2m}\langle\psi_{\mathrm{g}\alpha}(\tilde{p}), \sigma\psi_{\mathrm{g}\beta}(\tilde{p})\rangle_{\mathbb{C}^2\otimes\mathcal{F}} \\ & - \frac{\mathrm{i}}{2}e\langle\nabla_p\psi_{\mathrm{g}\alpha}(\tilde{p}), \times(H(\tilde{p}) - E(\tilde{p}))\nabla_p\psi_{\mathrm{g}\beta}(\tilde{p})\rangle_{\mathbb{C}^2\otimes\mathcal{F}}\Big) \\ & + e\big(-\nabla_q\phi_{\mathrm{ex}}(q) + v \times B_{\mathrm{ex}}(q)\big) \cdot \langle\psi_{\mathrm{g}\alpha}(\tilde{p}), \mathrm{i}\nabla_p\psi_{\mathrm{g}\beta}(\tilde{p})\rangle_{\mathbb{C}^2\otimes\mathcal{F}} \end{aligned} \tag{16.96}$$

with the velocity $v = \nabla_p E(\tilde{p})$ and $\tilde{p} = p - eA_{\mathrm{ex}}(q)$. The spin Hamiltonian has a simple interpretation: through the coupling to the field the electron acquires the effective magnetic moment

$$\begin{aligned} \langle\alpha|M_{\mathrm{m}}(\tilde{p})|\beta\rangle = & \frac{e}{2m}\langle\psi_{\mathrm{g}\alpha}(\tilde{p}), \sigma\psi_{\mathrm{g}\beta}(\tilde{p})\rangle_{\mathbb{C}^2\otimes\mathcal{F}} \\ & - \frac{\mathrm{i}}{2}e\langle\nabla_p\psi_{\mathrm{g}\alpha}(\tilde{p}), \times(H(\tilde{p}) - E(\tilde{p}))\nabla_p\psi_{\mathrm{g}\beta}(\tilde{p})\rangle_{\mathbb{C}^2\otimes\mathcal{F}} \end{aligned} \tag{16.97}$$

and the effective electric moment

$$\langle\alpha|M_{\mathrm{e}}(\tilde{p})|\beta\rangle = -e\langle\psi_{\mathrm{g}\alpha}(\tilde{p}), \mathrm{i}\nabla_p\psi_{\mathrm{g}\beta}(\tilde{p})\rangle_{\mathbb{C}^2\otimes\mathcal{F}}. \tag{16.98}$$

They are operators on spin space depending on the kinetic momentum $\tilde{p}$. The spin Hamiltonian then reads

$$H_{\mathrm{sp}} = -B_{\mathrm{ex}} \cdot M_{\mathrm{m}} - F_{\mathrm{L}} \cdot M_{\mathrm{e}} \tag{16.99}$$

with the Lorentz force $F_{\mathrm{L}} = -\nabla_q \phi_{\mathrm{ex}}(q) + v \times B_{\mathrm{ex}}(q)$. Note that on top of the obvious magnetic splitting, the effective moments are determined through geometric phases.

The semiclassical analysis of (16.90) together with (16.96) was discussed in the previous section. Of particular interest is the case of a small uniform magnetic field $B$, i.e. $\phi_{\mathrm{ex}} = 0$, $A_{\mathrm{ex}}(q) = \frac{1}{2} B \times q$. For small velocities the orbital motion is then governed by

$$m_{\mathrm{eff}} \frac{\mathrm{d}}{\mathrm{d}t} v_t = e v_t \times B; \tag{16.100}$$

see (15.23) for the definition of the effective mass, which yields the cyclotron frequency

$$\omega_{\mathrm{c}} = e|B|/m_{\mathrm{eff}}. \tag{16.101}$$

Since $\tilde{p} = 0$, we may pick arbitrarily the $J_3$-basis with eigenvectors $\psi_{\mathrm{g}\pm} = \psi_{\mathrm{g}\pm}(0)$ determined through $H(0)\psi_{\mathrm{g}\pm} = E(0)\psi_{\mathrm{g}\pm}$, $J_3 \psi_{\mathrm{g}\pm} = \pm\frac{1}{2}\psi_{\mathrm{g}\pm}$. Using first-order perturbation theory for $\nabla_p \psi_{\mathrm{g}\pm}(0)$, the spin Hamiltonian simplifies to

$$\begin{aligned}\langle \alpha | H_{\mathrm{sp}} | \beta \rangle = & -\frac{e}{2m} B \cdot \langle \psi_{\mathrm{g}\alpha}, \sigma \psi_{\mathrm{g}\beta} \rangle_{\mathbb{C}^2 \otimes \mathcal{F}} \\ & + \frac{\mathrm{i}}{2} e B \cdot \langle \psi_{\mathrm{g}\alpha}, \frac{1}{m}(P_{\mathrm{f}} + eA_\varphi) \times \frac{1}{H(0) - E(0)} \frac{1}{m}(P_{\mathrm{f}} + eA_\varphi) \psi_{\mathrm{g}\beta} \rangle_{\mathbb{C}^2 \otimes \mathcal{F}}.\end{aligned} \tag{16.102}$$

$H(0)$ is rotation invariant; see the discussion leading to (16.94). Therefore $H_{\mathrm{sp}}$ is necessarily of the form

$$H_{\mathrm{sp}} = -\frac{e}{2m} \frac{\widetilde{g}}{2} B \cdot \sigma, \tag{16.103}$$

which yields $\widetilde{g}$ as

$$\begin{aligned}\frac{1}{2}\widetilde{g} = & \langle \psi_{\mathrm{g}+}, \sigma_3 \psi_{\mathrm{g}+} \rangle_{\mathbb{C}^2 \otimes \mathcal{F}} \\ & - \frac{2}{m} \mathrm{Im} \langle \psi_{\mathrm{g}+}, (P_{\mathrm{f}} + eA_\varphi)_2 \frac{1}{H(0) - E(0)} (P_{\mathrm{f}} + eA_\varphi)_1 \psi_{\mathrm{g}+} \rangle_{\mathbb{C}^2 \otimes \mathcal{F}}.\end{aligned} \tag{16.104}$$

Note that $H_{\rm sp}$ does not depend on the choice of the phase $e^{-i\theta_+(p)}$ for $\psi_{g+}(p)$. In our approximation, the spin motion is governed by

$$\frac{d}{dt}\sigma(t) = -\frac{e}{2m}\widetilde{g}B \times \sigma(t), \tag{16.105}$$

from which the frequency of spin precession

$$\omega_s = e|B|\widetilde{g}/2m \tag{16.106}$$

follows.

The conventional definition of the gyromagnetic factor is

$$g = 2\omega_s/\omega_c. \tag{16.107}$$

Comparing (16.100) and (16.105) yields

$$g = \frac{m_{\rm eff}}{m}\widetilde{g}. \tag{16.108}$$

We stress that Eq. (16.108) is nonperturbative in the sense that it is valid for any coupling strength $e$. In the derivation it is assumed that the external magnetic field is weak, an assumption which certainly holds, since experimentally the radius of gyration is of the order of meters. Equation (16.108) is the $g$-factor at $p = 0$. At $p \neq 0$, since the Pauli–Fierz model is nonrelativistic, there is a $p$-dependent $g$-factor with components parallel and transverse to $p$.

Under our standard assumptions, $g$ depends analytically on the coupling strength $e$ and it is of interest to obtain the order $e^2$ correction to $g = 2$ at $e = 0$. For this purpose it is convenient to switch to the dimensionless units of section 19.3. The effective mass is defined through (15.23). Compared to (15.36) there is an extra contribution from the fluctuating magnetic field and one obtains

$$\begin{aligned}\frac{m_{\rm eff}}{m} = 1 &+ \frac{2}{3}e^2 \int d^3k|\widehat{\varphi}(k/\lambda_c)|^2\Big[k^2\Big(1 + \frac{1}{2}|k|\Big)\Big]^{-1} \\ &+ \frac{1}{6}e^2 \int d^3k|\widehat{\varphi}(k/\lambda_c)|^2\Big[\Big(1 + \frac{1}{2}|k|\Big)^3\Big]^{-1} + \mathcal{O}(e^4). \end{aligned} \tag{16.109}$$

Next we have to determine $\widetilde{g}$, which is the sum $\widetilde{g}_1 + \widetilde{g}_2$. $H(0)$ is written as $H(0) = H_0 + eH_1 + \frac{1}{2}e^2H_2$. At $e = 0$, $\psi_{g+} = \chi_+ \otimes \Omega$, $\sigma_3\chi_+ = \chi_+$, and $\widetilde{g}_1 = 2$, $\widetilde{g}_2 = 0$. Expanding $\psi_{g+}$ to first order in $e$ as $\psi_{g+} = \chi_+ \otimes \Omega + (e/2)H_0^{-1}\sigma \cdot B_\varphi\chi_+ \otimes \Omega + \mathcal{O}(e^2)$, we insert in (16.104). For $\widetilde{g}_1$ there is a contribution from the normalization of $\psi_{g+}$ and one contribution involving $(e^2/4)\langle\chi_+ \otimes \Omega, \sigma \cdot B_\varphi H_0^{-1}\sigma_3 H_0^{-1}\sigma \cdot$

$B_\varphi \chi_+ \otimes \Omega\rangle_{\mathbb{C}^2\otimes\mathcal{F}}$. The net result is

$$\frac{1}{2}\widetilde{g}_1 = 1 - \frac{1}{4}e^2 \int d^3k |\widehat{\varphi}(k/\lambda_c)|^2 \Big[|k|\Big(1+\frac{1}{2}|k|\Big)^2\Big]^{-1} - \frac{1}{12}e^2 \int d^3k |\widehat{\varphi}(k/\lambda_c)|^2 \Big[|k|\Big(1+\frac{1}{2}|k|\Big)^2\Big]^{-1} + \mathcal{O}(e^4). \quad (16.110)$$

For $\widetilde{g}_2$ only one of the two ground states is expanded to order $e$. Hence one has a contribution proportional to $\langle \chi_+ \otimes \Omega, (A_{\varphi 2} H_0^{-1} P_{\mathrm{f}1} H_0^{-1} \sigma \cdot B_\varphi - \sigma \cdot B_\varphi H_0^{-1} P_{\mathrm{f}2} H_0^{-1} A_{\varphi 1}) \chi_+ \otimes \Omega\rangle_{\mathbb{C}^2\otimes\mathcal{F}}$. The net result is

$$\frac{1}{2}\widetilde{g}_2 = -\frac{1}{3}e^2 \int d^3k |\widehat{\varphi}(k/\lambda_c)|^2 \Big[|k|\Big(1+\frac{1}{2}|k|\Big)^2\Big]^{-1}. \quad (16.111)$$

Adding up (16.109), (16.110), and (16.111), the $g$-factor to order $e^2$ is given by

$$g = 2\Big(1 + \frac{2}{3}e^2 \int d^3k |\widehat{\varphi}(k/\lambda_c)|^2 \Big[k^2\Big(1+\frac{1}{2}|k|\Big)^3\Big]^{-1}\Big) + \mathcal{O}(e^4). \quad (16.112)$$

In Heaviside–Lorentz units $e^2 = 4\pi\alpha$. We also set the sharp cutoff $\widehat{\varphi}(k) = (2\pi)^{-3/2}$ for $|k| \le \Lambda$, $\widehat{\varphi}(k) = 0$ for $|k| > \Lambda$. Then

$$g = 2\Big(1 + \frac{8}{3}\Big(\frac{\alpha}{2\pi}\Big)(1 - (1 + (\Lambda/2\lambda_c))^{-2}) + \mathcal{O}(\alpha^2)\Big). \quad (16.113)$$

Clearly $g > 2$, as observed experimentally. It is remarkable that $g$ stays bounded in the limit $\Lambda \to \infty$ and

$$g_\infty = 2\Big(1 + \frac{8}{3}\Big(\frac{\alpha}{2\pi}\Big)\Big) + \mathcal{O}(\alpha^2), \quad (16.114)$$

which is to be compared with $2(1 + (\alpha/2\pi)) + \mathcal{O}(\alpha^2)$ from fully relativistic QED. Evidently the nonrelativistic Pauli–Fierz model overestimates the contribution from large wave numbers by a factor 8/3. The result (16.114) is satisfactory, since it nourishes the hope that the Pauli–Fierz model makes reasonable predictions even when the ultraviolet cutoff $\Lambda$ is removed.

## Notes and references

### *Section 16.1*

In the old quantum theory classical adiabatic invariants were associated with good quantum numbers (Ehrenfest 1916). Thus the time-adiabatic theorem was an important consistency check of the Heisenberg–Schrödinger quantum mechanics (Born 1926; Born and Fock 1928). Kato (1958) proves the adiabatic theorem under the condition that the relevant subspace has finite dimension and is separated by a spectral gap. In fact, the theorem holds in much greater generality than explained in the text. Only a corridor separating the relevant energy band from the rest is needed. The spectrum inside the band can be arbitrary. The error in (16.8) may be improved to any order at the expense of a slight tilt of the subspace $P(\varepsilon t)\mathcal{H}$, as first recognized by Lenard (1959) and further refined by Garrido (1965), Berry (1990), Joye *et al.* (1991), Nenciu (1993), and Joye and Pfister (1994). We refer also to the interesting collection of articles by Shapere and Wilczek (1989). Sjöstrand (1993) discusses the higher-order corrections from the point of view of pseudodifferential operators; compare with section 16.4 and Panati *et al.* (2003a). If $H(t)$ depends analytically on $t$, the error becomes $\mathrm{e}^{-1/\varepsilon}$, which complements the Landau–Zener formula for almost crossing of eigenvalues (Joye and Pfister 1993). If there is no gap, but a smooth $t$-dependence as before, the adiabatic theorem still holds (Avron and Elgart 1999; Bornemann 1998; Teufel 2001). The error depends on the context. It can be as small as in (16.8), but in general it will be larger.

### *Section 16.2*

Our discussion of the space-adiabatic limit ignores technical details on purpose. They are supplied in Teufel and Spohn (2002), Spohn and Teufel (2001), and Teufel (2003). Most importantly, since $p_c < \infty$, one needs a local version of the result explained in the text in the following sense. In the limit $\varepsilon \to 0$ the initial state defines a classical probability measure $\rho_{cl}(\mathrm{d}^3q\mathrm{d}^3p)$ on phase space $\mathbb{R}^6$; compare with section 16.5. $\rho_{cl}$ is transported by the classical flow $\Phi_t$ with Hamiltonian (16.22) as $\rho_{cl} \circ \Phi_{-t}$. If $\rho_{cl}$ is supported in $\mathbb{R}^3 \times \{p \,|\, |p| < p_c\}$, then there is a first time $t_{hit}$ at which the support of $\rho_{cl} \circ \Phi_{-t}$ hits the boundary $\mathbb{R}^3 \times \{p \,|\, |p| = p_c\}$. The approximation through an effective Hamiltonian is valid for times $0 \le t < \varepsilon^{-1} t_{hit}$.

### *Section 16.3*

Weyl quantization, the Moyal product, and matrix-valued symbols are discussed in Robert (1987, 1998), Dimassi and Sjöstrand (1999), Martinez (2002), and Panati *et al.* (2003a). The Moyal product is introduced in Moyal (1949).

### *Section 16.4*

The methods explained in this section have a rich history with motivations ranging from singular partial differential equations and Fourier integral operators to the motion of electrons in solids subject to a small magnetic field. Blount (1962a, b, c) develops a similar scheme for computing effective Hamiltonians and applies it to Bloch electrons and to the Dirac equation. In particular, he computes the second-order symbol $h_2$. In the solid state physics literature his work is a standard reference, but his method is hardly applied to concrete problems. We refer to the discussion in Panati *et al.* (2003b) for an example in the case of magnetic Bloch bands. Starting from coupled wave equations Littlejohn and Flynn (1991) and Littlejohn and Weigert (1993) develop the technique of unitary operators close to the identity on the level of symbols in the case where the principal symbol is a nondegenerate matrix. They apply their scheme to Born–Oppenheimer-type problems, where $H_0(q, p) = p^2\mathbb{1} + V(q)$ with $V(q)$ an $n \times n$ matrix. On an abstract level the Born–Oppenheimer approximation is similar to the Pauli–Fierz model with a slowly varying external electrostatic potential only. The role of the invariant subspace is emphasized by Nenciu (1993). The formal power series for the projector $\pi(q, p)$ is constructed by Brummelhuis and Nourrigat (1999) for the Dirac equation, by Martinez and Sordoni (2002) for Born–Oppenheimer-type Hamiltonians and in the general matrix-valued case by Nenciu and Sordoni (2001). Our discussion is based on Panati *et al.* (2003a). The lecture notes by Teufel (2003) give detailed coverage with many examples, including the case of Bloch electrons (Panati *et al.* 2003b). There also a more complete listing of the literature can be found.

### *Section 16.5*

There is a vast literature on semiclassical methods, both on the theoretical physics and on the mathematical side; to mention only a few representatives: Maslov and Fedoriuk (1981), Gutzwiller (1990), and Robert (1987, 1998). These works are mostly concerned with the scalar case. An alternative technique is to employ matrix-valued Wigner functions (Gérard *et al.* 1997; Spohn 2000b). In this approach the adiabatic and semiclassical limits are fused, which is conceptually misleading. Also higher-order corrections are not accessible. An important example is the Dirac equation which has matrix dimension $n = 4$ and degeneracy $\ell = 2$ of, for example, the electron subspace. The adiabatic limit yields the BMT equation of chapter 10, as discussed in Panati *et al.* (2003a). Blount (1962c) computes the next-order correction. It seems to be of interest in accelerator physics (Heinemann and Barber 1999), despite its fairly complicated structure. Yajima (1992) studies the derivation of the BMT equation using WKB methods, which are rather difficult to handle because of the necessity to switch coordinate systems on the Lagrangian manifold.

The classical limit of the free Maxwell field with classical sources is regarded as sort of obvious. An instructive discussion is Thirring (1958) and Sakurai (1986). Photon counting statistics is covered by Carmichael (1999).

### *Section 16.6*

The gyromagnetic ratio of the electron is the most famous and precise prediction of QED with the current value $g_{\text{theor}}/2 = 1.001\,159\,652\,459\,(135)$ as based on an eight-loop computation, see Kinoshita and Sapirstein (1984) for a review. This result compares extraordinarily well with the experimental value $g_{\text{exp}}/2 = 1.001\,159\,652\,193\,(4)$ of van Dyck, Schwinberg and Dehmelt (1986) based on measurements on a single electron in a Penning trap, see also Brown and Gabrielse (1986), and Dehmelt (1990). The nonrelativistic theory yields $g_{\text{non}}/2 = 1.0031$, with no cutoffs. The nonperturbative formula (16.108) seems to be novel and is described in Panati *et al.* (2002b). A rough approximation is provided by Welton (1948). Grotch and Kazes (1977) discuss the $g$-factor for the Pauli–Fierz model and obtain the second-order result (16.113) through computing energy shifts; compare with section 19.3.5. Surprisingly, they do not stress the obvious point: the $g$-factor is not too far off the truth even in the limit $\Lambda \to \infty$. After all, the mistrust in QED up to the early 1940s was based mainly on the results being cutoff-dependent and diverging as $\Lambda \to \infty$; see Schweber (1994).

# 17
# Radiation

The theoretical understanding of the emission of light from atoms is inseparably linked with the development of quantum mechanics – the first glimpse of the full answer unraveled by P. A. M. Dirac in February 1927. A minimal model for radiation has to consist of at least one atom and the photons. Thus we fix an infinitely heavy nucleus at the origin, say, and describe the motion of a single electron by the spinless Pauli–Fierz Hamiltonian

$$H = \frac{1}{2m}(p - eA_{\varphi}(x))^2 + V_{\varphi\mathrm{coul}}(x) + H_{\mathrm{f}} \tag{17.1}$$

with $V_{\varphi\mathrm{coul}}(x) = -e^2 \int \mathrm{d}^3x_1 \mathrm{d}^3x_2 \varphi(x_1)\varphi(x_2)(4\pi|x + x_1 - x_2|)^{-1}$, the smeared Coulomb potential. Besides radiation, (17.1) describes a multitude of physical processes of interest. If the electron is free, i.e. far away from the nucleus, photons scatter off the electron (Compton effect). As the electron approaches the nucleus it will be scattered under the emission of bremsstrahlung (Rutherford scattering). In contrast, in this chapter we are interested in processes where the electron remains tightly bound to the nucleus. Of course, these two worlds are not strictly separated. The electron might be captured by the nucleus at the expense of radiated energy. Conversely, the atom may become ionized by hitting it with sufficiently energetic radiation (photoelectric effect). Even in the realm of a bound electron, several processes should be distinguished. The most basic one is spontaneous emission, through which the atom in an excited state loses energy and ends up in the radiationless ground state. A photon may be scattered by the atom leaving the atom behind in either its ground state (elastic Rayleigh scattering) or in an excited state (inelastic Rayleigh scattering) which is then followed by spontaneous emission. Both processes will be discussed in separate sections.

Under usual circumstances the wavelength of emitted light is much larger than the size of an atom. In this case one can ignore the variation of the vector potential in (17.1) and replace $A_{\varphi}(x)$ by $A_{\varphi}(0)$, the so-called dipole approximation. In addition we want to restrict the electron Hilbert space to bound states only. Taking into account the first $N$ of them results in an $N$-level system coupled to the radiation field. We point out that an enormous effort has been invested precisely to avoid such a mutilation of the Pauli–Fierz Hamiltonian (17.1). Still, in the first round a simplified version will suffice.

Radiation as discussed here has no classical counterpart. Of course, as explained, in the context of the Abraham model a charge loses energy through radiation. Its analog would be an extension of the results given in the previous chapter. There one has to give up $m_{\mathrm{ph}} > 0$. Then the spectral gap closes and the strict adiabatic protection is lost. For example, (16.105) would have a dissipative correction at the next order associated with a gradual emission of photons. In contrast, for the radiation processes studied here the emission of photons occurs on the atomic scale.

## 17.1 $N$-level system in the dipole approximation

The dipole approximation reads

$$H = \frac{1}{2m}(p - eA_\varphi(0))^2 + V(x) + H_{\mathrm{f}}. \tag{17.2}$$

If in addition we were to choose $V$ to be harmonic, $V(x) = \frac{1}{2}m\omega_0^2 x^2$, then (17.2) is a quadratic Hamiltonian, as can be seen, if on top of the Bose fields $a(k, \lambda), a^*(k, \lambda)$ one introduces the annihilation and creation operators $b, b^*$ for the particle; compare with section 13.7(i). The analysis of this model can be reduced to a Hamiltonian on the one-particle space $\mathbb{C}^3 \oplus (L^2(\mathbb{R}^3) \otimes \mathbb{C}^2)$, where $\mathbb{C}^3$ corresponds to the $b, b^*$ degrees of freedom. While such an analysis is very instructive, we stick here to the more realistic Coulomb-type potential. We rewrite (17.2) as

$$H = \frac{1}{2m}p^2 + V(x) + H_{\mathrm{f}} - \frac{e}{m}p \cdot A_\varphi(0) + \frac{e^2}{2m}A_\varphi(0)^2, \tag{17.3}$$

drop the $A_\varphi(0)^2$ term, and expand in the eigenbasis of $\frac{1}{2m}p^2 + V(x)$ up to the $N$-th eigenvalue, including multiplicity. This results in

$$H_\lambda = H_{\mathrm{at}} \otimes 1 + 1 \otimes H_{\mathrm{f}} + \lambda\tilde{Q} \cdot A_\varphi(0). \tag{17.4}$$

Here $H_{\mathrm{at}}$ and $\tilde{Q} = (\tilde{Q}_1, \tilde{Q}_2, \tilde{Q}_3)$ are symmetric $N \times N$ matrices. In our representation $H_{\mathrm{at}}$ is diagonal with nondegenerate smallest eigenvalue $\varepsilon_1$ and $\tilde{Q}$ is proportional to the dipole moments

$$\tilde{Q}_{ij} = \langle\psi_i, p\psi_j\rangle = \mathrm{i}m(\varepsilon_i - \varepsilon_j)\langle\psi_i, x\psi_j\rangle, \tag{17.5}$$

$i, j = 1, \dots, N$, where we used the facts that $\mathrm{i}[\frac{1}{2m}p^2 + V(x), x] = \frac{1}{m}p$ and

$(\frac{1}{2m}p^2 + V(x))\psi_j = \varepsilon_j \psi_j$ counting eigenvalues and eigenfunctions including their multiplicity. We also introduced explicitly the dimensionless small coupling parameter $\lambda$. If one follows the conventions of section 13.4, then $\lambda = \alpha^{3/2}$.

Note that in the functional integral representation of $\mathrm{e}^{-tH_\lambda}$, $H_\lambda$ of (17.4), the effective action is quadratic with the interaction potential

$$W_{\mathrm{dip}}(t) = \lambda^2 \int \mathrm{d}^3k |\widehat{\varphi}|^2 \frac{1}{2\omega} \mathrm{e}^{-\omega|t|}, \tag{17.6}$$

which decays as $t^{-2}$ for large $t$. Thus (17.4) is marginally infrared divergent. Generically $H_\lambda$ will lose its ground state at strong enough coupling, in contrast to the full Pauli–Fierz model, and (17.4) can be trusted only at small coupling.

An alternative route to the $N$-level approximation is first to transform to the $x \cdot E_\varphi(0)$ coupling through the unitary transformation

$$U = \mathrm{e}^{\mathrm{i}ex\cdot A_\varphi(0)}. \tag{17.7}$$

Then

$$\begin{aligned} U^* p U &= p + eA_\varphi(0), \quad U^* x U = x, \\ U^* a(k,\lambda) U &= a(k,\lambda) + \mathrm{i}(e_\lambda(k)\cdot x) e\widehat{\varphi}(k)/\sqrt{2\omega(k)} \end{aligned} \tag{17.8}$$

and therefore

$$U^* H U = \frac{1}{2m}p^2 + V(x) + \Big(\frac{2}{3}e^2 \int \mathrm{d}^3k |\widehat{\varphi}|^2\Big)x^2 + H_{\mathrm{f}} - ex\cdot E_\varphi(0). \tag{17.9}$$

As before, we expand in the eigenbasis of $\frac{1}{2m}p^2 + V(x)$ up to the $N$-th eigenvalue. This results in the Hamiltonian

$$H_\lambda = H_{\mathrm{at}} + H_{\mathrm{f}} + \lambda Q \cdot E_\varphi \tag{17.10}$$

with the matrix of dipole moments $Q_{ij} = \langle \psi_i, x\psi_j\rangle$, $E_\varphi = E_\varphi(0)$, and $\lambda = -e$. Since now the coupling is to $E_\varphi(0)$, the effective action (17.6) gains an extra factor of $\omega^2$ and therefore has a decay as $t^{-4}$ in accordance with the full model.

For the remainder of the chapter, we take (17.10) as the starting point. The particular origin of $H_{\mathrm{at}}$ and $Q$ is of no importance. We only record that they satisfy $H_{\mathrm{at}}^* = H_{\mathrm{at}}$, $Q^* = Q$. $H_{\mathrm{at}}$ has the spectrum $\sigma(H_{\mathrm{at}}) \subset \mathbb{R}$. It consists of the eigenvalues labeled without multiplicity as $\varepsilon_1 < \varepsilon_2 < \cdots < \varepsilon_{\bar N}$, $\bar N \le N$. The corresponding spectral projections are denoted by $P_1, \ldots, P_{\bar N}$. Their degeneracies are $\mathrm{tr}[P_j] = m_j$ with $m_1 = 1$ and $\sum_{j=1}^{\bar N} m_j = N$. In particular one has the spectral representation

$$H_{\mathrm{at}} = \sum_{j=1}^{\bar N} \varepsilon_j P_j. \tag{17.11}$$

## 17.2 The weak coupling theory

We plan to study the emission of light from atoms. The atom is assumed to have already been prepared in an excited state and thus the initial state of the coupled system is of the form $\psi \otimes \Omega$ with the atomic wave function $\psi \in \mathbb{C}^N$. To determine the radiated field one has to understand the long-time asymptotics of the solution $e^{-iHt}\psi \otimes \Omega$ of the time-dependent Schrödinger equation. For small coupling, which is well satisfied physically, the dynamics approximately decouples: the atom is governed by an autonomous reduced dynamics and the field evolves with the decaying atom as a source term. In this section we will first study the reduced dynamics of the atom in the weak coupling regime with our results to be supported through a nonperturbative resonance theory in section 17.3. In a follow-up we discuss the spectral characteristics of the emitted light.

By definition, the reduced dynamics refers to the reduced state of the atom, which allows one to determine atomic observables such as the probability of being in the $n$-th level at time $t$. Although by assumption the initial state of the atom is pure, it will not remain so because of the interaction with the radiation field. Thus it will be more natural to work directly in the set of all density matrices. The initial state is then of the form $\rho \otimes P_\Omega$ with $\rho$ the atomic density matrix, and $P_\Omega$ the projection onto Fock vacuum. The time evolution is given through

$$e^{-iH_\lambda t}\rho \otimes P_\Omega e^{iH_\lambda t} = e^{-iL_\lambda t}\rho \otimes P_\Omega. \tag{17.12}$$

Here $L_\lambda W = [H_\lambda, W]$ is the Liouvillean as acting on $\mathcal{T}_1(\mathbb{C}^N \otimes \mathcal{F})$, the trace class over $\mathbb{C}^N \otimes \mathcal{F}$. To distinguish typographically, $L_\lambda$ is written as a slanted symbol, like other operators, sometimes called superoperators, which act either on $\mathcal{T}_1$ or on $B(\mathbb{C}^N \otimes \mathcal{F})$, the space of bounded operators on $\mathbb{C}^N \otimes \mathcal{F}$. Clearly, states evolve into states, i.e. if $W \in \mathcal{T}_1$ is positive and normalized, so is $e^{-iL_\lambda t}W$. Sometimes, it is convenient to think of (17.12) as a Schrödinger evolution in a Hilbert space. This can be done by adopting the space $\mathcal{T}_2(\mathbb{C}^N \otimes \mathcal{F})$ of Hilbert–Schmidt operators with inner product $\langle A|B\rangle = \mathrm{tr}[A^*B]$. In this space the Liouvillean $L_\lambda$ is a self-adjoint operator, which explains our sign convention in front of the commutator. A further choice comes from regarding $B(\mathbb{C}^N \otimes \mathcal{F})$ as the space dual to $\mathcal{T}_1(\mathbb{C}^N \otimes \mathcal{F})$ through the duality relation $W \mapsto \mathrm{tr}[AW]$, $W \in \mathcal{T}_1(\mathbb{C}^N \otimes \mathcal{F})$, $A \in B(\mathbb{C}^N \otimes \mathcal{F})$. Then the dual of $L_\lambda$ is $-[H_\lambda, \cdot]$, which generates the Heisenberg evolution of operators.

The reduced dynamics is defined through

$$T_t^\lambda \rho = \rho^\lambda(t) = \mathrm{tr}_{\mathcal{F}}[e^{-iL_\lambda t}\rho \otimes P_\Omega], \tag{17.13}$$

where $\mathrm{tr}_{\mathcal{F}}[\cdot]$ denotes the partial trace over Fock space. $T_t^\lambda$ acts on $B(\mathbb{C}^N)$. It is linear, preserves positivity and normalization. In fact, since it originates from a

Hamiltonian dynamics, the even stronger property of complete positivity is satisfied. In such generality, $T_t^\lambda$ is intractable. But scales become separated for small $\lambda$ into atomic oscillations of the uncoupled dynamics $e^{-iH_{at}t}$ and the weak radiative damping of order $\lambda^2$ ($= \alpha^3 = 1/137^3$). When viewed on the dissipative scale the atomic oscillations are very rapid and effectively time-averaged. For small $\lambda$ memory effects are negligible and $T_t^\lambda$ becomes a dissipative semigroup, which is the autonomous dynamics we are looking for.

To write a formal evolution equation for $\rho^\lambda(t)$ one employs the Nakajima–Zwanzig projection operator method. We define the Liouvilleans $L_{at} = [H_{at}, \cdot\,]$ as acting on $B(\mathbb{C}^N) = \mathcal{T}_1(\mathbb{C}^N)$, $L_f = [H_f, \cdot\,]$ as acting on $\mathcal{T}_1(\mathcal{F})$, and $L_{int} = [Q \cdot E_\varphi, \cdot]$ as acting on $\mathcal{T}_1(\mathbb{C}^N \otimes \mathcal{F})$. For an arbitrary density matrix $W$ on $\mathbb{C}^N \otimes \mathcal{F}$ the Nakajima–Zwanzig projection is

$$PW = (\mathrm{tr}_{\mathcal{F}} W) \otimes P_\Omega. \tag{17.14}$$

Clearly $P^2 = P$ and

$$Pe^{-iL_\lambda t} \rho \otimes P_\Omega = \rho^\lambda(t) \otimes P_\Omega\,. \tag{17.15}$$

Let $W(t) = e^{-iL_\lambda t} \rho \otimes P_\Omega$. Then

$$i\frac{d}{dt} PW(t) = PL_\lambda W(t) = PL_\lambda PW(t) + PL_\lambda(1 - P)W(t), \tag{17.16}$$

$$\begin{aligned} i\frac{d}{dt}(1 - P)W(t) &= (1 - P)L_\lambda W(t) = (1 - P)L_\lambda PW(t) \\ &\quad + (1 - P)L_\lambda(1 - P)W(t). \end{aligned} \tag{17.17}$$

Substituting (17.17) back in (17.16) and using $PL_{int}P = 0$, we obtain

$$\begin{aligned} \frac{d}{dt}\rho^\lambda(t) &= -iL_{at}\rho^\lambda(t) \\ &\quad -\lambda^2 \int_0^t ds\, \mathrm{tr}_{\mathcal{F}}[L_{int}(1 - P)e^{-i(1-P)L_\lambda(1-P)(t-s)}(1 - P)L_{int}P_\Omega]\rho^\lambda(s), \end{aligned} \tag{17.18}$$

which is an exact memory-type equation.

As argued traditionally, the memory decays rapidly on the time scale of the variation of $\rho^\lambda(t)$. For small $\lambda$ one may ignore the interaction and replace $L_\lambda$ by $L_{at} + L_f$ in the exponential. In this approximation for small $\lambda$

$$\frac{d}{dt}\rho(t) = (-iL_{at} + \lambda^2 K_0)\rho(t) \tag{17.19}$$

is obtained as reduced dynamics with

$$K_0\rho = -\int_0^\infty dt\, \mathrm{tr}_{\mathcal{F}}[L_{\mathrm{int}} e^{-i(L_{\mathrm{at}}+L_{\mathrm{f}})t} L_{\mathrm{int}} P_\Omega]\rho. \tag{17.20}$$

This argument misses the point that both $\rho^\lambda(t)$ and the memory kernel have oscillatory contributions from $e^{-iH_{\mathrm{at}}t}$. In general, their products cannot be approximated as in (17.19), (17.20). To subtract the oscillations from the memory kernel we rewrite (17.18) as an integral equation,

$$\rho^\lambda(t) = e^{-iL_{\mathrm{at}}t}\rho - \lambda^2 \int_0^t ds e^{-iL_{\mathrm{at}}(t-s)} \\ \times \int_0^s du\, \mathrm{tr}_{\mathcal{F}}[L_{\mathrm{int}}(1-P) e^{-i(1-P)L_\lambda(1-P)(s-u)}(1-P)L_{\mathrm{int}}P_\Omega]\rho^\lambda(u). \tag{17.21}$$

After the change of variables $v = s - u$, one has

$$\rho^\lambda(t) = e^{-iL_{\mathrm{at}}t}\rho - \lambda^2 \int_0^t du\; e^{-iL_{\mathrm{at}}(t-u)} \\ \times \Big\{ \int_0^{t-u} dv\; e^{iL_{\mathrm{at}}v} \mathrm{tr}_{\mathcal{F}}[L_{\mathrm{int}}(1-P]) e^{-i(1-P)L_\lambda(1-P)v}(1-P)L_{\mathrm{int}}P_\Omega] \Big\} \rho^\lambda(u). \tag{17.22}$$

Now in the memory kernel the fast oscillations are properly counterbalanced and to a good approximation $\rho^\lambda(t)$ is governed by

$$\frac{d}{dt}\rho(t) = (-iL_{\mathrm{at}} + \lambda^2 K)\rho(t), \tag{17.23}$$

where

$$K\rho = -\int_0^\infty dt\; e^{iL_{\mathrm{at}}t} \mathrm{tr}_{\mathcal{F}}[L_{\mathrm{int}} e^{-i(L_{\mathrm{at}}+L_{\mathrm{f}})t} L_{\mathrm{int}} P_\Omega]\rho. \tag{17.24}$$

We state our result as

**Theorem 17.1** (Weak coupling quantum master equation). *Let*

$$e^2\langle \Omega, E_{\varphi\alpha} e^{-iH_{\mathrm{f}}t} E_{\varphi\beta}\Omega\rangle_{\mathcal{F}} = h_{\alpha\beta}(t) = \delta_{\alpha\beta} h(t), \tag{17.25}$$

$$h(t) = \frac{e^2}{3}\int d^3k |\widehat{\varphi}|^2 \omega(k) e^{-i\omega(k)t}, \tag{17.26}$$

$\alpha, \beta = 1, 2, 3$. *If*

$$\int_0^\infty dt |h(t)|(1+t)^\delta < \infty \tag{17.27}$$

*for some $\delta > 0$, then*

$$\lim_{\lambda\to 0} \sup_{0\le t\le \lambda^{-2}\tau} \| T_t^\lambda \rho - e^{(-iL_{at}+\lambda^2 K)t}\rho\| = 0. \tag{17.28}$$

$\tau$ is on the dissipative time scale. Thus in (17.28) a, possibly long, time interval on the dissipative time scale is fixed. Over that time span the true reduced dynamics is well approximated by a Markovian dynamics consisting of fast atomic oscillations, $-iL_{at}$, and slow dissipation, $K$.

The integrability condition (17.27) is seen to hold by transforming back to position space. Then

$$\begin{aligned}\int d^3k|\widehat{\varphi}|^2\omega e^{-i\omega t} &= \int d^3k|\widehat{\varphi}|^2\omega(\cos\omega t - i\sin\omega t)\\ &= -\partial_t^3\int d^3x d^3x' d^3y\varphi(x')|x-x'|^{-2}4\pi\frac{1}{t}\delta(|x-y|-t)\varphi(y)\\ &\quad + i\partial_t^2\int d^3x d^3y\varphi(x)\frac{1}{4\pi t}\delta(|x-y|-t)\varphi(y),\end{aligned} \tag{17.29}$$

which decays as fast as $t^{-4}$, since $\varphi$ is localized.

We still have to carry through properly the time-averaging, accounting for the fast oscillations of $e^{-iL_{at}t}$. We claim that, without further error, $K$ can be replaced by its time average

$$K^\natural\rho = \lim_{T\to\infty}\frac{1}{2T}\int_{-T}^{T} dt\, e^{iL_{at}t}Ke^{-iL_{at}t}, \tag{17.30}$$

as can be seen from going to the slow time scale and considering the interaction representation

$$\begin{aligned}e^{i\lambda^{-2}L_{at}\tau}e^{(-i\lambda^{-2}L_{at}+K)\tau}\rho = \rho + \int_0^\tau du\{e^{i\lambda^{-2}L_{at}u}Ke^{-i\lambda^{-2}L_{at}u}\}\\ \times e^{i\lambda^{-2}L_{at}u}e^{(-i\lambda^{-2}L_{at}+K)u}\rho.\end{aligned} \tag{17.31}$$

The term inside { } is rapidly oscillating and we are allowed to replace it by $K^\natural$. Theorem 17.1 remains valid when $K$ is replaced by $K^\natural$.

In conclusion, we have arrived at the approximate reduced dynamics of the atom:

$$\frac{d}{dt}\rho(t) = -i[H_{at}, \rho(t)] + \lambda^2 K^\natural\rho(t). \tag{17.32}$$

To understand the properties of this dynamics, the dissipative generator $K^\natural$ must be worked out more concretely. It is time-averaged with respect to the Liouvillean $L_{at} = [H_{at}, \cdot]$ and thus depends on the spectrum of $L_{at}$, which is given by

$\{\varepsilon_i - \varepsilon_j | i, j = 1, \ldots, \bar{N}\} = \sigma(L_{\text{at}})$. Accordingly we define

$$Q(\omega) = \sum_{i,j=1,\, \varepsilon_i - \varepsilon_j = \omega \in \sigma(L_{\text{at}})}^{\bar{N}} P_j Q P_i. \tag{17.33}$$

The degeneracy of $H_{\text{at}}$ enters through the projections $P_j$ whereas the degeneracy of the Liouvillean is reflected by the sum in (17.33). For instance, a harmonic oscillator has a nondegenerate Hamiltonian but a highly degenerate Liouvillean. The strength of the various transitions is determined by the one-sided Fourier transform of the field correlation (17.25). We decompose it into real and imaginary parts as

$$\int_0^\infty \mathrm{d}t\, \mathrm{e}^{-\mathrm{i}\omega t} h(t) = \frac{1}{2}\Gamma(\omega) - \mathrm{i}\Delta(\omega), \tag{17.34}$$

which gives

$$\frac{1}{2}\Gamma(\omega) = \frac{e^2}{3}\int \mathrm{d}^3k |\widehat{\varphi}(k)|^2 \omega(k) \pi \delta(\omega(k) + \omega), \tag{17.35}$$

$$\Delta(\omega) = \frac{e^2}{3}\,\mathrm{PV}\int \mathrm{d}^3k |\widehat{\varphi}(k)|^2 \omega(k) \frac{1}{\omega(k) + \omega}, \tag{17.36}$$

PV denoting the principal value of the integral. Using this notation, after working out the oscillatory integrals in (17.30), one obtains

$$K^\natural \rho = \sum_{\alpha=1}^{3} \sum_{\omega \in \sigma(L_{\text{at}})} \Big\{ -\mathrm{i}\Delta(\omega)[Q_\alpha(\omega) Q_\alpha^*(\omega), \rho] + \frac{1}{2}\Gamma(\omega)\big([Q_\alpha^*(\omega)\rho, Q_\alpha(\omega)] + [Q_\alpha^*(\omega), \rho Q_\alpha(\omega)]\big)\Big\}, \tag{17.37}$$

where the $\omega$-sum runs over all eigenvalues of the Liouvillean $L_{\text{at}}$.

The first term in (17.37) merely adds an extra term of order $\lambda^2$ to the atomic Hamiltonian $H_{\text{at}}$. Thereby the eigenvalues $\varepsilon_j$ are shifted and their degeneracy is possibly lifted. The second term represents the radiation damping. It is of Lindblad form which ensures that $T_t = \exp[(-\mathrm{i}[H_{\text{at}}, \cdot] + \lambda^2 K^\natural)t]$ is completely positive and in particular preserves positivity. For the nonaveraged variant $K$ such a property is in general not valid.

The details of the damping mechanism depend on $H_{\text{at}}$, $Q$, and $\widehat{h}$. Since $\Gamma(\omega) = 0$ for $\omega \geq 0$, only transitions to energetically lower levels are possible. Thus generically we expect that in the long-time limit the atom reaches its ground state,

$$\lim_{t\to\infty} T_t \rho = P_1 \tag{17.38}$$

independently of the initial state. Basically, there are two obstructions to (17.38). The analog of the classical Wiener condition (5.4) could be violated in the sense that $\Gamma(\varepsilon_i - \varepsilon_j) = 0$ for some $\varepsilon_i < \varepsilon_j$. Even if we assume $\Gamma(\omega) > 0$ for $\omega < 0$, $H_{\mathrm{at}}$ and $Q$ could be too commutative. For instance, in the extreme case $[H_{\mathrm{at}}, Q] = 0$, the damping vanishes and $K^\natural \rho = -\mathrm{i}\Delta(0)[Q^2, \rho]$. Under the Wiener condition a sufficient criterion for (17.38) to hold is $\{H_{\mathrm{at}}, Q_\alpha, \alpha = 1, 2, 3\}' = \mathbb{C}1$, i.e. the commutant of $\{H_{\mathrm{at}}, Q_\alpha, \alpha = 1, 2, 3\}$ (all operators which commute with $H_{\mathrm{at}}$ and $Q$) consists only of multiples of the unit matrix.

If the spectrum of $H_{\mathrm{at}}$ is nondegenerate, then the set of density matrices commuting with $H_{\mathrm{at}}$ is left invariant by $T_t$. We set $T_t \rho = \sum_{n=1}^{N} \rho_n(t) P_n$, $\mathrm{tr}[P_n] = 1$. The probabilities $\rho_n(t)$ are governed by the Pauli master equation

$$\frac{\mathrm{d}}{\mathrm{d}t}\rho_n(t) = \sum_{m=1}^{N} \big( w_{mn}\rho_m(t) - w_{nm}\rho_n(t) \big), \tag{17.39}$$

where

$$w_{mn} = \sum_{\alpha=1}^{3} \Gamma(\varepsilon_n - \varepsilon_m)\mathrm{tr}[P_m Q_\alpha P_n Q_\alpha] \tag{17.40}$$

is the transition rate from level $m$ to level $n$. Thus the coupling to the radiation field induces a Markov jump process on diagonal density matrices with transition rates given through Fermi's golden rule. The ground state is an absorbing state of the Markov chain. If every other state can be linked to the ground state by a sequence of jumps with nonzero rates, then $\lim_{t\to\infty} \rho_1(t) = 1$ and $\lim_{t\to\infty} \rho_n(t) = 0$ for $n \geq 2$ exponentially fast.

A much-studied variation is to immerse the atom in a black-body cavity at some temperature $T$. Based on rather general principles of statistical mechanics, Einstein came up with a phenomenological description of the atomic transitions in terms of his $A$, $B$-coefficients. Thereby he completely circumvented the yet nonexistent quantum statistical mechanics. Given such historical importance, we violate for a moment our principle of "zero temperature only", to provide a more fully fledged theory in chapter 18. Since we have already used density matrices, in the definition of the reduced dynamics we only have to replace $P_\Omega$ by the thermal state $Z^{-1}\mathrm{e}^{-H_{\mathrm{f}}/k_{\mathrm{B}}T}$. The physically correct procedure is to first enclose the radiation field in the cavity $[-\ell, \ell]^3$, i.e. the $k$-integration is to be replaced by a $k$-sum over the momentum lattice $((\pi/\ell)\mathbb{Z})^3$, followed by the infinite-volume limit $\ell \to \infty$. In the weak coupling approximation, as the only difference to the zero-temperature case, the time-correlation $h_{\alpha\beta}(t)$ for the field is to be computed from the thermal average. Explicitly, with $\langle\cdot\rangle_{k_{\mathrm{B}}T}$ denoting thermal average,

$$
\begin{aligned}
h_{\alpha\beta}(t) &= e^2\langle \mathrm{e}^{\mathrm{i}H_{\mathrm{f}}t} E_{\varphi\alpha}\mathrm{e}^{-\mathrm{i}H_{\mathrm{f}}t} E_{\varphi\beta}\rangle_{k_{\mathrm{B}}T} \\
&= e^2 \sum_{\lambda=1}^{2}\int \mathrm{d}^3k\widehat{\varphi}\sqrt{\omega/2}\, e_{\lambda\alpha}(k) \sum_{\lambda'=1}^{2}\int \mathrm{d}^3k'\widehat{\varphi}\sqrt{\omega/2} e_{\lambda'\beta}(k') \\
&\quad \mathrm{i}^2\langle\big(\mathrm{e}^{-\mathrm{i}\omega(k)t}a(k,\lambda) - \mathrm{e}^{\mathrm{i}\omega(k)t}a^*(k,\lambda)\big)\big(a(k',\lambda') - a^*(k',\lambda')\big)\rangle_{k_{\mathrm{B}}T} \\
&= \delta_{\alpha\beta}h(t), \qquad (17.41)\\
h(t) &= \frac{e^2}{3}\int \mathrm{d}^3k|\widehat{\varphi}|^2\omega(k)\big(\mathrm{e}^{-\mathrm{i}\omega(k)t} \\
&\quad + (\mathrm{e}^{\omega(k)/k_{\mathrm{B}}T} - 1)^{-1}(\mathrm{e}^{-\mathrm{i}\omega(k)t} + \mathrm{e}^{\mathrm{i}\omega(k)t})\big). \qquad (17.42)
\end{aligned}
$$

The friction coefficient, $\Gamma_{k_{\mathrm{B}}T}$, and the level shifts are still defined through (17.34). $\Gamma_{k_{\mathrm{B}}T}$ satisfies the condition of detailed balance as

$$\Gamma_{k_{\mathrm{B}}T}(\omega) = \Gamma_{k_{\mathrm{B}}T}(-\omega)\mathrm{e}^{-\omega/k_{\mathrm{B}}T}. \qquad (17.43)$$

At nonzero temperatures $\Gamma_{k_{\mathrm{B}}T}(\omega) > 0$ for all $\omega$, except for accidental zeros, and the energy can flow either way between atom and thermal bath. If the atom is well coupled to the black-body radiation, in the sense that $\Gamma_{k_{\mathrm{B}}T}(\omega) > 0$ and $\{H_{\mathrm{at}}, Q_\alpha, \alpha = 1, 2, 3\}' = \mathbb{C}1$, then the $N$-level system relaxes to the thermal state $Z^{-1}\mathrm{e}^{-H_{\mathrm{at}}/k_{\mathrm{B}}T}$ in the long-time limit. This is most easily seen in case all $\varepsilon_j$ are nondegenerate. Then the off-diagonal elements of $T_t\rho$ decay exponentially while the diagonal elements are still governed by the Pauli master equation (17.39), in which the transition rates now satisfy

$$w_{mn} = w_{nm}\mathrm{e}^{-(\varepsilon_n - \varepsilon_m)/k_{\mathrm{B}}T} \qquad (17.44)$$

as a result of the detailed balance (17.43). Under "good coupling" (17.44) ensures that the thermal state is the only invariant state for (17.39) and therefore

$$\lim_{t\to\infty} T_t\rho = Z^{-1}\mathrm{e}^{-H_{\mathrm{at}}/k_{\mathrm{B}}T}. \qquad (17.45)$$

As will be explained in chapter 18 the relaxation to thermal equilibrium can be established also for small, but fixed coupling strength and in fact should hold at arbitrary $\lambda$.

We note that in (17.41) there are two terms inside the big round bracket with the first one being temperature independent. This is the Einstein $A$-coefficient which regulates the spontaneous emission of a photon. The second term in (17.41) is the $B$-coefficient of stimulated emission and adsorption of a photon. It dominates for $\hbar|\varepsilon_i - \varepsilon_j| \ll k_{\mathrm{B}}T$. From the point of view of the atom, there is no way to distinguish the two emission processes.

## 17.3 Resonances

The virtue of the weak coupling theory consists in yielding a concise dynamical scenario with level shifts and lifetimes computed in terms of the microscopic Hamiltonian. High-precision experiments, e.g. of the Lamb shift in the hydrogen atom, show small deviations from the prediction of the theory, which however should not be regarded as a failure of the weak coupling theory. Rather, it is a failure of the Pauli–Fierz model at relativistic energies. Barring such fine details the weak coupling theory is the standard tool in atomic physics and there seems to be little incentive to go beyond. Still, we have not yet developed a firm link with the Hamiltonian. Are there corrections to the predicted exponential decay? Can one, at least in principle, obtain systematic corrections of higher order in $\lambda$? What is the long-time limit for small, but fixed $\lambda$? To answer such questions one has to go beyond perturbation theory and simple resummations. At present there is only one sufficiently powerful technique available, which is complex dilation. We explain this method first for the standard example of the Friedrichs–Lee model. The extension to the Pauli–Fierz model requires rather complex technical machinery, certainly beyond the present scope. We will, however, use complex dilations to study the return to equilibrium at nonzero temperatures in chapter 18, which turns out to be much simpler since the spectrum is the full real line and is translated rather than rotated.

We imagine a single energy level $\varepsilon > 0$, coupled to the continuum, which is labeled by $x \geq 0$, and should be thought of as energy. The Hilbert space of wave functions is then $\mathbb{C} \oplus L^2(\mathbb{R}_+, \mathrm{d}x)$ and the Hamiltonian reads

$$H_\lambda = H_0 + \lambda H_{\mathrm{int}} = \begin{pmatrix} \varepsilon & 0 \\ 0 & x \end{pmatrix} + \lambda \begin{pmatrix} 0 & \langle\varphi| \\ |\varphi\rangle & 0 \end{pmatrix} \tag{17.46}$$

in Dirac notation. $H_\lambda$ is known as the Friedrichs–Lee model. For some time we choose to denote by $H_\lambda$ the Hamiltonian of (17.46) and will give a warning to the reader when we return to the Hamiltonian (17.10). One needs $\varphi \in L^2$ to have $H_\lambda$ well defined and $\langle\varphi, x^{-1}\varphi\rangle < \infty$ for $\lambda H_{\mathrm{int}}$ to be form-bounded with respect to $H_0$. With no loss one can choose $\varphi$ to be real. For $\lambda = 0$ the eigenvalue $\varepsilon$ is embedded in the continuum and we want to understand its fate for small $\lambda$.

From scattering theory and the stability of the essential spectrum under rank-one perturbations it can be seen that the absolutely continuous spectrum of $H_\lambda$ is $[0, \infty)$ for all $\lambda$. In addition, there exists a critical $\lambda_c$ such that for $|\lambda| < \lambda_c$ there is no further spectrum, whereas for $|\lambda| > \lambda_c$ the eigenvalue $\varepsilon(\lambda) < 0$ gets expelled from the continuum. We are interested here in small $\lambda$ only, i.e. $|\lambda| \ll \lambda_c$, but, beyond mere spectral information, we want to know the decay of the survival amplitude

$$G(t) = \langle \psi_0, \mathrm{e}^{-\mathrm{i}H_\lambda t} \psi_0 \rangle \tag{17.47}$$

of the unperturbed eigenstate $\psi_0 = \binom{1}{0}$.

$G(t)$ has the spectral decomposition

$$G(t) = \int \mathrm{d}\omega g(\omega) \mathrm{e}^{-\mathrm{i}\omega t}, \tag{17.48}$$

$g(\omega) \geq 0$ for $\omega \geq 0$, $g(\omega) = 0$ for $\omega < 0$, and $\int \mathrm{d}\omega g(\omega) = 1$. Thus, $|G(t)|^2 \cong 1 - t^2$ for small $t$ and $G(t) \to 0$ as $t \to \infty$ by the Riemann–Lebesgue lemma. On the other hand, $G(t)$ cannot decay exponentially, for this would imply $g(\omega)$ to be analytic in a strip around the real axis and thus $g \equiv 0$, by the reasoning of Paley and Wiener. Since $H_{\mathrm{int}}$ is a one-dimensional projection, $g(\omega)$ is in fact easily computed. First, the resolvent is determined as

$$\widehat{G}(z) = \langle \psi_0, (z - H_\lambda)^{-1} \psi_0 \rangle = \Big[ z - \varepsilon - \lambda^2 \langle \varphi, \frac{1}{z - x} \varphi \rangle \Big]^{-1}, \tag{17.49}$$

$z \in \mathbb{C} \setminus \mathbb{R}_+$. Then

$$g(\omega) = (2\pi \mathrm{i})^{-1} \lim_{\eta \to 0+} [\widehat{G}(\omega + \mathrm{i}\eta) - \widehat{G}(\omega - \mathrm{i}\eta)]. \tag{17.50}$$

Since

$$\lim_{\eta \to 0+} \langle \varphi, (\omega \pm \mathrm{i}\eta - x)^{-1} \varphi \rangle = \Delta(\omega) \mp \mathrm{i}\Gamma(\omega)/2 \tag{17.51}$$

with

$$\Gamma(\omega)/2 = \pi |\varphi(\omega)|^2, \quad \Delta(\omega) = \mathrm{PV}\!\!\int_0^\infty \mathrm{d}x |\varphi(x)|^2 (x - \omega)^{-1}, \tag{17.52}$$

one has

$$g(\omega) = \frac{1}{2\pi} \frac{\lambda^2 \Gamma(\omega)}{(\omega - \varepsilon - \lambda^2 \Delta(\omega))^2 + (\lambda^2 \Gamma(\omega)/2)^2} \tag{17.53}$$

for $\omega \geq 0$, and $g(\omega) = 0$ for $\omega < 0$. For small $\lambda$, $g(\omega)$ has a huge bump located near $\omega = \varepsilon$. In the weak coupling theory, one ignores the variation of $\Gamma$ and $\Delta$ and approximates $g(\omega)$ for all $\omega$ by

$$g_{\mathrm{w}}(\omega) = \frac{1}{2\pi} \frac{\lambda^2 \Gamma(\varepsilon)}{(\omega - \varepsilon - \lambda^2 \Delta(\varepsilon))^2 + (\lambda^2 \Gamma(\varepsilon)/2)^2}, \tag{17.54}$$

which corresponds to the survival amplitude

$$G_{\mathrm{w}}(t) = \mathrm{e}^{-(\lambda^2 \Gamma(\varepsilon)/2)|t|} \mathrm{e}^{-\mathrm{i}(\varepsilon + \lambda^2 \Delta(\varepsilon))t}. \tag{17.55}$$

For the true survival amplitude one still obtains the bound

$$|G(t) - G_{\mathrm{w}}(t)| \leq c\lambda^2 \tag{17.56}$$

uniformly in $t$, provided $\varphi$ has some smoothness. The errors in (17.56) come from very short times, $\lambda^2 t \ll 1$, and very long ones, $\lambda^2 t \gg 1$. In the intermediate regime $G_{\mathrm{w}}(t)$ does very well.

For models like the Pauli–Fierz model one cannot hope for such explicit formulas. Instead, for the purpose of computing $g(\omega)$, the strategy is to continue the resolvent $\widehat{G}(z)$ from the upper half of the complex plane across $\mathbb{R}_+$ into the second Riemann sheet. Ideally, one should discover a simple pole, the resonance, located at $z_{\mathrm{r}}(\lambda) = \varepsilon + \lambda^2\Delta - \mathrm{i}\lambda^2\Gamma/2$ with $\Gamma > 0$. For small $\lambda$ one expects $\Delta \cong \Delta(\varepsilon)$, $\Gamma \cong \Gamma(\varepsilon)$, but as $\lambda$ is increased the pole $z_{\mathrm{r}}(\lambda)$ will move further away from the real axis. The resonance pole is responsible for the exponential decay as in (17.55) with $\Delta(\varepsilon)$, $\Gamma(\varepsilon)$ replaced by the true $\Delta$, $\Gamma$. The error, as in (17.56), comes from the background spectrum of $\widehat{G}(z)$ on the second Riemann sheet, unavoidable due to the branch cut at $z = 0$.

One would hope that $z_{\mathrm{r}}(\lambda)$ is an intrinsic property of $H_\lambda$ and not merely of the particular matrix element under study. Of course, we can always pick a bad coupling function $\varphi$ such that $\langle\varphi, (z - x)^{-1}\varphi\rangle$ cannot be analytically continued across $\mathbb{R}_+$ or for a nice coupling $\varphi$, we could pick a bad wave function $\psi$ such that $\langle\psi, (z - H_\lambda)^{-1}\psi\rangle$ cannot be analytically continued across $\mathbb{R}_+$. Thus the best we can expect is that for a given sufficiently smooth $\varphi$ the location of the resonance pole is independent of the choice of $\psi$ within a reasonably large set. To accomplish the desired analytic continuation we will implement a complex dilation of $H_\lambda$.

For real $\theta$ a dilation is defined by

$$U(\theta)\psi(x) = \mathrm{e}^{-\theta/2}\psi(\mathrm{e}^{-\theta}x). \tag{17.57}$$

$U(\theta)$ is unitary and $H_\lambda$ transforms under $U(\theta)$ as

$$\begin{aligned} U(\theta)H_\lambda U(\theta)^{-1} &= H_\lambda(\theta) = H_0(\theta) + \lambda H_{\mathrm{int}}(\theta) \\ &= \begin{pmatrix} \varepsilon & 0 \\ 0 & \mathrm{e}^{-\theta}x \end{pmatrix} + \lambda \begin{pmatrix} 0 & \langle\varphi_\theta| \\ |\varphi_\theta\rangle & 0 \end{pmatrix}, \end{aligned} \tag{17.58}$$

where $\varphi_\theta(x) = \mathrm{e}^{-\theta/2}\varphi(\mathrm{e}^{-\theta}x)$.

We want to extend (17.57), (17.58) to complex $\theta$ with $\theta$ inside the strip $\mathcal{S}_\beta = \{\theta \,|\, |\mathrm{Im}\theta| < \beta\}$ with some $\beta > 0$. $\mathrm{e}^{-\theta}$ is clearly analytic. For $\varphi$ we require that $\varphi_\theta$ extends as an analytic function to $\mathcal{S}_\beta$ such that $\int_0^\infty \mathrm{d}x\, |\mathrm{e}^{-\theta/2}\varphi(\mathrm{e}^{-\theta}x)|^2 < \infty$. Then $H_\lambda(\theta)$ is an analytic family of operators of type A in the sense of Kato, separately for $\theta \in \mathcal{S}_\beta$ and $|\lambda|$ sufficiently small. Note that $H_\lambda(\theta)^* = H_\lambda(\theta^*)$ for real $\lambda$, since $\varphi$ is real. The point of our construction is that for purely imaginary $\theta$, $\theta = \mathrm{i}\vartheta$,

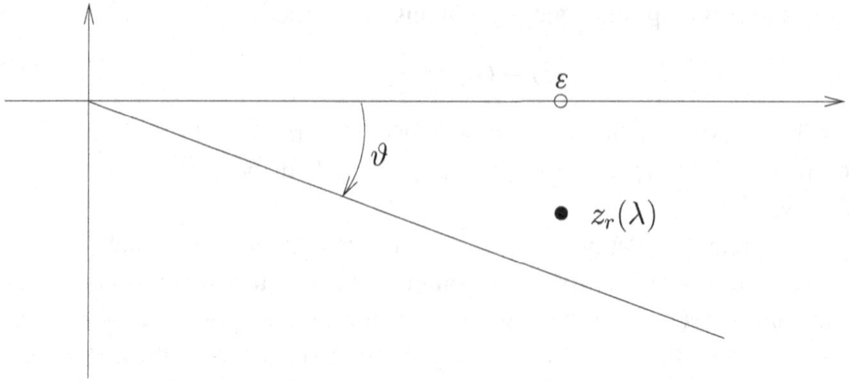


Figure 17.1: Spectrum of the rotated Hamiltonian $H_\lambda(\mathrm{i}\vartheta)$ for small coupling $\lambda$.

$0 < \vartheta < \beta$, the continuous spectrum of $H_0(\mathrm{i}\vartheta)$ rotates clockwise by the angle $\vartheta$, see figure 17.1. Thereby the previously embedded eigenvalue $\varepsilon$ becomes isolated and we can use ordinary perturbation theory to show that it shifts downwards to become the resonance pole $z_\mathrm{r}(\lambda)$ on the second Riemann sheet. $z_\mathrm{r}(\lambda)$ is analytic in $\theta$ as long as it remains isolated. If one sets $\theta = \kappa + \mathrm{i}\vartheta$, $\kappa, \vartheta \in \mathbb{R}$, then $H_\lambda(\kappa + \mathrm{i}\vartheta)$ is unitarily equivalent to $H_\lambda(\kappa' + \mathrm{i}\vartheta)$. Therefore $z_\mathrm{r}(\lambda)$ is constant along lines of fixed $\mathrm{i}\vartheta$ and by analyticity independent of $\theta$. As the continuous spectrum rotates clockwise, the resonance pole is uncovered and stays put. We summarize as

**Theorem 17.2** (Analytic continuation of the resolvent). *For $\lambda$ sufficiently small, there exists a dense set $D \subset \mathcal{H} = \mathbb{C} \oplus L^2(\mathbb{R}_+, \mathrm{d}x)$ such that for $\psi_1, \psi_2 \in D$ the resolvent $\langle \psi_1, (z - H_\lambda)^{-1}\psi_2\rangle$ has an analytic continuation from $\mathbb{C}_+$ across $\mathbb{R}_+$ into the second Riemann sheet. $\langle \psi_1, (z - H_\lambda)^{-1}\psi_2\rangle$ has a simple pole at $z_\mathrm{r}(\lambda)$, $\mathrm{Im} z_\mathrm{r}(\lambda) < 0$, with the property that $\lim_{\lambda\to 0} z_\mathrm{r}(\lambda) = \varepsilon$. $z_\mathrm{r}(\lambda)$ does not depend on the choice of $\psi_1, \psi_2$.*

*Proof*: Let $D \subset \mathcal{H}$ be the set of all vectors such that $\theta \mapsto U(\theta)\psi$ is an analytic vector-valued function on $S_\beta$. $D$ is dense in $\mathcal{H}$. For $\psi_1, \psi_2 \in D$ we have

$$\langle \psi_1, (z - H_\lambda)^{-1}\psi_2\rangle = \langle U(-\theta)^*\psi_1, (z - H_\lambda(\theta))^{-1}U(\theta)\psi_2\rangle. \quad (17.59)$$

For given $\theta$ with $\mathrm{Im}\theta > 0$, we can choose $\delta_0$ sufficiently small such that inside the open disc $|z - \varepsilon| \le \delta_0$ the location $z_\mathrm{r}(\lambda)$ of the pole is an analytic curve starting at $z_\mathrm{r}(0) = \varepsilon$. □

Let us follow the first step of the perturbation expansion. We fix $\theta = \mathrm{i}\vartheta$, $0 < \vartheta < \beta$. For $\lambda = 0$, $H_0(\theta)$ has the eigenvalue $\varepsilon$ with corresponding projector $|\psi_0\rangle\langle\psi_0|$. The eigenvalue persists for small $\lambda$ and we expand in $\lambda$. The first-order term vanishes and to second order we have

$$
\begin{aligned}
z_r(\lambda) &= \varepsilon + \lambda^2 \langle \psi_0, H_{\text{int}}(\theta)(\varepsilon - H_0(\theta))^{-1} H_{\text{int}}(\theta)\psi_0 \rangle \\
&= \varepsilon + \lim_{\eta \to 0_+} \lambda^2 \langle \psi_0, H_{\text{int}}(\varepsilon + \mathrm{i}\eta - H_0)^{-1} H_{\text{int}}\psi_0 \rangle \\
&= \varepsilon + \lambda^2 \Delta(\varepsilon) - \mathrm{i}\lambda^2 \Gamma(\varepsilon)/2. \qquad (17.60)
\end{aligned}
$$

No surprise, we recover the result from the weak coupling theory. We will see that this is a rather general fact and argue that the master equation (17.32) can be understood as arising from the resonances of the Liouvillean to lowest order. If the expansion in (17.60) is continued, the next order is $\lambda^4$ and the eigenprojection of the resonance will be slightly tilted.

With the Friedrichs–Lee model as a blueprint in hand we plan to implement complex dilation for the Pauli–Fierz model in the $N$-level approximation (17.10). As in the example above the complex dilation acts only on the photon degrees of freedom. For an $n$-photon vector we define

$$
U_f(\theta)\psi_n(k_1, \lambda_1, \ldots, k_n, \lambda_n) = \mathrm{e}^{-3n\theta/2}\psi_n(\mathrm{e}^{-\theta}k_1, \lambda_1, \ldots, \mathrm{e}^{-\theta}k_n, \lambda_n) \quad (17.61)
$$

for $\theta \in \mathbb{R}$. In particular

$$
U_f(\theta)a^*(f)U_f(\theta)^{-1} = a^*(f_\theta), \quad f_\theta(k, \lambda) = \mathrm{e}^{-3\theta/2} f(\mathrm{e}^{-\theta}k, \lambda). \quad (17.62)
$$

Then for the field energy

$$
U_f(\theta)H_f U_f(\theta)^{-1} = H_f(\theta) = \mathrm{e}^{-\theta} H_f \quad (17.63)
$$

and for the electric field

$$
\begin{aligned}
U_f(\theta)E_\varphi U_f(\theta)^{-1} &= E_\varphi(\theta) \\
&= \sum_{\lambda=1,2} \int \mathrm{d}^3k \mathrm{e}^{-3\theta/2} \widehat{\varphi}(\mathrm{e}^{-\theta}k) \mathrm{e}^{-\theta/2} \sqrt{\omega(k)/2} \\
&\quad \times e_\lambda(k) \mathrm{i}\big(a(k, \lambda) - a^*(k, \lambda)\big). \qquad (17.64)
\end{aligned}
$$

We want to extend (17.63), (17.64) to complex $\theta \in \mathcal{S}_\beta$. Clearly $H_f(\theta)$ is analytic in $\theta$. For the charge distribution we require that $\widehat{\varphi}_\theta(k)$ extends as an analytic function to $\mathcal{S}_\beta$ and

$$
\int \mathrm{d}^3k |\widehat{\varphi}_\theta|^2 \big(\omega(k) + \omega(k)^{-1}\big) < \infty. \quad (17.65)
$$

Then $E_\varphi(\theta)$ is bounded relative to $H_f(\theta)$ and

$$
H_\lambda(\theta) = H_{\text{at}} + H_f(\theta) + \lambda Q \cdot E_\varphi(\theta) \quad (17.66)
$$

is an analytic family of operators of type A separately in $\theta \in \mathcal{S}_\beta$ and $\lambda$, with $|\lambda| < \lambda_0$ and $\lambda_0$ sufficiently small. Thus we have established the abstract framework needed for complex dilation.

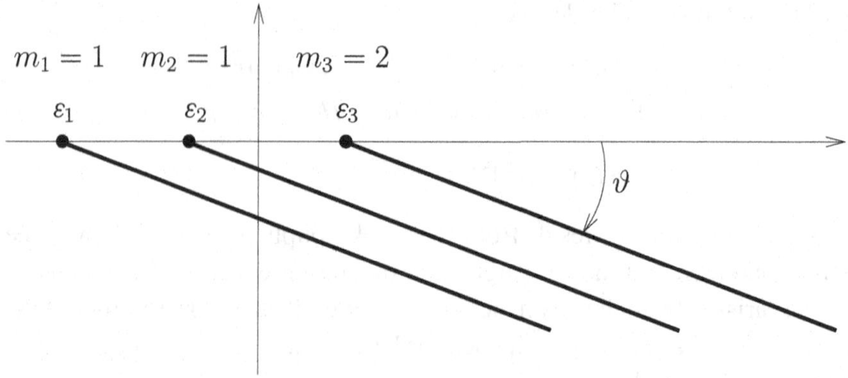


Figure 17.2: Spectrum of the rotated Hamiltonian $H_0(\mathrm{i}\vartheta)$ at zero coupling.

The difficulty already becomes apparent when the case of zero coupling is considered, i.e. $\lambda = 0$ ($= -e$). $H_\mathrm{f}$ has a zero eigenvalue and the continuous spectrum $\mathbb{R}_+$ is of infinite multiplicity. If $\theta = \mathrm{i}\vartheta$, the continuous spectrum rotates by the angle $\vartheta$. Thus for $H_\mathrm{at} + H_\mathrm{f}(\theta)$ we have a spectrum as shown in figure 17.2, where the eigenvalues $\varepsilon_j$, $j = 1, \ldots, \bar{N}$, are at the tip of the continuous spectrum. In contrast to the Friedrichs–Lee model, they are not isolated. We can make them become isolated by giving the photons a small mass $m_\mathrm{ph}$. Then $\omega(k) = (m_\mathrm{ph}^2 + k^2)^{1/2}$ which becomes $\omega_\theta(k) = (\mathrm{e}^{-2\theta}k^2 + m_\mathrm{ph}^2)^{1/2}$ when complex dilated. The eigenvalues are now isolated provided they do not lie in the set of thresholds $\{\varepsilon_j + nm_\mathrm{ph} | \, j = 1, \ldots, \bar{N}, n = 1, 2, \ldots\}$. Our previous arguments apply, but the range of allowed $\lambda$ is bounded by $m_\mathrm{ph}$.

In a beautiful piece of analysis V. Bach, J. Fröhlich, and I. M. Sigal succeed in controlling the situation depicted in figure 17.3. They prove that for sufficiently small $\lambda$ and a dense set $D$ of vectors the resolvent $\langle \psi, (z - H_\lambda)^{-1}\varphi\rangle$, $\psi, \varphi \in D$, can be analytically continued into a domain, schematically drawn in figure 17.3. For $\lambda = 0$ the eigenvalues are $\varepsilon_j$ with multiplicity $m_j$. Except for $j = 1$, for small $\lambda$ they turn into a group of resonances $z_{jm}(\lambda)$, $m = 1, \ldots, m_j$, with the property that $\lim_{\lambda\to 0} z_{jm}(\lambda) = \varepsilon_j$. The ground state energy $\varepsilon_1$ is nondegenerate and $z_1(\lambda)$ stays on the real axis. $z_1(\lambda)$ is the ground state energy of the coupled system. The $z_{jm}(\lambda)$ are eigenvalues of the complex dilated Hamiltonian $H_\lambda(\theta)$. The resonances are located at the apex of a cone, which is tilted by the angle $\theta$ and has a square root singularity at its tip.

To ensure that the resonances are strictly below the real axis we use the condition from second-order perturbation and require that

$$\frac{1}{2}\lambda^2 \sum_{\alpha=1}^{3} \sum_{i=1}^{j-1} \Gamma(\varepsilon_i - \varepsilon_j) P_j Q_\alpha P_i Q_\alpha P_j > 0 \tag{17.67}$$

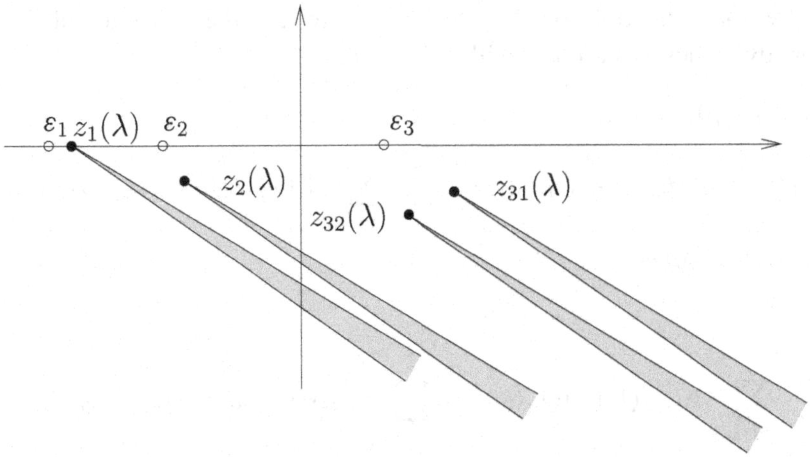


Figure 17.3: Spectrum of the rotated Hamiltonian $H_\lambda(\mathrm{i}\vartheta)$ at small nonzero coupling. The domain of analyticity is $\{z|\mathrm{Im}z > -b_0\}$ with small $b_0 > 0$ and away from the shaded regions.

as an $m_j \times m_j$ matrix for $j = 1, \ldots, \bar{N}$. The eigenvalues of this matrix are: $\mathrm{Im}z_{jm}(\lambda)$, $m = 1, \ldots, m_j$, to order $\lambda^2$. To second-order the imaginary part of the resonance poles agrees with the decay rates from the quantum master equation (17.32). Their real part coincides with the eigenvalues of $H_{\mathrm{at}}$ corrected by the Hamiltonian part of $K^\natural$ from (17.37). To obtain the full generator $K^\natural$ one has to study the resonances of the Liouvillean as will be discussed in chapter 18.

## 17.4 Fluorescence

We have described in considerable detail how the atom decays to its ground state, at least for small coupling. So what then are the spectral characteristics of the fluorescent light? How does the theory account for the experimental fact that the line shapes differ for equally and for unequally spaced unperturbed energy levels? We will address such questions only within the weak coupling theory.

The initial state of the atom is chosen to be a pure state $\psi \in \mathbb{C}^N$ and that of the field to be the vacuum. We want to determine $\mathrm{e}^{-\mathrm{i}H_\lambda t}\psi \otimes \Omega$ for small $\lambda$ and large $t$, order of $\lambda^{-2}$. One method is to use second-order perturbation theory for the resonance poles of the resolvent, as explained for a particular case in the previous section. Another method is to expand the resolvent $\langle \varphi \otimes \prod_{j=1}^{n} a^*(k_j, \lambda_j)\Omega, (z - H_\lambda)^{-1}\psi \otimes \Omega\rangle$ and to resum all nonoverlapping internal photon lines lying in between either the external photon legs or the atom legs. The results turn out to be identical and have a simple physical interpretation.

To state the approximation to $e^{-iH_\lambda t}\psi \otimes \Omega$, we rewrite the generator of the reduced atom dynamics as, compare with (17.37), (17.32),

$$\begin{aligned} L_D\rho &= [H_{at}, \rho] + i\lambda^2 K^\natural \rho \\ &= [H_{at} + \lambda^2 H_\Delta - i\lambda^2 H_\Gamma, \rho] + i\sum_{\alpha=1}^{3}\sum_{\omega\in\sigma(L_{at})} \Gamma(\omega) Q_\alpha(\omega)^* \rho Q_\alpha(\omega) \\ &= L_{D0}\rho + L_{D1}\rho. \end{aligned} \tag{17.68}$$

Then

$$H_\Delta = \sum_{\alpha=1}^{3}\sum_{\omega\in\sigma(L_{at})} \Delta(\omega) Q_\alpha(\omega) Q_\alpha(\omega)^* = \sum_{\alpha=1}^{3}\sum_{i,j=1}^{\bar N} \Delta(\varepsilon_i - \varepsilon_j) P_j Q_\alpha P_i Q_\alpha P_j, \tag{17.69}$$

and

$$H_\Gamma = \sum_{\alpha=1}^{3}\sum_{\omega\in\sigma(L_{at})} \Gamma(\omega) Q_\alpha(\omega) Q_\alpha(\omega)^* = \sum_{\alpha=1}^{3}\sum_{i<j=1}^{\bar N} \Gamma(\varepsilon_i - \varepsilon_j) P_j Q_\alpha P_i Q_\alpha P_j, \tag{17.70}$$

where we used the relation $\Gamma(\omega) = 0$ for $\omega \geq 0$. We introduce the convenient shorthand

$$H_d = H_{at} + \lambda^2 H_\Delta - i\lambda^2 H_\Gamma. \tag{17.71}$$

Note that $H_d$ is not symmetric. As a photon is emitted, the energy of the atom decreases by at least one level, which is described by the atom lowering part of the interaction Hamiltonian,

$$Q^- \cdot E_\varphi^+ = -i \sum_{i<j=1}^{\bar N} P_i Q P_j \otimes \Big( \sum_{\lambda=1,2} \int d^3k \widehat{\varphi}(k) \sqrt{\omega/2} \cdot e_\lambda(k) a^*(k,\lambda) \Big). \tag{17.72}$$

With this notation the approximate solution is

$$\begin{aligned} e^{-iH_\lambda t}\psi \otimes \Omega \cong{}& e^{-iH_d t}\psi \otimes \Omega + \sum_{n=1}^{\bar N - 1} (-i)^n \int_{0\leq t_1 \leq \ldots \leq t_n \leq t} dt_n \ldots dt_1 \\ &\times e^{-i(H_d + H_f)(t - t_n)} \lambda Q^- \cdot E_\varphi^+ \ldots e^{-i(H_d+H_f)(t_2 - t_1)} \\ &\times \lambda Q^- \cdot E_\varphi^+ e^{-i(H_d + H_f)t_1} \psi \otimes \Omega. \end{aligned} \tag{17.73}$$

The sum is finite, since $(Q^-)^{\bar N}\psi = 0$.

Taking in (17.73) the trace over the atom results in the reduced state of the photon field. Taking the trace over the field yields the reduced state of the atom. But this state was already determined in section 17.2. To be consistent with it we must have

$$\mathrm{tr}_{\mathcal{F}}[|\mathrm{e}^{-\mathrm{i}H_\lambda t}\psi \otimes \Omega\rangle\langle \mathrm{e}^{-\mathrm{i}H_\lambda t}\psi \otimes \Omega|] = \mathrm{e}^{-\mathrm{i}Lt} P_\psi, \tag{17.74}$$

at least for small $\lambda$, $P_\psi$ the projection onto $\psi$. If (17.74) holds, the case of an arbitrary initial density matrix follows by linearity.

To prove (17.74) we insert (17.73) and obtain

$$\begin{aligned}
&\mathrm{tr}_{\mathcal{F}}[|\mathrm{e}^{-\mathrm{i}H_\lambda t}\psi \otimes \Omega\rangle\langle \mathrm{e}^{-\mathrm{i}H_\lambda t}\psi \otimes \Omega|] \\
&= \mathrm{e}^{-\mathrm{i}H_{\mathrm{d}}t} P_\psi \mathrm{e}^{\mathrm{i}H_{\mathrm{d}}^* t} + \sum_{n=1}^{\bar{N}-1} \lambda^{2n} \int_{0\le t_1\le \ldots \le t_n \le t} \mathrm{d}t_n \ldots \mathrm{d}t_1 \\
&\quad \times \int_{0\le s_1\le \ldots \le s_n \le t} \mathrm{d}s_n \ldots \mathrm{d}s_1 \sum_{\alpha_1,\beta_1=1}^{3} \ldots \sum_{\alpha_n,\beta_n=1}^{3} \\
&\quad \times \prod_{j=1}^{n} \delta_{\alpha_j\beta_j} h(s_j - t_j) \mathrm{e}^{-\mathrm{i}H_{\mathrm{d}}(t-t_n)} Q_{\alpha_n}^- \mathrm{e}^{-\mathrm{i}H_{\mathrm{d}}(t_n - t_{n-1})} Q_{\alpha_{n-1}}^- \ldots Q_{\alpha_1}^- \mathrm{e}^{-\mathrm{i}H_{\mathrm{d}}t_1} P_\psi \\
&\quad \times \mathrm{e}^{\mathrm{i}H_{\mathrm{d}}^* s_1} Q_{\beta_1}^{-*} \ldots Q_{\beta_{n-1}}^{-*} \mathrm{e}^{\mathrm{i}H_{\mathrm{d}}^*(s_n - s_{n-1})} Q_{\beta_n}^{-*} \mathrm{e}^{\mathrm{i}H_{\mathrm{d}}^*(t-s_n)}.
\end{aligned} \tag{17.75}$$

Since $[H_{\mathrm{at}}, H_\Delta] = 0 = [H_{\mathrm{at}}, H_\Gamma]$, one can use the spectral representation

$$\mathrm{e}^{-\mathrm{i}H_{\mathrm{d}}t} = \sum_{j=1}^{\bar{N}} \mathrm{e}^{-\mathrm{i}\varepsilon_j t} P_j \mathrm{e}^{-\mathrm{i}\lambda^2 H_\Delta t - \lambda^2 H_\Gamma t} \tag{17.76}$$

and insert it for each propagator in (17.75). On the time scale $\lambda^{-2}\tau$, $\tau = \mathcal{O}(1)$, $h(t)$ decays quickly and the factors $\mathrm{e}^{-\mathrm{i}\varepsilon t}$ are rapidly oscillating. The generic integral in (17.75) is of the form

$$\int_0^\tau \mathrm{d}t \int_0^{\tau'} \mathrm{d}s \lambda^{-2} h(\lambda^{-2}(s-t)) \mathrm{e}^{\mathrm{i}(\varepsilon_i - \varepsilon_j)t/\lambda^2} \mathrm{e}^{-\mathrm{i}(\varepsilon_m - \varepsilon_n)s/\lambda^2}. \tag{17.77}$$

In the limit $\lambda \to 0$ it converges to

$$\min(\tau, \tau') \int \mathrm{d}t\, h(t) \mathrm{e}^{-\mathrm{i}(\varepsilon_m - \varepsilon_n)t} \delta_{\varepsilon_i - \varepsilon_j, \varepsilon_m - \varepsilon_n} = \min(\tau, \tau') \Gamma(\varepsilon_m - \varepsilon_n) \delta_{\varepsilon_i - \varepsilon_j, \varepsilon_m - \varepsilon_n}. \tag{17.78}$$

Using (17.78) and (17.76), the small-$\lambda$ limit of the expression in (17.75) is given by

$$\begin{aligned} &\mathrm{e}^{-\mathrm{i}L_{\mathrm{D}0}t} P_\psi + \sum_{n=1}^{\bar{N}-1} (-\mathrm{i})^n \int_{0\le t_1 \le \ldots \le t_n \le t} \mathrm{d}t_n \ldots \mathrm{d}t_1 \\ &\quad \times \mathrm{e}^{-\mathrm{i}L_{\mathrm{D}0}(t-t_n)} L_{\mathrm{D}1} \mathrm{e}^{-\mathrm{i}L_{\mathrm{D}0}(t_n - t_{n-1})} \ldots L_{\mathrm{D}1} \mathrm{e}^{-\mathrm{i}L_{\mathrm{D}0}t_1} P_\psi \\ &\quad = \mathrm{e}^{-\mathrm{i}L_{\mathrm{D}}t} P_\psi \end{aligned} \tag{17.79}$$

as was to be shown.

The approximate solution (17.73) describes the decay of the atom and the build-up of photons. Such details are experimentally inaccessible. However, what can be easily seen are the spectral characteristics of the fluorescent light, which are obtained from (17.73) in the limit $t \to \infty$ (on the time scale $\lambda^{-2}$). Then the atom is in its ground state and

$$\mathrm{e}^{-\mathrm{i}H_\lambda t} \psi \otimes \Omega \cong \mathrm{e}^{-\mathrm{i}(\varepsilon_1 + \lambda^2 \Delta_1)t} \psi_1 \otimes \mathrm{e}^{-\mathrm{i}H_{\mathrm{f}}t} \phi, \tag{17.80}$$

where $\phi$ is a photon state propagating freely to infinity through $\mathrm{e}^{-\mathrm{i}H_{\mathrm{f}}t}$. $\phi$ can be read off from (17.73) as

$$\begin{aligned} \psi_1 \otimes \phi = P_1 \psi \otimes \Omega + \sum_{n=1}^{\bar{N}-1} (-\mathrm{i})^n \int_{0\le t_1 \le \ldots \le t_n < \infty} \mathrm{d}t_n \ldots \mathrm{d}t_1 \, P_1 \mathrm{e}^{\mathrm{i}(H_{\mathrm{d}} + H_{\mathrm{f}})t_n} \\ \times \lambda Q^- \cdot E_\varphi^+ \ldots \mathrm{e}^{-\mathrm{i}(H_{\mathrm{d}} + H_{\mathrm{f}})(t_2 - t_1)} \lambda Q^- \cdot E_\varphi^+ \mathrm{e}^{-\mathrm{i}(H_{\mathrm{d}} + H_{\mathrm{f}})t_1} \psi \otimes \Omega. \end{aligned} \tag{17.81}$$

The projection $P_1$ comes in, since states in (17.73) which are orthogonal to the uncoupled ground state $\psi_1$ decay exponentially and only the piece parallel to $\psi_1$ persists in the long-time limit.

To see how (17.80) translates to the spectrum of the emitted light, it might be useful to work out two concrete cases.

*(i) Two-level atom.* We consider two nondegenerate levels $|1\rangle$, $|2\rangle$ with resonance poles $z_j = \varepsilon_j + \Delta_j - \mathrm{i}\Gamma_j/2$, $j = 1, 2$, $\Gamma_1 = 0$. Initially the atom is in state $|2\rangle$. Then the scattering state $\phi$ of (17.80) has only one photon, $\phi = (0, \phi_1, 0, \ldots)$, with wave function

$$\phi_1(k_1, \lambda_1) = \big((\Gamma_2/2) + \mathrm{i}(\varepsilon_2 + \Delta_2 - \varepsilon_1 - \Delta_1 - \omega(k_1))\big)^{-1} f_{12}(k_1, \lambda_1), \tag{17.82}$$

where

$$f_{12}(k, \lambda) = e\widehat{\varphi}(k) \langle 1|x|2\rangle \cdot e_\lambda(k) \sqrt{\omega(k)/2}. \tag{17.83}$$

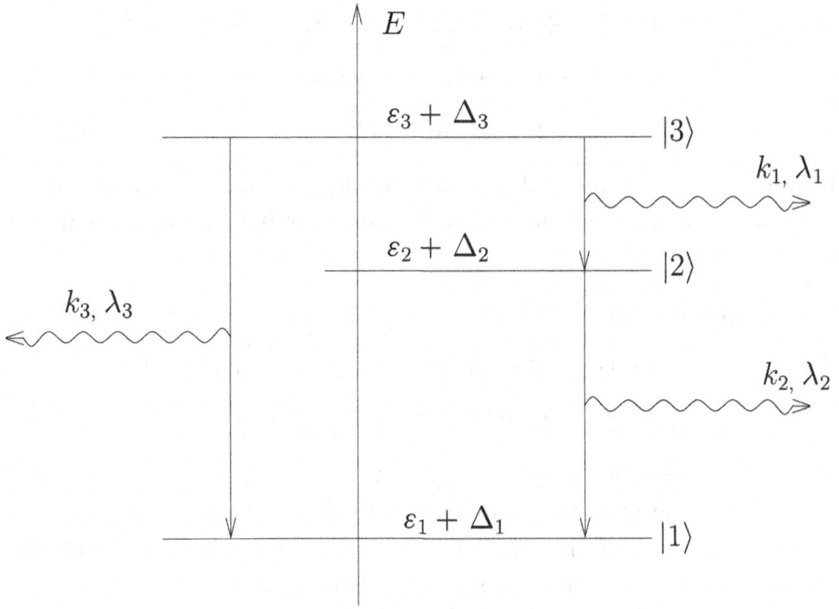


Figure 17.4: Radiation cascade for a three-level atom.

The spectral distribution is $|\phi_1|^2$. Since $\Gamma_2$ is small, the variation from $f_{12}$ can be ignored to obtain

$$|\phi_1(k,\lambda)|^2 \cong C\big[(\varepsilon_2+\Delta_2-\varepsilon_1-\Delta_1-\omega)^2+(\Gamma_2/2)^2\big]^{-1} = I(\omega,\widehat{k},\lambda) \tag{17.84}$$

with the constant $C = |f_{12}(k,\lambda)|^2$ evaluated at $\omega(k) = \varepsilon_2 - \varepsilon_1$, $C = e^2|\widehat{\varphi}(\varepsilon_2 - \varepsilon_1)|^2((\varepsilon_2-\varepsilon_1)/2)|\langle 1|x|2\rangle \cdot e_\lambda(k)|^2$, which depends on the direction of emission, $\widehat{k} = k/|k|$, and on the polarization. As a function of the frequency $\omega$ of the emitted light, the line shape is Lorentzian of natural width $\Gamma_2$ and centered at $\varepsilon_2 + \Delta_2 - \varepsilon_1 - \Delta_1$, differing from the bare line $\varepsilon_2 - \varepsilon_1$ by the shift $\Delta_2 - \Delta_1$.

If the initial state of the atom is $c_1|1\rangle + c_2|2\rangle$, normalized as $|c_1|^2 + |c_2|^2 = 1$, then $\phi = (c_1\Omega, c_2\phi_1(k_1,\lambda_1), 0, \ldots)$. With probability $|c_1|^2$ no photon is emitted and with probability $|c_2|^2$ the line shape is that of (17.84).

*(ii) Three-level atom.* We consider three nondegenerate levels $|1\rangle, |2\rangle, |3\rangle$ with resonance poles $z_j = \varepsilon_j + \Delta_j - \mathrm{i}\Gamma_j/2$, $j = 1, 2, 3$, $\Gamma_1 = 0$. The initial state of the atom is $|3\rangle$. There is a direct transition $|3\rangle \to |1\rangle$ as in case (i). In addition we have the cascade $|3\rangle \to |2\rangle \to |1\rangle$. Therefore the scattering state is $\phi = (0, \phi_1, \phi_2, 0, \ldots)$, see figure 17.4. $\phi_1$ is as in (17.82) with label 2 replaced by label 3. For the cascade one obtains

$$\begin{aligned}\phi_2(k_1,\lambda_1,k_2,\lambda_2) = \mathsf{S}\big\{&\big(\Gamma_3/2 + \mathrm{i}(\varepsilon_3 + \Delta_3 - \varepsilon_1 - \Delta_1 - \omega(k_1) - \omega(k_2))\big)^{-1}\\ &\times\big(\Gamma_2/2 + \mathrm{i}(\varepsilon_2 + \Delta_2 - \varepsilon_1 - \Delta_1 - \omega(k_2))\big)^{-1}\\ &\times\sqrt{2} f_{12}(k_2,\lambda_2) f_{23}(k_1,\lambda_1)\big\}\end{aligned} \tag{17.85}$$

with $\mathsf{S}$ denoting symmetrization. If the variation and the direction dependence from $f_{12}$, $f_{23}$ are ignored, the intensity distribution for the two photons in the cascade is

$$\begin{aligned}I(\omega_1,\omega_2) = C\big(&4(\Gamma_2/2)^2 + (\delta_2 - \omega_1 + \delta_2 - \omega_2)^2\big)\\ &\times\big[\big((\Gamma_3/2)^2 + (\delta_2 + \delta_3 - \omega_1 - \omega_2)^2\big)\big((\Gamma_2/2)^2\\ &+ (\delta_2 - \omega_1)^2\big)\big((\Gamma_2/2)^2 + (\delta_2 - \omega_2)^2\big)\big]^{-1}\end{aligned} \tag{17.86}$$

with the shorthand $\delta_3 = \varepsilon_3 + \Delta_3 - \varepsilon_2 - \Delta_2$, $\delta_2 = \varepsilon_2 + \Delta_2 - \varepsilon_1 - \Delta_1$. If $\varepsilon_3 - \varepsilon_2 \neq \varepsilon_2 - \varepsilon_1$, then in the frequency spectrum one will observe a Lorentzian at $\delta_2$ with natural width $\Gamma_2$ and a Lorentzian at $\delta_3$ with natural width $\Gamma_2 + \Gamma_3$. On the other hand if $\varepsilon_3 - \varepsilon_2 = \varepsilon_2 - \varepsilon_1$ and, just as an example, also $\Delta_2 = \Delta_3$, $2\Gamma_2 = \Gamma_3$, then

$$I(\omega_1,\omega_2) = C\big((\Gamma_2/2)^2 + (\delta_2 - \omega_1)^2\big)^{-1}\big((\Gamma_2/2)^2 + (\delta_2 - \omega_2)^2\big)^{-1}, \tag{17.87}$$

which corresponds to a single Lorentzian at $\delta_2 = \delta_3$ of natural width $\Gamma_2$ with double intensity. The two photons interfere when emitted. Otherwise, the intensity would be the sum of a Lorentzian of natural width $\Gamma_2$ and one of natural width $\Gamma_2 + \Gamma_3 = 3\Gamma_2$. If $\varepsilon_3 - \varepsilon_2 \cong \varepsilon_2 - \varepsilon_1$, the exact intensity distribution (17.86) has to be analyzed anew.

## 17.5 Scattering theory

From a very general perspective scattering theory is a comparison between an interacting dynamics and a simplified "free" dynamics in the limit of long times. In our context this means a study of

$$\mathrm{e}^{-\mathrm{i}Ht}\psi \quad \text{as } t \to \infty \tag{17.88}$$

for an arbitrary initial state $\psi \in \mathcal{H} = \mathbb{C}^N \otimes \mathcal{F}$. We stay within the dipole approximation and consider

$$H = H_{\mathrm{at}} + H_{\mathrm{f}} - eQ \cdot E_\varphi, \tag{17.89}$$

$E_\varphi = E_\varphi(0)$. Since the coupling is fixed, we omit the index $\lambda$ and return to $e = -\lambda$, see (17.10). Also, $\langle\cdot,\cdot\rangle$ always denotes the scalar product in $\mathcal{H}$. From the outset we state

**Condition 17.3** (Uniqueness and localization of the ground state). *$H$ has a unique ground state $\psi_g$, $H\psi_g = E_g\psi_g$ with the property that $\langle \psi_g, e^{\delta N_f}\psi_g\rangle < \infty$ for some $\delta > 0$. $H$ has no other eigenvalues.*

On physical grounds it is easy to conjecture the limit in (17.88). Photons are traveling outwards according to a scattering state $\phi$ and the atom decays to its ground state $\psi_g$. Thus for given $\psi \in \mathbb{C}^N \otimes \mathcal{F}$, there exists a $\phi \in \mathcal{F}$ such that

$$e^{-iHt}\psi \cong e^{-iE_g t}\psi_g \otimes_s e^{-iH_f t}\phi \quad \text{as } t \to \infty. \tag{17.90}$$

In rough terms, the state $e^{-iH_f t}\phi$ lives far away from the ground state $\psi_g$. Still, the bound photons of $\psi_g$ must be properly symmetrized with the freely propagating photons of $e^{-iH_f t}\phi$. This is achieved by the symmetrization $\otimes_s$ as defined in (17.91), (17.92) below. We note that in the previous sections we have discussed an initial state of the particular form $\chi \otimes \Omega$. The relation (17.90) constitutes a vast generalization thereof. Of course, the limit (17.90) can be considered also for $t \to -\infty$. Combining both limits then yields the $S$-matrix for Rayleigh scattering of photons from an atom.

To establish the limit (17.90) in this generality is a tough analytical problem, since no exceptions are allowed. The limit is supposed to hold for all states $\psi \in \mathcal{H}$. We will only outline the general framework, in particular the proper definition of the wave operators and their intertwining between the free and interacting dynamics. As an easy step a Cook-type argument is established ensuring (17.90) at least for a large class of states. One important consequence of the limit (17.90) is the relaxation of the atom to its ground state without taking recourse to weak coupling, respectively resonance theory. As will be explained, such a relaxation holds also for local field observables.

Let us first have a look at the right-hand side of (17.90). The symmetrization $\otimes_s$ can be defined for two arbitrary states in Fock space. We consider the Fock space $\mathcal{F} = \mathcal{F}(\mathfrak{h})$ over the one-particle space $\mathfrak{h}$. Then $\mathcal{F}(\mathfrak{h} \oplus \mathfrak{h}) = \mathcal{F}(\mathfrak{h}) \otimes \mathcal{F}(\mathfrak{h})$. On the one-particle space we define the map

$$(u_1, u_2) \mapsto u_1 + u_2 \in \mathfrak{h}. \tag{17.91}$$

The second quantization of this map defines $\psi_1 \otimes \psi_2 \in \mathcal{F}(\mathfrak{h}) \otimes \mathcal{F}(\mathfrak{h}) \mapsto \psi_1 \otimes_s \psi_2 \in \mathcal{F}(\mathfrak{h})$. More explicitly, one has

$$\Big(\prod_{j=1}^{n} a^*(f_j)\Omega\Big) \otimes_s \Big(\prod_{i=1}^{m} a^*(g_i)\Omega\Big) = \prod_{j=1}^{n} a^*(f_j) \prod_{i=1}^{m} a^*(g_i)\Omega. \tag{17.92}$$

In our case one factor is the ground state $\psi_{\mathrm{g}}$ which can be thought of as a spinor-valued vector in $\mathcal{F}$. We then define $J : \mathcal{F} \to \mathbb{C}^N \otimes \mathcal{F}$ through

$$J\phi = \psi_{\mathrm{g}} \otimes_{\mathrm{s}} \phi, \tag{17.93}$$

since $\psi_{\mathrm{g}}$ is considered as given. If $\phi$ is an $n$-photon vector, $\phi = (0, \ldots, \phi_n, 0, \ldots)$, then

$$(\psi_{\mathrm{g}} \otimes_{\mathrm{s}} \phi)_{n+j} = \binom{n+j}{n}^{1/2} \mathsf{S}\psi_{\mathrm{g}j}\phi_n, \tag{17.94}$$

with $\mathsf{S}$ denoting the symmetrizer.

As can be seen from (17.91), the symmetrization $\otimes_{\mathrm{s}}$ is unbounded. In particular,

$$\begin{aligned} \|J\phi\|^2 &= \sum_{n=0}^{\infty}\sum_{i=0}^{n}\sum_{j=0}^{n} \binom{n}{i}^{1/2}\binom{n}{j}^{1/2} \langle \mathsf{S}\psi_{\mathrm{g}i}\phi_{n-i}, \mathsf{S}\psi_{\mathrm{g}j}\phi_{n-j}\rangle \\ &\le \sum_{n=0}^{\infty}\Big(\sum_{j=0}^{n}\binom{n}{j}^{1/2} \|\psi_{\mathrm{g}j}\|\|\phi_{n-j}\|\Big)^2 \\ &\le \sum_{n=0}^{\infty}\sum_{j=0}^{n}\binom{n}{j}\|\phi_{n-j}\|^2 \mathrm{e}^{-\delta j}\sum_{i=0}^{n}\|\psi_{\mathrm{g}i}\|^2\mathrm{e}^{\delta i}. \end{aligned} \tag{17.95}$$

Let us define $D_\delta = \{\phi \mid \|\phi_n\| \le c(1-\mathrm{e}^{-\delta/2})^n\}$. Then for $\phi \in D_\delta$, we have $\|J\phi\| < \infty$, which is the reason for assuming the exponential bound in condition 17.3. Without it, we would have to go into details in what sense $\phi$ is far away from the atom.

If the state $\phi$ shifted to infinity, either by the spatial shift $\mathrm{e}^{-\mathrm{i}w\cdot P_{\mathrm{f}}}$ or by the time shift $\mathrm{e}^{-\mathrm{i}H_{\mathrm{f}}t}$, then only the coupled ground state remains in focus. To see this on a more formal level, we introduce the strictly local Weyl algebra $\mathcal{W}_R$ consisting of operators of the form $W(f) = \exp[a^*(f) - a(f)]$ with $f(x,\lambda) = 0$ for $|x| \ge R$. The quasi-local Weyl algebra $\mathcal{W}$ is the norm closure of $\cup_{R>0}\mathcal{W}_R$. The local character is of importance, e.g. $g(H_{\mathrm{f}})$ with $g$ bounded is obviously a bounded operator, but $g(H_{\mathrm{f}})$ does not lie in $\mathcal{W}$. Let $A \in B(\mathbb{C}^N) \otimes \mathcal{W}$. Shifting to infinity then

$$\lim_{|w|\to\infty} \langle \psi_{\mathrm{g}} \otimes_{\mathrm{s}} \mathrm{e}^{-\mathrm{i}w\cdot P_{\mathrm{f}}}\phi, A\psi_{\mathrm{g}} \otimes_{\mathrm{s}} \mathrm{e}^{-\mathrm{i}w\cdot P_{\mathrm{f}}}\phi\rangle = \langle \psi_{\mathrm{g}}, A\psi_{\mathrm{g}}\rangle\langle\phi,\phi\rangle_{\mathcal{F}}, \tag{17.96}$$

$$\lim_{|t|\to\infty} \langle \psi_{\mathrm{g}} \otimes_{\mathrm{s}} \mathrm{e}^{-\mathrm{i}H_{\mathrm{f}}t}\phi, A\psi_{\mathrm{g}} \otimes_{\mathrm{s}} \mathrm{e}^{-\mathrm{i}H_{\mathrm{f}}t}\phi\rangle = \langle \psi_{\mathrm{g}}, A\psi_{\mathrm{g}}\rangle\langle\phi,\phi\rangle_{\mathcal{F}} \tag{17.97}$$

for all $\phi \in D_\delta$.

To prove (17.96), (17.97), we choose an $n$-photon state of the form $\phi = (0, \ldots, \phi_n, 0, \ldots)$ with $\phi_n(x_1, \lambda_1, \ldots, x_n, \lambda_n) = \mathsf{S}\prod_{j=1}^{n} f_j(x_j, \lambda_j)$, in other words $\phi = (n!)^{-1/2} \prod_{j=1}^{n} a^*(f_j)\Omega$. We set $f_{jw}(x_j, \lambda_j) = f_j(x_j - w, \lambda_j)$ and similarly $\widehat{f}_{jt}(k_j, \lambda_j) = \mathrm{e}^{-\mathrm{i}\omega(k_j)t} \widehat{f}_j(k_j, \lambda_j)$. Equations (17.96) and (17.97) go in parallel and we consider only the latter. Then, for $M \in B(\mathbb{C}^N)$, $W(f) \in \mathcal{W}$, and since $W(f)a^*(f_j) = a^*(f_j)W(f) - \langle f, f_j\rangle_{\mathfrak{h}} W(f)$, we get

$$\begin{aligned}
&\langle J\mathrm{e}^{-\mathrm{i}H_{\mathrm{f}t}}\phi, M \otimes W(f) J \mathrm{e}^{-\mathrm{i}H_{\mathrm{f}t}}\phi\rangle \\
&= \frac{1}{n!} \langle \prod_{j=1}^{n} a^*(f_{jt})\psi_{\mathrm{g}}, M \otimes W(f) \prod_{j=1}^{n} a^*(f_{jt})\psi_{\mathrm{g}}\rangle \\
&= \frac{1}{n!} \langle \prod_{j=1}^{n} a^*(f_{jt})\psi_{\mathrm{g}}, \prod_{j=1}^{n} a^*(f_{jt}) M \otimes W(f)\psi_{\mathrm{g}}\rangle \\
&+ \frac{1}{n!} \sum_{\Lambda \subset \{1,\dots,n\}, \Lambda \neq \emptyset} \prod_{j \in \Lambda} (-\langle f, f_{jt}\rangle_{\mathfrak{h}}) \langle \prod_{j=1}^{n} a^*(f_{jt})\psi_{\mathrm{g}}, \prod_{j \in \Lambda^c} a^*(f_{jt}) M \psi_{\mathrm{g}}\rangle.
\end{aligned} \tag{17.98}$$

Since $f$ is local, by the Riemann–Lebesgue lemma, $\lim_{t\to\infty}\langle f, f_{jt}\rangle_{\mathfrak{h}} = 0$. Similarly, for space translations, $\lim_{|w|\to\infty}\langle f, f_{jw}\rangle_{\mathfrak{h}} = 0$. Therefore each term having at least one contraction vanishes in the limit $t \to \infty$, respectively $|w| \to \infty$. We still have to discuss the first summand corresponding to zero contraction which written out explicitly is

$$\sum_{j=0}^{\infty} \binom{n+j}{n} \langle \mathsf{S}\psi_{\mathrm{g}j}\phi_{nt}, \mathsf{S}(M \otimes W(f)\psi_{\mathrm{g}})_j \phi_{nt}\rangle, \tag{17.99}$$

by using (17.94) and setting $\phi_{nt} = (\mathrm{e}^{-\mathrm{i}H_{\mathrm{f}t}}\phi)_n$. There are two types of terms in the scalar product. If a $\phi_{nt}$ is integrated either against $\psi_{\mathrm{g}j}$ or against $(M \otimes W(f)\psi_{\mathrm{g}})_j$, then all such terms vanish as $t \to \infty$, again by the Riemann–Lebesgue lemma. The only terms which survive in the limit are of the form $\langle \psi_{\mathrm{g}j}, (M \otimes W(f)\psi_{\mathrm{g}})_j\rangle \langle \phi_{nt}, \phi_{nt}\rangle_{\mathcal{F}} = \langle \psi_{\mathrm{g}j}, (M \otimes W(f)\psi_{\mathrm{g}})_j\rangle \langle \phi_n, \phi_n\rangle_{\mathcal{F}}$ by unitarity. We conclude that the limit $t \to \infty$ in (17.99) equals

$$\sum_{j=0}^{\infty} \langle \psi_{\mathrm{g}j}, (M \otimes W(f)\psi_{\mathrm{g}})_j\rangle \langle \phi_n, \phi_n\rangle_{\mathcal{F}} = \langle \psi_{\mathrm{g}}, M \otimes W(f)\psi_{\mathrm{g}}\rangle \langle \phi, \phi\rangle_{\mathcal{F}}, \tag{17.100}$$

as claimed. To cover the general case one has to take suitable linear combinations and uniform limits.

With these preparations the limit in (17.90) can be formulated more concisely. We define the wave operators $\Omega^{\mp}$ through the strong limit

$$\Omega^{\mp}\phi = s - \lim_{t\to\pm\infty} e^{i(H-E_g)t} J e^{-iH_f t}\phi. \qquad (17.101)$$

The existence of this limit will be shown for all $\phi \in D_\delta$ by a Cook estimate in Proposition 17.6 below. But we first want to explore some consequences of our definition.

In the usual definition of wave operators one projects onto the scattering states of the comparison dynamics $e^{-iH_f t}$. This is not needed here because for $\phi = \Omega$, the limit in (17.101) equals $\psi_g$. The formulation (17.101) assumes that $H$ has no other bound state. If this had been the case, one would have to allow in (17.101) for several atomic channels, corresponding to the possibility that the atom remains in an excited state forever.

The wave operators $\Omega^{\pm}$ are isometries from $\mathcal{F}$ to $\mathbb{C}^N \otimes \mathcal{F}$, as can be seen from

$$\langle \Omega^{\mp}\phi, \Omega^{\mp}\phi\rangle = \lim_{t\to\pm\infty} \langle J e^{-iH_f t}\phi, J e^{-iH_f t}\phi\rangle = \langle \phi,\phi\rangle_{\mathcal{F}} \qquad (17.102)$$

by (17.97) for $\phi \in D_\delta$. By continuity this property extends to all of $\mathcal{F}$. $\Omega^{\pm}$ intertwines between the free and interacting dynamics as

$$e^{-i(H-E_g)t}\Omega^{\pm} = \Omega^{\pm} e^{-iH_f t}, \qquad (17.103)$$

which is an immediate consequence of the definition: for $\phi \in D_\delta$ one has $e^{-iH_f t}\phi \in D_\delta$ and

$$\begin{aligned}\Omega^{-}\phi &= \lim_{s\to\infty} e^{i(H-E_g)(t+s)} J e^{-iH_f(t+s)}\phi \\ &= \lim_{s\to\infty} e^{i(H-E_g)t} e^{i(H-E_g)s} J e^{-iH_f s} e^{-iH_f t}\phi \\ &= e^{i(H-E_g)t}\Omega^{-} e^{-iH_f t}\phi. \end{aligned} \qquad (17.104)$$

Since $D_\delta$ is dense in $\mathcal{F}$, (17.103) holds. As a consequence, Ran $\Omega^{\mp}$ are reducing subspaces for $H$ and $H - E_0$ restricted to Ran $\Omega^{\mp}$ is unitarily equivalent to $H_f$ on $\mathcal{F}$.

As emphasized, the limit in (17.90) should not only hold for some states but for *all* $\psi \in \mathbb{C}^N \otimes \mathcal{F}$. It is useful to have a name for such a property.

**Definition 17.4** $\Omega^{\pm}$ *are called asymptotically complete if*

$$\text{Ran}\,\Omega^{\pm} = \mathbb{C}^N \otimes \mathcal{F}. \qquad (17.105)$$

If $\Omega^{\pm}$ are asymptotically complete, then they are unitary and diagonalize $H$ as

$$(\Omega^{\pm})^{-1}(H - E_g)\Omega^{\pm} = H_f. \qquad (17.106)$$

In particular, $H$ has the absolutely continuous spectrum $[E_{\mathrm{g}}, \infty)$ of infinite multiplicity.

Under asymptotic completeness the long-time dynamics is fully characterized through

**Proposition 17.5** (Relaxation to the ground state). *Let $A$ be local in the sense that $A \in \mathcal{B}(\mathbb{C}^N) \otimes \mathcal{W}$. Then for every $\psi \in \operatorname{Ran} \Omega^-$ with $\|\psi\| = 1$ we have*

$$\lim_{t\to\infty} \langle \mathrm{e}^{-\mathrm{i}Ht}\psi, A\mathrm{e}^{-\mathrm{i}Ht}\psi\rangle = \langle \psi_{\mathrm{g}}, A\psi_{\mathrm{g}}\rangle. \tag{17.107}$$

*In particular, if asymptotic completeness holds, then the limit (17.107) is valid for all $\psi \in \mathbb{C}^N \otimes \mathcal{F}$.*

*Proof*: Let $\psi = \Omega^-\phi$ with $\phi \in D_\delta$. By (17.101) one has

$$\lim_{t\to\infty} \langle \mathrm{e}^{-\mathrm{i}Ht}\psi, A\mathrm{e}^{-\mathrm{i}Ht}\psi\rangle = \lim_{t\to\infty} \langle J\mathrm{e}^{-\mathrm{i}H_{\mathrm{f}}t}\phi, AJ\mathrm{e}^{-\mathrm{i}H_{\mathrm{f}}t}\phi\rangle, \tag{17.108}$$

which converges to the limit (17.107) as is seen by the argument explained in (17.98). Any $\psi \in \operatorname{Ran}\Omega^-$ can be approximated through states of the form $\Omega^-\phi$ with $\phi \in D_\delta$. □

**Proposition 17.6** (Cook estimate). *Let the integrability condition (17.27) be satisfied. Then for all $\phi \in D_\delta$ the strong limit*

$$\lim_{t\to\infty} \mathrm{e}^{\mathrm{i}(H-E_{\mathrm{g}})t} J\mathrm{e}^{-\mathrm{i}H_{\mathrm{f}}t}\phi = \Omega^-\phi \tag{17.109}$$

*exists.*

*Proof*: If $\phi = \Omega$, the limit exists and is $\psi_{\mathrm{g}}$. Let then $\langle\Omega, \phi\rangle = 0$ and $\phi \in D_\delta \cap D(H_{\mathrm{f}})$. Then $J\mathrm{e}^{-\mathrm{i}H_{\mathrm{f}}t}\phi \in D(H)$ and we have

$$\begin{aligned}\frac{\mathrm{d}}{\mathrm{d}t}\mathrm{e}^{\mathrm{i}(H-E_{\mathrm{g}})t} J\mathrm{e}^{-\mathrm{i}H_{\mathrm{f}}t}\phi &= \mathrm{i}\mathrm{e}^{\mathrm{i}(H-E_{\mathrm{g}})t}(HJ - E_{\mathrm{g}}J - JH_{\mathrm{f}})\mathrm{e}^{-\mathrm{i}H_{\mathrm{f}}t}\phi \\ &= -\mathrm{e}^{\mathrm{i}(H-E_{\mathrm{g}})t} eQ\psi_{\mathrm{g}} \otimes_{\mathrm{s}} \cdot E_{\varphi}^{-}\mathrm{e}^{-\mathrm{i}H_{\mathrm{f}}t}\phi.\end{aligned} \tag{17.110}$$

Here

$$E_{\varphi}^{-} = \mathrm{i}\sum_{\lambda=1,2}\int \mathrm{d}^3k\widehat{\varphi}(k)\sqrt{\omega(k)/2}e_\lambda(k)a(k,\lambda) \tag{17.111}$$

and we used

$$a(f)(\psi_{\mathrm{g}} \otimes_{\mathrm{s}} \phi) = a(f)\psi_{\mathrm{g}} \otimes_{\mathrm{s}} \phi + \psi_{\mathrm{g}} \otimes_{\mathrm{s}} a(f)\phi, \tag{17.112}$$

$$a^*(f)(\psi_{\mathrm{g}} \otimes_{\mathrm{s}} \phi) = a^*(f)\psi_{\mathrm{g}} \otimes_{\mathrm{s}} \phi, \tag{17.113}$$

which follow from the definition (17.92). Thus

$$e^{i(H-E_g)t} J e^{-iH_{ft}t} \phi = J\phi - e \int_0^t ds\, e^{i(H-E_g)s} Q\psi_g \otimes_s \cdot E_\varphi^- e^{-iH_{fs}s} \phi \quad (17.114)$$

and it is to be shown that $t \to \|Q\psi_g \otimes_s \cdot E_\varphi^- e^{-iH_{ft}t}\phi\|$ is integrable for a dense set of $\phi$'s. For this purpose we define $L_\varphi \subset L_\perp^2(\mathbb{R}^3, \mathbb{R}^3)$ to be the linear subspace spanned by the set $\{e^{-i\omega t}\widehat{\varphi}\sqrt{\omega/2}e_\lambda \,|\, t \in \mathbb{R}\}$. We choose an $n$-photon vector in product form, $\phi = (0, \ldots, \phi_n, 0, \ldots)$, $\phi_n = \mathrm{S}\prod_{j=1}^n \widehat{f}_j$ with each factor being a sum

$$\widehat{f}_j(k, \lambda) = \sum_{\ell=1}^{\ell_j} \alpha_{j\ell} e^{-i\omega(k)t_{j\ell}} \widehat{\varphi}(k)\sqrt{\omega(k)/2}e_\lambda(k) + \widehat{f}_j^\perp(k, \lambda) \quad (17.115)$$

with $\widehat{f}_j^\perp$ orthogonal to $L_\varphi$. Then

$$\|Q\psi_g \otimes_s \cdot E_\varphi^- e^{-iH_{ft}t}\phi\| \le \sum_{j=1}^n \Big| \sum_{\lambda=1,2} \int d^3k \widehat{\varphi}(k)\sqrt{\omega(k)/2}e_\lambda(k) \cdot \widehat{f}_j(k, \lambda) e^{-i\omega(k)t} \Big|. \quad (17.116)$$

Inserting from (17.115) yields a finite sum of terms of the form

$$\int d^3k |\widehat{\varphi}(k)|^2 \omega(k) Q_\perp(k) e^{-i\omega(k)(t+s)} \quad (17.117)$$

which are integrable, either by assumption or as a matter of fact for the Pauli–Fierz model, cf. the remark below Theorem 17.1.

Our argument establishes the limit (17.109) for a dense set of vectors in the $n$-photon subspace. By linearity and by taking uniform limits, this extends to all of $D_\delta$. □

For $\psi \in \mathrm{Ran}\, \Omega^\pm$ one has all the desired properties, relaxation to the ground state as in Proposition 17.5, long-time asymptotics as in (17.90), and spectral measures which are absolutely continuous except for a possible mass at $E_g$ with weight $\langle \psi, \psi_g \rangle$. Asymptotic completeness, i.e. the property $\mathrm{Ran}\, \Omega^\pm = \mathbb{C}^N \otimes \mathcal{F}$, ensures that there are no states with unphysical dynamics.

## Notes and references

### *Section 17.1*

The dipole approximation in conjunction with the $N$-level approximation is common practice in atomic physics, for example Agarwal (1974), Cohen-Tannoudji *et al.* (1992). The unitary transformation (17.7) is linked with Power and Zienau (1959). It is also used by Bloch and Nordsieck (1937). For the special case of $N = 2$ the transition from Fock to non-Fock ground state is studied in considerable detail by Leggett *et al.* (1987), Spohn (1989), Amann (1991), and Weiss (1999). The corresponding Hamiltonian, $H_{\mathrm{sb}}$, is known as the spin-boson model. The vector character of the Bose field is ignored and one sets $H_{\mathrm{at}} = \varepsilon\sigma_z$, $Q \cdot A_\varphi = \sigma_x \int \mathrm{d}^3k(\widehat{g}(k)a^*(k) + \widehat{g}(k)^* a(k))$. The $t^{-2}$-decay of (17.6) is the so-called Ohmic case, which is marginal for the transition to non-Fock. For small coupling $H_{\mathrm{sb}}$ has a unique ground state in $\mathbb{C}^2 \otimes \mathcal{F}$, whereas for large coupling $H_{\mathrm{sb}}$ acquires an infinite number of bosons, which leads to a two-fold degenerate ground state, both lying outside Fock space. Form factors with a decay different from $t^{-2}$ have been also investigated.

## Section 17.2

Landau (1927) uses density matrices in the description of the reduced state of the atom. He arrives at a variant of the master equation (17.32). Its diagonal part is often referred to as the Pauli master equation (Pauli 1928). A further influential work is Bloch (1928). The systematic weak coupling theory goes back to van Hove (1955, 1957) and has been further developed in response to the theoretical challenges in quantum optics. Just to remind the reader: In theoretical models of the laser one has to include dissipation for the field modes of the cavity to account for lossy reflection at the walls. For photon counting statistics one has to devise a simple model of a detector. An interesting exchange is Srinivas and Davies (1981) and Mandel (1981). For laser cooling and trapping the spontaneous emission and its associated recoil must be described in a concise way (Metcalf and van der Straten 1999). Thus the general problem of how to model open quantum systems necessarily comes into focus. On the classical level the addition of friction forces and possibly of noise serves well. But quantum mechanics poses constraints which are still of current research interest. As a short sample out of a large body of literature we refer to Lax (1968), Glauber (1969), Kossakowski (1972), Haake (1973), Spohn (1980), Carmichael (1999), Weiss (1999), and Breuer and Petruccione (2002). Our presentation here is based on Davies (1974, 1975, 1976a). He emphasizes time-averaging which has been overlooked mostly, but is done correctly in Cohen-Tannoudji *et al.* (1992) and Breuer and Petruccione (2002). The various generators of the dissipative evolution in the weak coupling limit are compared in Dümcke and Spohn (1979). In the text we discussed only single-time statistics. Stationary two-time statistics appear frequently in applications. Multi-time statistics are studied by Dümcke (1983) within the presented framework.

Although even the most simplistic theory yields a shift of the spectral line, such predictions were not taken seriously. The rough estimate of Bethe (1947) and the more sophisticated computation of Grotch (1981) resulting in a cutoff-independent shift could have been done as early as 1930. It is only through the war-related research on radar that experimental techniques became available to measure such fine effects. The theory followed soon; see Schweber (1994) for an excellent account.

The weak coupling theory is also a useful tool in studying decoherence. In essence one starts the dynamics with a coherent superposition of two spatially well-separated wave packets. According to the appropriate quantum master equation such a coherence is destroyed on a time scale which is much, much shorter than the friction time scale. Properly speaking the master equation should not be used on such short time scales. When decoherence is due to the coupling to the quantized rediation field, Dürr and Spohn (2002) provide an analysis based on the dipole approximation. A complete discussion, avoiding the dipole approximation, is given by Breuer and Petruccione (2001), who also list references to earlier work.

The weak coupling theory had a mathematical spin-off, going way beyond the specific application at hand. The basic observation is that the dissipative semigroup $T_t$ is the classical analog of the transition probability of a classical Markov process, the Markov character being embodied in the semigroup property $T_t T_s = T_{t+s}$, $t, s \geq 0$. $T_t$ is positivity and normalization preserving, in the sense that if $\rho$ is a density matrix so is $T_t \rho$. As recognized by Lindblad (1976) the stronger notion of complete positivity is very natural. It means that if $\mathcal{H}$ is extended to $\mathcal{H} \otimes \mathbb{C}^n$ and $T_t$ in the trivial way to $T_t \otimes 1$, then $T_t \otimes 1$ is positivity preserving for every $n$. In this framework the possible types of generators are classified by Lindblad (1976). He also characterizes dissipation through the decrease of relative entropy (Lindblad 1975). Mixing and the long-time limit $t \to \infty$ are studied by Spohn (1976), Frigerio (1978), and Frigerio and Verri (1982). The generalization of the notion of detailed balance to the quantum context is discussed by Gorini *et al.* (1984). Most recommended introductions are Davies (1976b) and Alicki and Lendi (1987). Clearly the next level is to inquire about multitime statistics and their build-up from the semigroup $T_t$. This is a fairly straightforward step for classical Markov processes through the concept of conditional independence of past and future. No such thing seems to exist on the quantum level and the theory of quantum stochastic processes tries to provide a consistent framework, possibly guided by specific model systems, that can be analyzed in detail. We refer to Accardi, Frigerio and Lewis (1982), Lindblad (1983), Hudson and Parthasarathy (1984), Accardi *et al.* (1991), and Parthasarathy (1992), and the recent monographs by Alicki and Fannes (2001) and by Accardi *et al.* (2001).

### *Section 17.3*

Already in his first work on radiation theory Dirac (1927) simplifies the problem to a single level coupled to a continuum of modes. A two-level atom coupled to the radiation field in the rotating wave approximation also reduces to a Friedrich–Lee-type Hamiltonian.

Complex dilations were investigated in connection with the study of Regge poles, cf. Reed and Simon (1978) for references. The mathematical framework is developed by Aguilar and Combes (1971) and Balslev and Combes (1971). A beautiful survey is Simon (1978). For an introduction we refer to Cycon *et al.* (1987). Hunziker (1990) focuses on the question of how to translate the results on the resolvent back to the real time-domain. Okamoto and Yajima (1985) observe that the dilation of the massive photon field can be used to unfold resonances. Resonances of the Pauli–Fierz model are studied in Bach, Fröhlich and Sigal (1995, 1998a, 1998b, 1999). They develop a renormalization-type iterative procedure to pin down the domain of analyticity of the complex dilated resolvent. This method is refined by Bach *et al.* (2002). An infinitesimal version based on Mourre-type estimates and the Feshbach method is Dereziński and Jakšić (2001).

### *Section 17.4*

Our discussion is based on Davies (1976a). $L_{\mathrm{D}}$ is the Davies generator in the weak coupling theory. Line shapes are discussed by Weisskopf and Wigner (1930). Our examples for the spectral characteristics of the emitted light are taken from Cohen-Tannoudji *et al.* (1992), Chapter IIIC and Exercise 15.

### *Section 17.5*

Potential scattering is discussed in Reed and Simon (1979) and $N$-body scattering in Cycon *et al.* (1987). We follow the presentation in Hübner and Spohn (1995a). The Cook argument is based on Høegh-Krohn (1970) who also studies the asymptotic electromagnetic fields; for a complete discussion see (Fröhlich, Griesemer and Schlein 2001). In the meantime the mathematical investigation of scattering of photons from an atom has flourished. For simplicity often the scalar field model of section 19.2 is studied. An important step is Dereziński and Gérard (1999) who establish asymptotic completeness in the case of massive photons, $\omega(k) = (k^2 + m_{\mathrm{ph}}^2)^{1/2}$, $m_{\mathrm{ph}} > 0$, and a strictly confining potential. Earlier work restricted to an $N$-level atom is Gérard (1996) and Skibsted (1998). An extension to massless photons under the condition $\widehat{\varphi}(0) = 0$ is Gérard (2002). For $m_{\mathrm{ph}} > 0$ Fröhlich, Griesemer and Schlein (2001, 2002) allow for potentials which are not strictly confining, like the Coulomb potential. Thereby the channel for a freely propagating electron is opened up as it occurs in the description of the photoelectric effect (Bach, Klopp and Zenk 2002). Ammari (2000) establishes asymptotic completeness for the Nelson model of section 19.2 with ultraviolet cutoff removed. In these works asymptotic completeness is defined in such a way that $H$ could have other eigenvalues besides its ground state. To exclude them one has to resort to Bach, Fröhlich and Sigal (1998a) and Fröhlich, Griesemer and Schlein (2002).

A different approach is to take the dipole approximation with harmonic confining potential. Since the Hamiltonian is quadratic, the scattering theory can be reduced to one-particle scattering with a finite rank perturbation. Maassen (1984) notices that for a weakly anharmonic confining potential the time-dependent perturbation series can be controlled uniformly in time. His estimates are improved and optimized in Maassen, Gută and Botvich (1999) and Fidaleo and Liverani (1999). With this input one can prove asymptotic completeness in the strong sense of Definition 17.4. The perturbing potential must be bounded and so small that the confining potential remains convex (Spohn 1997). The harmonic case is investigated by Arai (1983b).

# 18

# Relaxation at finite temperatures

The weak coupling theory of chapter 17 is the workhorse of quantum optics and serves very well in practice, also at nonzero temperatures. From the viewpoint of the theory one might wonder about the structure at fixed small, but nonzero, coupling strength, which needs to go beyond the analysis of the weak coupling theory. Much effort has been invested to achieve this goal, basically by trying to identify corrections within time-dependent perturbation theory. Unfortunately, since the long-time behavior must be extracted, the details quickly become unwieldy and one has to rely on ad hoc approximations.

Over recent years a novel approach has been pursued which investigates the pole structure of the analytic continuation of the resolvent of $H_\lambda$ across the real axis through complex dilations; compare with section 17.3. The techniques are demanding but simplify substantially at finite temperatures when the Hamiltonian is replaced by the Liouvillean and, since its spectrum is the full real line, complex dilations are replaced by complex translations which can be handled more easily. From the pole structure a fairly complete picture of the long-time dynamics can be extracted with the potential of computing systematically higher corrections to the weak coupling theory.

The finite temperature relaxation is a digression into the realm of time-dependent statistical mechanics with small deviations from thermal equilibrium. While this is of independent interest and has important applications in quantum optics and condensed matter, our goal is merely to illustrate the power of complex translations and make the connection to the weak coupling theory.

For completeness we recall once again the set-up. In the dipole approximation and $N$-level approximation the Hamiltonian is

$$H_\lambda = H_{\mathrm{at}} + H_{\mathrm{f}} + \lambda Q \cdot E_\varphi = H_0 + \lambda H_{\mathrm{int}}, \tag{18.1}$$

see (17.10). $H_{\mathrm{at}}$ acts on $\mathbb{C}^N$. Diverging from our previous convention, the energies $\varepsilon_j$ are labeled as $\varepsilon_1 \leq \varepsilon_2 \ldots \leq \varepsilon_N$ allowing for possible degeneracies. $Q$ is the dipole operator as an $N \times N$ matrix. The field energy $H_{\mathrm{f}}$, and the electric field $E_\varphi = E_\varphi(0)$ are operators on Fock space $\mathcal{F}$. For finite temperatures we need some extra structure, which will be explained below.

The photon field is at temperature $T > 0$. It will be more convenient to work with the inverse temperature and set $\beta = 1/k_{\mathrm{B}}T$. In the initial state the atom and its nearby photons are out of equilibrium and the goal is to understand how the coupled system relaxes back to global equilibrium. In the weak coupling theory one disentangles the effective dynamics of the atom and regards the field as driven by the atomic source. At fixed $\lambda$ such a distinction becomes hazy and a more global view is adopted with the separation deduced in the small-$\lambda$ limit.

The analysis of thermal relaxation relies on the following strategy. One introduces coordinates which encode the finite energy excitations away from equilibrium. Doing this properly relies on tools from the representation theory of $C^*$-algebras which for our context was mostly developed in the 1960–70s. For noninteracting photons the representation of Araki and Woods (1963) is of a sufficiently concrete form and also allows the incorporation of the coupling to the atom. Note that at zero coupling the spectrum of finite energy excitation covers the full real axis $\mathbb{R}$, since $\omega(k) = |k|$ and energy can be either below or above its equilibrium value. The energy differences of the atom are embedded in this spectrum as discrete eigenvalues. As the coupling is turned on, they become resonances which are uncovered by a complex downward translation of the photon excitation spectrum. The location of the resonance poles and the corresponding eigenspaces can be handled through standard analytic perturbation theory.

To convince the reader that the Araki–Woods Liouvillean correctly describes the finite energy excitation, we need some background material on quantum systems at finite temperature. We conform with established notation which to some extent deviates from our previous conventions.

## 18.1 Bounded quantum systems, Liouvillean

We start with an abstract quantum system on a separable Hilbert space $\mathcal{H}$ equipped with the scalar product $\langle \cdot, \cdot \rangle$. We assume that the Hamiltonian $H$ is bounded from below and has a purely discrete spectrum such that

$$\mathrm{tr}[\mathrm{e}^{-\beta H}] < \infty \tag{18.2}$$

for arbitrary $\beta > 0$. The algebra of observables, $\mathcal{A}$, is the set of all bounded operators on $\mathcal{H}$, denoted by $B(\mathcal{H})$. A general quantum state is given through the density matrix $\rho$, satisfying $\rho \geq 0$, $\mathrm{tr}\rho = 1$. In particular $\rho \in \mathcal{T}_1(\mathcal{H})$, denoting the two-sided ideal of trace class operators on $\mathcal{H}$. In the Heisenberg picture the time evolution is given through

$$\alpha_t(a) = e^{itH} a e^{-itH} \tag{18.3}$$

as acting on $a \in B(\mathcal{H})$. The dual Schrödinger picture provides the time evolution of states as

$$\rho \mapsto \rho_t = \alpha_{-t}(\rho) = e^{-itH} \rho e^{itH}. \tag{18.4}$$

We want the evolution of density matrices to look like the evolution of vectors on a Hilbert space and, for this purpose, introduce the two-sided ideal of Hilbert–Schmidt operators $\mathcal{T}_2(\mathcal{H})$. A bounded operator $a$ belongs to $\mathcal{T}_2(\mathcal{H})$ if and only if $\mathrm{tr}[a^*a] < \infty$. $\mathcal{T}_2(\mathcal{H})$ becomes a Hilbert space under the scalar product

$$\langle a|b\rangle = \mathrm{tr}[a^*b], \quad a, b \in \mathcal{T}_2(\mathcal{H}). \tag{18.5}$$

It will be useful to represent $\mathcal{A} = B(\mathcal{H})$ as an algebra of operators on $\mathcal{T}_2(\mathcal{H})$. $\mathcal{T}_2(\mathcal{H})$ carries a left representation through

$$\ell(a)\kappa = a\kappa \in \mathcal{T}_2(\mathcal{H}). \tag{18.6}$$

Later on we will need also the right antirepresentation defined through

$$r(a)\kappa = \kappa a^* \in \mathcal{T}_2(\mathcal{H}). \tag{18.7}$$

This representation is antilinear since $r(za)\kappa = z^* r(a)\kappa$ for $z \in \mathbb{C}$.

We transcribe states and dynamics to $\mathcal{T}_2(\mathcal{H})$. To every element $\kappa \in \mathcal{T}_2(\mathcal{H})$ a state $\rho$ is associated through

$$\rho = \langle\kappa|\kappa\rangle^{-1}\kappa\kappa^*. \tag{18.8}$$

The expectation of $a \in B(\mathcal{H})$ is given by

$$\langle a\rangle_\rho = \mathrm{tr}[\rho a] = \langle\kappa|\kappa\rangle^{-1}\langle\kappa|\ell(a)\kappa\rangle. \tag{18.9}$$

The time evolution becomes

$$\langle\alpha_t(a)\rangle_\rho = \langle\kappa|\kappa\rangle^{-1}\mathrm{tr}[\kappa^*\alpha_t(a)\kappa] = \langle\kappa|\kappa\rangle^{-1}\langle\kappa|\ell(\alpha_t(a))\kappa\rangle \tag{18.10}$$

and for $\kappa, \sigma \in \mathcal{T}_2(\mathcal{H})$

$$\begin{aligned}\langle\kappa|\ell(\alpha_t(a))\sigma\rangle &= \mathrm{tr}[\kappa^*\alpha_t(a)\sigma] = \mathrm{tr}[(e^{-itH}\kappa e^{itH})^* a (e^{-itH}\sigma e^{itH})] \\ &= \langle\alpha_{-t}(\kappa)|\ell(a)\alpha_{-t}(\sigma)\rangle = \langle e^{-it\mathcal{L}}\kappa|\ell(a)e^{-it\mathcal{L}}\kappa\rangle. \end{aligned} \tag{18.11}$$

The last identity defines the Liouvillean $\mathcal{L}$. Clearly

$$\mathcal{L}\kappa = [H, \kappa] \tag{18.12}$$

and $\kappa_t = \alpha_{-t}(\kappa)$ is governed by the Schrödinger-like equation

$$\mathrm{i}\frac{\mathrm{d}}{\mathrm{d}t}\kappa_t = \mathcal{L}\kappa_t. \tag{18.13}$$

The Liouvillean is a symmetric operator as can be seen from

$$\langle\kappa|\mathcal{L}\sigma\rangle = \mathrm{tr}[\kappa^*[H,\sigma]] = \mathrm{tr}[([H,\kappa])^*\sigma] = \langle\mathcal{L}\kappa|\sigma\rangle. \tag{18.14}$$

To work concretely with the left, respectively right, representation of $\mathcal{A}$ and the Liouvillean $\mathcal{L}$ it is convenient to identify $\mathcal{T}_2(\mathcal{H})$ with $\mathcal{H}\otimes\mathcal{H}$ through the isomorphism

$$I_C : \mathcal{T}_2(\mathcal{H}) \to \mathcal{H}\otimes\mathcal{H}. \tag{18.15}$$

In a suitable basis $C$ is simply complex conjugation. More abstractly $C$ is an anti-unitary involution on $\mathcal{H}$, i.e.

$$C^2 = 1 \quad \text{and} \quad \langle C\psi, C\varphi\rangle = \langle\varphi,\psi\rangle. \tag{18.16}$$

Then with $\kappa$ defined through $\kappa\psi = \psi_1\langle\psi_2,\psi\rangle$ one sets

$$I_C\kappa = \psi_1\otimes C\psi_2 \in \mathcal{H}\otimes\mathcal{H} \tag{18.17}$$

which extends by linearity. Note that

$$I_C\ell(a)\kappa = I_Ca\kappa = a\psi_1\otimes C\psi_2 = (a\otimes 1)I_C\kappa. \tag{18.18}$$

Thus $I_C$ intertwines with the left representation $\ell$ of $\mathcal{A}$ on $\mathcal{H}\otimes\mathcal{H}$ given by

$$\ell(a) = a\otimes 1. \tag{18.19}$$

Similarly

$$I_Cr(a)\kappa = I_C(\kappa a^*) = \psi_1\otimes Ca\psi_2 = 1\otimes CaCI_C\kappa \tag{18.20}$$

and

$$r(a) = 1\otimes CaC \quad \text{on} \quad \mathcal{H}\otimes\mathcal{H}. \tag{18.21}$$

In particular for the Liouvillean

$$\begin{aligned} I_C\mathcal{L}\kappa &= H\psi_1\otimes C\psi_2 - \psi_1\otimes CH\psi_2 = H\psi_1\otimes C\psi_2 - \psi_1\otimes(CHC)C\psi_2 \\ &= (H\otimes 1 - 1\otimes CHC)(\psi_1\otimes C\psi_2) = (H\otimes 1 - 1\otimes CHC)I_C\kappa \end{aligned} \tag{18.22}$$

and

$$L = I_C\mathcal{L}I_C^{-1} = H\otimes 1 - 1\otimes CHC \quad \text{on} \quad \mathcal{H}\otimes\mathcal{H}. \tag{18.23}$$

If $H$ is invariant under time-reversal, we may choose $C = T$, with time-reversal $T$, and $CHC = THT = H$. Then the Liouvillean is given by

$$L = H \otimes 1 - 1 \otimes H \tag{18.24}$$

as an operator on $\mathcal{H} \otimes \mathcal{H}$. Clearly the spectrum of $L$ consists of the energy differences $\{E_i - E_j |\, i, j = 0, 1, \dots\}$, where $E_i$, $i = 0, 1, \dots,$ are the eigenvalues of $H$.

## 18.2 Equilibrium states and their perturbations, KMS condition

Of the many possible quantum states thermal equilibrium plays a special role. It is defined by the density matrix

$$\rho_\beta = Z^{-1} \mathrm{e}^{-\beta H}, \quad Z = \mathrm{tr}[\mathrm{e}^{-\beta H}]. \tag{18.25}$$

As an element of $\mathcal{T}_2(\mathcal{H})$ we set

$$\kappa_\beta = Z^{-1/2} \mathrm{e}^{-\beta H/2}. \tag{18.26}$$

Then $\rho_\beta = \kappa_\beta \kappa_\beta^*$. Since $\rho_\beta$ is strictly positive, $\langle a^* a \rangle_\beta = \mathrm{tr}[\rho_\beta\, a^* a] = 0$ for $a \in \mathcal{A}$ implies $a = 0$. Equivalently,

$$\ell(a)\kappa_\beta = 0 \quad \text{implies} \quad a = 0, \tag{18.27}$$

which means that $\kappa_\beta$ is separating for the algebra $\ell(\mathcal{A})$. In principle, one should allow for additional conservation laws like total charge or total number of particles. However, this is ignored here since the Hamiltonian (18.1) does not have such a structure.

For the photon field in infinite space the spectrum of $H$ is continuous and $Z^{-1}\mathrm{e}^{-\beta H}$ as such makes no sense. On the other hand, the atom is a small perturbation. Thus the equilibrium state of the coupled system relative to that of the uncoupled system remains meaningful even at infinite volume and is the object of thermal perturbation theory.

We consider

$$H = H_0 + I. \tag{18.28}$$

$H_0$ is the unperturbed reference system and $I$ is the perturbation, assumed to be bounded, $\|I\| < \infty$. By the Golden–Thompson inequality

$$\begin{aligned} Z_\beta &= \mathrm{tr}[\mathrm{e}^{-\beta H}] = \mathrm{tr}[\mathrm{e}^{-\beta(H_0 + I)}] \le \mathrm{tr}[\mathrm{e}^{-\beta H_0} \mathrm{e}^{-\beta I}] \\ &\le \mathrm{e}^{\beta \|I\|} \mathrm{tr}[\mathrm{e}^{-\beta H_0}] = \mathrm{e}^{\beta \|I\|} Z_\beta^0. \end{aligned} \tag{18.29}$$

Thus if $Z_\beta^0 < \infty$, as assumed, then $\rho_\beta = Z_\beta^{-1} \mathrm{e}^{-\beta H} \in \mathcal{T}_1(\mathcal{H})$.

The Liouvillean of the reference system is given by

$$\mathcal{L}_0 = \ell(H_0) - r(H_0) \tag{18.30}$$

and the Liouvillean of the perturbed system by

$$\mathcal{L} = \ell(H) - r(H). \tag{18.31}$$

Under the isomorphism $I_C$ the Liouvilleans become

$$\begin{aligned} L &= I_C \mathcal{L} I_C^{-1} = (H_0 + I) \otimes 1 - 1 \otimes C(H_0 + I)C = L_0 + W, \\ L_0 &= H_0 \otimes 1 - 1 \otimes C H_0 C, \quad W = I \otimes 1 - 1 \otimes CIC. \end{aligned} \tag{18.32}$$

We also define the Radon–Nikodym operators, $\mathcal{L}_\ell$ and $\mathcal{L}_r$, through

$$\mathcal{L}_\ell = \mathcal{L}_0 + \ell(I), \quad \mathcal{L}_r = \mathcal{L}_0 - r(I). \tag{18.33}$$

Then, with $\kappa_\beta = (Z_\beta)^{-1/2} \mathrm{e}^{-\beta H/2}$, $\kappa_\beta^0 = (Z_\beta^0)^{-1/2} \mathrm{e}^{-\beta H_0/2}$, we have

$$\begin{aligned} \mathrm{e}^{-\beta \mathcal{L}_\ell/2} \kappa_\beta^0 &= \mathrm{e}^{-\beta(\ell(H_0) + \ell(I) - r(H_0))/2} \kappa_\beta^0 = \mathrm{e}^{-\beta(H_0 + I)/2} \kappa_\beta^0 \mathrm{e}^{\beta H_0/2} \\ &= (Z_\beta / Z_\beta^0)^{1/2} (Z_\beta)^{-1/2} \mathrm{e}^{-\beta H/2} = (Z_\beta / Z_\beta^0)^{1/2} \kappa_\beta \end{aligned} \tag{18.34}$$

and by a similar calculation

$$\mathrm{e}^{\beta \mathcal{L}_r/2} \kappa_\beta^0 = (Z_\beta / Z_\beta^0)^{1/2} \kappa_\beta. \tag{18.35}$$

$\kappa_\beta^0$ is in the domain of the operators $\mathrm{e}^{-\beta \mathcal{L}_\ell/2}$ and $\mathrm{e}^{\beta \mathcal{L}_r/2}$ and their action maps unperturbed to perturbed equilibrium,

$$\kappa_\beta = (Z_\beta^0 / Z_\beta)^{1/2} \mathrm{e}^{-\beta \mathcal{L}_\ell/2} \kappa_\beta^0, \quad \kappa_\beta = (Z_\beta^0 / Z_\beta)^{1/2} \mathrm{e}^{\beta \mathcal{L}_r/2} \kappa_\beta^0. \tag{18.36}$$

The thermal state $\mathrm{e}^{-\beta H}$ is related to the unitary time evolution $\mathrm{e}^{-\mathrm{i}tH}$ through analytic continuation to $\beta = \mathrm{i}t$, which gives rise to a very powerful analytic structure of equilibrium time correlations known as the Kubo–Martin–Schwinger (KMS) boundary condition. We define the time correlations as

$$\langle a\alpha_t(b) \rangle_\beta = F_{ab}(t), \quad \langle \alpha_t(b) a \rangle_\beta = G_{ab}(t). \tag{18.37}$$

They are linked through

$$\begin{aligned} \langle a\alpha_t(b) \rangle_\beta &= Z_\beta^{-1} \mathrm{tr}[\mathrm{e}^{-\beta H} a \mathrm{e}^{\mathrm{i}tH} b \mathrm{e}^{-\mathrm{i}tH}] = Z_\beta^{-1} \mathrm{tr}[\mathrm{e}^{-\beta H} \mathrm{e}^{(\beta + \mathrm{i}t)H} b \mathrm{e}^{-(\beta + \mathrm{i}t)H} a] \\ &= \langle \alpha_{-\mathrm{i}\beta + t}(b) a \rangle_\beta, \end{aligned} \tag{18.38}$$

which is the *KMS condition*. It states that $F_{ab}(t)$ is the boundary value of a function $G_{ab}(z)$ which is analytic in the strip $\mathcal{S}_{-\beta} = \{z \mid -\beta < \mathrm{Im} z < 0\}$ such that

$$\lim_{\eta \uparrow \beta} G_{ab}(t - \mathrm{i}\eta) = F_{ab}(t). \tag{18.39}$$

Equivalently, $G_{ab}(t)$ is the boundary value of a function $F_{ab}(z)$ analytic in the strip $\mathcal{S}_\beta$ such that

$$\lim_{\eta \uparrow \beta} F_{ab}(t + \mathrm{i}\eta) = G_{ab}(t). \tag{18.40}$$

A state which satisfies either of these boundary conditions is called a KMS state with respect to the time evolution $\alpha_t$. In our set-up, the only KMS state is $\rho_\beta$. The KMS condition is used as a defining property for equilibrium states in infinitely extended systems. In general, for the same group of automorphisms there could then be several KMS states. Physically they represent distinct thermodynamic phases.

## 18.3 Spectrum of the Liouvillean and relaxation

As discussed in section 17.5, at zero temperature the relaxation to the ground state can be reduced to a scattering problem, see Proposition 17.5. As a simplification, at finite temperature it suffices to have sufficiently strong spectral properties of the Liouvillean. We have in mind now a situation where the size of the black-body cavity is huge on the atomic scale. Therefore the relevant mathematical idealization is to have the photon field infinitely extended. The algebra $B(\mathcal{H})$ must be replaced then by a suitable algebra $\mathcal{A}$ of quasi-local observables. Its construction will be explained in the following. At the moment we focus on the abstract structure. Thus we have given the $C^*$-algebra $\mathcal{A}$ and a one-parameter group $\alpha_t$ of $*$-automorphisms as the dynamics. The distinguished state on $\mathcal{A}$ is the KMS state $\omega_\beta$ at inverse temperature $\beta$. Its time correlations are defined by

$$F_{ab}(t) = \omega_\beta(a\alpha_t(b)), \quad G_{ab}(t) = \omega_\beta(\alpha_t(b)a) \tag{18.41}$$

and they satisfy the KMS boundary condition

$$G_{ab}(t - \mathrm{i}\beta) = F_{ab}(t), \quad F_{ab}(t + \mathrm{i}\beta) = G_{ab}(t)\,; \tag{18.42}$$

compare with (18.39), (18.40). Note that $\omega_\beta$ is necessarily time-invariant, since $\omega_\beta(1\alpha_t(b)) = \omega_\beta(\alpha_t(b)1)$ and by the KMS condition

$$F_{1b}(t) = F_{1b}(t + \mathrm{i}\beta). \tag{18.43}$$

Let us define the $*$-algebra $\mathcal{A}^0$ through smoothing in time with a test function of compact support in Fourier space,

$$\mathcal{A}^0 = \big\{a_f = \int \mathrm{d}t f(t)\alpha_t(a) \,|\, a \in \mathcal{A}, \widehat{f} \in C_{00}(\mathbb{R})\big\}. \tag{18.44}$$

For $b \in \mathcal{A}^0$, $z \mapsto F_{1b}(z)$ is an entire function bounded as $|F_{1b}(z)| \le \|\alpha_z(b)\| \le \|\alpha_{i\mathrm{Im}z}(b)\|$. By (18.43) $F_{1b}$ is periodic with period $\mathrm{i}\beta$. Hence $F_{1b}$ is bounded and thus constant by Liouville's theorem, which implies

$$\omega_\beta(\alpha_t(b)) = \omega_\beta(b) \tag{18.45}$$

for all $t \in \mathbb{R}$.

We assume that $\mathcal{A}$ is a simple $C^*$-algebra, which means that the only two-sided $*$-ideals of $\mathcal{A}$ are either $\{0\}$ or $\mathcal{A}$ itself. The KMS condition then ensures that for every $a \in \mathcal{A}$

$$\omega_\beta(a^*a) = 0 \quad \text{implies} \quad a = 0. \tag{18.46}$$

To prove (18.46) we define $\mathcal{N} = \{a \in \mathcal{A} | \, \omega_\beta(a^*a) = 0\}$ with the goal of establishing that $\mathcal{N}$ is a two-sided $*$-ideal. Clearly, if $\omega_\beta(a^*a) = 0$ and $b \in \mathcal{A}$, then

$$\omega_\beta(a^*b^*ba) \leq \omega_\beta(a^*b^*bb^*ba)^{1/2}\omega_\beta(a^*a)^{1/2} = 0 \tag{18.47}$$

by the Schwarz inequality. Hence $\mathcal{A}\mathcal{N} \subset \mathcal{N}$. To show the converse one chooses $b \in \mathcal{A}$. By the KMS condition

$$\omega_\beta(b^*a^*ab) = \omega_\beta((b^*a^*a)b) = \omega_\beta(\alpha_{-\mathrm{i}\beta}(b)b^*a^*a) = 0 \tag{18.48}$$

by the Schwarz inequality as before. Thus $\mathcal{N}\mathcal{A} \subset \mathcal{N}$ and $\mathcal{N}$ is a two-sided $*$-ideal. Since $\mathcal{A}$ is simple, (18.46) follows.

Next we need the analog of $\mathcal{T}_2(\mathcal{H})$ and of the Liouvillean, which is the content of the Gelfand–Naimark–Segal (GNS) construction. The GNS Hilbert space $\mathcal{H}_\beta$ is defined as the completion of $\mathcal{A}$ equipped with the scalar product

$$\langle a|b\rangle = \omega_\beta(a^*b). \tag{18.49}$$

By the argument above $\langle a|a\rangle = 0$ implies $a = 0$, as it should. In our context $\mathcal{H}_\beta$ is a separable Hilbert space. We set $\Omega_\beta = 1$ and define the left representation of $\mathcal{A}$ through

$$\ell(a)b = ab. \tag{18.50}$$

Thereby $\ell(\mathcal{A}) \subset B(\mathcal{H}_\beta)$. In addition we define

$$\mathrm{e}^{-\mathrm{i}t\mathcal{L}}a = \alpha_t a \tag{18.51}$$

on $\mathcal{H}_\beta$. Since $\langle b|\mathrm{e}^{-\mathrm{i}t\mathcal{L}}a\rangle = \omega(b^*\alpha_t a) = \omega(\alpha_{-t}b^*a) = \langle \mathrm{e}^{\mathrm{i}t\mathcal{L}}b|a\rangle$ and since $\langle \mathrm{e}^{-\mathrm{i}t\mathcal{L}}b|\mathrm{e}^{-\mathrm{i}t\mathcal{L}}a\rangle = \omega(\alpha_t(b^*a)) = \langle b|a\rangle$ by time-invariance of $\omega_\beta$, $\mathrm{e}^{-\mathrm{i}t\mathcal{L}}$ is a strongly continuous unitary group on $\mathcal{H}_\beta$. By Stone's theorem it has a self-adjoint generator, which by definition is the Liouvillean $\mathcal{L}$.

The initial state of interest is a local perturbation of the equilibrium state $\omega_\beta$. It can be written as

$$\rho(a) = \omega_\beta(b^*ab), \quad b \in \mathcal{A}, \quad \omega_\beta(b^*b) = 1. \tag{18.52}$$

In the GNS representation $\rho$ corresponds to the state given by the vector $b \in \mathcal{H}_\beta$. More generally, a perturbed state can be written as

$$\rho(a) = \sum_{n=1}^{\infty} p_n \omega_\beta(b_n^* a b_n) \tag{18.53}$$

with $b_n \in \mathcal{A}$, $\omega(b_n^* b_n) = 1$, $p_n \geq 0$, $\sum_{n=1}^{\infty} p_n = 1$. States of the form (18.53) are called normal. A state not covered by this class would be a two-temperature state of the photon gas, for example, where the temperature to the far right differs from that to the far left. In fact, its long-time behavior would be rather different from that of the initial states discussed here.

Let $\rho$ be a normal state with time evolved $\rho_t(a) = \rho(\alpha_t(a))$. By relaxation to equilibrium we mean

$$\lim_{t\to\infty} \rho_t(a) = \omega_\beta(a) \tag{18.54}$$

for all $a \in \mathcal{A}$.

**Proposition 18.1** (Relaxation to equilibrium as a spectral property). *Suppose the Liouvillean $\mathcal{L}$ has a purely absolutely continuous spectrum except for a nondegenerate eigenvalue at 0. Then for all $a \in \mathcal{A}$*

$$\lim_{t\to\pm\infty} \rho_t(a) = \omega_\beta(a). \tag{18.55}$$

*Proof*: Since $\omega_\beta$ is time invariant, the (unique) zero eigenvector of $\mathcal{L}$ is $\Omega_\beta$. By assumption the spectral measure of $\langle\psi|\mathrm{e}^{-\mathrm{i}t\mathcal{L}}\varphi\rangle$ has the point mass $\langle\psi|\Omega_\beta\rangle\langle\Omega_\beta|\varphi\rangle$ at zero and is otherwise absolutely continuous. Therefore by the Riemann–Lebesgue lemma

$$\lim_{t\to\infty} \langle\psi|\mathrm{e}^{-\mathrm{i}t\mathcal{L}}\varphi\rangle = \langle\psi|\Omega_\beta\rangle\langle\Omega_\beta|\varphi\rangle \tag{18.56}$$

for all $\psi, \varphi \in \mathcal{H}_\beta$.

In view of the structure of normal states it suffices to study

$$\begin{aligned}\omega_\beta(b^*\alpha_t(a)c) &= \omega_\beta\big(\alpha_{-i\beta}(c)b^*\alpha_t(a)\big)\\ &= \langle\ell(b)\ell(\alpha_{i\beta}(c^*))\Omega_\beta|\ell(\alpha_t(a))\Omega_\beta\rangle\\ &= \langle\ell(b)\ell(\alpha_{i\beta}(c^*))\Omega_\beta|\mathrm{e}^{-\mathrm{i}t\mathcal{L}}\ell(a)\Omega_\beta\rangle.\end{aligned} \tag{18.57}$$

We assume that $a, b, c \in \mathcal{A}^0$, see (18.44). Then $\ell(b)\ell(\alpha_{i\beta}(c^*))\Omega_\beta, \ell(a)\Omega_\beta \in \mathcal{H}_\beta$. Therefore from (18.56)

$$\begin{aligned}\lim_{t\to\infty} \omega_\beta(b^*\alpha_t(a)c) &= \langle\ell(b)\ell(\alpha_{i\beta}(c^*))\Omega_\beta|\Omega_\beta\rangle\langle\Omega_\beta|\ell(a)\Omega_\beta\rangle\\ &= \omega_\beta(\alpha_{-i\beta}(c)b^*)\omega_\beta(a)\\ &= \omega_\beta(b^*c)\omega_\beta(a),\end{aligned} \tag{18.58}$$

which, upon inserting in (18.53), implies the limit (18.55). Note that the KMS condition is used twice, in the first identity of (18.57) and in the last identity of (18.58). □

Proposition 18.1 suggests that relaxation to equilibrium can be established in two steps: (i) One has to find for the equilibrium state a sufficiently concrete representation of the algebra of local observables and of the Liouvillean. (ii) The spectral properties of the Liouvillean must be studied. For (i) the natural representation is the Araki–Woods representation of the free photon gas in infinite volume. It will be taken up in the following section. The coupled system is constructed through perturbation series. For the dynamics the time-dependent Dyson series is used and for the thermal state the thermal perturbation theory of section 18.2. Of course, the convergence of both series relies on the atom being modeled as an $N$-level system and on the explicit control of the free photon gas. Only through the convergence of the perturbation series are we assured of the correct representation spaces for the interacting system. Nevertheless, we skip this important point completely and jump to the spectral analysis of the interacting Liouvillean.

### 18.4 The Araki–Woods representation of the free photon field

For photons in a cavity $\Lambda$, the spectrum of allowed momenta is discrete, $\mathrm{tr}[\exp[-\beta H_{\mathrm{f},\Lambda}]] < \infty$, and the rules of thermal equilibrium for bounded quantum systems are applicable, through which the time-correlations of local observables in the form $\omega_\beta^\Lambda(a\alpha_t^\Lambda(b))$ are defined. A macroscopic cavity with its surface kept at a uniform temperature is extremely well approximated by the infinite-volume limit $\Lambda \uparrow \mathbb{R}^3$. For the Hamiltonian (18.1) the infinite-volume limit of time-correlations can be established. Rather than going through the construction, we merely state the final answer, which will serve as a basis for the study of relaxation.

We work in the momentum space representation. Without risk of confusion we set $k = (k, \lambda) \in \mathbb{R}^3 \times \{1, 2\}$ and $\sum_{\lambda=1,2} \int \mathrm{d}^3k = \int \mathrm{d}k$, $\delta(k - k') = \delta_{\lambda,\lambda'}\delta(k - k')$. The bosonic field operators are

$$a(f) = \int \mathrm{d}k f(k)a(k) = \sum_{\lambda=1,2} \int \mathrm{d}^3k f(k,\lambda)a(k,\lambda), \quad a^*(f) = \int \mathrm{d}k f(k)a^*(k) \tag{18.59}$$

with $f \in \mathcal{S}_0(\mathbb{R}^3 \times \{1, 2\})$, the Schwartz space of functions that decrease rapidly and vanish at $k = 0$. Observe that our convention for the complex conjugation of the test function $f$ differs from that in (13.59), (13.60). Let us also introduce the complex conjugation $\tau f(k) = f(k)^* = \big(f(k,1)^*, f(k,2)^*\big)$. Its second quantization is the anti-unitary time-reversal operator $T$ on $\mathcal{F}$ with the properties

$$T\Omega = \Omega, \quad Ta^\sharp(f)T = a^\sharp(\tau f), \quad T = T^* = T^{-1}. \tag{18.60}$$

Note that $(a^*(f))^* = a(\tau f)$ and $\langle f, g\rangle_{\mathfrak{h}} = \int \mathrm{d}k(\tau f)g$. The boson fields satisfy the canonical commutation relations (CCR),

$$[a^*(f), a^*(g)] = 0 = [a(f), a(g)], \tag{18.61}$$

$$[a(\tau f), a^*(g)] = \langle f, g\rangle_{\mathfrak{h}}. \tag{18.62}$$

Let $\mathcal{P}$ denote the polynomial $*$-algebra generated by

$$\{a^*(f), a(g) \mid f, g \in \mathcal{S}_0(\mathbb{R}^3)^2\}. \tag{18.63}$$

On $\mathcal{P}$ the time evolution $\alpha_t^{\mathrm{f}}$ is defined through

$$\alpha_t^{\mathrm{f}}(a^*(k)) = \mathrm{e}^{\mathrm{i}t\omega(k)}a^*(k), \quad \alpha_t^{\mathrm{f}}(a(k)) = \mathrm{e}^{-\mathrm{i}t\omega(k)}a(k). \tag{18.64}$$

The equilibrium state $\omega_\beta^{\mathrm{f}}$ of the photon field at inverse temperature $\beta$ is a quasi-free state on $\mathcal{P}$. Set

$$\rho_\beta(k) = \frac{1}{\mathrm{e}^{\beta\omega(k)} - 1}. \tag{18.65}$$

Then the two-point function is given by

$$\omega_\beta^{\mathrm{f}}\big(a^*(\tau f)a(g)\big) = \langle f, \rho_\beta g\rangle_{\mathfrak{h}} \tag{18.66}$$

and all other moments by

$$\omega_\beta^{\mathrm{f}}\Big(\prod_{i=1}^{n} a^*(\tau f_i)\prod_{j=1}^{m} a(g_j)\Big) = \delta_{mn}\det\{\langle f_i, \rho_\beta g_j\rangle_{\mathfrak{h}}\}_{i,j=1,\ldots,n}. \tag{18.67}$$

$\omega_\beta^{\mathrm{f}}$ satisfies the KMS condition as can be seen directly from

$$\begin{aligned}\omega_\beta^{\mathrm{f}}\big(a(k)a^*(k')\big) &= \delta(k-k') + \omega_\beta^{\mathrm{f}}\big(a^*(k')a(k)\big)\\ &= \mathrm{e}^{\beta\omega(k)}\rho_\beta(k)\delta(k-k')\\ &= \omega_\beta^{\mathrm{f}}\big(\alpha_{-i\beta}^{\mathrm{f}}(a^*(k'))a(k)\big).\end{aligned} \tag{18.68}$$

Through the GNS construction the data $(\mathcal{P}, \alpha_t^{\mathrm{f}}, \omega_\beta^{\mathrm{f}})$ determine a separable Hilbert space $\mathcal{H}_\beta^{\mathrm{f}}$, a left representation $\ell$ of $\mathcal{P}$ on $\mathcal{H}_\beta^{\mathrm{f}}$, a vector $\Omega_\beta^{\mathrm{f}} \in \mathcal{H}_\beta^{\mathrm{f}}$ cyclic for $\ell(\mathcal{P})$, and a unitary one-parameter group $\mathrm{e}^{-\mathrm{i}t\mathcal{L}_{\mathrm{f}}}$, $t \in \mathbb{R}$, such that

$$\begin{aligned}\omega_\beta^{\mathrm{f}}(a) &= \langle\Omega_\beta^{\mathrm{f}}|\ell(a)\Omega_\beta^{\mathrm{f}}\rangle\\ \ell(\alpha_t(a)) &= \mathrm{e}^{\mathrm{i}t\mathcal{L}_{\mathrm{f}}}\ell(a)\mathrm{e}^{-\mathrm{i}t\mathcal{L}_{\mathrm{f}}}, \quad a \in \mathcal{P}.\end{aligned} \tag{18.69}$$

We follow Araki and Woods to construct, as for a bounded quantum system, an isomorphism $I_T$ between $\mathcal{H}^{\mathrm{f}}_\beta$ and $\mathcal{F}\otimes\mathcal{F}$. On $\mathcal{F}\otimes\mathcal{F}$ we introduce the Bose fields

$$a^\sharp_\ell(f) = a^\sharp(f)\otimes 1, \quad a^\sharp_r(f) = 1\otimes a^\sharp(\tau f). \tag{18.70}$$

Note that $a^\sharp_r$ is an antilinear representation of the CCR. The isomorphism $I_T$ is then defined through the following relations,

$$I_T\Omega^{\mathrm{f}}_\beta = \Omega\otimes\Omega, \tag{18.71}$$

$$I_T\ell(a(f))I_T^{-1} = a_\ell(\sqrt{1+\rho_\beta}f) + a^*_r(\sqrt{\rho_\beta}f), \tag{18.72}$$

$$I_Tr(a(f))I_T^{-1} = a^*_\ell(\sqrt{\rho_\beta}\tau f) + a_r(\sqrt{1+\rho_\beta}\tau f). \tag{18.73}$$

As it should be, (18.72) is linear and (18.73) is antilinear in $f$.

$I_T\ell(a^\sharp(f))I_T^{-1}$ and $I_Tr(a^\sharp(f))I_T^{-1}$ satisfy the CCR and one only has to check that the two-point function is properly transported,

$$\begin{aligned}
&\langle\Omega\otimes\Omega|I_T\ell(a^*(f))\ell(a(g))I_T^{-1}\Omega\otimes\Omega\rangle \\
&= \langle\Omega\otimes\Omega|a_r(\sqrt{\rho_\beta}f)a^*_r(\sqrt{\rho_\beta}g)\Omega\otimes\Omega\rangle = \langle\tau f, \rho_\beta g\rangle_{\mathfrak{h}} \\
&= \omega^{\mathrm{f}}_\beta(a^*(f)a(g)) = \langle\Omega^{\mathrm{f}}_\beta|\ell(a^*(f))\ell(a(g))\Omega^{\mathrm{f}}_\beta\rangle
\end{aligned} \tag{18.74}$$

and likewise for the right representation. We conclude that, indeed, $I_T:\mathcal{H}^{\mathrm{f}}_\beta\to\mathcal{F}\otimes\mathcal{F}$ is an isometry.

The Liouvillean is transported as $L_{\mathrm{f}} = I_T\mathcal{L}_{\mathrm{f}}I_T^{-1}$. From the bounded systems one would expect that

$$L_{\mathrm{f}} = \int \mathrm{d}k\omega(k)\big(a^*_\ell(k)a_\ell(k) - a^*_r(k)a_r(k)\big). \tag{18.75}$$

Then indeed, as required,

$$\mathrm{e}^{\mathrm{i}tL_{\mathrm{f}}}a_\ell(k)\mathrm{e}^{-\mathrm{i}tL_{\mathrm{f}}} = \mathrm{e}^{-\mathrm{i}t\omega(k)}a_\ell(k), \quad \mathrm{e}^{\mathrm{i}tL_{\mathrm{f}}}a_r(k)\mathrm{e}^{-\mathrm{i}tL_{\mathrm{f}}} = \mathrm{e}^{\mathrm{i}t\omega(k)}a_r(k) \tag{18.76}$$

and

$$\mathrm{e}^{\mathrm{i}tL_{\mathrm{f}}}I_T\ell(a(k))I_T^{-1}\mathrm{e}^{-\mathrm{i}tL_{\mathrm{f}}} = \mathrm{e}^{-\mathrm{i}t\omega(k)}I_T\ell(a(k))I_T^{-1} = I_T\ell(\alpha_t(a(k)))I_T^{-1}; \tag{18.77}$$

similarly for the right representation.

## 18.5 Atom in interaction with the photon gas

The atomic Hamiltonian $H_{\mathrm{at}}$ has $N$, possibly degenerate, eigenvalues, $\varepsilon_1\le\varepsilon_2\le\cdots\le\varepsilon_N$, and the atomic Hilbert space is $\mathcal{H}_{\mathrm{at}} = \mathbb{C}^N$. We fix a corresponding eigenbasis, $H_{\mathrm{at}}\varphi_j = \varepsilon_j\varphi_j$, $j = 1,\ldots,N$. The algebra of observables is the $N\times N$ complex matrices $\mathcal{M}_N$ and it carries the thermal state $\omega^{\mathrm{at}}_\beta = Z^{-1}\mathrm{e}^{-\beta H_{\mathrm{at}}}$. As long as there is no interaction we merely tensor the atom with the photon field. The algebra of observables is $\mathcal{M}_N \otimes \mathcal{P}$, the thermal state is

$$\omega_\beta^0 = \omega_\beta^{\mathrm{at}} \otimes \omega_\beta^{\mathrm{f}}, \tag{18.78}$$

and the dynamics is generated by $\alpha_t^0 = \alpha_t^{\mathrm{at}} \otimes \alpha_t^{\mathrm{f}}$. As before, the GNS construction determines a separable Hilbert space $\mathcal{H}_\beta^0$ with cyclic vector $\Omega_\beta^0$ and a unitary time evolution $\mathrm{e}^{-\mathrm{i}t\mathcal{L}_0}$. With $C$ denoting complex conjugation in the given basis of $\mathcal{H}_{\mathrm{at}}$, the map $I_0 = I_C \otimes I_T : \mathcal{H}_\beta^0 \to \mathcal{H}_{\mathrm{at}} \otimes \mathcal{H}_{\mathrm{at}} \otimes \mathcal{F} \otimes \mathcal{F} = \widehat{\mathcal{H}}_\beta$ is an isomorphism. In particular, $I_0 \Omega_\beta^0 = \widehat{\Omega}_\beta^0$ with

$$\widehat{\Omega}_\beta^0 = \sum_{j=1}^{N} \mathrm{e}^{-\beta\varepsilon_j} \varphi_j \otimes \varphi_j \otimes \Omega \otimes \Omega. \tag{18.79}$$

The Liouvillean is mapped as

$$I_0 \mathcal{L}_0 I_0^{-1} = L_0 = L_{\mathrm{at}} \otimes 1 + 1 \otimes L_{\mathrm{f}}, \quad L_{\mathrm{at}} = H_{\mathrm{at}} \otimes 1 - 1 \otimes H_{\mathrm{at}}. \tag{18.80}$$

The real task is to find out how the interaction is mapped to the Araki–Woods representation space. According to (18.1) one has

$$H_{\mathrm{int}} = Q \cdot E_\varphi = \sum_{\lambda=1,2} \int \mathrm{d}^3k\, \widehat{\varphi}(k) \sqrt{\omega/2}\, Q \cdot e_\lambda \big(\mathrm{i}a(k,\lambda) - \mathrm{i}a^*(k,\lambda)\big). \tag{18.81}$$

It is more convenient to slightly generalize from (18.81) as

$$H_{\mathrm{int}} = \int \mathrm{d}k \big(G(k) \otimes a^*(k) + G(k)^* \otimes a(k)\big), \tag{18.82}$$

where $G : \mathbb{R}^3 \times \{1,2\} \to \mathcal{M}_N$ as a matrix-valued function, with the memo that some specific features of the coupling in (18.81) will be used in the spectral analysis below.

If $G \in \mathcal{S}_0(\mathbb{R}^3 \times \{1,2\}, \mathcal{M}_N)$ as matrix-valued function, one has $H_{\mathrm{int}} \in \mathcal{M}_N \otimes \mathcal{P}$. Thus the Liouvillean in the GNS space necessarily takes the form

$$\mathcal{L}_\lambda = \mathcal{L}_0 + \lambda \ell(H_{\mathrm{int}}) - \lambda r(H_{\mathrm{int}}) = \mathcal{L}_0 + \lambda \mathcal{L}_{\mathrm{int}} \tag{18.83}$$

and only the transformations (18.72), (18.73) have to be applied, resulting in

$$\begin{aligned} L_{\mathrm{int}} &= I_0 \mathcal{L}_{\mathrm{int}} I_0^{-1} \\ &= \int \mathrm{d}k \big\{ \big(\sqrt{1+\rho_\beta}\, G_\ell(k) - \sqrt{\rho_\beta}\, \overline{G}_r^*(k)\big) a_\ell^*(k) \\ &\quad + \big(\sqrt{1+\rho_\beta}\, G_\ell^*(k) - \sqrt{\rho_\beta}\, \overline{G}_r(k)\big) a_\ell(k) \\ &\quad + \big(\sqrt{\rho_\beta}\, G_\ell^*(k) - \sqrt{1+\rho_\beta}\, \overline{G}_r(k)\big) a_r^*(k) \\ &\quad + \big(\sqrt{\rho_\beta}\, G_\ell(k) - \sqrt{1+\rho_\beta}\, \overline{G}_r^*(k)\big) a_r(k) \big\}. \end{aligned} \tag{18.84}$$

Here $G^\sharp_\ell = G^\sharp \otimes 1$, $G^\sharp_r = 1 \otimes G^\sharp$, $\overline{G}^\sharp = I_C G^\sharp I_C^{-1}$. Extending the test function notation to matrix-valued test functions, $L_{\text{int}}$ may be written more concisely as

$$L_{\text{int}} = a^*_\ell\big(\sqrt{1+\rho_\beta}G_\ell - \sqrt{\rho_\beta}\,\overline{G}^*_r\big) + a_\ell\big(\sqrt{1+\rho_\beta}G^*_\ell - \sqrt{\rho_\beta}\,\overline{G}_r\big)$$
$$+ a^*_r\big(\sqrt{\rho_\beta}G^*_\ell - \sqrt{1+\rho_\beta}\,\overline{G}_r\big) + a_r\big(\sqrt{\rho_\beta}G_\ell - \sqrt{1+\rho_\beta}\,\overline{G}^*_r\big). \quad (18.85)$$

With some effort we thus achieved our goal of writing the Liouvillean for the excitations away from equilibrium. Note that through $\rho_\beta$ the interaction is temperature-dependent and becomes singular as $\beta \to \infty$, which only reflects the fact that the ground state does not fall into the scheme explained before.

Two problems remain to be sorted out. First, $L_\lambda = L_0 + \lambda L_{\text{int}}$ must generate a unitary time evolution on $\widehat{\mathcal{H}}_\beta$. If

$$\int \mathrm{d}k\big(\omega(k) + \omega(k)^{-3}\big)\|G(k)\|^2 < \infty, \quad (18.86)$$

then the self-adjointness of $L_\lambda$ follows from the Nelson commutator theorem. Note that for the physical case (18.81) the condition (18.86) translates to $\int \mathrm{d}^3k|\widehat{\varphi}|^2(\omega^2 + \omega^{-2}) < \infty$, which is satisfied.

Secondly, the equilibrium state of the interacting system must be represented by a vector in $\widehat{\mathcal{H}}_\beta$. The thermal perturbation theory of section 18.2 tells us that this new vector is formally given by

$$\widehat{\Omega}^\lambda_\beta = (Z_\beta)^{-1/2}\mathrm{e}^{-\beta L_\ell/2}\widehat{\Omega}^0_\beta = (Z_\beta)^{-1/2}\mathrm{e}^{\beta L_r/2}\widehat{\Omega}^0_\beta, \quad (18.87)$$

where, according to (18.33), (18.85)

$$L_\ell = L_0 + \lambda L_{\text{int}\ell}, \quad L_r = L_0 - \lambda L_{\text{int}r} \quad (18.88)$$

and

$$L_{\text{int}\ell} = a^*_\ell(\sqrt{1+\rho_\beta}G_\ell) + a_\ell(\sqrt{1+\rho_\beta}G^*_\ell) + a^*_r(\sqrt{\rho_\beta}G^*_\ell) + a_r(\sqrt{\rho_\beta}G_\ell),$$
$$L_{\text{int}r} = a^*_\ell(\sqrt{\rho_\beta}\,\overline{G}^*_r) + a_\ell(\sqrt{\rho_\beta}\,\overline{G}_r) + a^*_r(\sqrt{1+\rho_\beta}\,\overline{G}_r) + a_r(\sqrt{1+\rho_\beta}\,\overline{G}^*_r). \quad (18.89)$$

$(Z_\beta)^{-1/2}$ normalizes the vector to one. It can be shown that $Z_\beta < \infty$ provided

$$\int \mathrm{d}k(1 + \omega^{-1})\|G(k)\|^2 < \infty. \quad (18.90)$$

Therefore under the condition (18.86), $\widehat{\Omega}^\lambda_\beta \in \widehat{\mathcal{H}}_\beta$.

By construction $L_\lambda\widehat{\Omega}^\lambda_\beta = 0$. Thus $L_\lambda$ has a zero eigenvector, which does not change under the dynamics and represents the state of global equilibrium. According to Proposition 18.1, we have to make sure that $\widehat{\Omega}^\lambda_\beta$ is the only eigenvector of $L_\lambda$ and that apart from the zero eigenvalue, the spectrum is purely absolutely continuous.

## 18.6 Complex translations

$L_{\mathrm{f}}$ has the full real line as spectrum. Its structure is more easily investigated by switching to spherical coordinates in momentum space and to the corresponding Bose field denoted here by $b(\omega, \widehat{k})$. We set

$$(k, \lambda) = (\omega, \widehat{k}), \quad \mathrm{d}k = \omega^2 \mathrm{d}\omega \mathrm{d}\widehat{k}, \tag{18.91}$$

where $\widehat{k} = (k/|k|, \lambda)$. The right representation in $L_{\mathrm{f}}$ has negative excitation energies, which we associate with $\omega < 0$. Thus the Bose field $b^\sharp(\omega, \widehat{k})$ lives on $\mathbb{R} \times S^2 \times \{1, 2\}$ and is defined by

$$b^\sharp(\omega, \widehat{k}) = \begin{cases} \omega a_\ell^\sharp(k) & \text{for} \quad \omega = |k|, \\ \omega a_r^\sharp(k) & \text{for} \quad \omega = -|k|. \end{cases} \tag{18.92}$$

From the definition of $a_\ell^\sharp$, $a_r^\sharp$ one confirms that $b, b^\sharp$ satisfy the CCR as

$$[b(\omega, \widehat{k}), b(\omega', \widehat{k}')] = 0 = [b^*(\omega, \widehat{k}), b^*(\omega', \widehat{k}')] \tag{18.93}$$

and

$$[b(\omega, \widehat{k}), b^*(\omega', \widehat{k}')] = \delta(\omega - \omega')\delta(\widehat{k} - \widehat{k}'). \tag{18.94}$$

In the new coordinates the Liouvillean becomes

$$L_{\mathrm{f}} = \int_{\mathbb{R}} \mathrm{d}\omega \int_{S^2 \times \{1,2\}} \mathrm{d}\widehat{k}\, \omega b^*(\omega, \widehat{k}) b(\omega, \widehat{k}). \tag{18.95}$$

We rewrite the interaction. Let us define the matrix-valued functions

$$F_\ell(\omega, \widehat{k}) = \begin{cases} \omega^{-1/2} G_\ell(k) & \text{for} \quad \omega = |k|, \\ -(-\omega)^{-1/2} G_\ell^*(k) & \text{for} \quad \omega = -|k|, \end{cases} \tag{18.96}$$

$$F_r(\omega, \widehat{k}) = \begin{cases} \omega^{1/2} C G_r^*(k) C & \text{for} \quad \omega = |k|, \\ -(-\omega)^{-1/2} C G_r(k) C & \text{for} \quad \omega = -|k|, \end{cases} \tag{18.97}$$

$$\begin{aligned} F_\ell^{(\beta)}(\omega, \widehat{k}) &= \big(\omega(1 - \mathrm{e}^{-\beta\omega})^{-1}\big)^{1/2} F_\ell(\omega, \widehat{k}), \\ F_r^{(\beta)}(\omega, \widehat{k}) &= \big(-\omega(1 - \mathrm{e}^{\beta\omega})^{-1}\big)^{1/2} F_r(\omega, \widehat{k}). \end{aligned} \tag{18.98}$$

Then

$$L_{\text{int}} = L_{\text{int}\ell} - L_{\text{int}\,r} \tag{18.99}$$

with

$$L_{\text{int}\sharp} = \int_{\mathbb{R}} \mathrm{d}\omega \int_{S^2\times\{1,2\}} \mathrm{d}\widehat{k}\big(F_\sharp^{(\beta)}(\omega,\widehat{k})b^*(\omega,\widehat{k}) + F_\sharp^{(\beta)}(\omega,\widehat{k})^* b(\omega,\widehat{k})\big). \tag{18.100}$$

With (18.95) and the definitions (18.99), (18.100) one concludes

$$L_\lambda = L_{\text{at}} + L_{\text{f}} + \lambda L_{\text{int}} = L_0 + \lambda L_{\text{int}}. \tag{18.101}$$

Since $L_{\text{f}}$ has the real line for its continuous spectrum, it is natural to try to move it through a downward translation. The generator $T$ of translations along the $\omega$-axis is given by

$$T = \int \mathrm{d}\omega \int \mathrm{d}\widehat{k} b^*(\omega,\widehat{k})(-\mathrm{i}\partial_\omega)b(\omega,\widehat{k}). \tag{18.102}$$

Let $\theta \in \mathbb{C}$. Then

$$L_0(\theta) = \mathrm{e}^{-\mathrm{i}\theta T} L_0 \mathrm{e}^{\mathrm{i}\theta T} = L_{\text{at}} + L_{\text{f}} - \theta N_{\text{f}} \tag{18.103}$$

with the number operator

$$N_{\text{f}} = \int \mathrm{d}\omega \int \mathrm{d}\widehat{k} b^*(\omega,\widehat{k})b(\omega,\widehat{k}). \tag{18.104}$$

We set $\theta = \mathrm{i}\vartheta$, $\vartheta > 0$. Then $L_0(\theta)$ has $\mathbb{R} - \mathrm{i}\vartheta$ as continuous spectrum and the isolated eigenvalues $\{\varepsilon_i - \varepsilon_j \,|\, i, j = 1, \dots, N\}$ on the real axis.

To be able to apply the theory of complex deformations $\theta \mapsto \mathrm{e}^{-\mathrm{i}\theta T} L_{\text{int}} \mathrm{e}^{\mathrm{i}\theta T} = L_{\text{int}}(\theta)$ has to be analytic in a strip around the real axis. $L_{\text{int}}(\theta)$ is obtained by shifting $F_\sharp^{(\beta)}(\omega,\widehat{k})$ in (18.100) to $F_\sharp^{(\beta)}(\omega+\theta,\widehat{k})$. Thus the issue is whether $F_\sharp^{(\beta)}(\omega,\widehat{k})$ extends to an analytic function near the real axis. For the physical coupling

$$G(k,\lambda) = -\mathrm{i}Q \cdot e_\lambda \sqrt{\omega/2}\widehat{\varphi}(k). \tag{18.105}$$

By assumption $\varphi$ is radial and has compact support in position space. Thus $\widehat{\varphi}_{\text{r}}$ is an analytic function on $\mathbb{C}$. Therefore $F_\ell$, $F_r$ of (18.96), (18.97) are analytic in $\omega$. The prefactors in (18.98) have simple poles at $\pm 2\pi\mathrm{i}\beta n$, $n = 1, 2, \dots$. We conclude that $F_\sharp^{(\beta)}(\omega,\widehat{k})$ are analytic in $\omega$ in the strip $\mathcal{S}_{2\pi/\beta} = \{\theta \mid |\mathrm{Im}\theta| < 2\pi/\beta\}$. $L_\lambda(\theta) = L_0(\theta) + \lambda L_{\text{int}}(\theta)$ is jointly analytic in $\lambda$ and $\theta \in \mathcal{S}_{2\pi/\beta}$. To derive this result we used the assumption that the photons have zero mass. Otherwise $L_{\text{f}}$ would have a spectral gap and complex translations could not be implemented. We also assumed that there is a $\sqrt{\omega}$ prefactor in the physical coupling. Both assumptions could be

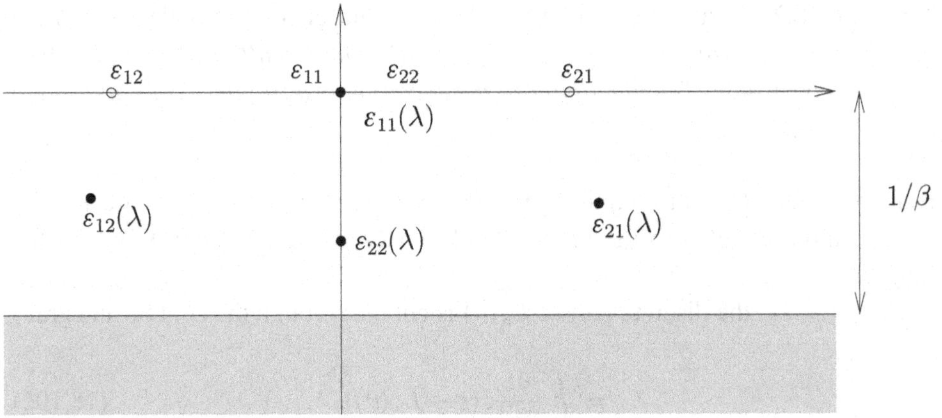


Figure 18.1: Spectrum of the complex translated Liouvillean in the case of a two-level atom for zero and nonzero coupling.

avoided at the expense of a considerably more involved analysis. Note that the width of the strip of analyticity decreases as $1/\beta$ which indicates that our estimates worsen as zero temperature is approached.

We are now in a position to use the considerations from section 17.3 and choose $\theta = \mathrm{i}\vartheta$ with $\vartheta$ close to the optimal value $2\pi/\beta$. For zero coupling the eigenvalues of $L_0(\theta)$ are $\varepsilon_{ij} = \varepsilon_i - \varepsilon_j, i, j = 1, \ldots, N$, see figure 18.1. The zero eigenvalue is at least $N$-fold degenerate. As the coupling is turned on, $\lambda \neq 0$, the eigenvalues $\varepsilon_{ij}(\lambda)$ move. From the general theory, there is a dense set of vectors, $\mathcal{E} \subset \widehat{\mathcal{H}}_\beta$, such that for $\psi, \varphi \in \mathcal{E}$ the resolvent $\langle\psi|(z - L_\lambda)^{-1}\varphi\rangle$ can be continued analytically from the upper complex plane to $\{z \mid \mathrm{Im} z > -\vartheta\}$. In this domain $\langle\psi|(z - L_\lambda)^{-1}\varphi\rangle$ is analytic except for poles at $z = \varepsilon_{ij}(\lambda)$. Thus $\varepsilon_{ij}(\lambda)$ are the resonance poles of the resolvent. These assertions remain valid up to the first $\lambda$ when a resonance hits the line $\{z = -\mathrm{i}\vartheta\}$. Thereby the theory is restricted up to some maximal coupling $\lambda_0$, $|\lambda| < \lambda_0$.

In our convention $\mathrm{Im}\,\varepsilon_{ij}(\lambda) < 0$ corresponds to exponential decay. Thus, since expectation values remain bounded in $t$, the resonances cannot move in the upper half complex plane. From the thermal perturbation theory we know that at least one eigenvalue remains at 0. Somewhat arbitrarily we label this eigenvalue by $\varepsilon_{11}(\lambda)$. To prove relaxation to equilibrium, according to Proposition 18.1, it must be ensured that all other resonances acquire a strictly negative imaginary part for $\lambda \neq 0$. At this point, second-order perturbation theory comes in handy. We require that the dissipative part $K^\natural$ in (17.32) has a nondegenerate eigenvalue 0. Then $\mathrm{Im}\,\varepsilon_{ij}(\lambda) = \mathcal{O}(\lambda^2)$ and, possibly further reducing $\lambda_0$, the second order controls the higher orders, which implies $\mathrm{Im}\,\varepsilon_{ij} < 0$ for $|\lambda| < \lambda_0$, except for $\varepsilon_{11}(\lambda)$.

**Theorem 18.2** (Absolute continuity of the spectrum of the Liouvillean). *If $K^\natural$ has a simple eigenvalue 0 and if $|\lambda| < \lambda_0$ for sufficiently small $\lambda_0$, then $L_\lambda$ has 0 as a simple eigenvalue. The remainder of the spectrum is absolutely continuous and covers the real line.*

From Theorem 18.2 in conjunction with Proposition 18.1 we conclude that an $N$-level atom coupled to the photon field relaxes to thermal equilibrium in the long-time limit.

For $|\lambda| < \lambda_0$, the discrete part of $L_\lambda(\theta)$ is cut out through the contour integral

$$\Sigma_\lambda = \oint_\gamma \frac{\mathrm{d}z}{2\pi\mathrm{i}} z(z - L_\lambda(\theta))^{-1}, \tag{18.106}$$

where $\gamma$ is a contour in the complex plane which encircles all eigenvalues $\varepsilon_{ij}(\lambda)$ and stays away from the half-space $\{z \mid \mathrm{Im} z \leq -\vartheta\}$. $\Sigma_\lambda$ remains unchanged under small shifts of $\vartheta$. By the same token one can construct two maps $W_\lambda^\pm \mathcal{E} \to \mathcal{M}_N$ such that

$$\langle \phi | \mathrm{e}^{-\mathrm{i}tL_\lambda} \psi \rangle = \langle W_\lambda^- \phi | \mathrm{e}^{-\mathrm{i}t\Sigma_\lambda} W_\lambda^+ \psi \rangle + \mathcal{O}(\mathrm{e}^{-\vartheta t}) \tag{18.107}$$

for $t \geq 0$. Equation (18.107) defines the level shift operator $\Sigma_\lambda$. Its eigenvalues are $\varepsilon_{ij}(\lambda)$, $i, j = 1, \ldots, N$. Thus (18.107) establishes *exponentially fast* relaxation to equilibrium for a large class of initial states and of observables.

Our scheme leaves somewhat open how rapidly specific expectation values decay. For example one could prepare the atom in the $n$-th level and ask how the probability of survival decays as $t \to \infty$. If $P_n$ denotes the projection on the $n$-th eigenstate $\varphi_n$, then the observable under consideration is $P_n \otimes 1$. As initial state one could take the uncorrelated state $P_n \otimes \omega_\beta^{\mathrm{f}}$. A physically more realistic choice would be the state $\omega^{(n)}(a) = \omega_\beta(P_n \otimes 1 a P_n \otimes 1)/\omega_\beta(P_n \otimes 1)$ with $\omega_\beta$ the equilibrium state of the coupled system. The issue is to compute the decay of $\omega^{(n)}(\alpha_t(P_n \otimes 1))$. Equation (18.107) suggests that $\omega^{(n)}(\alpha_t(P_n \otimes 1)) - \omega_\beta(P_n \otimes 1)$ decays exponentially to zero. To verify this one has to find the representation vectors and determine their analytic continuation in $\theta$. In our example the observables do not depend on the field and therefore the representation vectors are in $\mathcal{E}$, which ensures exponential decay.

In specific systems, say only two levels, also the order $\lambda^4$ could be computed. Up to errors from $\mathcal{O}(\mathrm{e}^{-\vartheta t})$ the line shape is still a Lorentzian, whose location and width are given with a precision superior to the weak coupling theory.

## 18.7 Comparison with the weak coupling theory

The weak coupling theory of section 17.2 predicts the decay of atomic expectations in the form

$$\mathrm{tr}[A\mathrm{e}^{-\mathrm{i}L_{\mathrm{D}}t}(B\rho_\beta B^*)], \tag{18.108}$$

which is written somewhat differently than before to ease comparison. The trace is over $\mathbb{C}^N$, $A = A^*$ is some atomic observable, $\rho_\beta = Z^{-1}\mathrm{e}^{-\beta H_{\mathrm{at}}}$, $B\rho_\beta B^*$ is the initial density matrix of the atom normalized as $\mathrm{tr}[\rho_\beta B^* B] = 1$, and $L_{\mathrm{D}}$ is the Davies generator of (17.68). We assume $\{H, Q_\alpha, \alpha = 1, 2, 3\}' = \mathbb{C}1$ and the Wiener condition $\Gamma(\omega) > 0$ for all $\omega$. Then (18.108) converges exponentially fast to the thermal equilibrium expectation of $A$, $\mathrm{tr}[\rho_\beta A]$, independently of the choice of $B$.

In the full microscopic theory the expectation in spirit closest to (18.108) is given by

$$\omega_\beta(B^* \otimes 1\alpha_t^\lambda(A \otimes 1)B \otimes 1), \tag{18.109}$$

where, as before, $\omega_\beta$ is the thermal state of the coupled system and $\alpha_t^\lambda$ is the time evolution with Liouvillean (18.83). From the thermal perturbation theory one concludes that there exists a local operator $c$ such that $\omega_\beta(B^* \otimes 1aB \otimes 1) = \omega_\beta(ca)$ for all $a \in \mathcal{M}_N \otimes \mathcal{A}$. Thus

$$\omega_\beta(B^* \otimes 1\alpha_t^\lambda(A \otimes 1)B \otimes 1) = \omega_\beta(c\alpha_t^\lambda(A \otimes 1)). \tag{18.110}$$

Since $B \otimes 1$, $A \otimes 1$ are atomic observables, in the GNS representation $c$ and $A \otimes 1$ become vectors in $\mathcal{E}$. Therefore the long-time behavior of the expectation in (18.109) is determined by the resonances $\varepsilon_{ij}(\lambda)$. If $|\lambda| < \lambda_0$, then $\mathrm{Im}\,\varepsilon_{ij}(\lambda) < 0$ except for $ij = 11$ when $\varepsilon_{11}(\lambda) = 0$. Thus also the expectation value in (18.109) decays exponentially fast to its equilibrium value $\omega_\beta(A \otimes 1)$.

Optimally, one would like to compare (18.108) and (18.109) for small $\lambda$. The form (18.108) is a sum of $N^2$ exponentials, decaying except for one constant term. Likewise, (18.109) is a sum of $N^2$ exponentials plus an error which has an even faster exponential decay independent of $\lambda$ and can be neglected for small $\lambda$. Most naturally, amplitudes and decay rates are compared. The amplitudes differ by order $\lambda^2$, since from the thermal perturbation theory $\omega_\beta(A \otimes 1) = \mathrm{tr}[\rho_\beta A] + \mathcal{O}(\lambda^2)$ using that $\omega_\beta^{\mathrm{f}}(E) = 0$.

The decay rates for (18.108) are the eigenvalues of $L_{\mathrm{D}}$. The eigenvalues of $L_{\mathrm{at}} = [H_{\mathrm{at}}, \cdot]$ are $\varepsilon_i - \varepsilon_j = \varepsilon_{ij}$, $i, j = 1, \ldots, N$. Since $L_{\mathrm{at}}$ and $\mathrm{i}K^\natural$ commute, $L_{\mathrm{D}}$ is block diagonal with respect to the eigenvalues of $L_{\mathrm{at}}$. The eigenvalues $\varepsilon_{ij}^{\mathrm{D}}$ of $L_{\mathrm{D}}$

are then necessarily of the form

$$\varepsilon_{ij}^{\mathrm{D}} = \varepsilon_{ij} + \lambda^2 \varepsilon_{ij}^{\natural} \tag{18.111}$$

with $\varepsilon_{ij}^{\natural}$ the eigenvalues of $\mathrm{i}K^{\natural}$ and they cluster at the eigenvalues of $L_{\mathrm{at}}$. $\mathrm{i}K^{\natural}$ decomposes as

$$\mathrm{i}K^{\natural}\rho = [H_{\Delta}, \rho] + \mathrm{i}K_{\mathrm{d}}^{\natural}\rho \tag{18.112}$$

with $H_{\Delta}$ given by (17.69). $[H_{\Delta}, H_{\mathrm{at}}] = 0$ by construction. $H_{\Delta}$ shifts the atomic levels and lifts possible degeneracies of $H_{\mathrm{at}}$. $K_{\mathrm{d}}^{\natural}$ and $L_{\mathrm{at}}$ also commute. By detailed balance $K_{\mathrm{d}}^{\natural}$ is symmetric with respect to the weighted inner product $\mathrm{tr}[\rho_{\beta} A^* B]$. Thus the eigenvalues of $K_{\mathrm{d}}^{\natural}$ are negative, real, and with a nondegenerate eigenvalue at 0. In general, $[H_{\Delta}, \cdot]$ and $\mathrm{i}K_{\mathrm{d}}^{\natural}$ do not commute. If, however, the eigenvalues of $H_{\mathrm{at}}$ are nondegenerate, then they do and the eigenvalues of $[H_{\Delta}, \cdot]$ and $\mathrm{i}K_{\mathrm{d}}^{\natural}$ can simply be added.

As explained the decay rates for (18.109) are determined by the resonances $\varepsilon_{ij}(\lambda)$. As a basic result one obtains that

$$|\varepsilon_{ij}(\lambda) - \varepsilon_{ij}^{\mathrm{D}}| = \mathcal{O}(\lambda^3), \tag{18.113}$$

where the naive error $\mathcal{O}(\lambda^4)$ is reduced because of possible crossings of eigenvalues. In the weak coupling theory there is some freedom in choosing the generator. For example, $K$ and $K^{\natural}$ cannot be distinguished, see (17.28), (17.31). The non-perturbative theory of resonances identifies $K^{\natural}$ as the optimal small-$\lambda$ limit. Any other version, like $K$, would have eigenvalues in general different from $K^{\natural}$, and its eigenvalues could thus not satisfy the bound (18.113).

## Notes and references

### *Sections 18.1–18.6*

These sections are based on the first part of Bach, Fröhlich and Sigal (2000). Jakšić and Pillet (1995, 1996a, 1996b, 1997) establish the relaxation to thermal equilibrium with the help of complex translations of the Liouvillean. Their method can be extended to the case when the small system is coupled to several reservoirs at distinct temperatures (Jakšić and Pillet 2002). By more sophisticated techniques one can control the analytic continuation of the resolvent uniformly in $\beta$ (Bach, Fröhlich and Sigal 2000). Dereziński and Jakšić (2003a) use an infinitesimal version based on Mourre-type estimates. Such a technique has been used before in the simplification of the spin-boson Hamiltonian (Hübner and Spohn 1995b). Positive commutator techniques are employed by Merkli (2001). Dereziński, Jakšić and Pillet (2003) systematically develop the $W^*$-algebraic approach.

The standard reference on the algebraic formulation of quantum statistical mechanics is Bratteli and Robinson (1987, 1997); see also Sewell (1986) for a more gentle introduction. The representation theory for the free Bose gas is due to Araki and Woods (1963). A very readable introduction to free quantum gases in the frame of the algebraic approach is Dubin (1974).

Within the thermal context also the translation-invariant model (15.15) is of considerable interest. The initial state can be taken to be factorized as $\rho \otimes \omega_\beta^{\mathrm{f}}$, with $\rho$ some density matrix of the electron. For small coupling, the electron has a rate proportional to $\lambda^2$ to be scattered by the photons. The collisions are approximately independent and result in a finite energy and momentum transfer. Between consecutive collisions the electron travels freely. Such a situation is well approximated by a classical linear Boltzmann equation. Only the jump rates know about the quantum nature of the electron. We refer to Spohn (1978), Erdös and Yau (1998, 2000), and Erdös (2002). Transport of independent electrons by scattering either through phonons or through impurities is discussed in Fujita (1966) and Vollhardt and Wölfle (1980).

### *Section 18.7*

Jakšić and Pillet (1997) and Dereziński and Jakšić (2003a) introduce the level shift operator $\Sigma_\lambda$. Dereziński and Jakšić (2003b) discuss in more detail the relation to the weak coupling theory. If one defines $\tilde{L}_D A = (\rho_\beta)^{-1/2} L_D^*((\rho_\beta)^{1/2} A)$, then they establish that $\|\Sigma_\lambda - \tilde{L}_D\| = \mathcal{O}(\lambda^3)$.

# 19

# Behavior at very large and very small distances

For the classical Abraham model, and its relativistic generalization, we had to accept a phenomenological charge distribution. The physically appealing idea to let this charge distribution shrink to a point charge failed because the charged particle acquires a mass which grows beyond any limit. There is simply no bare parameter in the model which would balance the divergence in a meaningful way. Nevertheless the situation is much less dramatic than it sounds. When probed over distances that are large compared to the size of the charge distribution and correspondingly long times, only global properties of the charge distribution, like total charge and total electrostatic energy, are needed, thereby greatly reducing the dependence on the choice of the form factor. In the quantized version one has to investigate the problem anew, which requires the study of the properties of the Pauli–Fierz Hamiltonian at very small distances. The form factor $\widehat{\varphi}$ cuts off the interaction with the Maxwell field at large wave numbers. The point-charge limit thus means removing this ultraviolet cutoff. If it could be done, we would be in the very satisfactory position of having the empirical masses and empirical charges of the quantum particles as the only model parameters. Of course, the validity of the theory would not extend beyond what we have discussed already. In particular, relativistic corrections are not properly accounted for.

As we will see, the ultraviolet behavior of the Pauli–Fierz model is not so well understood. If the Maxwell field is replaced by a scalar Bose field, the ultraviolet divergencies simplify considerably and have been studied by E. Nelson in detail. To have a sort of blueprint we therefore include a section on the scalar field model.

Since the photons have zero mass, the Coulomb potential decreases as $-e^2/4\pi|x|$. In a quantized field theory one has to check whether states which have such a slow decay for the average fields still lie in Fock space, the Hilbert space which we used throughout to develop our theory. This issue leads to a study of the infrared behavior of the Pauli–Fierz Hamiltonian. Note that for this purpose the dispersion relation $\omega(k) = |k|$ is crucial, whereas an ultraviolet cutoff in the interaction can be accommodated without harm. On the other hand, for the point-charge limit we may assign the photons a small mass. The infrared and ultraviolet behavior appear as disjoint properties.

## 19.1 Infrared photons

A classical charge traveling at constant velocity $v$ carries with it the electric field $E_v^{\mathrm{cl}}$ and the magnetic field $B_v^{\mathrm{cl}}$, see Eq. (4.5), where we omitted the boldface and added the superscript "cl" to distinguish from the quantized sister. One would expect that the quantized theory reproduces these fields on the average, at least very far away from the charge. Thus we are led to consider states $\psi$ in Fock space such that

$$\langle \psi, E_\varphi(x)\psi\rangle_{\mathcal{F}} = E_{v\perp}^{\mathrm{cl}}(x), \quad \langle \psi, B_\varphi(x)\psi\rangle_{\mathcal{F}} = B_v^{\mathrm{cl}}(x). \tag{19.1}$$

Under these constraints the average number of photons is minimal for the coherent state $\psi_v^{\mathrm{coh}}$ having averages (19.1) and the minimum is given by

$$\begin{aligned}\langle \psi_v^{\mathrm{coh}}, N_{\mathrm{f}}\psi_v^{\mathrm{coh}}\rangle_{\mathcal{F}} = \frac{e^2}{2}\int \mathrm{d}^3k|\widehat{\varphi}(k)|^2(k^2-(v\cdot k)^2)^{-2}\\ \times\omega(1+\omega^{-2}(v\cdot k)^2)(v\cdot Q^\perp v).\end{aligned} \tag{19.2}$$

If $\widehat{\varphi}(0)=(2\pi)^{-3/2}$ and $\omega(k)=|k|$, then the integrand diverges as $|k|^{-3}$ for small $k$ which makes the integral in (19.2) logarithmically infrared divergent. There is no vector in Fock space which satisfies (19.1), unless $v=0$.

A natural consequence is to take $\psi_v^{\mathrm{coh}}$ as the basic object and to build the Fock space $\mathcal{F}_v$ out of finite photon excitations away from it. If in $\mathcal{F}_v$ one searches for a vector reproducing the classical fields at velocity $u$ on the average, then the constraint (19.1) becomes

$$\langle \psi, E_\varphi(x)\psi\rangle_{\mathcal{F}_v} = E_{u\perp}^{\mathrm{cl}}(x)-E_{v\perp}^{\mathrm{cl}}(x), \quad \langle \psi, B_\varphi(x)\psi\rangle_{\mathcal{F}_v} = B_u^{\mathrm{cl}}(x)-B_v^{\mathrm{cl}}(x). \tag{19.3}$$

The minimal photon number consistent with (19.3) is

$$\begin{aligned}&\frac{1}{2}\int \mathrm{d}^3k|\widehat{\varphi}(k)|^2\omega\Big((v\widehat{\phi}_v-u\widehat{\phi}_u)\cdot Q^\perp(v\widehat{\phi}_v-u\widehat{\phi}_u)\\ &\quad+\big(v\widehat{\phi}_v\frac{1}{\omega}(v\cdot k)-u\widehat{\phi}_u\frac{1}{\omega}(u\cdot k)\big)\cdot Q^\perp\big(v\widehat{\phi}_v\frac{1}{\omega}(v\cdot k)-u\widehat{\phi}_u\frac{1}{\omega}(u\cdot k)\big)\Big),\end{aligned} \tag{19.4}$$

$\widehat{\phi}_v(k)$ from (4.6), which again diverges logarithmically for small $k$, unless $u=v$. The family of coherent states $\{\psi_v^{\mathrm{coh}}\,|\,|v|<1\}$ leads to mutually inequivalent representations of the canonical commutation relations. Mathematically it is bad news, since there is no single Hilbert space which can accommodate states corresponding to the electron freely traveling at arbitrary uniform velocity.

To probe the subject further let us consider the scattering of photons where, to simplify matters, it is assumed that the motion of the quantized particle is replaced by a classical current. To figure out the Hamiltonian we return to (13.47) and regard $j(x,t)$ as a given current. In the Coulomb gauge Eq. (13.47) reads

$$\partial_t A = -E, \quad \partial_t E = -\Delta A - j, \tag{19.5}$$

where it is understood that (19.5) refers to the transverse components only. The longitudinal piece of $E$ is determined through the Poisson equation. Equations (19.5) are the Heisenberg equations of motion for the time-dependent Hamiltonian

$$H(t) = H_{\mathrm{f}} - \int \mathrm{d}^3 x j(x,t) A(x) \tag{19.6}$$

acting on $\mathcal{F}$. Since $H(t)$ is quadratic in $a, a^*$, its unitary propagator can be computed explicitly. For $t \geq 0$ one obtains, with time ordering denoted by $\mathcal{T}$,

$$\begin{aligned} U(t,0) &= \mathcal{T}\exp\Big[-\mathrm{i}\int_0^t \mathrm{d}s\; H(s)\Big] \\ &= \mathrm{e}^{-\mathrm{i}H_{\mathrm{f}}t} \exp\Big[\mathrm{i}\int_0^t \mathrm{d}s \sum_{\lambda=1,2} \int \mathrm{d}^3k (2\omega)^{-1/2} \big(e_\lambda \cdot \widehat{j}(k,s)^* \mathrm{e}^{-\mathrm{i}\omega s} a(k,\lambda) \\ &\quad + e_\lambda \cdot \widehat{j}(k,s) \mathrm{e}^{\mathrm{i}\omega s} a^*(k,\lambda)\big) + \frac{1}{2}\mathrm{i}\mathrm{Im} \int_0^t \mathrm{d}s \int_0^t \mathrm{d}s'\; \Theta(s-s') \\ &\quad \times \sum_{\lambda=1,2} \int \mathrm{d}^3k (2\omega)^{-1} (e_\lambda \cdot \widehat{j}(k,s))(e_\lambda \cdot \widehat{j}(k,s'))^* \mathrm{e}^{\mathrm{i}\omega(s-s')}\Big] \end{aligned} \tag{19.7}$$

with $\Theta(s) = 1$ for $s \geq 0$, $\Theta(s) = -1$ for $s < 0$.

Let us first examine the case where the charge travels at constant velocity, i.e. $j(x,t) = e\varphi(x - vt)v$, $|v| < 1$, and the initial $\psi = \Omega$. Classically, the current would build up the charge soliton; compare with (4.31), (4.32). There is no accompanying radiation. The quantum wave function $\psi(t) = U(t,0)\Omega$ is a coherent state of the Maxwell field. This implies that $N_{\mathrm{f}}$ has a Poisson distribution with average

$$
\begin{aligned}
&\langle \psi(t), N_{\mathrm{f}}\psi(t)\rangle_{\mathcal{F}} \\
&= e^2 \sum_{\lambda=1,2} \int \mathrm{d}^3k |\widehat{\varphi}|^2 \omega^{-1} (e_\lambda \cdot v)^2 (\omega - k \cdot v)^{-2} \big(1 - \cos((\omega - v \cdot k)t)\big) \\
&\cong v^2 \log t
\end{aligned} \tag{19.8}
$$

for large $t$. On the other hand, $\langle \psi(t), H_{\mathrm{f}}\psi(t)\rangle_{\mathcal{F}}$ stays bounded because of the extra factor of $\omega$ from the definition of the energy. Also, in every bounded region in position space and in any region in momentum space avoiding the origin, the number of photons is Poisson-distributed with a finite mean. The photons in (19.8) are bound to the charge, i.e. virtual in the usual parlance. For the Pauli–Fierz Hamiltonian virtual photons can be probed only indirectly, e.g. through the effective dynamics discussed in chapter 16. As long as the energy remains finite no qualitative changes are expected, as confirmed by the fact that the $g$-factor and the effective mass are infrared convergent at least to order $e^2$.

As a second example let us study the generation of photons through accelerated motion. We prescribe the trajectory $q_t$ with velocity $v_t$ of the classical charge, and thus the current $j(x,t) = e\varphi(x - q_t)v_t$. The scattering process is captured most conveniently through the $S$-matrix defined by

$$
S = \lim_{t \to \infty} U(t,0)^* \mathrm{e}^{2\mathrm{i}H_{\mathrm{f}}t} U(0,-t). \tag{19.9}
$$

From (19.7) we conclude

$$
\begin{aligned}
S = \exp\Big[ &- \mathrm{i} \int_{-\infty}^{\infty} \mathrm{d}t \sum_{\lambda=1,2} \int \mathrm{d}^3k (2\omega)^{-1/2} e_\lambda \cdot v_t \big( e\widehat{\varphi}^* \mathrm{e}^{-\mathrm{i}(\omega t - k\cdot q_t)} a(k,\lambda) \\
&+ e\widehat{\varphi} \mathrm{e}^{\mathrm{i}(\omega t - k\cdot q_t)} a^*(k,\lambda) \big) - \frac{1}{2}\mathrm{i}\mathrm{Im} \int_{-\infty}^{\infty} \mathrm{d}s \int_{-\infty}^{\infty} \mathrm{d}s' \Theta(s - s') \\
&\times \sum_{\lambda=1,2} \int \mathrm{d}^3k e^2 |\widehat{\varphi}|^2 (2\omega)^{-1} (e_\lambda \cdot v_s)(e_\lambda \cdot v_{s'}) \mathrm{e}^{\mathrm{i}(\omega s - \omega s' - k\cdot q_s + k \cdot q_{s'})} \Big].
\end{aligned} \tag{19.10}
$$

Note that for constant velocity, $v_t = v$, $|v| < 1$, the time-integration yields the $\delta$-function $\delta(\omega - k \cdot v)$ and therefore the $S$-matrix is trivial, $S = 1$. For the sake of an example let us assume that there are no incoming photons. Then the scattering state of interest is $S\Omega$, which is a coherent state with average number of photons

$$\langle S\Omega, N_{\mathrm{f}} S\Omega\rangle_{\mathcal{F}} = e^2 \sum_{\lambda=1,2} \int \mathrm{d}^3k |\widehat{\varphi}|^2 (2\omega)^{-1} \Big| \int \mathrm{d}t\, (e_\lambda \cdot v_t) \mathrm{e}^{\mathrm{i}(\omega t - k\cdot q_t)} \Big|^2. \quad (19.11)$$

In standard scattering $v_t \to v_\pm$ for $t \to \pm\infty$. If $v_+ = v_-$, from the previous argument one concludes that $\langle S\Omega, N_{\mathrm{f}} S\Omega\rangle_{\mathcal{F}} < \infty$. However if $v_+ \neq v_-$, then from the time-integration a factor $|k|^{-2}$ appears which together with the factor $1/\omega$ makes the integral in (19.11) logarithmically divergent at small $k$. As before, $\langle S\Omega, H_{\mathrm{f}} S\Omega\rangle_{\mathcal{F}} < \infty$. Also the number of photons is finite in any region of the form $\{k \,|\, |k| > \delta\}$ with $\delta > 0$.

If an electron is scattered by, say, a short-range electrostatic potential then in the collision process a large number of infrared photons is generated. Strictly speaking, there is no channel with elastic scattering. Since the total energy of scattered photons is bounded, the collision cross-section is slightly modified but remains finite. These infrared photons are however somewhat elusive objects. For example, for the state $S\Omega$ the photon density in position space decays as $|x|^{-3}$ for large $|x|$, which means that there is a small probability for the photons to have been created very far away from the source. A real detector necessarily makes a cutoff in the energy range and in position, thus necessarily misses the infrared part.

## 19.2 Energy renormalization in Nelson's scalar field model

On the classical level we consider a scalar wave field and couple it to a mechanical particle in such a way that the interaction is linear in the field, local, and translation invariant. This fixes the Hamiltonian function to be of the form

$$H = \frac{1}{2m} p^2 + \frac{1}{2} \int \mathrm{d}^3x \big(\pi(x)^2 + (\nabla\phi(x))^2 + m_{\mathrm{ph}}^2 \phi(x)^2\big) + e\phi_\varphi(q). \quad (19.12)$$

Here $q, p$ are the position and momentum of the particle with bare mass $m$ and $\pi(x)$ is the momentum field canonically conjugate to the scalar wave field $\phi(x)$. The wave speed $c$ is set equal to one. $e$ is the coupling strength, and $m_{\mathrm{ph}} \geq 0$ is the mass of the bosons. The equations of motion read

$$\partial_t^2 \phi(x,t) = (\Delta - m_{\mathrm{ph}}^2)\phi(x,t) - e\varphi(x - q_t), \quad (19.13)$$

$$m\ddot{q}_t = -e\nabla\phi_\varphi(q_t). \quad (19.14)$$

The solutions to (19.13) and (19.14) bear a fair qualitative similarity to the Abraham model, in particular, our discussion of the energy–momentum relation, the radiation reaction, and the center manifold could be repeated almost word for word.

The quantization of (19.12) is straightforward. $\pi(x)$ and $\phi(x)$ become a scalar Bose field with commutation relations

$$[\phi(x), \pi(x')] = \mathrm{i}\delta(x - x'), \tag{19.15}$$

setting $\hbar = 1$. It is convenient to introduce the scalar creation and annihilation operators $a^*(k)$, $a(k)$ in momentum space. Then

$$\phi(x) = \int \mathrm{d}^3k \frac{1}{\sqrt{2\omega}}(2\pi)^{-3/2}\big(\mathrm{e}^{\mathrm{i}k\cdot x}a(k) + \mathrm{e}^{-\mathrm{i}k\cdot x}a^*(k)\big), \tag{19.16}$$

$$\pi(x) = \int \mathrm{d}^3k\sqrt{\omega/2}(2\pi)^{-3/2}\big( - \mathrm{i}\mathrm{e}^{\mathrm{i}k\cdot x}a(k) + \mathrm{i}\mathrm{e}^{-\mathrm{i}k\cdot x}a^*(k)\big) \tag{19.17}$$

with $\omega(k) = (k^2 + m_{\mathrm{ph}}^2)^{1/2}$. The quantized Hamiltonian reads

$$H = \frac{1}{2m}p^2 + H_{\mathrm{f}} + e\phi_\varphi(x), \tag{19.18}$$

where the momentum operator $p = -\mathrm{i}\nabla_x$ is canonically conjugate to the position $x$ and

$$\phi_\varphi(x) = \int \mathrm{d}^3k\widehat{\varphi}(k)\frac{1}{\sqrt{2\omega}}\big(\mathrm{e}^{\mathrm{i}k\cdot x}a(k) + \mathrm{e}^{-\mathrm{i}k\cdot x}a^*(k)\big) \tag{19.19}$$

with $\widehat{\varphi}$ assumed to be real. $H$ acts on $L^2(\mathbb{R}^3) \otimes \mathcal{F}$; we call it the *Nelson Hamiltonian*. If $\int \mathrm{d}^3k|\widehat{\varphi}|^2(\omega^{-2} + 1) < \infty$, then the interaction $e\phi_\varphi(x)$ is infinitesimally bounded with respect to $\frac{1}{2m}p^2 + H_{\mathrm{f}}$ and, by the Kato–Rellich theorem, $H$ is self-adjoint with domain $D((p^2/2m) + H_{\mathrm{f}})$.

Since $H$ is invariant under translations, the total momentum

$$P = p + P_{\mathrm{f}}, \quad P_{\mathrm{f}} = \int \mathrm{d}^3k k a^*(k)a(k), \tag{19.20}$$

is conserved. As in section 15.2, $H$ can be unitarily transformed to fixed total momentum with the result

$$H(P) = \frac{1}{2m}(P - P_{\mathrm{f}})^2 + H_{\mathrm{f}} + e\phi_\varphi \tag{19.21}$$

and $\phi_\varphi = \phi_\varphi(0)$. The ground state energy of (19.21) defines the energy–momentum relation $E(P)$. If one sets

$$H(0) = \frac{1}{2m}P_{\mathrm{f}}^2 + H_{\mathrm{f}} + e\phi_\varphi, \tag{19.22}$$

then the effective mass is given by

$$\frac{m}{m_{\mathrm{eff}}} = 1 - \frac{2}{3m}\langle\psi_{\mathrm{g}}, P_{\mathrm{f}} \cdot \frac{1}{H(0) - E(0)}P_{\mathrm{f}}\psi_{\mathrm{g}}\rangle_{\mathcal{F}}, \tag{19.23}$$

where $\psi_{\mathrm{g}}$ is the ground state of $H(0)$, i.e. $H(0)\psi_{\mathrm{g}} = E(0)\psi_{\mathrm{g}}$.

With this somewhat rapid introduction the problem under consideration is whether the Nelson Hamiltonian (19.18) remains well-defined in the point-charge limit $\varphi(x) \to \delta(x)$. Following the usual convention to denote the ultraviolet cutoff in momentum space by $\Lambda$, the point-charge limit means scaling the form factor as

$$\varphi_\Lambda(x) = \Lambda^3 \varphi(\Lambda x), \quad \text{respectively} \quad \widehat{\varphi}_\Lambda(k) = \widehat{\varphi}(k/\Lambda) \tag{19.24}$$

with $\Lambda \to \infty$.

The interaction $\phi_\varphi(x)$ is bounded relative to $H_\mathrm{f}$ only if $\int \mathrm{d}^3k |\widehat{\varphi}|^2/\omega^2 < \infty$. At $\Lambda = \infty$ this condition is violated indicating that the limit $\Lambda \to \infty$ is singular. To find out how singular we compute the ground state energy to second order in $e^2$, regarding in (19.22) $e\phi_\varphi$ as a perturbation. Then

$$E(0) = -e^2 \int \mathrm{d}^3k |\widehat{\varphi}_\Lambda(k)|^2 \frac{1}{2\omega}\Big(\omega + \frac{1}{2m}k^2\Big)^{-1} + \mathcal{O}(e^4) \tag{19.25}$$

which diverges as $-\log\Lambda$ for $\Lambda \to \infty$. Physically only energy differences count and one may want to subtract $E(0)$ from $H(0)$. After all, in the definition of $H_\mathrm{f}$ an infinite zero-point energy was already subtracted. There are two caveats to this. First, $E(0)$ from (19.25) is only a second-order perturbation and a priori one does not know which energy to subtract. More importantly, it must be ensured that physical properties are not distorted as $\Lambda \to \infty$. In the classical model the effective mass is the relevant indicator and we adopt the same criterion here. From (19.23) we compute, compare with (15.36),

$$\frac{m}{m_\mathrm{eff}} = 1 - \frac{2}{3m} e^2 \int \mathrm{d}^3k |\widehat{\varphi}_\Lambda(k)|^2 \frac{1}{2\omega} k^2 \Big(\omega + \frac{1}{2m}k^2\Big)^{-3} + \mathcal{O}(e^4), \tag{19.26}$$

which stays finite as $\Lambda \to \infty$, at least to second order, fostering our hope that $H(0) - E(0)$ is a well-defined Hamiltonian as $\Lambda \to \infty$.

The Nelson model has the simplifying feature that the energy renormalization can be made explicit through a unitary transformation originally introduced by E. P. Gross. It is constructive to work out the case of $N$ charges coupled to the Bose field. The Hamiltonian (19.18) then generalizes to

$$H_N = \sum_{j=1}^N \frac{1}{2m_j} p_j^2 + H_\mathrm{f} + \sum_{j=1}^N e_j \phi_\varphi(x_j). \tag{19.27}$$

Here the $j$-th particle has position $x_j$, momentum $p_j = -\mathrm{i}\nabla_{x_j}$, mass $m_j$, and charge $e_j$. We define

$$T = -\sum_{j=1}^N e_j \int \mathrm{d}^3k \widehat{\varphi} \frac{1}{\sqrt{2\omega}} \beta_j \big(\mathrm{e}^{\mathrm{i}k\cdot x_j} a(k) - \mathrm{e}^{-\mathrm{i}k\cdot x_j} a^*(k)\big) \tag{19.28}$$

with $\beta_j = (\omega + \frac{1}{2m_j}k^2)^{-1}$. $\mathrm{e}^{-T}$ is the Gross transformation. Since $\int \mathrm{d}^3k\beta_j^2/\omega < \infty$ provided $m_{\mathrm{ph}} > 0$, $\mathrm{e}^{-T}$ is unitary and well defined in $\mathcal{H}$ even at $\Lambda = \infty$. Let us set

$$A_{\varphi j}^{-}(x) = -\int \mathrm{d}^3k\widehat{\varphi}\frac{1}{\sqrt{2\omega}}k\beta_j \mathrm{e}^{\mathrm{i}k\cdot x}a(k)\,,$$
$$A_{\varphi j}^{-}(x)^* = A_{\varphi j}^{+}(x)\,, \quad A_{\varphi j}(x) = A_{\varphi j}^{+}(x) + A_{\varphi j}^{-}(x)\,. \tag{19.29}$$

◇ We here use on purpose the same notation as for the transverse vector potential, since through the Gross transformation $A_{\varphi j}(x)$ appears in the Hamiltonian in the same way as the transverse vector potential does for the Pauli–Fierz model. However $A_{\varphi j}(x)$ is longitudinal and $[p, A_{\varphi j}(x)] \neq 0$. It is better behaved at small $x$ because the factor $\beta_j$ gains one extra power in decay at large $k$. *Only* for section 19.2, $A_\varphi$ is defined through (19.29). ◇

$\mathrm{e}^{-T}$ acts as

$$\mathrm{e}^{T} p_j \mathrm{e}^{-T} = p_j - e_j A_{\varphi j}^{-}(x_j) - e_j A_{\varphi j}^{+}(x_j), \quad \mathrm{e}^{T} x_j \mathrm{e}^{-T} = x_j,$$
$$\mathrm{e}^{T} a(k)\mathrm{e}^{-T} = a(k) - \sum_{j=1}^{N} e_j\widehat{\varphi}\frac{1}{\sqrt{2\omega}}\beta_j \mathrm{e}^{-\mathrm{i}k\cdot x_j},$$
$$\mathrm{e}^{T} a^*(k)\mathrm{e}^{-T} = a^*(k) - \sum_{j=1}^{N} e_j\widehat{\varphi}\frac{1}{\sqrt{2\omega}}\beta_j \mathrm{e}^{\mathrm{i}k\cdot x_j}. \tag{19.30}$$

When normally ordered, the Gross-transformed Hamiltonian becomes

$$\begin{aligned}\mathrm{e}^{T} H_N \mathrm{e}^{-T} &= \sum_{j=1}^{N} \frac{1}{2m_j}\big(p_j^2 - 2e_j p_j\cdot A_{\varphi j}^{-}(x_j) - 2e_j A_{\varphi j}^{+}(x_j)\cdot p_j \\ &\quad + e_j^2 A_{\varphi j}^{-}(x_j)^2 + e_j^2 A_{\varphi j}^{+}(x_j)^2 + 2e_j^2 A_{\varphi j}^{+}(x_j)\cdot A_{\varphi j}^{-}(x_j)\big) + H_{\mathrm{f}} \\ &\quad - \sum_{i\neq j=1}^{N} e_i e_j \int \mathrm{d}^3k|\widehat{\varphi}|^2\frac{1}{2\omega}(\beta_i + \beta_j - \omega\beta_i\beta_j)\mathrm{e}^{\mathrm{i}k\cdot(x_i - x_j)} \\ &\quad - \sum_{j=1}^{N} e_j^2 \int \mathrm{d}^3k|\widehat{\varphi}|^2\frac{1}{2\omega}\beta_j.\end{aligned} \tag{19.31}$$

Note that $A_{\varphi j}(x_j)$ does not commute with $p_j$. The last term in (19.31) is the energy renormalization, granted for a moment that the remainder is a well-defined Hamiltonian with energy bounded from below. The energy renormalization coincides with $E(0)$ as computed from second-order perturbation theory, compare with (19.25), and diverges as $-N\log\Lambda$.

The next to last term in (19.31) is the instantaneous interaction between the particles which dominates their dynamics at small velocities; see section 20.2. Let us set $m_{\mathrm{ph}} = 0$ and $\Lambda = \infty$. Then the interaction potential for particles $i$ and $j$ is

$$V_{ij}(x) = -e_i e_j \int \mathrm{d}^3k \frac{1}{\omega}(\beta_i + \beta_j - \omega\beta_i\beta_j)\mathrm{e}^{\mathrm{i}k\cdot x} \tag{19.32}$$

as a function of their relative distance. $V_{ij}(x) \cong -e_i e_j/4\pi|x|$ for large $|x|$, and $V_{ij}(x) \cong e_i e_j \log|x|$ for small $|x|$. Even in the point-charge limit the interaction deviates from a strict Coulomb law at distances on the scale of the Compton wavelength for particles $i$, $j$. This confirms our previous findings that it is natural to regard the Compton wavelength as an effective size of the charged particles in the quantized theory. Even more importantly, the sign of the interaction is $-e_i e_j$. In the scalar theory particles of equal charge attract, those of opposite charge repel each other. Thus particles of opposite charge tend to segregate and a big cluster of one sign would be separated from a big cluster of the opposite sign. There could not be the delicate balance between nuclei (ions) and electrons which is responsible for the formation of atoms and molecules. If the photons were spinless, the world would have no similarity to the one we know.

We are left with the first piece of (19.31). Since it is additive in the particles, for notational simplicity we return to $N = 1$ and rewrite it as

$$\begin{aligned}
&\lim_{\Lambda\to\infty} \mathrm{e}^{T}\Big(H + e^2 \int \mathrm{d}^3k|\widehat{\varphi}_\Lambda|^2 \frac{1}{2\omega}\beta\Big)\mathrm{e}^{-T} \\
&= \frac{1}{2m}p^2 - \frac{e}{m}(p\cdot A^-(x) + A^+(x)\cdot p) \\
&\quad + \frac{e^2}{2m}\big(A^-(x)^2 + A^+(x)^2 + 2A^+(x)\cdot A^-(x)\big) + H_{\mathrm{f}} \\
&= \widetilde{H}_{\mathrm{ren}}.
\end{aligned} \tag{19.33}$$

Here, using $\widehat{\varphi}_\Lambda(0) = (2\pi)^{-3/2}$, we have

$$\begin{aligned}
&A^-(x) = -\int \mathrm{d}^3k \frac{1}{\sqrt{2\omega}} k\beta(2\pi)^{-3/2}\mathrm{e}^{\mathrm{i}k\cdot x}a(k)\,, \\
&A^+(x) = A^-(x)^*, \quad A(x) = A^+(x) + A^-(x).
\end{aligned} \tag{19.34}$$

$\widetilde{H}_{\mathrm{ren}}$ is the physical Hamiltonian in the point-charge limit. The splitting into $A^-$ and $A^+$ results from normal ordering. The $A$-field is longitudinal but otherwise plays a role very similar to the vector potential in the Pauli–Fierz model.

In the following it will be convenient to rewrite $\widetilde{H}_{\mathrm{ren}}$ in dimensionless form. Through the canonical transformation (13.88) one obtains

$$\begin{aligned}\widetilde{H}_{\mathrm{ren}} &= mH_{\mathrm{ren}}\\ &= m\Big(\frac{1}{2}p^2 - e(p\cdot A^-(x) + A^+(x)\cdot p)\\ &\quad + \frac{1}{2}e^2\big(A^-(x)^2 + A^+(x)^2 + 2A^+(x)\cdot A^-(x)\big) + H_{\mathrm{f}}\Big)\\ &= m(H_0 + H_{\mathrm{int}})\end{aligned} \tag{19.35}$$

with $\omega = (k^2 + (m_{\mathrm{ph}}/m)^2)^{1/2}$, $\beta = (\omega + \frac{1}{2}k^2)^{-1}$, and $A^-(x)$ as in (19.34). We repeat the relative form bound estimates from section 13.3 with the result

$$|\langle\psi, H_{\mathrm{int}}\psi\rangle| \le \Big(3e^2(2\pi)^{-3}\int \mathrm{d}^3k k^2\beta^2\omega^{-2}\Big)\langle\psi, H_0\psi\rangle. \tag{19.36}$$

If

$$3e^2(2\pi)^{-3}\int \mathrm{d}^3k k^2\beta^2\omega^{-2} < 1, \tag{19.37}$$

then $H_{\mathrm{int}}$ is $H_0$-form bounded with a bound less than 1, which implies that $H_{\mathrm{ren}}$ is a self-adjoint operator bounded from below.

The total momentum transforms as $\mathrm{e}^T(p + P_{\mathrm{f}})\mathrm{e}^{-T} = p + P_{\mathrm{f}} = P$ and

$$[H_{\mathrm{ren}}, P] = 0, \tag{19.38}$$

as can also be checked directly. For fixed total momentum $H_{\mathrm{ren}}$ becomes

$$\begin{aligned}H_{\mathrm{ren}}(P) &= \frac{1}{2}(P - P_{\mathrm{f}})^2 - e\big((P - P_{\mathrm{f}})\cdot A^- + A^+\cdot(P - P_{\mathrm{f}})\big)\\ &\quad + \frac{1}{2}e^2\big(A^+\cdot A^+ + A^-\cdot A^- + 2A^+\cdot A^-\big) + H_{\mathrm{f}}\end{aligned} \tag{19.39}$$

as acting on $\mathcal{F}$ with the shorthand $A = A(0)$.

The expression in (19.37) is finite also for $m_{\mathrm{ph}} = 0$. Thus $H_{\mathrm{ren}}$ and $H_{\mathrm{ren}}(P)$ are well-defined Hamiltonians even for massless bosons with infrared and ultraviolet cutoffs removed. However, $\mathrm{e}^{-T}$ is unitarily implemented only for $m_{\mathrm{ph}} > 0$. At $m_{\mathrm{ph}} = 0$, $H$ and $H_{\mathrm{ren}}$ are not unitarily equivalent. As can be seen from (19.30) the Gross-transformed $\phi$-field has a vacuum expectation which decays as $-e/4\pi|x|$ for large $x$ and thus singles out the $P = 0$ representation; compare with our discussion in section 19.1. $H_{\mathrm{ren}}(0)$ has a ground state in Fock space, whereas $H_{\mathrm{ren}}(P)$, $P \neq 0$, has no ground state in Fock space, just as is the case for the Pauli–Fierz model.

$H_{\mathrm{ren}}$ is the result of a mathematical limit procedure and it is not automatically guaranteed that the limit Hamiltonian inherits the physically desired properties.

For a modest check we compute the self-energy, the effective mass, and the binding energy for hydrogen-like atoms in low-order perturbation theory. While these quantities are well defined, it is not known whether they can be expanded around $e = 0$. Only if the bosons had the strictly positive mass $m_{\mathrm{ph}} > 0$, $H_{\mathrm{ren}}(P)$ for small $|P|$ and $H_{\mathrm{ren}} - Ze^2 V_{\mathrm{coul}}$ have a gap between their ground state and the continuous spectrum, which implies a convergent Taylor expansion at $e = 0$.

*(i) Self-energy*

We expand $E_{\mathrm{ren}}(0)$ in powers of $e^2$. $H_{\mathrm{ren}}(0)$ is split as

$$H_{\mathrm{ren}}(0) = H_0 + eH_1 + \frac{1}{2}e^2 H_2 \tag{19.40}$$

with $H_0 = \frac{1}{2}P_{\mathrm{f}}^2 + H_{\mathrm{f}}$, $H_1 = P_{\mathrm{f}} \cdot A^- + A^+ \cdot P_{\mathrm{f}}$, $H_2 = A^- \cdot A^- + A^+ \cdot A^+ + 2A^+ \cdot A^-$. The unperturbed ground state is $\Omega$ with energy 0. The expansion is written as

$$E_{\mathrm{ren}}(0) = \frac{1}{2}e^2 E^{(2)} + \frac{1}{4!}e^4 E^{(4)} + \frac{1}{6!}e^6 E^{(6)} + \mathcal{O}(e^8). \tag{19.41}$$

$E^{(2)} = 0$, since $H_1\Omega = 0$. The next order is

$$\begin{aligned}\frac{1}{4!}e^4 E^{(4)} &= -e^4 \frac{1}{2}\langle \Omega, A^- \cdot A^- \frac{1}{H_0} A^+ \cdot A^+ \Omega\rangle_{\mathcal{F}} \\ &= -e^4 (2\pi)^{-6} \int \mathrm{d}^3k_1 \int \mathrm{d}^3k_2 \frac{1}{2\omega_1}\beta_1^2 \frac{1}{2\omega_2}\beta_2^2 (k_1 \cdot k_2)^2 \frac{1}{E_{12}}\end{aligned} \tag{19.42}$$

with $\omega_i = \omega(k_i)$, $\beta_i = \big(\omega(k_i) + \frac{1}{2}k_i^2\big)^{-1}$, $i = 1, 2$, and $E_{12} = \omega_1 + \omega_2 + \frac{1}{2}(k_1 + k_2)^2$.

For the discussion below we still need the sixth order, which is given by

$$\begin{aligned}e^6 \frac{1}{6!}E^{(6)} = e^6 \frac{1}{4}\Big( &- \langle \Omega, A^- \cdot A^- \frac{1}{H_0} P_{\mathrm{f}} \cdot A^- \frac{1}{H_0} A^+ \cdot P_{\mathrm{f}} \frac{1}{H_0} A^+ \cdot A^+ \Omega\rangle_{\mathcal{F}} \\ &- \langle \Omega, A^- \cdot A^- \frac{1}{H_0} A^+ \cdot P_{\mathrm{f}} \frac{1}{H_0} P_{\mathrm{f}} \cdot A^- \frac{1}{H_0} A^+ \cdot A^+ \Omega\rangle_{\mathcal{F}} \\ &+ \langle \Omega, A^- \cdot A^- \frac{1}{H_0} A^+ \cdot A^- \frac{1}{H_0} A^+ \cdot A^+ \Omega\rangle_{\mathcal{F}} \Big).\end{aligned} \tag{19.43}$$

The integrals appearing in the expressions for $E^{(4)}$ and $E^{(6)}$ are convergent.

*(ii) Effective mass*

From the definition (15.23) and (19.39) one concludes

$$\frac{m}{m_{\mathrm{eff}}} = 1 - \frac{2}{3}\langle \psi_{\mathrm{g}}, (P_{\mathrm{f}} + eA) \cdot (H_{\mathrm{ren}}(0) - E_{\mathrm{ren}}(0))^{-1} (P_{\mathrm{f}} + eA)\psi_{\mathrm{g}}\rangle_{\mathcal{F}} \tag{19.44}$$

with $H_{\mathrm{ren}}(0)\psi_{\mathrm{g}} = E_{\mathrm{ren}}(0)\psi_{\mathrm{g}}$. $m/m_{\mathrm{eff}}$ is even in $e$ and, provided $m_{\mathrm{ph}} > 0$, analytic for small $e$. Expanding in (19.44) to order $e^2$ by the scheme already explained, one finds

$$\frac{m_{\mathrm{eff}}}{m} = 1 + \frac{2}{3}e^2(2\pi)^{-3}\int \mathrm{d}^3k(k^2\beta^3/2\omega) + \mathcal{O}(e^4) \tag{19.45}$$

which agrees with (19.26) in the limit $\Lambda \to \infty$. For $m_{\mathrm{ph}} = 0$,

$$m_{\mathrm{eff}} = m\Big(1 + \frac{1}{6\pi^2}e^2 + \mathcal{O}(e^4)\Big) \tag{19.46}$$

is obtained. Since the mass renormalization is finite, the relation (19.46) allows us in principle to obtain the bare mass $m$ from an acceleration experiment at small velocities which measures $m_{\mathrm{eff}}$ according to our discussion in section 16.6.

*(iii) Binding energy*

We consider two charges, a nucleus of charge $Ze$ of infinite mass nailed down at the origin and a "meson" of charge $e$. According to (19.31) the renormalized Hamiltonian for $\Lambda \to \infty$ reads then

$$H = H_{\mathrm{ren}} - \frac{Ze^2}{4\pi|x|} \tag{19.47}$$

in units of $m$. For sufficiently small $e$, $e \neq 0$, $H$ has a ground state. Denoting its ground state energy by $E$, by definition the (positive) binding energy is

$$E_{\mathrm{bin}} = m\big(E_{\mathrm{ren}}(0) - E\big), \tag{19.48}$$

since $mE_{\mathrm{ren}}(0)$ is the energy of the meson far away from the nucleus. $E_{\mathrm{bin}}$ is even in $e$ and proportional to the bare mass $m$. Physically the natural units for $E_{\mathrm{bin}}$ are $m_{\mathrm{eff}}c^2$ and we write

$$E_{\mathrm{bin}} = m_{\mathrm{eff}}h_{\mathrm{bin}}(e^2), \tag{19.49}$$

which is regarded as a definition of $h_{\mathrm{bin}}$.

We expand $E$ in powers of $e^2$. To better follow the subtraction of the self-energy we first transform to the total momentum representation. Then the split-up for $H$ is

$$\begin{aligned} H &= H_{\mathrm{at}} + H_{\mathrm{f}} + \frac{1}{2}P_{\mathrm{f}}^2 - p\cdot P_{\mathrm{f}} + e\big((P_{\mathrm{f}} - p)\cdot A^- + A^+\cdot(P_{\mathrm{f}} - p)\big) \\ &\quad + \frac{1}{2}e^2(A^-\cdot A^- + A^+\cdot A^+ + 2A^+\cdot A^-) \\ &= H_0 + eH_1 + \frac{1}{2}e^2 H_2. \end{aligned} \tag{19.50}$$

The atomic Hamiltonian is $H_{\text{at}} = \frac{1}{2}p^2 - Ze^2/4\pi|x|$ with ground state energy $E_{\text{at}} = -\frac{1}{2}(Ze^2/4\pi)^2$ and ground state $\psi_{\text{at}}(x) = (\pi r_{\text{B}}^3)^{-1/2}e^{-|x|/r_{\text{B}}}$, $r_{\text{B}} = 4\pi/Ze^2$. The unperturbed ground state is $\psi_{\text{at}} \otimes \Omega$ with energy $E_{\text{at}}$. The perturbation expansion up to order $e^6$ is given as in (19.42), (19.43) with the corresponding substitutions for $H_0$, $H_1$. Let us first consider those terms not containing either $p \cdot A^-$ or $A^+ \cdot p$. The inverse operator $(H_{\text{at}} + H_{\text{f}} + \frac{1}{2}P_{\text{f}}^2 - p \cdot P_{\text{f}})^{-1}$ is expanded in $p \cdot P_{\text{f}}$. Since $\langle \psi_{\text{at}}, p^2\psi_{\text{at}}\rangle_{L^2} = -2E_{\text{at}}$, only the leading term contributes and all the self-energy terms cancel including order $e^6$. The only remaining contribution is

$$\begin{aligned} E_{\text{bin}} &= -E_{\text{at}} + e^2\langle \psi_{\text{at}} \otimes \Omega,\ p \cdot A^-(H_0)^{-1}A^+ \cdot p\psi_{\text{at}} \otimes \Omega\rangle_{\mathcal{H}} + \mathcal{O}(e^8) \\ &= -E_{\text{at}} + e^2(2\pi)^{-3}\int \mathrm{d}^3k\frac{1}{2\omega}\beta^2\langle\psi_{\text{at}}, p \cdot k\frac{1}{H_{\text{at}} - E_{\text{at}} + E_1 - p \cdot k}p \cdot k\psi_{\text{at}}\rangle_{L^2} \\ &\quad + \mathcal{O}(e^8) \end{aligned} \tag{19.51}$$

with $E_1 = \beta^{-1}$. Expanding in $p \cdot k$ and in $H_{\text{at}} - E_{\text{at}}$ yields

$$\begin{aligned} E_{\text{bin}} &= -E_{\text{at}} + \frac{1}{3}e^2(2\pi)^{-3}\int \mathrm{d}^3k\frac{1}{2\omega}\beta^2k^2\langle\psi_{\text{at}}, (E_1)^{-1}p^2\psi_{\text{at}}\rangle_{L^2} + \mathcal{O}(e^8) \\ &= -E_{\text{at}}\Big(1 + \frac{2}{3}e^2(2\pi)^{-3}\int \mathrm{d}^3k(k^2\beta^3/2\omega)\Big) + \mathcal{O}(e^8). \end{aligned} \tag{19.52}$$

Note that in (19.51) the Taylor coefficient of order $e^{10}$ is infrared divergent, which implies that $E_{\text{bin}}$ cannot be analytic at $e = 0$.

As a final step, we carry out the mass renormalization to order $e^2$ as required according to (19.49). The corrections $\mathcal{O}(e^6)$ cancel and

$$E_{\text{bin}} = -m_{\text{eff}}E_{\text{at}} + \mathcal{O}(e^8). \tag{19.53}$$

$h_{\text{bin}}$ acquires a radiative correction at least as small as $\mathcal{O}(e^8)$, which confirms the conventional picture. For small coupling the predictions of the one-particle theory are reliable. The coupling to the field generates to leading order the attractive Coulomb potential. Further effects of the interaction with the scalar field are small. Having no compelling incentive, the strong coupling regime of $H_{\text{ren}}$ is apparently little explored. It is conceivable that for large $e$ the kinetic energy of the meson cannot balance the singular Coulomb attraction. If so, $H$ of (19.50) would no longer be bounded from below.

## 19.3 Ultraviolet limit, energy and mass renormalization

The ultraviolet limit of the Pauli–Fierz model is a poorly understood subject. All we can do is to explain the few hints available, which in their optimistic interpretation indicate that the ultraviolet cutoff may be removed at the expense of a renormalization in energy and mass.

As we learned from the Nelson model, the indicative quantities are the self-energy, the effective mass, and the binding energy of the electron. To study these properties in the point-charge (= ultraviolet) limit, it is convenient to switch to relativistic units as explained at the end of section 13.4. To repeat, for constant total momentum $p$ one has

$$H(p) = \frac{1}{2}\big(p - P_{\mathrm{f}} - \sqrt{4\pi\alpha}A_{\varphi}\big)^2 + H_{\mathrm{f}}, \tag{19.54}$$

where

$$H_{\mathrm{f}} = \sum_{\lambda=1,2}\int \mathrm{d}^3k|k|a^*(k,\lambda)a(k,\lambda), \quad P_{\mathrm{f}} = \sum_{\lambda=1,2}\int \mathrm{d}^3kka^*(k,\lambda)a(k,\lambda),$$

$$A_{\varphi} = \sum_{\lambda=1,2}\int \mathrm{d}^3k\widehat{\varphi}(k/\Lambda\lambda_{\mathrm{c}})\frac{1}{\sqrt{2|k|}}\big(a(k,\lambda) + a^*(k,\lambda)\big). \tag{19.55}$$

Here $\alpha = e^2/4\pi\hbar c$ is the fine-structure constant written in Heaviside–Lorentz units, $\lambda_{\mathrm{c}} = \hbar/mc$ the Compton wavelength, and $\Lambda$ the large $k$ cutoff, $\Lambda \to \infty$ eventually. Energies are measured in units of $mc^2$, momenta in units of $mc$. In the case of the hydrogen atom with the nucleus pinned down at the origin, in relativistic units the Hamiltonian reads

$$H = \frac{1}{2}\big(-\mathrm{i}\nabla_x - \sqrt{4\pi\alpha}A_{\varphi}(x)\big)^2 + H_{\mathrm{f}} - \frac{\alpha Z}{|x|}, \tag{19.56}$$

where we ignored the smearing of the Coulomb potential by $\varphi$; compare with (13.89).

### *19.3.1 Self-energy*

Since $E(p)$ has its minimum at $p = 0$, the self-energy is given by

$$E_{\mathrm{self}} = mc^2E_{\Lambda}, \quad E_{\Lambda} = \inf_{\|\psi\|=1}\langle\psi, H(0)\psi\rangle_{\mathcal{F}}, \tag{19.57}$$

and the first task is to get some idea of how $E_{\Lambda}$ diverges as $\Lambda \to \infty$. Of course, the self-energy has no observable consequences. Still, it is a sort of theoretical test which must be passed before more difficult problems can be tackled. We normal-order $H(0)$ as

$$\begin{aligned}H(0) &= \frac{1}{2}P_{\mathrm{f}}^2 + H_{\mathrm{f}} + e(P_{\mathrm{f}} \cdot A_\varphi^- + A_\varphi^+ \cdot P_{\mathrm{f}}) + \frac{1}{2}e^2(A_\varphi^+ \cdot A_\varphi^+ + A_\varphi^- \cdot A_\varphi^- \\ &\quad + 2A_\varphi^+ \cdot A_\varphi^-) + 4\pi\alpha \int \mathrm{d}^3k |\widehat{\varphi}(k/\Lambda\lambda_{\mathrm{c}})|^2 \frac{1}{2|k|} \\ &= H_0 + eH_1 + \frac{1}{2}e^2 H_2 + E_0, \end{aligned} \tag{19.58}$$

where $A_\varphi$ is the transverse vector potential split as $A_\varphi = A_\varphi^+ + A_\varphi^-$, $A_\varphi^+ = (A_\varphi^-)^*$. $E_0$ is the lowest order of the self-energy and diverges as $\Lambda^2$. The next order is computed as in the Gross-transformed Nelson Hamiltonian with the result

$$\begin{aligned}E_\Lambda &= E_0 - (4\pi\alpha)^2 \frac{1}{2} \langle \Omega, A_\varphi^- \cdot A_\varphi^- \frac{1}{H_0} A_\varphi^+ \cdot A_\varphi^+ \Omega \rangle_{\mathcal{F}} \\ &= E_0 - (4\pi\alpha)^2 \int \mathrm{d}^3k_1 \int \mathrm{d}^3k_2 |\widehat{\varphi}(k_1/\Lambda\lambda_{\mathrm{c}})|^2 |\widehat{\varphi}(k_2/\Lambda\lambda_{\mathrm{c}})|^2 \big((2|k_1|)(2|k_2|) \\ &\quad \times 4(|k_1| + |k_2| + \frac{1}{2}(k_1 + k_2)^2)\big)^{-1} \big(1 + (\widehat{k}_1 \cdot \widehat{k}_2)^2\big) + \mathcal{O}(\alpha^3). \end{aligned} \tag{19.59}$$

The order $\alpha^2$ diverges also as $\Lambda^2$ with a negative prefactor, however. Thus in contrast to the Nelson model, here perturbation theory does not tell of the self-energy. If the electron spin were included, there are cancellations between $E_0$ and the spin contribution which yields a divergence proportional to $\Lambda$.

A second attempt is to guess a variational wave function. Variation over coherent states leads to the trivial minimizer $\psi = \Omega$, which reflects that for $p = 0$ the transverse vector field vanishes classically. A more ingenious approach is due to Lieb and Loss. They give up the zero total momentum restriction and consider

$$H = \frac{1}{2}\big(-\mathrm{i}\nabla_x - \sqrt{4\pi\alpha}A_\varphi(x)\big)^2 + H_{\mathrm{f}}. \tag{19.60}$$

The minimum of $H$ equals the self-energy $E_\Lambda$; see section 15.2. The variational wave function is taken to be of the Pekar form $\psi = \phi \otimes \Phi$, with $\Phi \in \mathcal{F}$ and $\phi(x)$ a real function. Therefore

$$\begin{aligned}E_\Lambda &\le \langle \psi, H\psi \rangle_{\mathcal{H}} \\ &= \frac{1}{2}\int \mathrm{d}^3x |\nabla\phi(x)|^2 + 2\pi\alpha \int \mathrm{d}^3x \phi(x)^2 \langle \Phi, A_\varphi(x)^2 \Phi \rangle_{\mathcal{F}} + \langle \Phi, H_{\mathrm{f}}\Phi \rangle_{\mathcal{F}}, \end{aligned} \tag{19.61}$$

since the cross-term has average zero. For $\Phi$ we choose the ground state of

$$H_\phi = 2\pi\alpha \int \mathrm{d}^3x \phi(x)^2 A_\varphi(x)^2 + H_{\mathrm{f}}. \tag{19.62}$$

$H_\phi$ is a quadratic Hamiltonian and thus its ground state energy is given by

$$E_\phi = \frac{1}{2}\mathrm{tr}\big[\big(Q_\perp\widehat{\varphi}(-\Delta + 4\pi\alpha\phi(x)^2)\widehat{\varphi}Q_\perp\big)^{1/2} - \big(Q_\perp\widehat{\varphi}(-\Delta)\widehat{\varphi}Q_\perp\big)^{1/2}\big]. \quad (19.63)$$

Here the trace is over $L^2(\mathbb{R}^3, \mathbb{R}^3)$, $Q_\perp$ is the projection onto transverse vector fields, $\widehat{\varphi}$ is regarded as a multiplication operator in momentum space, and $-\Delta + 4\pi\alpha\phi(x)^2$ is diagonal with respect to the vector indices. Combining (19.61) and (19.63) we obtain

$$E_\Lambda \leq \frac{1}{2}\int \mathrm{d}^3x|\nabla\phi(x)|^2 + E_\phi \quad (19.64)$$

as a nonlinear variational bound for $E_\Lambda$.

The difference of square roots is unpleasant. To simplify we use that $\mathrm{tr}[\sqrt{A+B} - \sqrt{A} - \sqrt{B}] \leq 0$. Then

$$E_\phi \leq \sqrt{\pi\alpha}\,\mathrm{tr}[(Q_\perp\widehat{\varphi}\phi^2\widehat{\varphi}Q_\perp)^{1/2}] \leq \sqrt{\pi\alpha}\,\mathrm{tr}[(\widehat{\varphi}\phi^2\widehat{\varphi})^{1/2}], \quad (19.65)$$

since the square root is increasing. In spirit, the bound (19.65) equals $\mathrm{tr}[\widehat{\varphi}\phi] = (2\pi)^{3/2}\varphi(0)\widehat{\phi}(0)$. To actually achieve it, one sets $\phi(x) = \phi_K(x) = K^{3/2}\phi_s(Kx)$ with scaling parameter $K \cong \Lambda^{6/7}$, such that $\widehat{\phi}_s$ has support in a ball of radius 1 and $\|\phi_s\| = 1$. Let us choose $\widehat{\varphi}_\Lambda(k) = \chi(|k|/\Lambda\lambda_c)$, $\chi(|k|) = (2\pi)^{-3/2}$ for $|k| \leq 1$, $\chi = 0$ for $|k| > 1$. If $K < \Lambda\lambda_c$, then $\widehat{\varphi}_{2\Lambda}\phi_K\widehat{\varphi}_\Lambda = \phi_K\widehat{\varphi}_\Lambda$. Thus

$$\begin{aligned}\mathrm{tr}[(\widehat{\varphi}_\Lambda\phi_K^2\widehat{\varphi}_\Lambda)^{1/2}] &= \mathrm{tr}[(\widehat{\varphi}_\Lambda\phi_K\widehat{\varphi}_{2\Lambda}\phi_K\widehat{\varphi}_\Lambda)^{1/2}] \leq \mathrm{tr}[(\widehat{\varphi}_{2\Lambda}\phi_K\widehat{\varphi}_{2\Lambda}\phi_K\widehat{\varphi}_{2\Lambda})^{1/2}]\\ &= \mathrm{tr}[\widehat{\varphi}_{2\Lambda}\phi_K] = (2\pi)^{-1/2}(2/3\pi^3)(\Lambda\lambda_c)^3\int \mathrm{d}^3x\,\phi_K(x). \quad (19.66)\end{aligned}$$

Hence

$$E_\Lambda \leq \frac{1}{2}\int \mathrm{d}^3x|\nabla\phi_K(x)|^2 + \sqrt{2\alpha}(2\Lambda\lambda_c)^3(1/3\pi^3)\int \mathrm{d}^3x\,\phi_K(x). \quad (19.67)$$

One can choose $\phi_s$ such that $\widehat{\phi}_s(0) > 0$. Then

$$E_\Lambda \leq c_1K^2 + c_2\sqrt{\alpha}(\Lambda\lambda_c)^3K^{-3/2} \quad (19.68)$$

with $c_1, c_2 > 0$. Optimizing with respect to $K$ yields, for $\Lambda\lambda_c$ sufficiently large,

$$E_{\mathrm{self}} \leq c_+(\Lambda\lambda_c)^{12/7}mc^2. \quad (19.69)$$

The guess is that 12/7 is the correct power. The best available lower bound is of order $(\Lambda\lambda_c)^{3/2}mc^2$.

### 19.3.2 Effective mass

We turn to the effective mass, which is defined by

$$\frac{m}{m_{\mathrm{eff}}} = 1 - \frac{2}{3}\langle \psi_{\mathrm{g}}, \big(P_{\mathrm{f}} + \sqrt{4\pi\alpha}A_{\varphi}\big) \cdot \frac{1}{H(0) - E(0)}\big(P_{\mathrm{f}} + \sqrt{4\pi\alpha}A_{\varphi}\big)\psi_{\mathrm{g}}\rangle_{\mathcal{F}}, \tag{19.70}$$

where $\psi_{\mathrm{g}}$ is the ground state of $H(0)$, $H(0)\psi_{\mathrm{g}} = E(0)\psi_{\mathrm{g}}$, setting $p = 0$ in (19.54). $m_{\mathrm{eff}}/m$ is an even function of $e$. The issue of interest is its cutoff dependence for fixed $e$. Clearly the right-hand side depends only on $\Lambda\lambda_{\mathrm{c}}$, compare with (19.55). This allows us to write

$$\frac{m_{\mathrm{eff}}}{m} = h_{\mathrm{mas}}(\hbar\Lambda/mc), \tag{19.71}$$

which defines $h_{\mathrm{mas}}$. $h_{\mathrm{mas}}$ depends on $\alpha$ with $h_{\mathrm{mas}} \geq 1$ and $h_{\mathrm{mas}}(0) = 1$.

If $h_{\mathrm{mas}}$ has a finite limit as $\Lambda \to \infty$, then

$$m^{\star}_{\mathrm{eff}} = m h_{\mathrm{mas}}(\infty), \tag{19.72}$$

where $m^{\star}_{\mathrm{eff}}$ is the effective mass in the model with removed ultraviolet cutoff. This situation is realized for the Nelson Hamiltonian (19.18). On the other hand, if asymptotically $h_{\mathrm{mas}}$ increases linearly in $\Lambda$, i.e. $h_{\mathrm{mas}}(\lambda) = b\lambda$, $b > 0$, for large $\lambda$, then

$$m^{\star}_{\mathrm{eff}} = \lim_{\Lambda\to\infty} m_{\Lambda} h_{\mathrm{mas}}(\hbar\Lambda/m_{\Lambda}c) = \infty \tag{19.73}$$

for any choice of $m = m_{\Lambda}$ as long as $m_{\Lambda} > 0$, which is required by a stable theory. Such a linear dependence we found for the classical Abraham model, where in the point-charge limit the electron becomes infinitely heavy with no counterbalancing mechanism.

The most intriguing case, presumably realized in the Pauli–Fierz model, is

$$h_{\mathrm{mas}}(\lambda) \cong b_0\lambda^{\gamma} \tag{19.74}$$

for large $\lambda$ with $0 < \gamma < 1$ and $\gamma$ possibly depending on $\alpha$. Then

$$m_{\mathrm{eff}} = b_0(c^{-1}\hbar\Lambda)^{\gamma} m^{1-\gamma}. \tag{19.75}$$

Setting now

$$m = (c^{-1}\hbar\Lambda)^{-\gamma/(1-\gamma)} b_1^{1/(1-\gamma)}, \tag{19.76}$$

we obtain

$$m^{\star}_{\mathrm{eff}} = b_0 b_1. \tag{19.77}$$

Thus, as $\Lambda \to \infty$ simultaneously we have to let $m \to 0$ in accordance with (19.76), recall that $\gamma < 1$. The effective mass $m^{\star}_{\text{eff}}$ stays finite in this limit. Such a limiting procedure is the standard *mass renormalization*. $b_0$ is dimensionless and defined through (19.70). $b_1$ has the dimension of mass and is a free scaling parameter adjustable to the effective mass $m^{\star}_{\text{eff}}$ as supplied from sources outside of theory, e.g. from an experiment. Note that, in contrast to a finite mass renormalization, the bare mass $m$ has disappeared from the scene.

At present the only way of deciding whether $\gamma < 1$ is a sort of consistency check by expanding $m_{\text{eff}}$ in powers of $\alpha$. We use the normal-ordered $H_0$ from (19.58) and follow the scheme outlined in the case of Nelson's model.

The order $\alpha$ is straightforward, since the approximations $\psi_{\text{g}} = \Omega$ and $E(0) = 0$ suffice, giving the result

$$\frac{m}{m_{\text{eff}}} = 1 - \frac{2}{3}(4\pi\alpha) \int \mathrm{d}^3k |\widehat{\varphi}(k/\lambda_{\text{c}})|^2 \Big(k^2\Big(1 + \frac{1}{2}|k|\Big)\Big)^{-1} + \mathcal{O}(\alpha^2). \quad (19.78)$$

The conventional sharp ultraviolet cutoff is made through the choice $\widehat{\varphi}(k) = (2\pi)^{-3/2}$ for $|k| \leq \Lambda$ and $\widehat{\varphi}(k) = 0$ for $|k| > \Lambda$. Inserting in (19.78) we obtain

$$\begin{aligned} \frac{m}{m_{\text{eff}}} &= 1 - \frac{4\alpha}{3\pi} \int_0^{\Lambda\lambda_{\text{c}}} \mathrm{d}k \Big(1 + \frac{1}{2}k\Big)^{-1} + \mathcal{O}(\alpha^2) \\ &= 1 - \frac{8\alpha}{3\pi} \log\Big(1 + \frac{1}{2}\Lambda\lambda_{\text{c}}\Big) + \mathcal{O}(\alpha^2). \end{aligned} \quad (19.79)$$

To order $\alpha$, $m_{\text{eff}}$ diverges as $\log \Lambda$ in contrast to the classical Abraham model which has a divergence proportional to $\Lambda$. Equation (19.79) suggests that

$$\frac{m_{\text{eff}}}{m} = (\Lambda\lambda_{\text{c}})^{8\alpha/3\pi} \quad (19.80)$$

for small $\alpha$ and large $\Lambda$. If so, $\gamma = 8\alpha/3\pi$. If the electron spin is included, then there is an extra contribution from the fluctuating magnetic field, see Eq. (15.68), and $8\alpha/3\pi$ is increased to $16\alpha/3\pi$.

The order $\alpha^2$ requires more effort. The normalized ground state is needed up to order $e^3$ and is given by

$$\psi_{\text{g}} = \Big(1 - e^2 \frac{1}{2}\frac{1}{H_0} A^{+}_{\varphi} \cdot A^{+}_{\varphi} + e^3 \frac{1}{H_0}(P_{\text{f}} \cdot A^{-}_{\varphi} + A^{+}_{\varphi} \cdot P_{\text{f}}) \frac{1}{H_0} A^{+}_{\varphi} \cdot A^{+}_{\varphi}\Big)\,\Omega. \quad (19.81)$$

Expanding $(H(0) - E(0))^{-1}$ results in six terms proportional to $\alpha^2$. The details are lengthy and not particularly illuminating. We obtain

$$\begin{aligned}\frac{m}{m_{\mathrm{eff}}} &= 1 - \frac{2}{3}(4\pi\alpha)\int \mathrm{d}^3k_1|\widehat{\varphi}(k_1/\lambda_c)|^2\frac{1}{2|k_1|}\frac{2}{E_1}\\ &\quad - \frac{2}{3}(4\pi\alpha)^2\int \mathrm{d}^3k_1|\widehat{\varphi}(k_1/\lambda_c)|^2\frac{1}{2|k_1|}\int \mathrm{d}^3k_2|\widehat{\varphi}(k_2/\lambda_c)|^2\frac{1}{2|k_2|}\\ &\quad \times\Big\{-\Big(\frac{1}{E_1}+\frac{1}{E_2}\Big)\frac{1}{E_{12}}(1+s)+\frac{1}{2(E_{12})^3}(k_1+k_2)^2(1+s)\\ &\quad \times\Big(\frac{1}{E_1}+\frac{1}{E_2}\Big)\frac{1}{(E_{12})^2}(k_1\cdot k_2)(-1+s)-\frac{1}{E_1}\frac{1}{E_2}(1+s)\\ &\quad +\Big(\frac{k_1^2}{E_1^2}+\frac{k_2^2}{E_2^2}\Big)\frac{1}{E_{12}}(1-s)+\frac{1}{E_1}\frac{1}{E_2}\frac{1}{E_{12}}(k_1\cdot k_2)(-1+s)\Big\}+\mathcal{O}(\alpha^3)\end{aligned} \tag{19.82}$$

with the shorthand

$$E_i = |k_i| + \frac{1}{2}k_i^2,\quad i = 1,2,\quad E_{12} = |k_1| + |k_2| + \frac{1}{2}(k_1+k_2)^2,\quad s = (\widehat{k}_1\cdot\widehat{k}_2)^2. \tag{19.83}$$

The conventional wisdom is to take the lowest-order approximation seriously and to make the ansatz

$$\frac{m_{\mathrm{eff}}}{m} = (\Lambda\lambda_c)^{((8\alpha/3\pi)+b\alpha^2)}. \tag{19.84}$$

Expanding in $\alpha$ yields

$$\frac{m_{\mathrm{eff}}}{m} = 1 + \frac{8\alpha}{3\pi}\log(\Lambda\lambda_c) + \frac{1}{2}\Big(\frac{8\alpha}{3\pi}\log(\Lambda\lambda_c)\Big)^2 + b\alpha^2\log(\Lambda\lambda_c) + \mathcal{O}(\alpha^3). \tag{19.85}$$

To be consistent, the $(\log(\Lambda\lambda_c))^2$ term must have the correct prefactor, whereas the $\log(\Lambda\lambda_c)$ term identifies the as yet unknown coefficient $b$. Indeed, inserting in (19.82) the sharp cutoff $\widehat{\varphi}$ results in terms which diverge as $\log(\Lambda\lambda_c)$ and $(\log(\Lambda\lambda_c))^2$. Only the second term inside the curly brackets diverges as $(\Lambda\lambda_c)^{1/2}$. This would suggest $h_{\mathrm{mas}}(\lambda) = \sqrt{\lambda}$ for large $\lambda$ and $\gamma = \frac{1}{2}$ independent of $\alpha$, at least for small $\alpha$. Whether this is an artifact of our method remains to be understood.

To have an intuitive picture why in the ultraviolet limit the Pauli–Fierz model can behave so differently from its classical relative, it is useful to turn to the functional integral (14.51) with the Maxwell field already integrated out. First note that the self-energy is automatically cancelled by the normalizing partition function. Also, since we study the ultraviolet limit, to be definite we may set $t = 1$, $V = 0$, and pin the Brownian motion at both ends, $q_{-1} = 0 = q_1$. $m \to 0$ means that in (14.51) the underlying Wiener measure has local fluctuations diverging as $1/\sqrt{m}$. They fight the singular behavior of $W(q_s - q_{s'}, s - s')$ near the diagonal $\{s = s'\}$. If successfully, the two effects balance each other such that the limit measure locally looks like Brownian motion with effective diffusivity $1/m^{\star}_{\text{eff}}$.

### 19.3.3 Binding energy

With $E^{\text{coul}}_{\Lambda}$ denoting the ground state energy of $H$ from (19.56), the binding energy is defined by

$$E_{\text{bin}} = mc^2\big(E_{\Lambda} - E^{\text{coul}}_{\Lambda}\big). \tag{19.86}$$

Since $m \to 0$, it is mandatory to take the binding energy in units of $m_{\text{eff}}$, and we write

$$E_{\text{bin}} = m_{\text{eff}}\frac{m}{m_{\text{eff}}}h_{\text{bin}}(\Lambda\lambda_{\text{c}}) = m_{\text{eff}}\Big(\frac{h_{\text{bin}}(\Lambda\lambda_{\text{c}})}{h_{\text{mas}}(\Lambda\lambda_{\text{c}})}\Big). \tag{19.87}$$

The scaling function $h_{\text{bin}}$ depends on $\alpha$, $h_{\text{bin}} \geq 0$. For the binding energy to remain finite (and nonzero) in the limit $\Lambda \to \infty$, assuming already the validity of (19.74), it is required that

$$h_{\text{bin}}(\lambda) = b_0'\lambda^{\gamma'} \quad \text{and} \quad \gamma = \gamma', \tag{19.88}$$

for large $\lambda$. If (19.88) holds, then

$$E_{\text{bin}} = m^{\star}_{\text{eff}}c^2(b_0'/b_0) \tag{19.89}$$

in the limit $\Lambda \to \infty$. The ratio $b_0'/b_0$ is a consequence of the theory. To have agreement with experiments, on top of (19.88) one should have

$$b_0'/b_0 \cong (\alpha Z)^2/2, \tag{19.90}$$

at least for small $\alpha$.

As before, a minimal control is provided by perturbation theory. The atomic Hamiltonian is

$$H_{\text{at}} = -\tfrac{1}{2}\Delta - \alpha Z/|x| \tag{19.91}$$

with eigenvalues $E^{\text{at}}_n$ and eigenfunctions $\psi_n$, $H_{\text{at}}\psi_n = E^{\text{at}}_n\psi_n$, $n = 1, 2, \dots$, ground state $\psi_1 = \psi_{\text{at}}$. $E^{\text{at}}_1 = E_{\text{at}} = -(\alpha Z)^2/2$ is the atomic ground state energy. The computation proceeds in perfect analogy with the Nelson model. Replacing $m$ by $m_{\text{eff}}(m/m_{\text{eff}})$ to order $\alpha$ removes the large $k$ divergence of the matrix element for the perturbed energy. In the limit $\Lambda \to \infty$ the net result is

$$E_{\mathrm{bin}} = -m_{\mathrm{eff}}c^2\Big[E_1^{\mathrm{at}} + 4\pi\alpha\frac{2}{3}(2\pi)^{-3}\int \mathrm{d}^3k\frac{1}{2\omega}\Big(\omega + \frac{1}{2}k^2\Big)^{-1} \times\langle p\psi_{\mathrm{at}}, (H_{\mathrm{at}} - E_1^{\mathrm{at}})\Big(H_{\mathrm{at}} - E_1^{\mathrm{at}} + \omega + \frac{1}{2}k^2\Big)^{-1} \cdot p\psi_{\mathrm{at}}\rangle_{L^2}\Big] + \mathrm{h.o.}, \tag{19.92}$$

where h.o. stands for higher orders in $\alpha$. Of course, the hope is that through mass renormalization the cancellation is so precise that h.o. really means smaller than the leading correction.

To compute the matrix element in (19.92) we switch to atomic units through the replacements $x \rightsquigarrow x/\alpha$, $p \rightsquigarrow p\alpha$, which implies $H \rightsquigarrow \alpha^2 H$. Let us denote by $\mu(\mathrm{d}\lambda)$ the spectral measure of $Z^{-1}(p^2)^{1/2}\psi_{\mathrm{at}}$ in atomic units. It is normalized as $Z^{-2}\langle\psi_{\mathrm{at}}, p^2\psi_{\mathrm{at}}\rangle_{L^2} = 1$ and has a support starting at $E_2^{\mathrm{at}} - E_1^{\mathrm{at}} = Z^2(1/2)(3/4)$. With this notation (19.92) becomes

$$E_{\mathrm{bin}} = -m_{\mathrm{eff}}c^2 E_1^{\mathrm{at}}\big[1 - \frac{8}{3\pi}\alpha^3\int_{3Z^2/8}^{\alpha} \mu(\mathrm{d}\lambda)\lambda \times \int_0^\infty \mathrm{d}k(2+k)^{-1}(\alpha^2\lambda + k + \frac{1}{2}k^2)^{-1}\big]. \tag{19.93}$$

Because of the coupling to the radiation field the binding energy is reduced. The shift is, however, rather small, $\alpha^3|\log\alpha|$ in relative and $\alpha^5|\log\alpha|$ in absolute order. Evaluating the integral in (19.93) yields a shift which is only a few percent away from the experimental value of 8173 MHz, which should be compared with the ionization energy of $3 \times 10^9$ MHz for the unperturbed hydrogen atom.

In addition the upper bound

$$\gamma' \le 6/7 \tag{19.94}$$

is available. While the bound could be far from truth, the crucial point is its being less than one. The proof of (19.94) is based on the operator bound

$$-\frac{1}{|x|} \ge -\kappa|-\mathrm{i}\nabla_x + A(x)| \tag{19.95}$$

which holds for any vector field $A$. The numerical coefficient is $\kappa = \pi Z/2 + 2.22Z^{2/3} + 1.04$. Setting $T = (-\mathrm{i}\nabla_x - \sqrt{4\pi\alpha}A_\varphi(x))^2/2$ we obtain for $H$ of (19.56) with $Z = 1$

$$H \ge T + H_{\mathrm{f}} - \kappa\alpha\sqrt{2T} \ge T + H_{\mathrm{f}} - \kappa\alpha\sqrt{2(T + H_{\mathrm{f}})}. \tag{19.96}$$

Let now $\psi$ be the ground state of $H$. Then by Jensen's inequality and with the abbreviation $f(x) = x - \kappa\alpha\sqrt{2x}$

$$E_\Lambda^{\mathrm{coul}} \ge \langle\psi, f(T + H_{\mathrm{f}})\psi\rangle \ge f(\langle\psi, (T + H_{\mathrm{f}})\psi\rangle). \tag{19.97}$$

$f$ attains its minimum at $x_{\min} = \frac{1}{2}(\kappa\alpha)^2$. If $E_\Lambda \geq x_{\min}$, which is the case for sufficiently large $\Lambda$, then

$$E_\Lambda - E_\Lambda^{\text{coul}} \leq \kappa\alpha\sqrt{2E_\Lambda}. \tag{19.98}$$

Therefore by (19.69)

$$E_{\text{bin}} \leq \widetilde{c}_+(\Lambda\lambda_c)^{6/7}mc^2 \tag{19.99}$$

and

$$h_{\text{bin}}(\lambda) \leq \widetilde{c}_+\lambda^{6/7} \tag{19.100}$$

for large $\lambda$.

### 19.3.4 Lamb shift and line width

As explained in chapter 17, through the coupling to the quantized radiation field the energy levels of the hydrogen atom are shifted and acquire a finite lifetime which is measured by the inverse width of the spectral line. The expressions (17.35), (17.36) are derived for an $N$-level atom in the dipole approximation. For the removal of the ultraviolet cutoff, retardation effects are of importance and the translation-invariant coupling must be used. Thus the arguments of chapter 17 have to be adapted to the Hamiltonian (19.56), which could be easily done. An alternative, for our purposes equivalent, route is to use perturbation theory for the level shift. The lifetime then follows from a Kramers–Kronig relation, since both quantities are linked to the same spectral measure; compare with Eq. (17.34).

We follow this second route. The computation is basically identical to that leading to (19.92) and uses the virial theorem $\langle\psi_n, p^2\psi_n\rangle = -2E_n^{\text{at}}$. If $\delta E_n$ denotes the level shift relative to $E_\Lambda$, the net result reads

$$\begin{aligned}\delta E_n = m_{\text{eff}}c^2\Big[E_n^{\text{at}} + 4\pi\alpha\int d^3k|\widehat{\varphi}(k/\Lambda\lambda_c)|^2\frac{1}{2\omega}\Big(\omega+\frac{1}{2}k^2\Big)^{-1}\langle\nabla_x\psi_n, (H_{\text{at}} - E_n^{\text{at}})\\ \times\Big(H_{\text{at}} - E_n^{\text{at}} + \omega + \frac{1}{2}k^2\Big)^{-1}\cdot Q_\perp(k)\nabla_x\psi_n\rangle_{L^2}\Big] + \text{h.o.}\end{aligned} \tag{19.101}$$

For large $k$ the matrix element decays as $|k|^{-2}$, which makes the integral (19.101) ultraviolet convergent. The Lamb shift refers to the frequency of emitted radiation and is therefore an energy difference. In fact, experimentally the splitting between the $2S_{1/2}$ and $2P_{1/2}$ levels is 1058 MHz, in comparison to the unperturbed ground state energy of $3 \times 10^9$ MHz, and is mostly due to the coupling to the quantized radiation field. Evaluating numerically the intergrals in (19.101) at $\Lambda = \infty$ and taking into account the coupling of the electron to the quantized magnetic field, a few percent effect only, yields a Lamb shift which is 2.5% lower than the true value.

### 19.3.5 g-factor of the electron

The gyromagnetic ratio was discussed in section 16.6 through investigating the motion of an electron in a homogeneous weak magnetic field. Here we point out that the $g$-factor can also be obtained from the Zeeman splitting of the ground state energy at total momentum $P = 0$, which is the basis of the high-precision Penning trap experiment. For a constant external magnetic field $B$, in relativistic units, the Hamiltonian reads

$$\begin{aligned} H_B = m\Big(\frac{1}{2}\big(\sigma \cdot (p - eA_\varphi(x))\big)^2 + H_{\mathrm{f}} \\ - \frac{e}{2m^2}\sigma \cdot B - \frac{e}{2m^2}\big((p - eA_\varphi(x)) \times x\big) \cdot B + \frac{e^2}{8m^2}(x \times B)^2\Big). \quad (19.102) \end{aligned}$$

Let $\psi$ be an approximate ground state for $H = H_{B=0}$. Then the linear Zeeman splitting, $\Delta E$, is given through first-order perturbation theory in $B$ as

$$\Delta E = \frac{1}{m}B \cdot \Big( - \frac{e}{2}\langle\psi, \sigma\psi\rangle_{\mathcal{H}} - \frac{e}{2}\langle\psi, (p - eA_\varphi(x)) \times x\psi\rangle_{\mathcal{H}}\Big). \quad (19.103)$$

Next we write

$$\big(H - E(0)\big)x\psi = [H - E(0), x]\psi = -\mathrm{i}(p - eA_\varphi(x))\psi. \quad (19.104)$$

In this form the total momentum can be fixed at $P = 0$. Then $H$ becomes

$$H(0) = \frac{1}{2}(P_{\mathrm{f}} + eA_\varphi)^2 - \frac{e}{2}\sigma \cdot B_\varphi + H_{\mathrm{f}} \quad (19.105)$$

with $\psi$ the ground state $\psi_{\mathrm{g}}$ of $H(0)$. Hence

$$\begin{aligned} \Delta E &= \frac{1}{m_{\mathrm{eff}}}\frac{m_{\mathrm{eff}}}{m}B \cdot \Big( - \frac{e}{2}\langle\psi_{\mathrm{g}}, \sigma\psi_{\mathrm{g}}\rangle_{\mathbb{C}^2\otimes\mathcal{F}} \\ &\quad + \frac{e}{2}\mathrm{i}\langle\psi_{\mathrm{g}}, (P_{\mathrm{f}} + eA_\varphi)\frac{1}{H(0) - E(0)} \times (P_{\mathrm{f}} + eA_\varphi)\psi_{\mathrm{g}}\rangle_{\mathbb{C}^2\otimes\mathcal{F}}\Big) \\ &= |B|\frac{e}{2m_{\mathrm{eff}}}g. \end{aligned} \quad (19.106)$$

We orient the $B$-field along the $z$-axis and take as ground state the one with total angular momentum pointing parallel to $B$; compare with section 16.6. Since

$(\sigma_3 + 2J_{f3} + 2S_{f3})\psi_{g+} = \psi_{g+}$, the $g$-factor is thus given by

$$\frac{1}{2}g = \Big(1 - \frac{2}{3}\langle\psi_{g+}, (P_f + eA_\varphi)(H(0) - E(0))^{-1}(P_f + eA_\varphi)\psi_{g+}\rangle_{\mathbb{C}^2\otimes\mathcal{F}}\Big)^{-1}$$
$$\times\Big(1 - 2\langle\psi_{g+}, (J_{f3} + S_{f3})\psi_{g+}\rangle_{\mathbb{C}^2\otimes\mathcal{F}}$$
$$- 2\mathrm{Im}\langle\psi_{g+}, (P_f + eA_\varphi)_2(H(0) - E(0))^{-1}(P_f + eA_\varphi)_1\psi_{g+}\rangle_{\mathbb{C}^2\otimes\mathcal{F}}\Big), \tag{19.107}$$

in agreement with (19.104).

We recall that, to second order and with no cutoffs, the $g$-factor is computed to

$$\frac{1}{2}g = 1 + \frac{8}{3}\Big(\frac{\alpha}{2\pi}\Big) + \mathcal{O}(\alpha^2), \tag{19.108}$$

which is 0.2% away from the true value. For fixed, small $e$ both numerator and denominator in (19.107) are expected to tend to 0 as $\Lambda \to \infty$ in such a way that their ratio is close to 1.

## Notes and references

### *Section 19.1*

The infrared behavior of radiative corrections to scattering was first studied by Bloch and Nordsieck (1937), Nordsieck (1937), and Pauli and Fierz (1938). Within the framework of the massless scalar Nelson model of section 19.2, Fröhlich (1973) constructs the one-particle shell and investigates the scattering theory. Further progress in this direction is Pizzo (2000). In his thesis Chen (2001) establishes the infrared limit of the energy–momentum relation. In contrast to the Pauli–Fierz model the scalar Nelson model is infrared divergent also at $p = 0$. The Gross transformation (19.28) switches to the representation corresponding to $p = 0$. In fact it implements the shift $\phi(x) - eV_{\varphi\mathrm{coul}}(x)$. We refer to Arai (2001), Lőrinczi, Minlos and Spohn (2002b) and Hirokawa, Hiroshima and Spohn (2002). The quantized Maxwell field coupled to a classical current is a standard textbook example (Kibble 1968; Thirring 1958). The representation theory for coherent states is developed by Klauder, McKenna and Woods (1966) with follow-ups within the algebraic framework (Emch 1972; Dubin 1974; Bratteli and Robinson 1987, 1997).

### *Section 19.2*

The scalar field model is studied in solid state physics and includes a large body of experimental work. It describes an electron coupled to the optical mode of a polar crystal and is known as a polaron (Landau 1933; Fröhlich 1954). In the standard approximation the dispersion of the field is $\omega(k) = \omega_0$ and the coupling function

$\widehat{\varphi}(k) = |k|^{-1}$. The ground state energy is well approximated by the variational approach of Feynman (1955). Upper and lower bounds are proved by Lieb and Yamazaki (1958). A large coupling theory is available (Pekar 1954). A rigorous proof of the Pekar limit can be found in Donsker and Varadhan (1983) and Lieb and Thomas (1997). The effective mass is studied in Spohn (1987) who also provides an extensive list of references. The Pekar limit of the effective mass still remains an open problem. Useful reviews are Devreese and Peeters (1984) and Gerlach and Löwen (1991). Gross (1976) develops systematic corrections to the large coupling theory. Nelson (1964a) studies the scalar model through functional integration; see chapter 14. Nelson (1964b) uses the transformation of Gross (1962), itself inspired by Lee, Low and Pines (1953), to control the removal of the ultraviolet cutoff. Nelson's analysis is pushed much further in Fröhlich (1973, 1974). The discussion of chapter 14 transcribes word for word to the Nelson model with the welcome simplification that stochastic Ito integrals become Riemann integrals. We refer to Lőrinczi and Minlos (2001), Lőrinczi *et al.* (2002a), and Betz *et al.* (2002). For sufficiently small coupling the existence of a ground state for $H$ of (19.18) is proved in Hirokawa *et al.* (2002). On the one-particle level, $H = \sqrt{p^2 + m^2} - e^2/4\pi|x|$ is not bounded from below for large $e$. Since for the Nelson model $E(p) \simeq |p|$ for large $p$, the same instability could be present for the Hamiltonian (19.18). Hainzl, Hirokawa and Spohn (2003) provide upper and lower bounds on $E_{\text{bin}}$ which establish (19.52) with an error $\mathcal{O}(e^7 \log e)$.

## *Section 19.3*

The estimates of the ground state energy are taken from Lieb and Loss (2000, 2002), who study in addition the case of many particles and semirelativistic models. The scaling (19.80) follows also from a perturbative one-loop renormalization (Chen 1996; Bugliaro *et al.* 1996). The effective mass to order $\alpha^2$ seems to be novel. Details of the perturbative computation leading to (19.82) can be found in Hiroshima and Spohn (2003). Fröhlich argues that the effective mass depends nonanalytically on $\alpha$ and therefore the interchange of limits, $\alpha \to 0$ and $\Lambda \to \infty$, leads to erroneous results. In a more proper treatment one should successively eliminate the interaction at high momenta. The resulting renormalization group flow equations yield a plausible outcome and, indeed, reflect the nonanalytic dependence in $\alpha$. Moniz and Sharp (1974, 1977) and Grotch *et al.* (1982) claim cutoff dependences of the effective mass which are in contradiction to our findings. The bound for $\gamma'$ is from Lieb and Loss (2002), which is based on the lower operator bound (19.95) for the Coulomb potential as proved in Lieb, Loss and Siedentop (1996). The famous calculation of the Lamb shift by Bethe (1947) is based on the dipole approximation and has a divergence as $\log \Lambda$. As pointed out immediately (Kroll and Lamb 1949), the shift becomes ultraviolet convergent in a relativistic theory which necessarily includes positrons. The role of retardation is mentioned by Kroll (1965). An accurate calculation is Au and Feinberg (1974), ignoring spin however. It is included in Grotch (1981), from which our numbers are taken. Quantum electrodynamic effects are most dominant for the 1S Lamb shift, which experimentally is determined with higher accuracy than the $2S_{1/2} - 2P_{1/2}$ splitting. We refer to Weitz *et al.* (1994). For small coupling quantitative estimates on the binding energy are available. One constructs upper and lower bounds with the leading terms given by formal perturbation theory, which is not directly applicable because of the missing spectral gap; see Catto and Hainzl (2004), Chen, Voulgater and Vulgater (2003), Hainzl (2002, 2003), Hainzl, Seiringer (2002), and Hainzl, Voulgater, Vulgater (2003). At present, one obstacle is that no corresponding result for the effective mass is available. Since physically energies are calibrated through $m_{\text{eff}}c^2$, such a bound is mandatory. The $g$-factor as based on the shift in energy is computed in Grotch and Kazes (1977) to second order in $e$. The derivation of the nonperturbative expression (19.107) seems to be new.

# 20

# Many charges, stability of matter

In the low-energy sector, to an excellent approximation, the world consists of photons, electrons, and nuclei. To simplify the forthcoming discussion, let us consider only one species of nuclei with charge $eZ$, $Z = 1, 2, \dots$. In fact, we also assume that the nuclei are infinitely heavy and located at positions $r_1, \dots, r_K \in \mathbb{R}^3$. This is hardly realistic, but not of central importance for the stability issues studied here. We also ignore nuclear spins. To include them would require yet another layer of considerations. With these assumptions we have an arbitrary number of photons, $N$ electrons, and $K$ nuclei governed by the Hamiltonian

$$H = \sum_{j=1}^{N} \frac{1}{2m} \big(\sigma_j \cdot (p_j - eA_\varphi(x_j))\big)^2 + H_\mathrm{f} + V_{\varphi\mathrm{coul}}\,, \tag{20.1}$$

compare with (13.39). $\sigma_j$ are the Pauli spin matrices for the $j$-th electron. Since electrons are fermions, the corresponding Hilbert space is

$$\mathcal{H} = P_\mathrm{a}\big(L^2(\mathbb{R}^3, \mathbb{C}^2)^{\otimes N}\big) \otimes \mathcal{F} \tag{20.2}$$

with $P_\mathrm{a}$ denoting the projection onto the subspace of antisymmetric wave functions. $V_{\varphi\mathrm{coul}}$ is the smeared Coulomb potential, cf. (13.17), which in the case considered here is given through

$$\begin{aligned} V_{\varphi\mathrm{coul}}(x_1, \dots, x_N) = e^2 \int \mathrm{d}^3k |\widehat{\varphi}(k)|^2 |k|^{-2} \Big( \sum_{1 \le i < j \le N} \mathrm{e}^{\mathrm{i}k\cdot(x_i - x_j)} \\ - Z \sum_{i=1}^{N} \sum_{j=1}^{K} \mathrm{e}^{\mathrm{i}k\cdot(x_i - r_j)} + Z^2 \sum_{1 \le i < j \le K} \mathrm{e}^{\mathrm{i}k\cdot(r_i - r_j)} \Big). \end{aligned} \tag{20.3}$$

One of the most basic facts about nature, which the Hamiltonian (20.1) should better explain, is the apparent stability of ordinary matter over extremely long periods of time. It has become customary to divide the issue roughly into

(i) atomic stability,
(ii) energy stability (or H-stability),
(iii) thermodynamic stability.

An atom is the special case of (20.1) with $K = 1$ (hence $r_1 = 0$) and $N$, $Z$ arbitrary. *Atomic stability* means that the ground state for $H$ looks like what we know from real atoms in nature. In particular, provided that $N < Z + 1$, or perhaps $N \leq Z + 1$ admitting a negatively charged ion, $H$ has a ground state eigenvector with an exponentially localized electronic density. Also the ultraviolet cutoff should not have to be fine-tuned. Our understanding of the stability of atoms and molecules within nonrelativistic QED has advanced spectacularly over the past few years. An overview is provided in section 20.1.

*Energy stability* and *thermodynamic stability* refer to the property that matter at the human scale is (volume) extensive: Adding two buckets of water of 10 liters each merely results in 20 liters of water. Since now many molecules are involved, (20.1) is to be considered for large $N$ with $N \cong KZ$, $Z \leq 100$. For an energy stable system, the volume of the combined system in its ground state is at least as large as the sum of the volumes of the subsystems. It is more convenient to re-express this property in energetic terms. If $E(N; K, r_1, \ldots, r_K)$ denotes the ground state energy of $H$ in (20.1), then for an H-stable system

$$E(N; K, r_1, \ldots, r_K) \geq -c_0\,(N + K) \tag{20.4}$$

with suitable $c_0 \geq 0$ independent of the location of the nuclei. In fact, such a bound obviously holds, since

$$H \geq V_{\varphi\mathrm{coul}} \geq -\frac{1}{2}\Big(\int \mathrm{d}^3k |\widehat{\varphi}(k)|^2 |k|^{-2}\Big)(e^2 N + e^2 Z^2 K)\,. \tag{20.5}$$

While correct, (20.5) teaches us little about the physics involved, since the bound is cutoff-dependent and is not of the order of one Rydberg, as expected.

The condition (20.4) overlooks the fact that even when the electrons are stripped off to infinity they still carry a self-energy. Denoting as before the self-energy of a single electron by $E_{\mathrm{self}}$, the sharper stability condition is

$$E(N; K, r_1, \ldots, r_K) - N E_{\mathrm{self}} \;\geq\; -c_1\,(N + K) \tag{20.6}$$

with some suitable constant $c_1$ independent of the location of the nuclei. Hopefully $c_1$ is of order of a Rydberg and less sensitive to the cutoff than $c_0$. Energy stability, as far as aspects of the quantized radiation field are involved, is discussed in section 20.3.

As the name indicates, thermodynamic stability means that the thermodynamic potentials are volume extensive. In particular, the thermodynamic pressure, i.e. the force per unit area on the confining container, is in essence size independent.

For proper statistical mechanics also the nuclei should have a finite mass. In our context a natural model would be to assume charge neutrality, i.e. $N = KZ$, and that the nuclei form a regular crystal lattice. Then one aspect of thermodynamic stability is a ground state energy proportional to the number of particles, which requires (20.6) to be augmented by an upper bound linear in $N + K$. We refer to the notes at the end of the chapter for further details.

No surprise, energy and thermodynamic stability are best understood in the case when the interaction with the radiation field is neglected. This raises the question: In what sense is the standard $N$-body Coulomb Hamiltonian a good approximation to (20.1)? In the classical context we discussed this problem rather exhaustively in section 11.2. Quantum mechanics adds a layer of difficulty, as will be explained in section 20.2.

## 20.1 Stability of atoms and molecules

The number of electrons, $N$, is regarded as fixed and the goal is to understand under what conditions, in their lowest-energy state, they are all bound to the nuclei. For this purpose the interaction between nuclei can be dropped. We also ignore the smearing of the Coulomb potential. On the other hand, we want to allow a variation in the nucleon charge, i.e. the $j$-th nucleus is located at $r_j$ and has charge $eZ_j$, $Z_j > 0$, $j = 1, \ldots, K$. With these modifications, the Hamiltonian reads

$$H^{\mathrm{V}}(N) = \sum_{j=1}^{N} \frac{1}{2m}\big(\sigma_j \cdot (p_j - eA_\varphi(x_j))\big)^2 + H_{\mathrm{f}} + \sum_{1 \le i < j \le N} e^2 (4\pi |x_i - x_j|)^{-1}$$
$$- \sum_{i=1}^{N}\sum_{j=1}^{K} e^2 Z_j (4\pi |x_i - r_j|)^{-1} . \tag{20.7}$$

The form factor ensures a smooth cutoff at large $k$, but $\widehat{\varphi}(0) = (2\pi)^{-3/2}$ as it should. The bottom of the spectrum for $H^{\mathrm{V}}(N)$ is

$$E^{\mathrm{V}}(N) = \inf \sigma(H^{\mathrm{V}}(N)) = \inf_{\psi, \|\psi\|_{\mathcal{H}}=1} \langle \psi, H^{\mathrm{V}}(N)\psi \rangle_{\mathcal{H}} . \tag{20.8}$$

We will have to compare with free electrons whose Hamiltonian is

$$H^{0}(N) = \sum_{j=1}^{N} \frac{1}{2m}\big(\sigma_j \cdot (p_j - eA_\varphi(x_j))\big)^2 + H_{\mathrm{f}} + \sum_{1 \le i < j \le N} e^2 (4\pi |x_i - x_j|)^{-1} . \tag{20.9}$$

Its lowest energy is denoted by $E^0(N)$. It is unlikely that the effective interaction induced by the photon cloud overrules the combined Coulomb repulsion and Fermi exclusion. Thus $E^0(N) = NE^0(1)$ is expected, but will not be assumed here.

The general strategy is to introduce a suitable notion of the ionization energy $E_{\text{ion}}(N)$. Then the binding energy is defined by

$$E_{\text{bin}}(N) = E_{\text{ion}}(N) - E^{\text{V}}(N)\,. \tag{20.10}$$

If $E_{\text{bin}}(N) > 0$, the energy interval $\Delta = [E^{\text{V}}(N), E_{\text{ion}}(N))$ is nonempty and states with an energy distribution supported by $\Delta$ should be well localized near the nuclei. Amongst them there will be the stable ground state.

It seems clear how to proceed. If one electron is moved to infinity it has energy $E^0(1)$ and the corresponding lowest-energy state of $H^{\text{V}}(N)$ has the energy $E^{\text{V}}(N-1) + E^0(1)$. Of course it could be energetically more favorable to move two electrons to infinity, etc. Thus

$$E_{\text{ion}}(N) = \min_{1 \le N' \le N} \left\{E^{\text{V}}(N-N') + E^0(N')\right\} \tag{20.11}$$

with the convention $E^{\text{V}}(0) = 0$. Note that if the interaction with the photon field is turned off, formally setting $\widehat{\varphi} = 0$, then $E^0(N) = 0$ and (20.11) agrees with the standard definition of the ionization energy for the Coulomb Hamiltonian.

There is a more direct way of moving electrons to infinity. As in chapter 16, we regard $\psi(x)$ as a $\mathbb{C}^{2N} \otimes \mathcal{F}$-valued wave function, $x = (x_1, \ldots, x_N)$. We define $P_R$ as the projection on the subspace of wave functions satisfying $\psi(x) = 0$ for $|x| < R$. Then the alternative definition is

$$E_{\text{ion}}(N) = \lim_{R \to \infty} \inf \sigma(P_R H^{\text{V}}(N) P_R)\,. \tag{20.12}$$

As proved by Griesemer (2002) the definitions (20.11) and (20.12) of the ionization energy agree in the context of the Pauli–Fierz Hamiltonian. Note that with (20.12) it is obvious that $E_{\text{bin}} \ge 0$. Also, if $H^{\text{V}}(N)$ admits surplus electrons, necessarily $E_{\text{bin}} = 0$.

Let us denote by $E_\lambda = E_\lambda(H^{\text{V}}(N))$ the spectral resolution of $H^{\text{V}}(N)$, i.e. $E_\lambda$ is the projection corresponding to the energy interval $(-\infty, \lambda]$.

**Theorem 20.1** (Exponential localization). *Let $E_{\text{bin}}(N) > 0$ and let us choose $\lambda, \beta$ such that $\lambda + (\beta^2/2m) < E_{\text{ion}}(N)$, $E^{\text{V}}(N) \le \lambda$, $\beta > 0$. If $E_\lambda \psi = \psi$, then*

$$\|e^{\beta|x|}\psi\|_{\mathcal{H}} \le c_0 \|\psi\|_{\mathcal{H}}\,. \tag{20.13}$$

The proof is due to Griesemer (2004). In fact, the proof exploits only properties of the Laplacian. As in chapter 16, we regard $\mathcal{H} = L^2(\mathbb{R}^n, \mathcal{H}_{\text{f}}) = L^2(\mathbb{R}^n, \text{d}^n x) \otimes \mathcal{H}_{\text{f}}$ with some Hilbert space $\mathcal{H}_{\text{f}}$ of "internal degrees of freedom". The operator

$H$ with domain $D(H)$ is self-adjoint on $\mathcal{H}$. Let $f \in C^\infty(\mathbb{R}^n, \mathbb{R})$ with $f$ and $\nabla f$ bounded. The crucial assumption concerns the double commutator

$$[[H, f], f] = -2|\nabla f|^2 \tag{20.14}$$

with $f$ regarded as a multiplication operator. Note that (20.14) holds in the case $\mathcal{F} = \mathbb{C}$ and $H = -\Delta$. But (20.14) holds also for $H = H^{\mathrm{V}}(N)$ setting $n = 3N$. As before the ionization threshold for $H$ is

$$E_{\mathrm{ion}} = \lim_{R\to\infty} \inf \sigma(P_R H P_R) . \tag{20.15}$$

**Proposition 20.2** *Let $H$ satisfy (20.14) and let $\lambda + \beta^2 < E_{\mathrm{ion}}$, $\beta > 0$. Then*

$$\|\mathrm{e}^{\beta|x|} E_\lambda(H)\|_{\mathcal{H}} < \infty . \tag{20.16}$$

Let us return to the existence of a ground state for $H^{\mathrm{V}}(N)$. If $E_{\mathrm{bin}} > 0$, the exponential localization is a favorable indication. But it could happen that more and more photons are bound by the electrons. Thus we need a soft photon bound of the type of Theorem 15.1. The proof is now considerably more demanding and established by Griesemer, Lieb and Loss (2001).

**Theorem 20.3** (Existence of a ground state). *If $E_{\mathrm{bin}}(N) > 0$, then $H^{\mathrm{V}}(N)$ has a ground state.*

Because of Pauli exclusion and spin, no obvious positivity is available which would ensure uniqueness. Note that there is no restriction on the coupling strength $e$. Also, by Proposition 20.2, the ground state is exponentially localized with length less than $1/\sqrt{2mE_{\mathrm{bin}}(N)}$.

The existence of a ground state is reduced to the issue of whether $E_{\mathrm{bin}}(N) > 0$. While the statement looks innocent and seems to require only the clever choice of a wave function, the actual construction is ingenious and has been achieved only very recently by Lieb and Loss (2003). The main obstacle is the, in position space, nonlocal nature of the photon kinetic energy.

**Theorem 20.4** (Strictly positive binding energy). *Let $eZ_{\mathrm{tot}}$ be the total nuclear charge, $Z_{\mathrm{tot}} = \sum_{j=1}^{K} Z_j$. If*

$$N < Z_{\mathrm{tot}} + 1 , \tag{20.17}$$

*then $E_{\mathrm{bin}}(N) > 0$.*

In nature ions carrying one, or perhaps two, extra electrons are rather common. Such fine chemical features are difficult to access. In fact, even on the level of the Coulomb Hamiltonian the excess charge for stable ions is poorly understood.

## 20.2 Quasi-static limit

We plan to investigate under what limiting conditions the many-particle Pauli–Fierz Hamiltonian can be approximated by the Coulomb Hamiltonian with possible corrections. The implementation of the limit (11.8) on the quantum level has not yet been attempted. Thus we have to be satisfied with the more down-to-earth limit $c \to \infty$ already discussed briefly at the beginning of section 11.2. $c \to \infty$ means that the interaction between the charges becomes instantaneous, a principle on which the Coulomb Hamiltonian is built. To study this limit we had better reintroduce the velocity of light, which amounts to

$$H(c) = \sum_{j=1}^{N} \frac{1}{2m_j}\Big(\sigma_j \cdot \Big(p_j - \frac{1}{\sqrt{c}} e_j A_\varphi(x_j)\Big)\Big)^2 + V_{\varphi\text{coul}} + cH_\mathrm{f}. \tag{20.18}$$

$A_\varphi(x)$, $V_{\varphi\text{coul}}$, and $H_\mathrm{f}$ do not depend on $c$. The prefactors as written result from reintroducing $\omega(k) = c|k|$. The masses and charges are arbitrary.

$c$ has a dimension. So what we really mean is $|v|/c \to 0$, where $v$ is some characteristic velocity of the charges. Thus either $c \to \infty$ at fixed $|v|$ or $|v| \to 0$ at fixed $c$. The latter can also be achieved by assuming the particles to be heavy and, hence, by replacing in (20.18) $m_j$ by $\varepsilon^{-2} m_j$, $\varepsilon \ll 1$. On the classical level the limits $c \to \infty$ and $\varepsilon \to 0$ are related through the time scale change $t$ to $\varepsilon t$, and thus are completely equivalent. Quantum mechanically the two Hamiltonians are not unitarily related, which reflects the additional scale coming from $\hbar$.

Let us first study the limit $c \to \infty$. Except for normal order the Hamiltonian (20.18) reads

$$\begin{aligned} H(c) = &\sum_{j=1}^{N} \frac{1}{2m_j} p_j^2 + V_{\varphi\text{coul}} - \frac{1}{\sqrt{c}} \sum_{j=1}^{N} \frac{e_j}{m_j} p_j \cdot A_\varphi(x_j) \\ &- \frac{1}{\sqrt{c}} \sum_{j=1}^{N} \frac{e_j}{2m_j} \sigma_j \cdot B_\varphi(x_j) + \frac{1}{c} \sum_{j=1}^{N} \frac{e_j^2}{2m_j} :A_\varphi(x_j)^2: + cH_\mathrm{f}. \end{aligned} \tag{20.19}$$

$H(c)$ should be compared with the weak coupling Hamiltonian (17.4), written for the long-time scale $\lambda^{-2}\tau$ and with the abbreviation $H_\text{int} = \tilde{Q} \cdot A_\varphi(0)$,

$$H_\lambda = \lambda^{-2} H_\text{at} + \lambda^{-1} H_\text{int} + \lambda^{-2} H_\mathrm{f}. \tag{20.20}$$

The interaction part $H_\text{int}$ satisfies $\langle \Omega, H_\text{int}\Omega\rangle_\mathcal{F} = 0$, which holds also for (20.19), since $\langle \Omega, A_\varphi(x_j)\,\Omega\rangle_\mathcal{F} = 0$, $\langle \Omega, B_\varphi(x_j)\,\Omega\rangle_\mathcal{F} = 0$, and $\langle \Omega, :A_\varphi(x_j)^2:\,\Omega\rangle_\mathcal{F} = 0$. The central insight of the weak coupling theory is that the correction to $H_\text{at}$ results from balancing $\lambda^{-2}(H_\text{int})^2$ with the time averaging due to $\lambda^{-2}H_\mathrm{f}$; compare with

(17.22). Clearly this balance can be achieved also in (20.19) by considering the long-time scale

$$t = c^2 \tau \quad \text{with} \quad \tau = \mathcal{O}(1) . \tag{20.21}$$

Then $H_\mathrm{f}$ has the prefactor $c^3$, $H_\mathrm{int}$ the prefactor $c^{3/2}$ with a subleading correction of order $c$, and $H_\mathrm{at}$ has the prefactor $c^2$. The analog of (17.24) becomes

$$\mathcal{K}\rho = -\int_0^\infty \mathrm{d}t \, \mathrm{e}^{\mathrm{i}\mathcal{L}_\mathrm{at} c^{-1} t} \mathrm{tr}_\mathcal{F} \mathcal{L}_\mathrm{int} \mathrm{e}^{-\mathrm{i}(c^{-1}\mathcal{L}_\mathrm{at} + \mathcal{L}_\mathrm{f})t} \mathcal{L}_\mathrm{int} P_\Omega ] \rho . \tag{20.22}$$

In the limit $c \to \infty$ the dependence on $\mathcal{L}_\mathrm{at}$ drops out. In particular, this implies that the correction term must be nondissipative; compare with (17.35) which is evaluated at $\omega = 0$.

Let us first write out the limiting objects. The analog of $H_\mathrm{at}$ is

$$H_{\varphi\mathrm{coul}} = \sum_{j=1}^N \frac{1}{2m_j} p_j^2 + V_{\varphi\mathrm{coul}} . \tag{20.23}$$

It is corrected by

$$(-i) V_{\varphi\mathrm{darw}} = \mathrm{i} \int_0^\infty \mathrm{d}t \, \langle \Omega, H_\mathrm{int} \mathrm{e}^{-\mathrm{i}t H_\mathrm{f}} H_\mathrm{int} \Omega \rangle_\mathcal{F} , \tag{20.24}$$

which upon working out the integrals becomes

$$\begin{aligned} V_{\varphi\mathrm{darw}} = &- \sum_{i,j=1}^N \frac{e_i e_j}{m_i m_j} \int \mathrm{d}^3 k |\widehat{\varphi}(k)|^2 \frac{1}{2k^2} \mathrm{e}^{\mathrm{i}k\cdot x_i} (p_i \cdot Q^\perp(k) p_j) \mathrm{e}^{-\mathrm{i}k\cdot x_j} \\ &- \sum_{i,j=1}^N \frac{e_i e_j}{12 m_i m_j} \sigma_i \cdot \sigma_j \int \mathrm{d}^3 k |\widehat{\varphi}(k)|^2 \mathrm{e}^{\mathrm{i}k\cdot(x_i - x_j)} , \end{aligned} \tag{20.25}$$

which is the Darwin correction. We set

$$H_{\varphi\mathrm{darw}} = H_{\varphi\mathrm{coul}} + c^{-2} V_{\varphi\mathrm{darw}} . \tag{20.26}$$

Note that the integrability condition (17.27) is satisfied, since the integrand in (20.24) is bounded by $(1 + t^2)^{-1}$. In contrast to section 17.2, $H_\mathrm{int}$ has an unbounded factor acting on $\mathcal{H}_\mathrm{p}$, which necessitates a restriction on the initial wave function. We summarize as

**Theorem 20.5** (Coulomb Hamiltonian and correction). *Let* $\psi \in L^2$ *with* $\langle \psi, H_{\mathrm{coul}}\psi\rangle_{L^2} < \infty$. *Then*

$$\lim_{c\to\infty} \|\big(\mathrm{e}^{-\mathrm{i}H(c)c^2 t} - \mathrm{e}^{-\mathrm{i}(H_{\varphi\mathrm{darw}}+cH_{\mathrm{f}})c^2 t}\big)\psi \otimes \Omega\| = 0\,. \tag{20.27}$$

Since the limit (20.27) is on the long-time scale $c^2$, the Darwin correction is meaningfully singled out.

Except for operator ordering, (20.27) is in accordance with the results in section 11.2. However in $L_{\mathrm{darw}}$ of (11.27) the kinetic energy is modified and the Coulomb potential is not smeared out, which reflects the fact that limits here and in section 11.2 differ somewhat.

For the limit $m_j \to \infty$ we can also rely on methods developed before. We start with the classical symbol

$$\begin{aligned} H(q,p) = \sum_{j=1}^{N}\Big(&\frac{1}{2m_j}p_j^2 - \varepsilon\frac{e_j}{m_j}p_j\cdot A_\varphi(q_j) \\ &- \varepsilon^2\frac{e_j}{2m_j}\sigma_j\cdot B_\varphi(q_j) + \varepsilon^2\frac{e_j^2}{2m_j}:A_\varphi(q_j)^2:\Big) + V_{\varphi\mathrm{coul}}(q) + H_{\mathrm{f}}\,, \end{aligned} \tag{20.28}$$

$q = (q_1, \ldots, q_N)$, $p = (p_1, \ldots, p_N)$. The Weyl quantization of $H(q,p)$ is $H(c)$ of (20.18) with $m_j$ replaced by $\varepsilon^{-2}m_j$, where for convenience we returned to $c = 1$. The leading symbol for $H(q,p)$ is

$$H_0(p,q) = \sum_{j=1}^{N}\frac{1}{2m_j}p_j^2 + V_{\varphi\mathrm{coul}}(q) + H_{\mathrm{f}}\,. \tag{20.29}$$

Its ground state band has the projection $P_0(q,p) = 1\otimes P_\Omega$, independent of $q, p$, and the eigenvalue $e_0(q,p) = \sum_{j=1}^{N}(p_j^2/2m_j) + V_{\varphi\mathrm{coul}}(q)$. Thus, following section 16.4, the Coulomb Hamiltonian can be understood as Peierls' substitution for (20.28). It approximates on the time scale $\varepsilon^{-1}t$ the true unitary evolution projected to $1\otimes P_\Omega$.

To obtain corrections we have to first compute $h_1$. Since $\langle\Omega, H_{\mathrm{int}}\Omega\rangle = 0$ and since $P_0$ does not depend on $p, q$, $h_1 = 0$, in accordance with the previous findings that the Darwin correction is of order $\varepsilon^{-2}$. Thus we need $h_2$. In section 16.4 no explicit formula was given, since it is already somewhat lengthy. In our particular case, many simplifications occur and as a net result one finds that

$$h_2(q, p) = -\sum_{i,j=1}^{N} \frac{e_i e_j}{m_i m_j}\Big(\int d^3k|\widehat{\varphi}(k)|^2 \frac{1}{2\omega^2}(p_i \cdot Q^{\perp}(k)p_j)e^{ik\cdot(q_i-q_j)} + \frac{1}{12}(\sigma_i \cdot \sigma_j)\int d^3k|\widehat{\varphi}(k)|^2 \frac{k^2}{\omega^2}e^{ik\cdot(q_i-q_j)}\Big). \tag{20.30}$$

While, since in agreement with the previous result, (20.30) is very satisfactory on a formal level, a complete proof has to deal with the fact that the ground state band is not separated by a gap from the remainder of the spectrum. If one is willing to impose a gap by hand through a massive dispersion $\omega$, then a suitable version of the results described in section 16.4 becomes available. The picture so derived is somewhat different from the $c \to \infty$ limit: the almost invariant subspace is tilted by order $\varepsilon$ relative to $(1 \otimes P_\Omega)\mathcal{H}$. Over the time scale $\varepsilon^{-2}t$ the motion in this subspace is governed by $h_0 + \varepsilon^2 h_2$.

## 20.3 H-stability

For the (no-cutoff) Coulomb Hamiltonian

$$H_{\text{coul}} = \sum_{j=1}^{N} \frac{1}{2}p_j^2 + V_{\text{coul}},$$

$$V_{\text{coul}} = \frac{e^2}{4\pi}\Big(\sum_{1\le i<j\le N} |x_i - x_j|^{-1} - Z\sum_{i=1}^{N}\sum_{j=1}^{K}|x_i - r_j|^{-1} + Z^2 \sum_{1\le i<j\le K} |r_i - r_j|^{-1}\Big), \tag{20.31}$$

the H-stability is a famous result by Dyson and Lenard. An independent proof was achieved by Lieb and Thirring, who succeeded in a fairly realistic estimate of the stability constant. For stability to hold the electrons must satisfy the Pauli exclusion principle, as they do in nature. For bosons the energy would decrease as $-N^{5/3}$. If the nuclei have a finite mass, for a H-stable system at least one of the two species must be fermions.

To extend H-stability to the realm of nonrelativistic quantum electrodynamics, one has to establish a lower bound on $H^{\text{V}}(N) = H$, see (20.7), linear in $K + N$. Note that for spinless electrons $H \ge H_{\text{coul}}$ by the diamagnetic inequality (14.69) and one is back to the H-stability in (20.31). Thus the difficult point is to deal with electron spin and the associated magnetic energy. The Schrödinger representation, as explained in chapter 14, suggests that for the purpose of a lower bound, $H_{\text{f}}$

could be substituted by the classical field energy stored in the $A$-field, i.e. by

$$E_{\mathrm{magn}} = \frac{1}{2}\int \mathrm{d}^3x\, B(x)^2\,. \tag{20.32}$$

If it could be established that $H_{\mathrm{f}} - E_{\mathrm{magn}} \geq 0$, then

$$H = H + E_{\mathrm{magn}} - E_{\mathrm{magn}} \geq \sum_{j=1}^{N} \frac{1}{2}\big(\sigma_j \cdot (p_j - eA_\varphi(x_j))\big)^2 + V_{\mathrm{coul}} + E_{\mathrm{magn}}\,. \tag{20.33}$$

H-stability of the Coulomb Hamiltonian with magnetic field energy added would have to be shown for an *arbitrary* external transverse vector potential.

To progress towards our goal we note that, for an arbitrary operator $A$, $|\langle\psi, A^2\psi\rangle| \leq \|A^*\psi\|\,\|A\psi\| \leq \frac{1}{2}\langle\psi, (AA^* + A^*A)\psi\rangle$ and therefore

$$(A + A^*)^2 \leq 2(AA^* + A^*A)\,. \tag{20.34}$$

We split the magnetic field as $B_\varphi(x) = B_\varphi^+(x) + B_\varphi^-(x)$ and apply (20.34),

$$B_\varphi(x)^2 \leq 4B_\varphi^+(x)B_\varphi^-(x) + 2[B_\varphi^+(x), B_\varphi^-(x)]\,, \tag{20.35}$$

which remains true when multiplied by $f(x) \geq 0$. Then

$$\frac{1}{2}\int \mathrm{d}^3x f(x) B_\varphi(x)^2 \leq \|f\|_\infty \sum_{\lambda=1,2}\int \mathrm{d}^3k (2\pi)^3 |\widehat{\varphi}(k)|^2 |k| a^*(k,\lambda) a(k,\lambda)$$
$$+ \|f\|_1 \int \mathrm{d}^3k |\widehat{\varphi}(k)|^2 |k|\,. \tag{20.36}$$

Let us assume that $|\widehat{\varphi}(k)| \leq (2\pi)^{-3/2}$ and $\int \mathrm{d}^3k |\widehat{\varphi}(k)|\,|k| = C_\Lambda < \infty$. For $f$ we choose $f(x) = 1$ if $|x - r_j| \leq 1$ for some $j$ and $f(x) = 0$ otherwise. Then

$$H \geq \sum_{j=1}^{N} \frac{1}{2}\big(\sigma_j \cdot (p_j - eA_\varphi(x_j))\big)^2 + V_{\mathrm{coul}} + \frac{1}{2}\int \mathrm{d}^3x f(x) B(x)^2 - KC_\Lambda\,. \tag{20.37}$$

The energy stability with an arbitrary external $B$-field is difficult, but has been done. Unfortunately the field energy balances the Coulomb attraction only for $|e|$ sufficiently small. To have H-stability for all $e$ one also has to include the $B$-field gradients. In addition, the choice of $f$ should be optimized. As one result we state

**Theorem 20.6** (H-stability of nonrelativistic QED). *Let $\widehat{\varphi}$ be the form factor with sharp cutoff at $\Lambda$. Then there exists a positive constant $C(e, Z)$ such that*

$$H \geq -C(e, Z)(\Lambda + 1)K \tag{20.38}$$

*independently of N.*

The proof of Theorem 20.6 relies on the H-stability of the Hamiltonian on the right hand side of (20.37) and thus requires the electrons to be fermions.

In the Pauli–Fierz Hamiltonian (20.7) the self-energy of the electrons is not subtracted. Thus, in principle, the stability bound (20.38) could exclusively be due to the positive contribution from the self-energy. To rule out such an unphysical mechanism we employ a technique, briefly touched upon already in section 19.3. The results available are sharper in the case of spinless electrons with Hamiltonian

$$H_N = \sum_{j=1}^{N} \frac{1}{2}\big(p_j - eA_\varphi(x_j)\big)^2 + H_{\mathrm{f}} + V_{\mathrm{coul}} = T_N + H_{\mathrm{f}} + V_{\mathrm{coul}}\,. \tag{20.39}$$

Since $N$ is the important parameter, it is displayed explicitly. The no-cutoff Coulomb potential carries the information on the $K$ nuclei located at $r_1, \ldots, r_K$. Let $E(N)$ be the bottom of the spectrum of $H_N$ and $E_0(N)$ that of $T_N + H_{\mathrm{f}}$. The binding energy for $H_N$ is defined as in (20.10) with the nucleon repulsion as an additive constant. Then, using (20.11) and assuming $E^0(N) = E_0(N)$,

$$E_{\mathrm{bin}}(N) \le E_0(N) - E(N)\,. \tag{20.40}$$

Similar to (19.95) the Coulomb energy is bounded from below as

$$V_{\mathrm{coul}} \ge -\kappa e^2 \sum_{j=1}^{N} |p_j - eA_\varphi(x_j)| \tag{20.41}$$

with $\kappa = \big((\pi/2)Z + (2.22)Z^{2/3} + 1.03\big)/4\pi$. Therefore, using Schwarz's inequality,

$$H_N \ge T_N + H_{\mathrm{f}} - \kappa e^2 \sqrt{2N}\sqrt{T_N + H_{\mathrm{f}}}\,. \tag{20.42}$$

The function $f(x) = x - \kappa e^2\sqrt{2N}\sqrt{x}$ takes its minimum at $x_{\mathrm{min}} = \frac{1}{2}(\kappa e^2)^2 N$, $f(x_{\mathrm{min}}) = -\frac{1}{2}(\kappa e^2)^2 N$. Thus, if

$$E_0(N) \;\le\; \frac{1}{2} N(\kappa e^2)^2 \quad \text{(case I)}\,, \tag{20.43}$$

then $H_N \ge -\frac{1}{2}(\kappa e^2)^2 N$ and $E(N) - E_0(N) \ge -(\kappa e^2)^2 N$. On the other hand, if

$$E_0(N) \;\ge\; \frac{1}{2} N(\kappa e^2)^2 \quad \text{(case II)}\,, \tag{20.44}$$

we can use the fact that $f$ is monotonically increasing to conclude that $H_N \ge f(E_0(N))$ and $E(N) \ge E_0(N) - (e^2\kappa)\sqrt{2N}\sqrt{E_0(N)}$. We summarize as

**Theorem 20.7** (Upper bound for $N$-particle binding energy). *For the Hamiltonian $H_N$ of (20.39), in case I*

$$E_{\mathrm{bin}}(N) \le (\kappa e^2)^2 N \tag{20.45}$$

*and in case II*

$$E_{\text{bin}}(N) \le (\kappa e^2)\sqrt{2N}\sqrt{E_0(N)}\,. \tag{20.46}$$

Note that energies are in units of $mc^2$.

The bound (20.46) is unexpected, since the binding energy is estimated in terms of the self-energy of a system of $N$ electrons without Coulomb repulsion. The Pauli exclusion principle has not yet been invoked.

To make further progress one needs a good estimate on $E_0(N)$. Fermions like to stay alone and the state of lowest energy should be achieved once they are infinitely separated.

**Conjecture 20.8** *For fermions*

$$E_0(N) = N E_0(1)\,. \tag{20.47}$$

If Conjecture 20.8 is assumed to hold, then the condition for the two cases reads

$$\text{(case I)}: \quad E_0(1) \le \frac{1}{2}(\kappa e^2)^2, \quad \text{(case II)}: \quad E_0(1) \ge \frac{1}{2}(\kappa e^2)^2\,. \tag{20.48}$$

As explained in section 19.3, $E_0(1) \le c_2 e^{4/7}(\Lambda\lambda_{\text{c}})^{12/7}$. Consequently

$$\begin{aligned} E_{\text{bin}} &\le (\kappa e^2)^2 N && \text{(case I)}, \\ E_{\text{bin}} &\le (\kappa e^2)\sqrt{2c_2}\,e^{2/7}(\Lambda\lambda_{\text{c}})^{6/7} N && \text{(case II)}. \end{aligned} \tag{20.49}$$

The binding energy is extensive. However, our estimate on the stability bound diverges with the cutoff $\Lambda$. Since energies are calibrated in units of $mc^2$, the folklore tells us that multiplying the true stability constant by $m/m_{\text{eff}}$ should result in a $\Lambda$-independent prefactor.

## Notes and references

Stability for the Coulomb Hamiltonian is covered extensively and excellently in survey articles. Particularly recommended are Lieb (1976, 1990), which have become classics. Some of the original articles are reprinted in the Lieb Selecta (2001), where the reader should in addition consult the introduction by Thirring, see also Thirring (2002). The first proof of stability is Dyson and Lenard (1968). The use of the Thomas–Fermi theory as a comparison standard is introduced in Lieb and Thirring (1975). Extensions to Coulomb systems with relativistic kinetic energy are investigated by Conlon (1984), Feffermann and de la Llave (1986), and were finally settled in Lieb and Yau (1988a, b). The basic discovery is that stability holds only under a smallness condition on $Z\alpha$ and $\alpha$. If electrons were bosons, they would cluster with a density increasing with $N$. In the ground state energy this can be seen in a faster than linear decrease with $N$. For bosons and fixed nuclei it is known that $E_N \simeq -N^{5/3}$ (Lieb 1979 and references therein), while with bosonic nuclei of finite mass, $E_N \cong -N^{7/5}$ (Dyson 1967; Conlon, Lieb and Yau 1988).

Thermodynamic stability for Coulomb systems is proved by Lieb and Lebowitz (1972).

If photons were scalar, then the Coulomb potential has the "wrong" sign; see section 19.2. This leads to instability, some partial aspects of which are studied in Gallavotti, Ginibre and Velo (1970).

### *Section 20.1*

For Schrödinger operators, $\mathcal{F} = \mathbb{C}$, the exponential localization of Proposition 20.2 goes back to Agmon (1982). Griesemer (2004) observes that it remains valid for general $\mathcal{F}$. Theorem 20.4 is proved by Lieb and Loss (2003). For the helium atom, $N = 2$, the strict positivity of the binding energy is established by Barbaroux *et al.* (2003).

### *Section 20.2*

For the Nelson model, i.e. a scalar Bose field, the limit $c \to \infty$ is studied by Davies (1979), see also Hiroshima (1997a), and the limit $m \to \infty$ by Teufel (2002). They prove that the dynamics is well-approximated through the Coulomb Hamiltonian. Our observation seems to be new, but could have been made already by Davies, if he had chosen the Gross-transformed Nelson Hamiltonian as a starting point.

### *Section 20.3*

The argument leading to Theorem 20.6 is taken from Fefferman, Fröhlich and Graf (1997), see also Bugliaro, Fröhlich and Graf (1996). The harder part is to establish stability for the Hamiltonian on the right hand side of (20.37), which is achieved by Feffermann (1996) with a "sufficiently small" constant and is subsequently improved and simplified by Lieb, Loss and Solovej (1995) to include the physical case. Theorem 20.7 is a result by Lieb and Loss (2002). They also establish that the self-energy for $N$ electrons is bounded as $c_1\alpha^{1/2}(\Lambda\lambda_c)^{3/2}N \leq E_0(N) \leq c_2\alpha^{2/7}(\Lambda\lambda_c)^{12/7}N$ with suitable constants $c_1, c_2$, which is somewhat weaker than our Conjecture 20.8. The discussion does not change; only the prefactors are less sharp. For bosons the bounds $c_3\alpha^{1/2}(\Lambda\lambda_c)^{3/2}N^{1/2} \leq E_0(N) \leq c_4\alpha^{2/7}(\Lambda\lambda_c)^{12/7}N^{5/7}$ are available, which together with Theorem 20.7 strongly indicate that, as to be expected, bosons remain unstable when the quantized radiation field is added. The basic inequality (20.41) holds also in the case where the electron spin is included, see Lieb and Loss (2002).

# References


Abraham M. (1903). Prinzipien der Dynamik des Elektrons, *Ann. Physik* **10**, 105–79.
(1904). Die Grundhypothesen der Elektronentheorie, *Physikalische Zeitschrift* **5**, 576–9.
(1905). *Theorie der Elektrizität*, vol. II. *Elektromagnetische Theorie der Strahlung*. Leipzig, Teubner; 2nd edition (1908).
Accardi L., Frigerio A. and Lewis J. T. (1982). Quantum stochastic processes, *Publ. RIMS Kyoto University* **18**, 97–133.
Accardi L., Lu Y. G., Alicki R. and Frigerio A. (1991). An invitation to the weak coupling and the low density limit. In *Quantum Probability and Applications*, vol. VI, pp. 3–62. Singapore: World Scientific.
Accardi L., Lu Y. G. and Volovich I. (2001). *Quantum Theory and its Stochastic Limit*. Berlin: Springer.
Agarwal G. S. (1974). *Quantum Statistical Theories of Spontaneous Emission and their Relation to Other Approaches*. Springer Tracts in Modern Physics, vol. 70, Berlin: Springer.
Agmon S. (1982). *Lectures on Exponential Decay of Solutions of Second Order Elliptic Equations: Bounds on Eigenfunctions of N-body Schrödinger Operators*. Mathematical Notes 29. Princeton N.J.: Princeton University Press.
Aguilar J. and Combes J. M. (1971). A class of analytic perturbations for one-body Schrödinger Hamiltonians, *Comm. Math. Phys.* **22**, 269–79.
Alastuey A. and Appel W. (2000). On the decoupling between classical Coulomb matter and radiation, *Physica* A **276**, 508–20.
Alicki R. and Fannes M. (2001). *Quantum Dynamical Systems*. Oxford: Oxford University Press.
Alicki R. and Lendi K. (1987). *Quantum Dynamical Semigroups and Applications*. Berlin: Springer.
Amann A. (1991). Molecules coupled to their environment. In *Large-Scale Molecular Systems: Quantum and Stochastic Aspects*, edited by W. Gans, A. Blumen and A. Amann. London: Plenum.
Ammari Z. (2000). Asymptotic completeness for a renormalized non-relativistic Hamiltonian in quantum field theory: the Nelson model, *Math. Phys. Anal. Geom.* **3**, 217–85.
Appel W. and Kiessling M. (2001). Mass and spin renormalization in Lorentz electrodynamics, *Ann. Phys.* **289**, 24–83.

(2002). Scattering and radiation damping in gyroscopic Lorentz electrodynamics, *Lett. Math. Phys.* **60**, 31–46.

Arai A. (1981). Self-adjointness and spectrum of Hamiltonians in nonrelativistic quantum electrodynamics, *J. Math. Phys.* **22**, 534–7.

(1983a). Rigorous theory of spectra and radiation for a model in quantum electrodynamics, *J. Math. Phys.* **24**, 1886–910.

(1983b). A note on scattering theory in nonrelativistic quantum electrodynamics, *J. Phys.* A **16**, 49–69.

(1990). An asymptotic analysis and its applications to the nonrelativistic limit of the Pauli–Fierz and a spin–boson model, *J. Math. Phys.* **31**, 2653–63.

(1991). Long-time behavior of an electron interacting with a quantized radiation field, *J. Math. Phys.* **32**, 2224–42.

(2001). Ground state of the massless Nelson model without cutoff in a non-Fock representation, *Rev. Math. Phys.* **13**, 1075–94.

Araki H. and Woods E. (1963). Representations of the canonical commutation relations describing a non-relativistic infinite free Bose gas, *J. Math. Phys.* **4**, 637–62.

Au C.-K. and Feinberg G. (1974). Effects of retardation on electromagnetic self-energy of atomic states, *Phys. Rev.* A **9**, 1794–800.

Avron J. and Elgart A. (1999). Adiabatic theorem without a gap condition, *Comm. Math. Phys.* **203**, 445–63.

Bach V., Chen T., Fröhlich J. and Sigal I. M. (2002). Smooth Feshbach map and operator-theoretic renormalization methods, *J. Funct. Anal.* **203**, 44–92.

Bach V., Fröhlich J. and Sigal I. M. (1995). Mathematical theory of non-relativistic matter and radiation, *Lett. Math. Phys.* **34**, 183–201.

(1998a). Quantum electrodynamics of confined non-relativistic particles, *Adv. Math.* **137**, 205–98.

(1998b). Renormalization group analysis of spectral problems in quantum field theory, *Adv. Math.* **137**, 299–395.

(1999). Spectral analysis for systems of atoms and molecules coupled to the quantized radiation field, *Comm. Math. Phys.* **207**, 249–90.

(2000). Return to equilibrium, *J. Math. Phys.* **41**, 3985–4060.

Bach V., Fröhlich J., Sigal I. M. and Soffer A. (1999). Positive commutators and the spectrum of Pauli-Fierz Hamiltonian of atoms and molecules, *Comm. Math. Phys.* **207**, 557–87.

Bach V., Klopp F. and Zenk H. (2002). Mathematical analysis of the photoelectric effect, *Adv. Theor. Math. Phys.* **5**, 969–99.

Bailey J. and Picasso E. (1970). The anomalous magnetic moment of the muon and related topics, *Progress in Nuclear Physics* **12**, 43–75.

Balslev E. and Combes J. M. (1971). Spectral properties of many-body Schrödinger operators with dilation-analytic interactions, *Comm. Math. Phys.* **22**, 280–94.

Bambusi D. (1994). A Nekhoroshev-type theorem for the Pauli–Fierz model of classical electrodynamics, *Ann. Inst. H. Poincaré, Phys. Théor.* **60**, 339–71.

(1996). A proof of the Lorentz–Dirac equation for charged point particles, mp_arc 96–118.

Bambusi D. and Galgani L. (1993). Some rigorous results on the Pauli–Fierz model of classical electrodynamics, *Ann. Inst. H. Poincaré, Phys. Théor.* **58**, 155–71.

Bambusi D. and Noja D. (1996). On classical electrodynamics of point particles and mass renormalization, some preliminary results, *Lett. Math. Phys.* **37**, 449–60.

Barbaroux J.-M., Chen T. and Vugalter S. (2003). Binding conditions for atomic $N$-electron systems in nonrelativistic QED, *Ann. H. Poincaré* **6**, 1101–36.

Bargmann V., Michel L. and Telegdi V. L. (1959). Precession of the polarization of particles moving in a homogeneous electromagnetic field, *Phys. Rev. Lett.* **2**, 435–6.

Barker B. M. and O'Connell R. F. (1980a). Removal of acceleration terms from the two-body Lagrangian to order $c^{-4}$ in electromagnetic theory, *Canad. J. Phys.* **58**, 1659–66.

(1980b). The post-Newtonian problem in classical electromagnetic theory, *Ann. Phys.* **129**, 358–77.

Barut A. O. (1964). *Electrodynamics and Classical Theory of Fields and Particles*. New York: Dover.

(ed.) (1980). *Foundations of Radiation Theory and Quantum Electrodynamics*. New York: Plenum.

Batt J. (2001). $N$-particle approximation to the nonlinear Vlasov–Poisson system, *Nonlinear Anal.* **47**, 1445–56.

Bauer G. (1997). Ein Existenzsatz für die Wheeler–Feynman Elektrodynamik. Dissertation, LMU München, unpublished.

Bauer G. and Dürr D. (2001). The Maxwell–Lorentz system of a rigid charge, *Ann. H. Poincaré* **2**, 179–96.

Bauer S. and Kunze M. (2003). The Darwin approximation of the relativistic Vlasov-Maxwell system, arXiv:math-ph/0401012.

Baylis W. E. and Huschilt J. (1976). Nonuniqueness of physical solutions to the Lorentz–Dirac equation, *Phys. Rev.* D **13**, 3237–9.

Baylis W.E. and Huschilt J. (2002). Energy balance with the Landau–Lifshitz equation, *Phys. Lett.* A **301**, 7–12.

Berry M. (1990). Histories of adiabatic quantum transitions, *Proc. Roy. Soc. London* A **429**, 61–72.

Bethe H. A. (1947). The electromagnetic shift of energy levels, *Phys. Rev.* **72**, 339–41.

Betz V., Hiroshima F., Lőrinczi J., Minlos R.A. and Spohn H. (2002). Ground state properties of the Nelson Hamiltonian – a Gibbs measure-based approach, *Rev. Math. Phys.* **14**, 173–98.

Bhabha H. J. (1939). Classical theory of electrons, *Proc. Indian Acad. Sci.* A **10**, 324–32.

Bhabha H. J. and Corben H. C. (1941). General classical theory of spinning particles in a Maxwell field, *Proc. Roy. Soc. Lond.* A **178**, 273–314.

Blanco R. (1995). Nonuniqueness of the Lorentz–Dirac equation with the free-particle asymptotic condition, *Phys. Rev.* E **51**, 680–9.

Bloch F. (1928). Zur Strahlungsdämpfung in der Quantenmechanik. *Physikalische Zeitschrift* **29**, 58–66.

Bloch F. and Nordsieck A. (1937). Notes on the radiation field of the electron, *Phys. Rev.* **52**, 54–9.

Blount E. I. (1962a). Formalism of band theory. In *Solid State Physics, Advances in Research and Applications*, edited by F. Seitz and D. Turnbull, Vol. 13. New York: Academic Press, pp. 305–73.

(1962b). Bloch electrons in a magnetic field, *Phys. Rev.* **126**, 1636–53.

(1962c). Extension of the Foldy–Wouthuysen transformation, *Phys. Rev.* **128**, 2454–8.

Bohm D. and Weinstein M. (1948). The self-oscillations of a charged particle, *Phys. Rev.* **74**, 1789–98.

Bolte J. and Keppeler S. (1999). A semiclassical approach to the Dirac equation, *Ann. Phys.* **274**, 125–62.

Bonnor W. B. (1974). A new equation of motion for a radiating charged particle, *Proc. Roy. Soc. Lond.* A **337**, 591–8.

Bordag M., Geyer B., Klimchitskaya G.L. and Mostepanenko V.M. (2000). Casimir force at both nonzero temperature and finite conductivity, *Phys. Rev. Lett.* **85**, 503–6.

Born M. (1909). Die Theorie des starren Elektrons in der Kinematik des Relativitätsprinzips, *Ann. Physik* **30**, 1–56.
(1926). Das Adiabatenprinzip der Quantenmechanik, *Z. Physik* **40**, 167–92.
(1933). Modified field equations with a finite radius of the electron, *Nature* **132**, 282.
Born M. and Fock V. (1928). Beweis des Adiabatensatzes, *Z. Physik* **51**, 165–80.
Born M., Heisenberg, W. and Jordan P. (1926). Zur Quantenmechanik, *Z. Physik* **35**, 557–615 (translated in van der Waerden (1968)).
Born M. and Infeld L. (1933). Electromagnetic mass, *Nature* **132**, 970.
Bornemann F. A. (1998). *Homogenization in Time of Singularly Perturbed Mechanical Systems*. Lecture Notes in Mathematics, vol. 1687, Berlin: Springer.
Brascamp H. J., Lieb E. H. and Lebowitz J. L. (1976). The statistical mechanics of anharmonic lattices. *Bulletin of the International Statistical Institute, Proceedings of the Fortieth Session* **1**, 393–404. ISI, the Netherlands.
Bratteli O. and Robinson D. (1987). *Operator Algebras and Quantum Statistical Mechanics*, vol. 1. Berlin: Springer.
(1997). *Operator Algebras and Quantum Statistical Mechanics*, vol. 2, 2nd edition, Berlin: Springer.
Braun W. and Hepp K. (1977). The Vlasov dynamics and its fluctuations in the $1/N$ limit of interacting classical particles, *Comm. Math. Phys.* **56**, 101–13.
Breit G. (1932). Quantum theory of dispersion, *Rev. Mod. Phys.* **4**, 504–76.
Breuer H.-P. and Petruccione F. (2001). Destruction of quantum coherence through emission of bremsstrahlung, *Phys. Rev.* A **36**, 032102 (18 pages).
(2002). *The Theory of Open Quantum Systems*. Oxford: Oxford University Press.
Brown L. S. and Gabrielse G. (1986). Geonium theory: physics of a single electron or ion in a Penning trap, *Rev. Mod. Phys.* **58**, 233–78.
Brummelhuis R. and Nourrigat J. (1999). Scattering amplitude for Dirac operators, *Commun. Part. Diff. Eqs.* **24**, 377–94.
Brydges D. and Martin Ph. (1999). Coulomb systems at low density: a review, *J. Stat. Phys.* **96**, 1163–330.
Bucherer A. H. (1904). *Mathematische Einführung in die Elektronentheorie*. Berlin: Teubner.
(1905). Das deformierte Elektron und die Theorie des Elektromagnetismus, *Physikalische Zeitschrift* **6**, 833–4.
(1908). Messungen an Becquerelstrahlen. Die experimentelle Bestätigung der Lorentz–Einsteinschen Theorie, *Physikalische Zeitschrift* **9**, 755–62.
(1909). Die experimentelle Bestätigung des Relativitätsprinzips, *Ann. Physik* **28**, 513–36.
Bugliaro L., Fröhlich J. and Graf G.-M. (1996). Stability of quantum electrodynamics with nonrelativistic matter, *Phys. Rev. Lett.* **77**, 3494–7.
Burke W. (1970). Runaway solutions: Remarks on the asymptotic theory of radiation damping, *Phys. Rev.* A **2**, 1501–5.
Burko L. (2000). Self-force approach to synchroton radiation, *Am. J. Phys.* **68**, 456–68.
Caldirola P. (1956). A new model of classical electron, *Nuov. Cim.* **3**, Supplemento 2, 297–343.
Carati A., Delzanno P., Galgani L. and Sassarini J. (1995). Nonuniqueness properties of the physical solutions of the Lorentz–Dirac equation, *Nonlinearity* **8**, 65–79.
Carati A. and Galgani L. (1993). Asymptotic character of the series of classical electrodynamics and an application to Bremsstrahlung, *Nonlinearity* **6**, 905–14.

Carmichael H. J. (1999). *Statistical Methods in Quantum Optics*, vol. 1: *Master Equations and Fokker–Planck Equations*. Berlin: Springer.
Carmona R. (1978). Pointwise bounds for Schrödinger operators, *Comm. Math. Phys.* **62**, 97–106.
Casimir H. B. G. (1948). On the attraction between two perfectly conducting plates, *Proc. K. Ned. Akad. Wet.* **51**, 793–6.
Casimir H. B. G. and Polder D. (1948). The influence of retardation on the London–van der Waals forces, *Phys. Rev.* **73**, 360–72.
Catto I. and Hainzl C. (2004). Self-energy of one electron in non-relativistic QED, *J. Funct. Anal.* **207**, 68–110.
Chen T. (1996). Aspects of the theory of non-relativistic matter coupled to the quantized radiation field. Diplomarbeit, ETH Zürich, unpublished.
(2001). Operator-theoretic infrared renormalization and construction of dressed one-particle states in non-relativistic QED. Ph.D. thesis, ETH Zürich, mp_arc 01–310.
Chen T., Vougalter V. and Vugalter S. (2003). The increase of binding energy and enhanced binding in nonrelativistic QED, *J. Math. Phys.* **44**, 1961–70.
Cohen-Tannoudji C., Dupont-Roc J. and Grynberg G. (1989). *Photons and Atoms. Introduction to Quantum Electrodynamics*. New York: John Wiley.
(1992). *Atom–Photon Interactions. Basic Processes and Applications*. New York: John Wiley.
Coleman S. (1982). Classical electron theory from a modern standpoint, chapter 6 in *Electromagnetism: Paths to Research*, edited by D. Teplitz. New York: Plenum.
Conlon J. (1984). The ground state energy of a classical gas, *Comm. Math. Phys.* **94**, 439–58.
Conlon J., Lieb E. H. and Yau H.-T. (1988). The $N^{7/5}$ law for charged bosons, *Comm. Math. Phys.* **116**, 417–88.
Corben H. C. (1961). Spin in classical and quantum theory, *Phys. Rev.* **121**, 1833–9.
Craig D. P. and Thirunamachandran T. (1984). *Molecular Quantum Electrodynamics*. New York: Dover.
Cushing J. T. (1981). Electromagnetic mass, relativity, and the Kaufmann experiments, *Am. J. Phys.* **49**, 1133–49.
Cycon H. L., Froese R. G., Kirsch W. and Simon B. (1987). *Schrödinger Operators*. Berlin: Springer.
Damour T. and Schäfer G. (1991). Redefinition of position variables and the reduction of higher order Lagrangians, *J. Math. Phys.* **22**, 127–34.
Davies E. B. (1974). Markovian master equations, *Comm. Math. Phys.* **39**, 91–110.
(1975). Markovian master equations III, *Ann. Inst. H. Poincaré, Phys. Théor.* **11**, 265–73.
(1976a). Markovian master equations II, *Math. Ann.* **219**, 147–58.
(1976b). *Quantum Theory of Open Systems*. London: Academic Press.
(1979). Particle-boson interactions and the weak coupling limit, *J. Math. Phys.* **20**, 345–51.
Dehmelt H. (1990). Experiments with an isolated subatomic particle at rest, *Rev. Mod. Phys.* **62**, 525–30.
Dereziński J. and Gérard C. (1999). Asymptotic completeness in quantum field theory. Massive Pauli-Fierz Hamiltonians, *Rev. Math. Phys.* **11**, 383–450.
(2003). Scattering theory of infrared divergent Pauli–Fierz Hamiltonians, mp-arc 03–363.
Dereziński J. and Jakšić V. (2001). Spectral theory of Pauli–Fierz operators, *J. Funct. Anal.* **180**, 243–327.

(2003a). Return to equilibrium for Pauli–Fierz systems, *Ann. H. Poincaré* **4**, 739–94.
(2003b). On the nature of the Fermi golden rule for open quantum systems, mp-arc 03–354, *J. Stat. Phys.*, to appear.
Dereziński J., Jakšić V. and Pillet C.-A. (2003). Perturbation theory of W*-dynamics, Liouvillians and KMS-states, *Rev. Math. Phys.* **15**, 447–89.
Devreese J. T. and Peeters F. M. (1984). *Polarons and Excitons in Polar Semiconductors and Ionic Crystals*. New York: Plenum.
Dimassi M. and Sjöstrand J. (1999). *Spectral Asymptotics in the Semi-Classical Limit*. Lond. Math. Soc. Lecture Note Series 268, Cambridge: Cambridge University Press.
Dirac P. A. M. (1927). The quantum theory of emission and dispersion, *Proc. Roy. Soc. Lond.* A **114**, 243–65. Reprinted in Schwinger (1958).
(1938). Classical theory of radiating electrons, *Proc. Roy. Soc. Lond.* A **167**, 148–69.
Donsker M. D. and Varadhan S. R. S. (1983). Asymptotic for the polaron, *Comm. Pure Appl. Math.* **36**, 505–28.
Dresden M. (1987). *H. A. Kramers. Between Tradition and Revolution*. New York: Springer.
Dubin D. A. (1974). *Solvable Models in Algebraic Statistical Mechanics*. Oxford: Clarendon Press.
Dümcke R. (1983). Convergence of multitime correlation functions in the weak and singular couplings limits, *J. Math. Phys.* **24**, 311–15.
Dümcke R. and Spohn H. (1979). The proper form of the generator in the weak coupling limit, *Z. Physik* B **34**, 419–22.
Dürr D. and Spohn H. (2000). Decoherence through coupling to the radiation field. In *Decoherence: Theoretical, Experimental, and Conceptual Problems*, edited by Ph. Blanchard, D. Guilini, E. Joos, C. Kiefer and I.-O. Stamatescu, Lecture Notes in Physics vol. 538, pp. 77–86. Heidelberg: Springer.
van Dyck R. S., Schwinberg P. B. and Dehmelt H. G. (1986). Electron magnetic moment from geonium spectra: early experiments and background concepts, *Phys. Rev.* D **34**, 722–86.
Dyson F. J. (1967). Ground state energy for a finite system of charged particles, *J. Math. Phys.* **8**, 1538–45.
Dyson F. J. and Lenard A. (1968). Stability of matter, I and II, *J. Math. Phys.* **8**, 423–34, *ibid.* **9**, 119–215.
Ehrenfest P. (1916). On adiabatic changes of a system in connection with the quantum theory, *Proc. Amsterdam Acad.* **19**, 576–97.
Einstein A. (1905a). Zur Elektrodynamik bewegter Körper, *Ann. Physik* **17**, 891–921, translation in: *The Principle of Relativity*. New York: Dover, 1952.
(1905b). Ist die Trägheit eines Körpers von seinem Energiegehalt abhängig?, *Ann. Physik* **17**, 639–41, translation in: *The Principle of Relativity*. New York: Dover, 1952.
Eliezer C. J. (1950). A note on electron theory, *Proc. Camb. Phil. Soc.* **46**, 199–201.
Emch G. G. (1972). *Algebraic Methods in Statistical Mechanics and Quantum Field Theory*. New York: Wiley Interscience.
Endres D. J. (1993). The physical solution to the Lorentz–Dirac equation for planar motion in a constant magnetic field, *Nonlinearity* **6**, 953–71.
Erber T. (1961). The classical theories of radiation reaction, *Fortschritte der Physik* **9**, 343–92.
(1971). Some external field problems in quantum electrodynamics, *Acta Physica Austriaca*, Suppl. VIII, 323–57. Proceedings X. Internationale Universitätswochen für Kernphysik, Schladming, ed. P. Urban.

Erdös L. (2002). Linear Boltzmann equation as the long time dynamics of an electron weakly coupled to a phonon field, *J. Stat. Phys.* **107**, 1043–127.
Erdös L. and Yau H. T. (1998). Linear Boltzmann equation as scaling limit of quantum Lorentz gas, *Advances in Differential Equations and Mathematical Physics, Contemp. Math.* **217**, 137–55.
(2000). Linear Boltzmann equation as the weak coupling limit of the random Schrödinger equation, *Comm. Pure Appl. Math.* **53**, 667–735.
Fefferman C. (1996). On electrons and nuclei in a magnetic field, *Adv. Math.* **124**, 100–53.
Fefferman C., Fröhlich J. and Graf G. M. (1997). Stability of ultraviolet-cutoff quantum electrodynamics with non-relativistic matter, *Comm. Math. Phys.* **190**, 309–30.
Fefferman C. and de la Llave R. (1986). Relativistic stability of matter. I, *Rev. Math. Iberoamericana* **2**, 119–215.
Feinberg J., Mann A. and Revzen M. (2000). Casimir effect: the classical limit, *Ann. Phys.* **288**, 103–36.
Fermi E. (1922). Über einen Widerspruch zwischen der elektrodynamischen und der relativistischen Theorie der elektromagnetischen Masse, *Physikalische Zeitschrift* **23**, 340–4.
(1930). Sopra l'ellettrodinamica quantistica, *Atti della Reale Accademia Nazionale dei Lincei* **12**, 431–5.
(1932). Quantum theory of radiation, *Rev. Mod. Phys.* **4**, 87–132.
Feynman R. P. (1948). Space-time approach to nonrelativistic quantum mechanics, *Rev. Mod. Phys.* **20**, 367–87.
(1955). Slow electrons in a polar crystal, *Phys. Rev.* **97**, 660–5.
Feynman R. P. and Hibbs A. (1965). *Quantum Mechanics and Path Integrals*. New York: McGraw-Hill.
Feynman R. P., Leighton R. B. and Sands M. (1963). *The Feynman Lectures in Physics*. Reading, MA: Addison-Wesley.
Fidaleo F. and Liverani C. (1999). Ergodic properties for quantum nonlinear dynamics, *J. Stat. Phys.* **97**, 957–1009.
Fokker A. D. (1929). Ein invarianter Variationssatz für die Bewegung mehrerer elektrischer Massenteilchen, *Z. Physik* **58**, 386–93.
Ford G. W., Kac M. and Mazur P. (1965). Statistical mechanics of assemblies of coupled oscillators, *J. Math. Phys.* **6**, 504–15.
Ford G. W., Lewis J. T. and O'Connell R. F. (1988a). Quantum Langevin equation, *Phys. Rev.* A **37**, 4419–28.
(1988b). Independent oscillator model of a heat bath: exact diagonalization of the Hamiltonian, *J. Stat. Phys.* **53**, 439–55.
Ford G. W. and O'Connell R. F. (1991). Radiation reaction in electrodynamics and the elimination of runaway solutions, *Phys. Lett.* A **157**, 217–20.
(1993). Relativistic form of radiation reaction, *Phys. Lett.* A **174**, 182–4.
Frenkel J. (1925). Zur Elektrodynamik punktförmiger Elektronen, *Z. Physik* **32**, 518–34.
(1926). Die Elektrodynamik des rotierenden Elektrons, *Z. Physik* **37**, 243–62.
Frigerio A. (1978). Stationary states of quantum dynamical semigroups, *Comm. Math. Phys.* **63**, 269–76.
Frigerio A. and Verri M. (1982). Long time asymptotic properties of dynamical semigroups on $W^*$-algebras, *Math. Z.* **180**, 257–86.
Fröhlich H. (1954). Electrons in lattice fields, *Advances in Physics* **3**, 325–62.
Fröhlich J. (1973). On the infrared problem in a model of scalar electrons and massless, scalar bosons, *Ann. Inst. H. Poincaré, Phys. Théor.* **19**, 1–103.

(1974). Existence of dressed one electron states in a class of persistent models, *Fortschritte der Physik* **22**, 159–89.
Fröhlich J., Griesemer M. and Schlein B. (2001). Asymptotic electromagnetic fields in models of quantum-mechanical matter interacting with the quantized radiation field, *Adv. Math.* **164**, 349–98.
(2002). Asymptotic completeness for Rayleigh scattering, *Ann. H. Poincaré* **3**, 107–70.
(2003). Asymptotic completeness for the Compton scattering, arXiv:math-ph/0111032 v3, *Comm. Math. Phys.*, to appear.
Fröhlich J. and Park Y. M. (1978). Correlation inequalities and the thermodynamic limit for classical and quantum continuous systems, *Comm. Math. Phys.* **59**, 235–66.
Fujita S. (1966). *Introduction to Non-Equilibrium Statistical Mechanics*. Philadelphia: Saunders.
Fulton T. and Rohrlich F. (1960). Classical radiation from a uniformly accelerated charge, *Ann. Phys.* **9**, 499–517.
Galgani L., Angaroni C., Forti L., Giorgilli A. and Guerra F. (1989). Classical electrodynamics as a nonlinear dynamical system, *Phys. Lett.* A **139**, 221–30.
Gallavotti G., Ginibre J. and Velo G. (1970). Statistical mechanics of the electron-phonon system, *Lett. Nuovo Cimento* **4**, 1293–7.
Garrido L. M. (1965). Generalized adiabatic invariance, *J. Math. Phys.* **5**, 355–62.
Georgescu V., Gérard C. and Møller J.S. (2004). Spectral theory of massless Pauli–Fierz models, *Comm. Math. Phys.*, to appear.
Gérard C. (1996). Asymptotic completeness for the spin-boson model with a particle number cut-off, *Rev. Math. Phys.* **8**, 549–89.
(2000). On the existence of ground states for massless Pauli–Fierz Hamiltonians, *Ann. H. Poincaré* **1**, 443–59.
(2002). On the scattering theory of massless Nelson models, *Rev. Math. Phys.* **14**, 1165–280.
Gérard P., Markowich P., Mauser N. and Poupaud F. (1997). Homogenization limits and Wigner transforms, *Comm. Pure Appl. Math.* **50**, 323–79.
Gerlach B. and Löwen H. (1991). Analytical properties of polaron systems or: do polaronic phase transitions exist or not?, *Rev. Mod. Phys.* **63**, 63–90.
Giacomin G., Olla S. and Spohn H. (2001). Equilibrium fluctuations for $\nabla\varphi$ interface model, *Ann. Probab.* **29**, 1138–72.
Glassey R. T. and Schaeffer J. (1991). Convergence of a particle method for the relativistic Vlasov-Maxwell system, *SIAM J. Num. Anal.* **28**, 1–25.
(1997). The "two and one-half" relativistic Vlasov–Maxwell system, *Comm. Math. Phys.* **185**, 257–84.
(2000). The relativistic Vlasov–Maxwell system in 2D and 2.5D, *Contemp. Math.* **263**, 61–9.
Glauber R. J. (1969). Coherence and quantum detection. In *Quantum Optics. Proceedings of the International School of Physics Enrico Fermi*, Course XLVII, p. 15, edited by R.J. Glauber. New York: Academic Press.
Glimm J. and Jaffe A. (1987). *Quantum Physics: A Functional Integral Point of View*, 2nd edition. Berlin: Springer.
Gorini V., Frigerio A. and Accardi L. (1984). *Quantum Probability and Applications to the Quantum Theory of Irreversible Processes*. New York: Springer.
Grabert H., Schramm P. and Ingold G.-L. (1988). Quantum Brownian motion: the functional integral approach, *Phys. Rep.* **168**, 115–207.

Grandy W. T. and Aghazadeh A. (1982). Radiative corrections for extended charged particles in classical electrodynamics, *Ann. Phys.* **142**, 284–98.

Griesemer M. (2004). Exponential decay and ionization thresholds in non-relativistic quantum electrodynamics, *J. Funct. Anal.* **210**, 321–40.

Griesemer M., Lieb E. and Loss M. (2001). Ground states in non-relativistic quantum electrodynamics, *Inv. Math.* **145**, 557–95.

Gross E. P. (1962). Particle-like solutions in field theory, *Ann. Phys.* **19**, 219–33.

(1976). Strong coupling polaron theory and translational invariance, *Ann. Phys.* **99**, 1–29.

Grotch H. (1981). Lamb shift in nonrelativistic quantum electrodynamics, *Am. J. Phys.* **49**, 48–51.

Grotch H. and Kazes E. (1977). Nonrelativistic quantum mechanics and the anomalous part of the electron *g* factor, *Am. J. Phys.* **45**, 618–23.

Grotch H., Kazes E., Rohrlich F. and Sharp D. H. (1982). Internal retardation, *Acta Phys. Austr.* **54**, 31–8.

Gutzwiller M. C. (1990). *Chaos in Classical and Quantum Mechanics.* New York: Springer.

Haag R. (1955). Die Selbstwechselwirkung des Elektrons, *Z. Naturforsch.* **10a**, 752–61.

Haake F. (1973). *Statistical Treatment of Open Systems by Generalized Master Equations.* Springer Tracts in Modern Physics **66**, Berlin: Springer.

Hainzl C. (2002). Enhanced binding through coupling to the photon field. In *Mathematical Results in Quantum Mechanics*, pp. 149–54, edited by R. Weder, P. Exner and B. Grebert, AMS Contemporary Mathematics Vol. 307.

(2003). One non-relativistic particle coupled to a photon field, *Ann. H. Poincaré* **4**, 217–37.

Hainzl C., Hirokawa M. and Spohn H. (2003). Binding energy for hydrogen-like atoms in the Nelson model without cutoffs, arXiv:math-ph/0312025.

Hainzl C. and Seiringer R. (2002). Mass renormalization and energy level shift in non-relativistic QED, *Adv. Theor. Math. Phys.* **6**, 847–69.

Hainzl C., Vougalter V. and Vugalter S. (2003). Enhanced binding in non-relativistic QED, *Comm. Math. Phys.* **233**, 13–26.

Haken H. (1983). *Advanced Synergetics. Instability Hierarchies of Self-Organizing Systems and Devices.* Berlin: Springer.

Healy W. P. (1982). *Non-Relativistic Quantum Electrodynamics.* London: Academic Press.

Heinemann K. and Barber D. P. (1999). The semiclassical Foldy–Wouthuysen transformation and the derivation of the Bloch equation for spin-1/2 polarised beams using Wigner functions, arXiv:phys 9901044.

Heitler W. (1936). *The Quantum Theory of Radiation.* Oxford: Clarendon Press; 3rd edition 1958.

Herglotz G. (1903). Zur Elektronentheorie, *Nachr. K. Ges. Wiss. Göttingen* (6), 357–82.

Hirokawa M., Hiroshima F. and Spohn H. (2002). The two-particle massless Nelson model without cutoffs, arXiv:math-ph/0211050. *Adv. Math.*, to appear.

Hiroshima F. (1996). Diamagnetic inequalities for systems of nonrelativistic particles with a quantized field, *Rev. Math. Phys.* **8**, 185–203.

(1997a). Scaling limit of a model of quantum electrodynamics with many nonrelativistic particles, *Rev. Math. Phys.* **9**, 201–25.

(1997b). Functional integral representation of a model in quantum electrodynamics, *Rev. Math. Phys.* **9**, 489–530.

(1999). Ground states of a model in nonrelativistic quantum electrodynamics I, *J. Math. Phys.* **40**, 6209–22.
(2000a). Ground states of a model in nonrelativistic quantum electrodynamics II, *J. Math. Phys.* **41**, 661–74.
(2000b). Essential self-adjointness of translation-invariant quantum field models for arbitrary coupling constants, *Comm. Math. Phys.* **211**, 585–613.
(2001). Spectral analysis of atoms interacting with a quantized radiation field, mp_arc 01–97.
(2002). Self-adjointness of the Pauli–Fierz Hamiltonian for arbitrary values of coupling constants, *Ann. H. Poincaré* **3**, 171–201.
Hiroshima F. and Spohn H. (2001). Enhanced binding through coupling to a quantum field, *Ann. H. Poincaré* **2**, 1159–87.
(2002). Two-fold degeneracy of the ground state band for the Pauli–Fierz model with spin, *Adv. Theor. Math. Phys.* **5**, 1091–104.
(2003). Mass renormalization in nonrelativistic QED, arXiv:math-ph/0310043.
Høegh-Krohn R. (1970). On the scattering theory for quantum fields, *Comm. Math. Phys.* **18**, 109–26.
Holley R. and Stroock D. W. (1978). Generalized Ornstein–Uhlenbeck processes and infinite branching Brownian motions, *RIMS Kyoto Publications* A **14**, 741–814.
Hönl H. (1952). Feldmechanik des Elektrons und der Elementarteilchen, *Ergeb. Exakt. Naturwiss.* **26**, 291–382.
van Hove L. (1955). Quantum-mechanical perturbations giving rise to a statistical transport equation, *Physica* **21**, 517–40.
(1957). The approach to equilibrium in quantum statistics, *Physica* **23**, 441–80.
Huang K. (1987). *Statistical Mechanics*, 2nd edition. New York: Wiley.
(1998). *Quantum Field Theory: From Operators to Path Integrals*. New York: Wiley.
Hübner M. and Spohn H. (1995a). Radiative decay: nonperturbative approaches, *Rev. Math. Phys.* **7**, 363–87.
(1995b). Spectral properties of the spin-boson Hamiltonian, *Ann. Inst. H. Poincaré, Phys. Théor.* **62**, 289–323.
Hudson R. L. and Parthasarathy K. R. (1984). Quantum Ito's formula and stochastic evolutions, *Comm. Math. Phys.* **93**, 301–23.
Hulse R. A. (1994). The discovery of the binary pulsar, *Rev. Mod. Phys.* **66**, 699–710.
Hunziker W. (1990). Resonances, metastable states and exponential decay laws in perturbation theory, *Comm. Math. Phys.* **132**, 177–88.
Huschilt J. and Baylis W. E. (1976). Numerical solutions to two-body problems in classical electrodynamics: head on collisions with retarded fields and radiation reaction, *Phys. Rev.* D **13**, 3256–61 and D **13**, 3262–8.
Imaikin V., Komech A. and Mauser N. (2003). Soliton type asymptotics for the coupled Maxwell–Lorentz equations, preprint.
Imaikin V., Komech A. and Spohn H. (2002). Soliton-type asymptotics and scattering for a charge coupled to the Maxwell field, *Russian J. Math. Phys.* **9**, 428–36.
(2004). Rotating charge coupled to the Maxwell field: scattering theory and adiabatic limit, *Monatsheft für Mathematik*, to appear.
Jackson J. D. (1999). *Classical Electrodynamics*, 3rd edition. New York: Wiley.
Jakšić V. and Pillet C. A. (1995). On a model of quantum friction, I: Fermi's golden rule and dynamics at zero temperature, *Ann. Inst. H. Poincaré, Phys. Théor.* **62**, 47–68.
(1996a). On a model of quantum friction, II: Fermi's golden rule and dynamics at positive temperature, *Comm. Math. Phys.* **176**, 619–44.

(1996b). On a model of quantum friction, III: Ergodic properties of the spin-boson system, *Comm. Math. Phys.* **178**, 627–51.

(1997). From resonances to master equations, *Ann. Inst. H. Poincaré, Phys. Théor.* **67**, 425–45.

(2002). Nonequilibrium steady states of finite quantum systems coupled to thermal reservoirs, *Comm. Math. Phys.* **226**, 131–62.

Jaranowski P. and Schäfer G. (1998). Third post-Newtonian order ADM Hamilton dynamics for two-body point mass systems, *Phys. Rev.* D **57**, 7274–90.

Jones C. (1995). Geometric singular perturbation theory. In *Dynamical Systems, Proceedings, Montecatini Terme 1994*, edited by R. Johnson, Lecture Notes in Mathematics 1609, pp. 44–118. New York: Springer.

Jordan P. and Pauli W. (1928). Zur Quantenelektrodynamik ladungsfreier Felder, *Z. Phys.* **47**, 151–73.

Joye A., Kunz H. and Pfister C.-E. (1991). Exponential decay and geometric aspect of transition probabilities in the adiabatic limit, *Ann. Phys.* **208**, 299–332.

Joye A. and Pfister C.-E. (1993). Superadiabatic evolution and adiabatic transition probability between two nondegenerate levels in the spectrum, *J. Math. Phys.* **34**, 454–73.

(1994). Quantum adiabatic evolution. In *On Three Levels*, ed. M. Fannes, C. Maes and A. Verbeure, pp. 139–48. New York: Plenum.

Kac M. (1950). On some connections between probability theory and differential equations, *Proc. 2nd Berk. Symp. Math. Statist. Probability*, pp. 189–215.

Kampen van N. (1951). Contribution to the quantum theory of light scattering, *Det Kongelige Danske Videns. Selskab, Matt. Fys. Medd.* **26**, 1–77.

Kato T. (1958). On the adiabatic theorem of quantum mechanics, *Phys. Soc. Jap.* **5**, 435–9.

Kaufmann W. (1901). Series of papers in *Nachr. K. Ges. Wiss. Göttingen*, (2), 143–55 (1901); (5), 291–96 (1902); (3), 90–103 (1903). *Physikalische Zeitschrift* **4**, 54–7 (1902). *Sitzungsber. K. Preuss. Akad. Wiss.* **2**, 949–56 (1905). *Ann. Phys.* **19**, 487–553 (1906).

Keitel C. H., Szymanowski C., Knight P. L. and Maquet A. (1998). Radiative reaction in ultraintense laser-atom interaction, *J. Phys. B: At. Mol. Opt. Phys.* **31**, L75–L83.

Kibble T. W. B. (1968). Coherent soft-photon states and infrared divergencies I. Classical currents, *J. Math. Phys.* **9**, 315–24.

Kiessling M. (1999). Classical electron theory and conservation laws, *Phys. Lett.* A **258**, 197–204.

(2003). Electromagnetic field theory without divergence problems: 1. The Born legacy, arXiv:math-ph 0306076, 2. A least invasively quantized theory, arXiv:math-ph 0311034. *J. Stat. Phys.*, to appear.

Kinoshita T. and Sapirstein J. (1984). New developments in QED. In *Atom Physics*, vol. 9, pp. 38–52, edited by R. S. van Dyck and E. N. Forston. Singapore: World Scientific.

Klauder J. R., McKenna J. and Woods E. J. (1966). Direct-product representation of the canonical commutation relations, *J. Math. Phys.* **7**, 822–8.

Kleinert H. (1995). *Path Integrals in Quantum Mechanics, Statistics and Polymer Physics*, 2nd edition. Singapore: World Scientific.

Komech A., Kunze M. and Spohn H. (1999). Effective dynamics of a mechanical particle coupled to a wave field, *Comm. Math. Phys.* **203**, 1–19.

Komech A. and Spohn H. (1998). Soliton-like asymptotics for a classical particle interacting with a scalar wave field, *Nonlin. Analysis* **33**, 13–24.

(2000). Long-time asymptotics for coupled Maxwell–Lorentz equations, *Comm. PDE* **25**, 559–84.

Komech A., Spohn H. and Kunze M. (1997). Long-time asymptotics for a classical particle interacting with a scalar wave field, *Comm. PDE* **22**, 307–35.

Kossakowski A. (1972). On quantum statistical mechanics of non-hamiltonian systems, *Rep. Math. Phys.* **3**, 247–74.

Kramers H. A. (1948). Nonrelativistic quantum electrodynamics and correspondence principle. *Collected Scientific Papers*, pp. 845–69. North-Holland, Amsterdam, 1956.

Kroll N. M. (1965). *Quantum Optics and Electronics*, pp. 51–3. New York: Gordon and Breach.

Kroll N. M. and Lamb E. (1949). On the self-energy of a bound electron, *Phys. Rev.* **75**, 388–98.

Kunze M. (1998). Instability of the periodic motion of a particle interacting with a scalar wave field, *Comm. Math. Phys.* **195**, 509–23.

Kunze M. and Rendall A. (2001). The Vlasov–Poisson system with radiation damping, *Ann. H. Poincaré* **2**, 857–86.

Kunze M. and Spohn H. (2000a). Radiation reaction and center manifolds, *SIAM J. Math. Analysis* **32**, 30–53.

(2000b). Adiabatic limit for the Maxwell–Lorentz equation, *Ann. H. Poincaré* **1**, 625–55.

(2000c). Slow motion of charges interacting through the Maxwell field, *Comm. Math. Phys.* **212**, 437–67.

(2001). Post-Coulombian dynamics at order $c^{-3}$, *J. Nonlinear Science* **11**, 321–96.

Lamoreaux S. K. (1997). Demonstration of the Casimir force in the 0.6$\mu$m to 6$\mu$m range, *Phys. Rev. Lett.* **78**, 5–8, Erratum, *ibid.* **81**, 5475–6.

Landau L. (1927). Das Dämpfungsproblem in der Wellenmechanik, *Z. Phys.* **45**, 430–41.

(1933). Electron motion in crystal lattices, *Phys. Z. Sowjet.* **3**, 664–665. In *Collected Papers*, pp. 67–8, New York: Gordon and Breach, 1965.

Landau L. D. and Lifshitz E. M. (1959). *The Classical Theory of Fields*. Reading, MA: Addison-Wesley, London: Pergamon Press.

Landau L. and Peierls R. (1930). Quantenelektrodynamik im Konfigurationsraum, *Z. Phys.* **62**, 188–200.

Langevin, P. (1905). La physique des électrons. Lecture delivered on 22 September 1904 at the International Congress of Arts and Science at St. Louis, Missouri, and published in *Rev. Générale Scis. Pures Appl.* **16**, 257–76.

Laue M. von (1909). Die Wellenstrahlung einer bewegten Punktladung nach dem Relativitätsprinzip, *Ann. Physik* **28**, 436–42.

(1911a). *Das Relativitätsprinzip*. Vieweg: Braunschweig.

(1911b). Zur Diskussion über den starren Körper in der Relativitätstheorie, *Physikalische Zeitschrift* **12**, 85–7.

Lax M. (1968). Fluctuations and coherence phenomena in classical and quantum physics. In *Brandeis Summer Institute Lectures, 1966*, vol. 2, p. 269, edited by M. Chrétien, E. P. Gross, and S. Dreser. New York: Gordon and Breach.

Lebowitz J. L. and Lieb E. H. (1969). Existence of thermodynamics for real matter with Coulomb forces, *Phys. Rev. Lett.* **22**, 631–4.

Lee T. D., Low F. E. and Pines D. (1953). The motion of slow electrons in a polar crystal, *Phys. Rev.* **90**, 297–302.

Leggett A. J., Chakravarty S., Dorsey A. T., Fisher M. P. A., Garg A. and Zwerger W. (1987). Dynamics of the dissipative two-state system, *Rev. Mod. Phys.* **59**, 1–85.

Lenard A. (1959). Adiabatic invariance to all orders, *Ann. Phys.* **6**, 261–76.

Levine H., Moniz E. J. and Sharp D. H. (1977). Motion of extended charges in classical electrodynamics, *Am. J. Phys.* **45**, 75–8.

Lieb E. H. (1976). The stability of matter, *Rev. Mod. Phys.* **48**, 553–69.

(1979). The $N^{5/3}$ law for bosons, *Phys. Lett.* A **70**, 71–3.

(1990). The stability of matter: from atoms to stars, *Bull. Amer. Math. Soc.* **22**, 1–49.

(2001). *The Stability of Matter: From Atoms to Stars*, edited by W. Thirring, 3rd edition. Berlin: Springer.

Lieb E. H. and Lebowitz J. L. (1972). The constitution of matter: existence of thermodynamics for systems composed of electrons and nuclei, *Adv. Math.* **9**, 316–98.

(1973). Lectures on the thermodynamic limit for Coulomb systems. In *Statistical Mechanics and Mathematical Problems*, Batelle 1971 Rencontres, Lecture Notes in Physics, vol. 20, pp. 136–61, Berlin: Springer.

Lieb E. H. and Loss M. (2000). Self-energy of electrons in non-perturbative QED. In *Differential Equations and Mathematical Physics*, edited by R. Weikard and G. Weinstein, pp. 279–93. Cambridge, MA: Amer. Math. Soc.

(2002). A bound on binding energies and mass renormalization in models of quantum electrodynamics, *J. Stat. Phys.* **108**, 1057–69.

(2003). Existence of atoms and molecules in non-relativistic quantum electrodynamics, *Adv. Theor. Math. Phys.* 7667–710.

(2004). A note on polarization vectors in quantum electrodynamics, arXiv:math-ph/0401016.

Lieb E. H., Loss M. and Siedentop H. (1996). Stability of relativistic matter via Thomas–Fermi theory, *Helv. Phys. Acta* **69**, 978–84.

Lieb E. H., Loss M. and Solovej J. P. (1995). Stability of matter in magnetic fields, *Phys. Rev. Lett.* **75**, 985–8.

Lieb E. H. and Thirring W. (1975). Bound for the kinetic energy of fermions which proves the stability of matter, *Phys. Rev. Lett.* **35**, 687–9, Erratum, *ibid.* **35**, 1116.

Lieb E. H. and Thomas L. E. (1997). Exact ground state energy of the strong-coupling polaron, *Comm. Math. Phys.* **183**, 511–19.

Lieb E. H. and Yamazaki K. (1958). Ground-state energy and effective mass of the polaron, *Phys. Rev.* **111**, 728–33.

Lieb E. H. and Yau H. T. (1988a). Many-body stability implies a bound on the fine structure constant, *Phys. Rev. Lett.* **61**, 1995–7.

(1988b). The stability and instability of relativistic matter, *Comm. Math. Phys.* **118**, 177–213.

Lindblad G. (1975). Completely positive maps and entropy inequalities, *Comm. Math. Phys.* **40**, 147–51.

(1976). On the generators of quantum dynamical semigroups, *Comm. Math. Phys.* **48**, 119–30.

(1983). *Non-Equilibrium Entropy and Irreversibility*. Dordrecht: D. Reidel.

Littlejohn R. G. and Flynn W. G. (1991). Geometric phases in the asymptotic theory of coupled wave equations, *Phys. Rev.* A **44**, 5239–55.

Littlejohn R. G. and Weigert S. (1993). Diagonalization of multicomponent wave equations with a Born–Oppenheimer example, *Phys. Rev.* A **47**, 3506–12.

Lorentz H. A. (1892). La théorie électromagnetique de Maxwell et son application aux corps mouvants, *Arch. Néerl. Sci. Exactes Nat.* **25**, 363–552.

(1904a). Electromagnetic phenomena in system moving with any velocity less than that of light, *Proceedings of the Academy of Sciences of Amsterdam* **6**, 809–31; contained in: Perret W. and Jeffery G. B., *The Principle of Relativity*. New York: Dover, 1952.

(1904b). Weiterbildung der Maxwell'schen Theorie: Elektronentheorie. *Enzyklopädie der Mathematischen Wissenschaften*, Band 14, 145–288.

(1915). *The Theory of Electrons and its Applications to the Phenomena of Light and Radiant Heat*, 2nd edition. Reprinted by New York: Dover, 1952.

Lőrinczi J. and Minlos R. (2001). Gibbs measures for Brownian paths under the effect of an external and a small pair potential, *J. Stat. Phys.* **105**, 607–49.

Lőrinczi J., Minlos R. and Spohn H. (2002a). The infrared behavior in Nelson's model of a quantum particle coupled to a massless scalar field, *Ann. H. Poincaré* **3**, 269–95.

(2002b). Infrared regular representation of the three-dimensional massless Nelson model, *Lett. Math. Phys.* **108**, 189–98.

Louisell W. H. (1973). *Quantum Statistical Properties of Radiation*. New York: Wiley-Interscience.

Maassen H. (1984). Return to thermal equilibrium by a solution of a quantum Langevin equation, *J. Stat. Phys.* **34**, 239–62.

Maassen H., Gută M. and Botvich D. (1999). Stability of Bose dynamical systems and branching theory, mp-arc 99–130.

McManus H. (1948). Classical electrodynamics without singularities, *Proc. Roy. Soc. Lond.* **195**, 323–36.

Mandel L. (1981). Comment on "photon counting probabilities in quantum optics", *Optica Acta* **28**, 1447–50.

Marino M. (2002). Classical electrodynamics of point charges, *Ann. Phys.* **301**, 85–127.

(2003). The unexpected flight of the electron in a classical hydrogen-like atom, *J. Phys.* A **36**, 11247–56.

Martinez A. (2002). *An Introduction to Semiclassical Analysis*. Berlin: Springer.

Martinez A. and Sordoni V. (2002). On the time-dependent Born–Oppenheimer approximation with smooth potential, *Comptes Rendus Acad. Sci. Paris* **334**, 51–85.

Maslov V. P. and Fedoriuk M. V. (1981). *Semi-classical Approximation in Quantum Mechanics*. Dordrecht: D. Reidel.

Mehra J. and Rechenberg H. (2000). *The Historical Development of Quantum Theory, vol. 6: The Completion of Quantum Mechanics, 1926–1941, Part 1: The Probabilistic Interpretation and the Empirical and Mathematical Foundation of Quantum Mechanics*, 1926–1936. New York: Springer.

Merkli M. (2001). Positive commutators in non-equilibrium quantum statistical mechanics, *Lett. Math. Phys.* **57**, 225–37.

Metcalf H. J. and van der Straten P. (1999). *Laser Cooling and Trapping*. New York: Springer.

Miller A. I. (1994). *Early Quantum Electrodynamics. A Source Book*. Cambridge: Cambridge University Press.

(1997). *Albert Einstein's Special Theory of Relativity*. New York: Springer.

Milonni P. W. (1994). *The Quantum Vacuum: An Introduction to Quantum Electrodynamics*. San Diego: Academic Press.

Misner C. W., Thorne K. S. and Wheeler J. A. (1973). *Gravitation*. New York: W. H. Freeman.

Mo T. C. and Papas C. H. (1971). New equation of motion for classical charged particles, *Phys. Rev.* D **4**, 3566–71.

Møller C. (1952). *The Theory of Relativity*. Oxford: Oxford University Press.

Moniz E. J. and Sharp D. H. (1974). Absence of runaways and divergent self-mass in nonrelativistic quantum electrodynamics, *Phys. Rev*. D **10**, 1113–6.

(1977). Radiation reaction in nonrelativistic quantum electrodynamics, *Phys. Rev.* D **15**, 2850–65.

Morse P. M. and Feshbach H. (1953). *Methods of Theoretical Physics, Part I*. New York: McGraw-Hill.

Moyal J. E. (1949). Quantum mechanics as a statistical theory, *Proc. Cambridge Phil. Soc.* **45**, 99–124.

Nelson E. (1964a). Schrödinger particles interacting with a quantized scalar field. In *Proceedings of a Conference on Analysis in Function Space*, edited by W. T. Martin and I. Segal. Cambridge, MA: MIT Press.

(1964b). Interaction of nonrelativistic particles with a quantized scalar field, *J. Math. Phys.* **5**, 1190–7.

(1966). A quartic interaction in two dimensions. In *Mathematical Theory of Elementary Particles*, edited by R. Goodman and I. Segal, pp. 69–73. Cambridge, MA: MIT Press.

(1973). Construction of quantum fields from Markoff fields, *J. Funct. Anal.* **12**, 97–112.

Nenciu G. (1993). Linear adiabatic theory. Exponential estimates, *Comm. Math. Phys.* **152**, 479–96.

Nenciu G. and Sordoni V. (2001). Semiclassical limit for multistate Klein–Gordon systems: almost invariant subspaces and scattering theory, mp_arc 01–36.

Neumann G. (1914). Die träge Masse schnell bewegter Elektronen, *Ann. Phys.* **45**, 529–79.

Neunzert H. (1975). Neuere qualitative und numerische Methoden in der Plasmaphysik, Gastvorlesung im Rahmen des Paderborner Ferienkurses in Angewandter Mathematik.

Nodvik J. S. (1964). A covariant formulation of classical electrodynamics for charges of finite extension, *Ann. Phys.* **28**, 225–319.

Noja D. and Posilicano A. (1998). The wave equation with one point interaction and the (linearized) classical electrodynamics of a point particle, *Ann. Inst. H. Poincaré, Phys. Théor.* **68**, 351–77.

(1999). On the point limit of the Pauli–Fierz model, *Ann. Inst. H. Poincaré, Phys. Théor.* **71**, 425–58.

Nordsieck A. (1937). The low frequency radiation of a scattered electron, *Phys. Rev.* **52**, 59–62.

Nyborg P. (1962). On classical theories of spinning particles, *Nuovo Cim.* **23**, 47–62.

Okamoto T. and Yajima K. (1985). Complex scaling technique in non-relativistic massive QED, *Ann. Inst. H. Poincaré, Phys. Théor.* **42**, 311–27.

Page L. (1918). Is a moving mass retarded by the reaction of its own radiation?, *Phys. Rev.* **11**, 377–400.

Pais A. (1972). The early history of the theory of the electron: 1897–1947. In *Aspects of Quantum Theory*, edited by A. Salam and E. P. Wigner. Cambridge: Cambridge University Press.

(1982). *'Subtle is the Lord'. The Science and Life of Albert Einstein*. Oxford: Oxford University Press.

Panati G., Spohn H. and Teufel S. (2002). Space-adiabatic perturbation theory in quantum dynamics, *Phys. Rev. Lett.* **88**, 250–405.

(2003a). Space-adiabatic perturbation theory, *Adv. Theor. Math. Phys.* **7**, 145–204.

(2003b). Effective dynamics for Bloch electrons: Peierls substitution and beyond, *Comm. Math. Phys.* **242**, 547–78.

Panofsky W. K. H. and Phillips M. (1962). *Classical Electricity and Magnetism*, 2nd edition. Reading, MA: Addison-Wesley.

Parrot S. (1987). *Relativistic Electrodynamics and Differential Geometry*. Berlin: Springer.

Parthasarathy K. (1992). *An Introduction to Quantum Stochastic Calculus*. Basel: Birkhäuser.

Pauli W. (1921). *Relativitätstheorie, Enzyklopädie der Mathematischen Wissenschaften*, vol. 19, pp. 543–775. Translated as: *Theory of Relativity*, New York: Pergamon, 1958.

(1928). Über das H-Theorem vom Anwachsen der Entropie vom Standpunkt der neuen Quantenmechanik. In *Probleme der modernen Physik, Arnold Sommerfeld zum 60. Geburtstage, gewidmet von seinen Schülern*, pp. 30–45. Leipzig: S. Hirzel-Verlag.

(1929). Theorie der schwarzen Strahlung. Chapter 27 in *Müller-Pouillets Lehrbuch*, vol. 2, part 2, 11th edition, pp. 1483–1553. Braunschweig: Friedrich Vieweg.

Pauli W. and Fierz M. (1938). Zur Theorie der Emission langwelliger Lichtquanten, *Il Nuovo Cimento* **15**, 167–88.

Pearle P. (1977). Absence of radiationless motions of relativistically rigid classical electron, *Found. Phys.* **7**, 931–45.

(1982). Classical electron models. Chapter 7 in *Electromagnetism: Paths to Research*, edited by D. Teplitz. New York: Plenum.

Peierls R. (1991). *More Surprises in Theoretical Physics*. Princeton, NJ: Princeton University Press.

Pekar S. I. (1954). *Untersuchungen zur Elektronentheorie der Kristalle*. Berlin: Akademie-Verlag.

Perret W. and Jeffery G. B. (1952). *The Principle of Relativity: A Collection of Original Memoirs on the Special and General Theories of Relativity by H. A. Lorentz, A. Einstein, H. Minkowski, and H. Weyl.* London: Dover; originally London: Methuen, 1923.

Pfaffelmoser K. (1992). Global classical solutions of the Vlasov–Poisson system in three dimensions for general initial data, *J. Diff. Eqn.* **95**, 281–303.

Pizzo A. (2000). One particle (improper) states and scattering states in Nelson's massless model, arXiv:math-ph/0010043.

Plass G. N. (1961). Classical electrodynamic equations of motion with radiative reaction, *Rev. Mod. Phys.* **33**, 37–62.

Poincaré H. (1905). Sur la dynamique de l'électron, *C. R. Acad. Sci.* **140**, 1504–8.

(1906). Sur la dynamique de l'électron, *Rendiconti del Circolo Matematico di Palermo* **21**, 129–76. Translated by H.M. Schwartz, *Am. J. Phys.* **39**, 1287–94, **40**, 862–72, and **40**, 1282–7.

Power E. A. (1964). *Introductory Quantum Electrodynamics*. London: Longmans.

Power E. A. and Zienau S. (1959). Coulomb gauge in non-relativistic quantum electrodynamics and the shape of spectral lines, *Phil. Trans. Roy. Soc. Lond.* **251**, 427–54.

Reed M. and Simon B. (1980). *Methods of Modern Mathematical Physics*, vol. I. New York: Academic Press; first edition 1972.

(1975). *Methods of Modern Mathematical Physics*, vol. II. New York: Academic Press.

(1979). *Methods of Modern Mathematical Physics*, vol. III. New York: Academic Press.

(1978). *Methods of Modern Mathematical Physics*, vol. IV. New York: Academic Press.

Richardson O. W. (1916). *The Electron Theory of Matter*. Cambridge: Cambridge University Press, 2nd edition.

Robert D. (1987). *Autour de l'approximation semi-classique*. Progress in Mathematics, vol. 68. Basel: Birkhäuser.

(1998). Semi-classical approximation in quantum mechanics. A survey of old and recent mathematical results, *Helv. Phys. Acta* **71**, 44–116.

Roepstorff G. (1994). *Path Integral Approach to Quantum Physics*. Berlin: Springer.

Rohrlich F. (1960). Self-energy and stability of the classical electron, *Am. J. Phys.* **28**, 639–43.

(1973). The electron: development of the first elementary particle theory. In *The Physicist's Conception of Nature*, edited by J. Mehra. Dordrecht: D. Reidel.
(1990). *Classical Charged Particles*, 2nd edition. Redwood City, CA: Addison-Wesley.
(1997). The dynamics of a charged sphere and the electron, *Am. J. Phys.* **65**, 1051–6.
Rubinow S. I. and Keller J. B. (1963). Asymptotic solution of the Dirac equation, *Phys. Rev.* **131**, 2789–96.
Rudin W. (1977). *Functional Analysis*. New York: McGraw–Hill.
Ruf B. and Srikanth P. N. (2000). The Lorentz–Dirac equation, *Rev. Math. Phys.* **12**, 657–86, and **12**, 1137–57.
Sakamoto K. (1990). Invariant manifolds in singular perturbation problems for ordinary differential equations, *Proc. Roy. Soc. Edinburgh, Sect.* A **116**, 45–78.
Sakurai J. J. (1986). *Advanced Quantum Mechanics*. Reading, MA: Addison-Wesley.
Schaeffer J. (1986). The classical limit of the relativistic Vlasov–Maxwell system, *Comm. Math. Phys.* **104**, 403–21.
(1991). Global existence of smooth solutions to the Vlasov–Poisson system in three dimensions, *Comm. PDE* **16**, 1313–35.
Scharf G. (1994). *From Electrostatics to Optics*. Heidelberg: Springer.
Schild A. (1963). Electromagnetic two-body problem, *Phys. Rev.* **131**, 2762–6.
Schott G. A. (1912). *Electromagnetic Radiation*. Cambridge: Cambridge University Press.
(1915). On the motion of the Lorentz electron, *Phil. Mag.* **29**, 49–62.
Schulman L. S. (1981). *Techniques and Applications of Path Integration*. New York: Wiley.
Schwabl F. and Thirring W. (1964). Quantum theory of laser radiation. *Erg. Ex. Naturw.* **36**, 219–42.
Schwarzschild K. (1903). Zur Elektrodynamik III. Über die Bewegung des Elektrons, *Göttinger Nachr.* 245–278.
Schweber S. (1994). *QED and the Men Who Made It: Dyson, Feynman, Schwinger, and Tomonaga*. Princeton, NJ: Princeton University Press.
Schwinger J. (1949). On the classical radiation of accelerated electrons, *Phys. Rev.* **75**, 1912–25.
(1958). *Selected Papers on Quantum Electrodynamics*. New York: Dover.
(1983). Electromagnetic mass revisited, *Found. Phys.* **13**, 373–83.
Schwinger J., De Raad L. L. and Milton K. A. (1978). Casimir effect in dielectrics, *Ann. Phys.* **115**, 1–23.
Senitzky I. R. (1960). Dissipation in quantum mechanics: the harmonic oscillator, *Phys. Rev.* **119**, 670–9.
Sewell G. L. (1986). *Quantum Theory of Collective Phenomena*. Oxford: Clarendon Press.
Shapere A. and Wilczek F. (eds.) (1989). *Geometric Phases in Physics*. Singapore: World Scientific.
Shen C. S. (1970). Synchrotron emission at strong radiative damping, *Phys. Rev. Lett.* **24**, 410–15.
(1972a). Magnetic Bremsstrahlung in an intense magnetic field, *Phys. Rev.* D **6**, 2736–54.
(1972b). Comment on the new equation of motion for classical charged particles, *Phys. Rev.* D **6**, 3039–40.
(1978). Radiation and acceleration of a relativistic charged particle in an electromagnetic field, *Phys. Rev.* D **17**, 434–45.
Simon B. (1974). *The* $P(\phi)_2$ *Euclidean (Quantum) Field Theory*. Princeton, NJ: Princeton University Press.

(1978). Resonances and complex scaling: a rigorous overview, *J. Quant. Chem.* **14**, 529–42.

(1979). *Functional Integrals and Quantum Physics*. New York: Academic Press.

Sjöstrand J. (1993). Projecteurs adiabatiques du point de vue pseudodifférentiel, *C. R. Acad. Sci. Paris Sér. I Math.* **317**, 217–20.

Skibsted E. (1998). Spectral analysis of $N$-body system coupled to a bosonic field, *Rev. Math. Phys.* **10**, 989–1026.

Sommerfeld A. (1904a). Zur Elektronentheorie: 1. Allgemeine Untersuchung des Feldes einer beliebig bewegten Ladung; 2. Grundlagen für eine allgemeine Dynamik des Elektrons, *Nachr. der Kgl. Ges. der Wiss. Göttingen, Math.-Phys. Klasse*, 99–130, 363–469.

(1904b). Grundlagen für eine allgemeine Dynamik des Elektrons, *Göttinger Nachr.*, 363–439.

(1904c). Simplified deduction of the field and the forces of an electron, moving in any given way, *Proc. R. Acad. Amsterdam* **7**, 346–67.

(1905). Über Lichtgeschwindigkeits- und Überlichtgeschwindigkeits-Elektronen, *Göttinger Nachr.*, 201–35. Reprinted in *Gesammelte Schriften*, II, 148–82.

(1968). *Gesammelte Schriften, Band I–IV*, edited by F. Sauter. Braunschweig: Vieweg.

Sparnaay M. J. (1958). Measurement of attractive forces between flat plates, *Physica* **24**, 751–64.

Spohn H. (1976). An algebraic condition for the approach to equilibrium of an open $N$-level system, *Lett. Math. Phys.* **2**, 33–8.

(1978). Derivation of the transport equation for electrons moving through random impurities, *J. Stat. Phys.* **60**, 277–90.

(1980). Kinetic equations from Hamiltonian dynamics: Markovian limits, *Rev. Mod. Phys.* **53**, 569–615.

(1987). Effective mass of the polaron: a functional integral approach, *Ann. Phys.* **175**, 278–304.

(1988). The polaron at large total momentum, *J. Phys.* A **21**, 1199–211.

(1989). Ground state(s) of the spin-boson Hamiltonian, *Comm. Math. Phys.* **123**, 277–304.

(1991). *Large Scale Dynamics of Interacting Particles*. Berlin: Springer.

(1997). Asymptotic completeness for Rayleigh scattering, *J. Math. Phys.* **38**, 2281–96.

(1998). Runaway charged particles and center manifolds, unpublished manuscript.

(2000a). The critical manifold of the Lorentz–Dirac equation, *Europhys. Lett.* **50**, 287–92.

(2000b). Semiclassical limit of the Dirac equation and spin precession, *Ann. Phys.* **282**, 420–31.

Spohn H. and Teufel S. (2001). Adiabatic decoupling and time-dependent Born–Oppenheimer theory, *Comm. Math. Phys.* **224**, 113–32.

Srinivas M. D. and Davies E. B. (1981). Photon counting probabilities in quantum optics, *Optica Acta* **28**, 981–96.

Stephas P. (1992). Analytic solutions for Wheeler–Feynman interaction, *J. Math. Phys.* **33**, 612–24.

Taylor J. H. (1994). Binary pulsars and relativistic gravity, *Rev. Mod. Phys.* **66**, 699–710.

Teitelbom C., Villarroel D. and van Weert Ch. G. (1980). Classical electrodynamics of retarded fields and point particles. *Rev. Nuovo Cim.* **3**, 1–64.

Templin, J. (1999). Radiation reaction and runaway solutions in acoustics, *Am. J. Phys.* **67**, 407–13.

Teufel S. (2001). A note on the adiabatic theorem without gap condition, *Lett. Math. Phys.* **58**, 261–6.

(2002). Effective $N$-body dynamics for the massless Nelson model and adiabatic decoupling without spectral gap, *Ann. H. Poincaré* **3**, 939–65.
(2003). *Adiabatic Perturbation Theory in Quantum Dynamics*. Springer Lecture Notes in Mathematics, vol. 1821. Berlin: Springer.
Teufel S. and Spohn H. (2002). Semiclassical motion of dressed electrons, *Rev. Math. Phys.* **14**, 1–28.
Thirring W. (1958). *Principles of Quantum Electrodynamics*. New York: Academic Press.
(1997). *Classical Mathematical Physics, Dynamical Systems and Field Theory*, 3rd edition. New York: Springer.
(2002). *Quantum Mathematical Physics, Atoms, Molecules, and Large Systems*, 2nd edition. New York: Springer.
Thomas L. H. (1926). The motion of the spinning electron, *Nature* **117**, 514–21.
(1927). On the kinematics of an electron with an axis, *Phil. Mag.* **3**, 1–22.
Thomson J. J. (1881). On the electric and magnetic effects produced by the motion of electrified bodies, *Phil. Mag.* **11**, 229–49.
(1897). Cathode rays, *Phil. Mag.* **44**, 294–316.
Ullersma P. (1966). An exactly solvable model for Brownian motion, *Physica* **32**, 27–55, 56–73, 74–89, 90–6.
Unruh W. G. and Zurek W. H. (1989). Reduction of a wave packet in quantum Brownian motion, *Phys. Rev.* D **40**, 1071–94.
Valentini A. (1988). Resolution of causality violation in the classical radiation reaction, *Phys. Rev. Lett.* **61**, 1903–5.
Van der Waerden B. L. (1908). *Sources of Quantum Mechanics*. New York: Dover.
Vlasov A. A. (1961). *Many-Particle Theory and its Applications to Plasma Physics*. New York: Gordon and Breach.
Vollhardt D. and Wölfle P. (1980). Diagrammatic, self-consistent treatment of the Anderson localization in $d \leq 2$ dimensions, *Phys. Rev.* B **22**, 4666–79.
Weiss U. (1999). *Quantum Dissipative Systems*. Singapore: World Scientific.
Weisskopf V. and Wigner E. (1930). Berechnung der natürlichen Linienbreite auf Grund der Diracschen Lichttheorie, *Z. Physik* **63**, 64–73.
Weitz M., Huber A., Schmidt-Kaler F., Leibford D. and Hänsch T.W. (1994). Precision measurement of the hydrogen and deuterium 1S ground state Lamb shift, *Phys. Rev. Lett.* **72**, 328–31.
Welton T. A. (1948). Some observable effects of the quantum-mechanical fluctuations of the electromagnetic field, *Phys. Rev.* **74**, 1157–67.
Wheeler J. A. (1998). *Geons, Black Holes, and Quantum Foam* (with K. Ford). New York: W. W. Norton.
Wheeler J. A. and Feynman R. P. (1945). Interaction with the absorber as the mechanism of radiation, *Rev. Mod. Phys.* **17**, 157–81.
(1949). Classical electrodynamics in terms of direct interparticle action, *Rev. Mod. Phys.* **21**, 425–33.
Wildermuth K. (1955). Zur physikalischen Interpretation der Elektronenselbstbeschleunigung, *Z. Naturf.* **10a**, 450–9.
Will C. M. (1999). Gravitational radiation and the validity of general relativity, *Physics Today*, October, 38–43.
Yaghjian A. D. (1992). *Relativistic Dynamics of a Charged Sphere*. Lecture Notes in Physics m11. Berlin: Springer.
Yajima K. (1992). The quasi-classical approximation to the Dirac equation, *J. Fac. Sci. Univ. Tokyo Sect. IA Math.* **29**, 161–94.
Zahn C. T. and Spees A. A. (1938). An improved method for the determination of the specific charge of beta-particles, *Phys. Rev.* **35**, 357–64.

# Index

Printed in the United States
by Baker & Taylor Publisher Services